T0324082

Process Safety Calculations

Process Safety Calculations

Renato Benintendi
MSc CEng FIChemE

Elsevier
Radarweg 29, PO Box 211, 1000 AE Amsterdam, Netherlands
The Boulevard, Langford Lane, Kidlington, Oxford OX5 1GB, United Kingdom
50 Hampshire Street, 5th Floor, Cambridge, MA 02139, United States

Library of Congress Cataloging-in-Publication Data
A catalog record for this book is available from the Library of Congress

British Library Cataloguing-in-Publication Data
A catalogue record for this book is available from the British Library

ISBN: 978-0-08-101228-4

Publisher Director: Joe Hayton
Acquisition Editor: Anita A. Koch
Editorial Project Manager: Ana Claudia A. Garcia
Production Project Manager: Sruthi Satheesh
Cover Designer: Victoria Pearson

Typeset by SPi Global, India

Professor Gianni Astarita (1933–97)

This book is dedicated to the memory of Gianni Astarita, professor of chemical engineering at Naples and Delaware universities and a foreign associate of the National Academy of Engineering. The author has collected many bricks from his temple of science to build up this small house in the chemical engineering village.

To my mother

Contents

Preface

The aim of this book is to provide the reader with some guidance on calculations in process safety. Accordingly, the intention of the author was not to duplicate or to emulate the many excellent literature works produced since the many years of study on process safety techniques and models, but rather to build-up a logical and fluid thread to overcome doubts, uncertainties, and difficulties often met in calculation exercises. The available literature sources offer either a broad range of different models and approaches or, even when they are calculations oriented, sometimes unavoidably and faultlessly leave some gaps in the calculation criteria; this is a fee to be paid to the richness and variety of data and information. This book has a different target: to provide a clear indication on where to go in practical applications when a crossroads is met, and when available data are difficult to be converted into figures and findings. Nevertheless, the theoretical and conceptual background is deemed to be effective in enabling the user to properly frame the topics and, to some extent, some aspects not included in the existing literature sources have also been dealt with, from principles to applications.

The book is the final step of a long trip the author started in 1988, when, more than 10 years after the incidents of Flixborough and Seveso, and some years after the unresolved tragedy of Bhopal, the Seveso I directive was actuated in Italy. It is doubtless that this European legislative act has given a tremendous impulse to the development of systematic methods and techniques in process safety engineering. In the nineties the author was involved as a teacher in the final part of the Chemical Plants course held at the Chemical Engineering faculty at the University of Salerno (Italy), providing some guidance about process risk assessment methodologies. In the same years, a long experience acquired as an instructor within the course for Risk Analysis, managed by the Italian Inspectorates engaged in the Seveso Directive safety reports assessment, clearly showed how difficult and challenging it was to relate theory to real cases. Specifically, even if chemical engineers, and engineers in general, should have a thorough knowledge of background concepts underpinning process safety studies, the experience has shown that cultural transition from process to process safety engineering is neither automatic nor easy. The author has analysed this aspect in a recent article (Benintendi, 2016), where he has pointed out that the effectiveness of *adding-on* basic process safety concepts to the university background is not always high. The combination of the experience acquired from process safety teaching, tutoring, and lecturing at several universities in the UK, Italy, Asia, and USA, with the professional expertise developed in almost 30 years' work, has suggested that the provision of basic concepts already calibrated on process safety is much more effective.

In this respect, this book includes a first part where basic concepts of chemistry, thermodynamics, reactor engineering, hydraulics, and fluid-dynamics are reviewed with a specific focus on process safety scenarios. Dozens of fully resolved examples focusing on process safety applications have been included. This *Fundamentals* section ends with one chapter dealing with structural analysis for process safety and another one including a statistics and reliability overview, aiming to provide the basic concepts to properly manage the probabilistic aspect of risk assessment studies. All these first chapters include many literature data, with the intention to provide the users with a complete tool for their calculations.

The *Consequence Assessment* section is organised according to the typical sequential outcomes following a release after loss of containment. Some efforts have been made to ensure that all potential gaps and uncertainties in the calculations were covered and overcome, based on the professional involvement of the author in many projects dealing with oil and gas, petrochemical, pharmaceutical, fine chemistry, food, and environmental subjects. In this respect, the users will be driven across a relatively simple and direct route, unlike what happens when they go to the literature, where obviously a much wider spread of methods is provided. Chapter 7 focuses on releases from containments and from pools: on the basis of the theoretical background provided in the *Fundamentals* section, a systematic analysis of possible scenarios has been carried out, with the support of many fully resolved examples. Release of carbon dioxide has been dealt with in detail, due to the relatively new hazardous scenarios presented after the introduction of Carbon Capture and Storage (CCS) process, and to the specific nature of this substance, which shows a solid-liquid equilibrium below the triple point and does not fully behave according to equilibrium thermodynamics. Chapter 8 presents dispersion models; in the author's intention, the effort has been made to resolve the various uncertainties met by process safety engineers on which model to adopt, which regime to select, which phase of the dispersion route to identify, and where to localise it. Key parameters have been identified to drive this approach with the support of many examples. Eisenberg's model for flash fire and Kalghatgi solid's flame model for jet fire have been selected for their simplicity, completeness, and robustness in Chapter 9, which covers fire. A specific focus has been made on ignition sources, according to the systematic BS EN 1127-1 standard, with the aim to reduce the incompleteness of the approach often followed. Chapter 10 deals with gas and vapour explosions, consisting of all of scenarios potentially resulting in significant overpressures, including BLEVE, Rapid Phase Transition, and thermal runaway. The Multi Energy Method (MEM) has been fitted with the findings of the GAME projects, and this has been very effective in removing the traditional large level of subjectivity and uncertainty in blast curve selection. A MEM detailed, and fully resolved, example has shown a very good consistency with the findings of the Baker, Strehlow, and Tang (BST) method. Chapter 11 has been included to cover dust explosions. In addition to the models describing the primary and the secondary explosions, some HAZID cases relating to dust processing equipment have been included, according to the great emphasis the machinery and the ATEX directives have put on this specific aspect. A case study dealing with the Imperial Sugar Company has been analysed and verified against some calculation findings.

Chapter 12 deals with QRA techniques, including the exceedance frequency curve build-up, the ALARP model demonstration, the FN curves, and the parts count. Some applications have also been given in this chapter.

In this book, unless otherwise specified, all units are expressed according to the International System (SI) or mks system. This book aims to support scientists and engineers working in process safety engineering. It is worth repeating that it is a book of calculations offering a large number of data useful for this purpose. The author guesses that it is not free from mistakes and defects, and the author apologies in advance for that. He will be grateful for any contributions readers will wish to give him, to ensure that the objectives the writer had in his mind can be fully achieved.

Reading (Berkshire), 30 April 2017

Reference

Benintendi, R., April 2016. The bridge link between university and industry: a key factor for achieving high performance in process safety. Educ. Chem. Eng. 15, 23–32. IChemE, Elsevier.

Acknowledgements

I wish to thank various professors, colleagues, and friends for their contribution to this book: Prof Gianni Astarita of University of Naples (Italy) and Delaware, who provided me with the mysterious codes of chemical engineering. This book is dedicated to his memory as an appreciation for the prestigious Chemical Engineering School he created in Naples, that I had the honour and the pleasure to attend. Simona Rega, for her precious support and for the contribution to the development of the Rapid Phase Transition Phase included in this book. My colleagues, Foster Wheeler and Amec Foster Wheeler, Reading office, who inspired this work through their joint activity and the commitment of the Process Safety Calculations course held in Reading in 2014. My students of the master in process safety engineering attended at Sheffield, Leeds, and Paris, whom I tutored, giving me the opportunity to make a much better focus on the subject from this standpoint. The team of the Project Evaluation Laboratory of the University of Salerno (Italy), with whom I am sharing and extending the risk assessment techniques in a much wider perspective, which has resulted in a sharper focus on methods and philosophy.

Finally, I would like to express my thanks, gratitude, and appreciation to the Elsevier team for their support and patience: Fiona Geraghty, Anita Koch, Renata Rodrigues, and Maria Convey, without whose advice these pages would have been neither written nor published.

Reading (Berkshire), 30 April 2017

PART 1

Fundamentals

CHAPTER 1

Chemistry of Process Safety

Nothing is lost, nothing is created. Everything is transformed.

(A. Lavoisier)

1.1 Stoichiometry and Mass Balances

Stoichiometry (from ancient Greek *στοιχεῖον element* and *μέτρον measure*) is a fundamental part of basic chemistry that can be defined as *the relationship between the relative quantities of substances taking part in a reaction or forming a compound, typically a ratio of whole integer.*

The stoichiometry of a chemical reaction can be expressed through the expression:

$$aA + bB + \cdots \rightarrow cC + dD + \cdots \tag{1.1}$$

or

$$\sum_{1}^{n} r_i R_i \rightarrow \sum_{1}^{m} q_j Q_j \tag{1.2}$$

where a, b, c, d are the reactants and products reaction coefficients, and A, B, C, D are the atoms or the molecules involved in the reaction. Eq. (1.2) is a generalised form of the same formula. It is worth noting that:

$$s_{ij} = \frac{r_i}{q_j} \tag{1.3}$$

is the stoichiometric ratio of R_i to Q_j and indicates the constant relationship between the concentrations of these chemical compounds throughout the reaction path.

Pure mixing and dispersion processes don't cause any modifications of the chemical identity of the substances because of the physical nature of these phenomena.

1.1.1 Mass Balances

Mass balances are broadly used in process engineering where process streams and configurations are well defined. Instead, process safety scenarios are variable and often very complex. Considering a portion of space (Fig. 1.1), the following general equation of mass balance applies:

$$W_{acc} = W_{in} - W_{out} + W_{gen} \tag{1.4}$$

Process Safety Calculations. https://doi.org/10.1016/B978-0-08-101228-4.00001-0

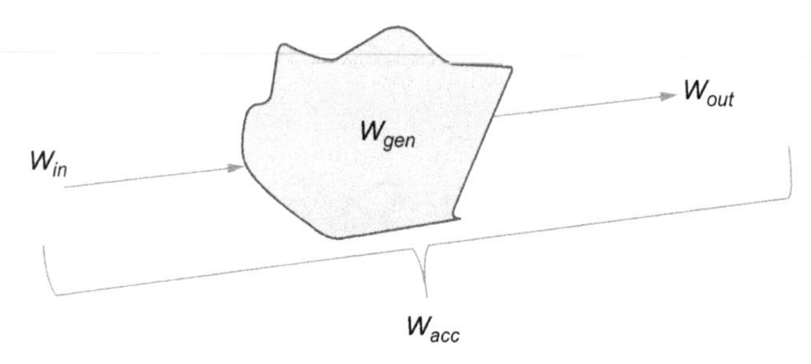

Fig. 1.1
Mass balance on a generic space domain.

where:

- W_{in} is the mass entering the space domain.
- W_{out} is the mass leaving the space domain.
- W_{gen} is the mass generated or converted.
- W_{acc} is the mass inventory variation.

According to Lavoiser' s principle, W_{gen} exists only for components which are transformed into others.

Process safety engineering entails a broad range of complex and variable scenarios where full understanding of stoichiometry and mass balances is necessary to properly analyse and assess the related process and plant configurations. Some cases are discussed here, and specific scenarios have been analysed in the next paragraphs.

1.1.2 Chemical Reactions

A low pressure vessel contains a stoichiometric mixture of carbon monoxide and pure oxygen at ambient temperature T_o (Fig. 1.2).

The system undergoes a chemical reaction that converts all carbon monoxide into carbon dioxide and is assumed to be at thermal equilibrium so that initial ambient temperature is attained. Application of ideal gases law, with V and T_o constant, gives:

$$P_1 = P_o \frac{N_1}{N_o} \tag{1.5}$$

where N_1 and N_o are the final and initial number of kmoles of product and reactants respectively, which in this specific case coincides with the reaction stoichiometric coefficient s_{rp}:

$$s_{rp} = \frac{N_1}{N_o} = \frac{1}{1.5} = 0.67 \tag{1.6}$$

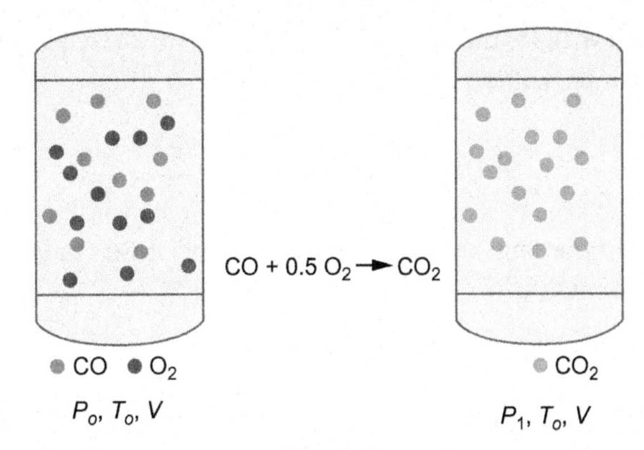

Fig. 1.2
Oxidation of carbon monoxide leading to vessel depressurisation.

It can be concluded the reaction, assuming an overall isothermal and isochoric condition, causes a 33% pressure drop, which could result in a catastrophic outcome for the vessel.

1.1.3 Jet Flows From Pressurised Systems

Jet flows from pressurised containments are frequent in process safety. The consequences of toxic or flammable compound dispersion strictly depend on the jet dynamics. The scenario shown in Fig. 1.3 illustrates the release of hydrogen sulphide from a pipeline.

The toxic gas is released with a mass flow rate of W_{H_2S}. Air is entrained into the jet as long as this is developed, resulting in a progressive dilution of H_2S. Depending on the effect of the entrainment, toxic concentrations are proportionally reduced, while flammability will be promoted by air mixing within a specific region of the jet. Assuming a steady state value of

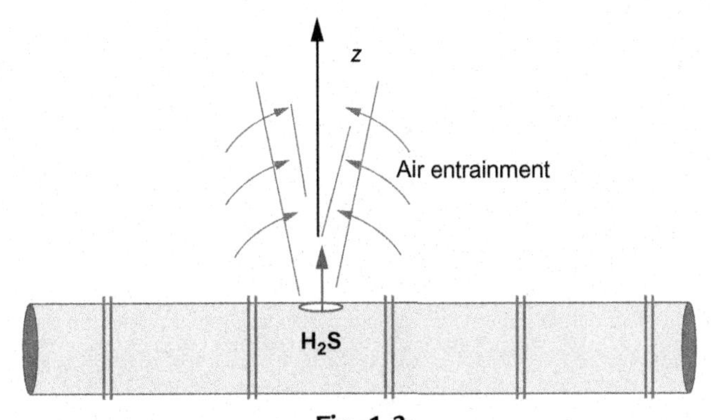

Fig. 1.3
Release of hydrogen sulphide from a pipeline.

W_{H_2S}, and indicating with $W_{AIR}(z)$ the air entrainment mass flow rate per length unit along z, the mass balance at $z = h$ can be written as:

$$W(h) = W_{H_2S} + \int_0^h W_{AIR}(z)dz \tag{1.7}$$

It has been shown how important the correct management of this balance is in jet flow consequence assessment studies.

1.1.4 Flash Flow

A flash flow is the release of a liquid from a containment where the operating temperature is significantly greater than its downstream boiling temperature, typically the normal boiling temperature. The liquid is forced to vaporise a fraction of it to reach the downstream equilibrium condition. This is the case with LPG stored at ambient temperature (Fig. 1.4).

The liquid mass W splits into the flashed vapour fraction X_V and the liquid fraction X_L. It is:

$$W = X_V + X_L \tag{1.8}$$

1.1.5 Absorption and Adsorption

Removal of toxic or dangerous compounds can be accomplished via mass transfer units, such as absorber or adsorption towers. A typical example is the amine treatment of sour gas (Fig. 1.5), or absorption of carbon dioxide with sodium hydroxide. For sour gas treatment, neglecting changes of flow rates Q and q, the mass balance of H_2S can be simplified considering the concentration of sulphur S:

$$Q \cdot S_{in} = Q \cdot S_{out} + q \cdot S_{am} \tag{1.9}$$

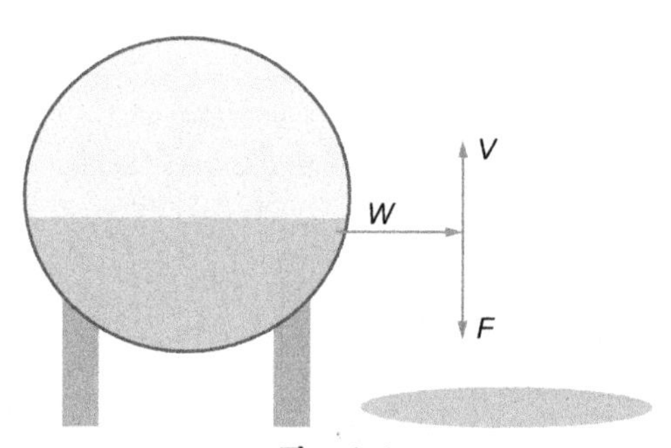

Fig. 1.4
Flash flow from pressurised containment.

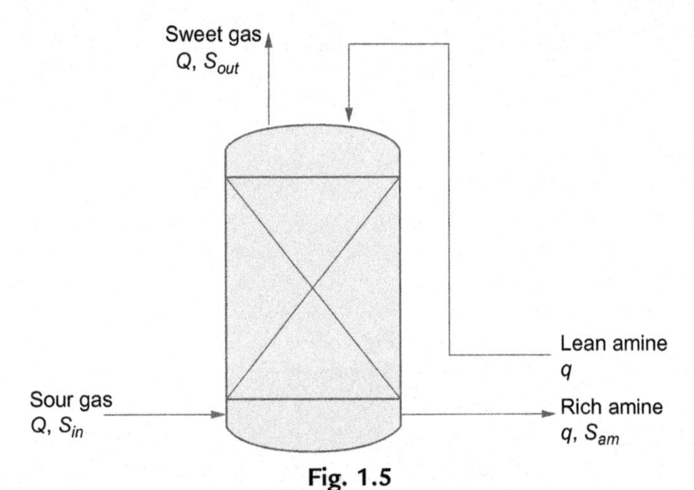

Fig. 1.5
Amine treatment of sour gas.

Example 1.1

How many kilograms of oxygen are required to enrich 500 kg of air to 50% of O_2 (molar basis)? (Air molecular weight = 29, oxygen molecular weight = 32)

Solution
500 kg of air correspond to the following kmoles:

$$N_{air} = \frac{500}{29} = 17.24 \quad (1.10)$$

$$N_{O_2} = \frac{500}{29} \times 0.21 = 3.62 \quad (1.11)$$

$$N_{N_2} = \frac{500}{29} \times 0.79 = 13.62 \quad (1.12)$$

The final amount of oxygen will be, as requested, the same as that of nitrogen, i.e. 13.62 kmol. Therefore, the oxygen to be added can be calculated by difference:

$$\Delta N_{O_2} = 13.62 - 3.62 = 10\,\text{kmoles} = 320\,\text{kg} \quad (1.13)$$

Example 1.2

A gas stream, other than air, containing a certain amount of hydrogen sulphide is sent to a burner that is fed with the necessary air to oxidise the sulphide to sulphur dioxide. Knowing that the off gas contains 464.8 kg/h of molecular nitrogen and 20.8 kg/h of molecular oxygen and that the sulphide is the only one oxidisable gas, determine:

(a) The mass flow rate of hydrogen sulphide sent to the burner
(b) The sulphide oxidation efficiency

(Hydrogen sulphide molecular weight = 34, nitrogen molecular weight = 28)

Fig. 1.6
H_2S oxidiser.

Solution

With reference to Fig. 1.6, the combustion process is represented by the following reaction:

$$2H_2S + 3O_2 + 3 \cdot \frac{0.79}{0.21} N_2 \rightarrow 2SO_2 + 2H_2O + 3 \cdot \frac{0.79}{0.21} N_2 \tag{1.14}$$

Nitrogen does not participate in the reaction, so being the stoichiometric coefficient:

$$s_{H_2S-N_2} = \frac{2}{3 \cdot \frac{0.79}{0.21}} = 0.177 \tag{1.15}$$

and the nitrogen molar flow rate $\frac{468.4}{28} = 16.6$ kmol/h, the hydrogen sulphide mass flow rate is:

$$W_{H_2S} = 0.177 \cdot 16.6 \cdot 34 = 100\,\text{kg/h} \tag{1.16}$$

Unreacted oxygen is 0.65 kmol/h. Oxygen fed to the burner is equal to $16.6 \cdot \frac{0.21}{0.79} = 4.4$ kmoles/h. The conversion efficiency is calculated considering the ratio of unreacted to supplied oxygen:

$$\eta_{H_2S} = 1 - \frac{0.65}{4.4} = 0.85 \tag{1.17}$$

Example 1.3

A storage tank contains a heavy hydrocarbon with a negligible vapour pressure at the operating temperature. The tank is nitrogen blanketed so that a positive overpressure of P_o is maintained. The tank emptying is started with a head free volume of V_o and a liquid flow rate equal to Q. Find the nitrogen mass to be provided with time by the controller PIC in order to ensure that positive overpressure P_o is maintained during all emptying phases (Fig. 1.7).

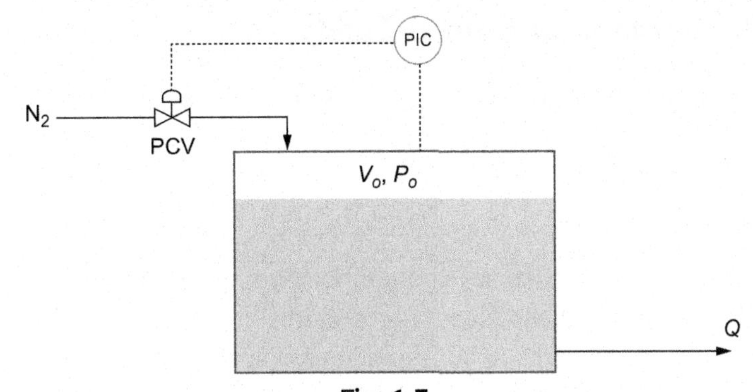

Fig. 1.7
Emptying of nitrogen blanketed tank.

Solution
The tank head space is assumed to be occupied by nitrogen only. Application of ideal gas laws:

$$P \cdot V = n_{N_2} \cdot R \cdot T \tag{1.18}$$

Imposing that pressure is maintained constant by the PCV. It is:

$$\frac{dP}{dt} = R \cdot T \cdot \left[n_{N_2} \cdot \frac{d\frac{1}{V}}{dt} + \frac{1}{V} \cdot \frac{dn_{N_2}}{dt} \right] = 0 \tag{1.19}$$

Simplifying and rearranging:

$$n_{N_2} \cdot \frac{1}{V}\frac{dV}{dt} = \frac{dn_{N_2}}{dt} \tag{1.20}$$

Considering that $\frac{dV}{dt} = Q$ and that $V = V_o + Q \cdot t$, Eq. (1.20) may be written as:

$$\frac{1}{V_o + Q \cdot t} \cdot Q = \frac{1}{n_{N_2}} \cdot \frac{dn_{N_2}}{dt} \tag{1.21}$$

Separating:

$$\frac{d(V_o + Q \cdot t)}{V_o + Q \cdot t} = \frac{dn_{N_2}}{n_{N_2}} \tag{1.22}$$

Solving:

$$W_{N_2}(t) = 28 \cdot n_{N_2}(t) = 28 \cdot n_{N_{2_o}} \cdot \left(\frac{V_o + Q \cdot t}{V_o} \right) \tag{1.23}$$

where $n_{N_{2_o}}$ is the molar nitrogen amount of V_o at $t = 0$. This result is intuitive but has been rigorously obtained here via the application of mass balances.

1.2 States of Substances in Process Safety

Substances in process safety can be present in the following forms:

1.2.1 Gases and Vapours

Gas is a fundamental state of substances at a temperature higher than their critical temperature. Hydrogen and methane have to be regarded as gases at ambient temperature, whereas propane and sulphur dioxide are vapours and can be condensed by compression.

1.2.2 Liquid

Liquids are the condensed phase of vapours. They can be in equilibrium with their vapours at any temperature, and vapour pressure is the equilibrium pressure exerted by vapour above the liquids. Liquids can be miscible or immiscible, polar or non-polar, and this behaviour strongly affects the release and dispersion scenarios.

1.2.3 Dusts

In addition to being combustible, dusts which are finely divided solid particles can be explosive. According to BS-EN 60079-10-2:2015, combustible dusts, 500 μm or less in nominal size, may form an explosive mixture with air at atmospheric pressure and normal temperatures. Particular types of solid particles, including fibres, are combustible flyings, greater than 500 μm in nominal size, which may form an explosive mixture with air at atmospheric pressure and normal temperatures (BS-EN 60079-10-2:2015).

As for gases and vapour, the mechanism of dust explosion consists of the rapid release of heat due to the chemical reaction:

$$\text{Fuel} + \text{oxygen} \rightarrow \text{oxides} + \text{heat}$$

Metal dusts can also exothermically react with nitrogen and carbon dioxide according to Eckhoff (2003), which classifies explosive dusts as follows:

- Natural organic matters
- Synthetic organic materials
- Coal and peat
- Metals such as aluminium, magnesium, zinc, and iron.

1.2.4 Hybrid Mixtures

A hybrid mixture is a combined mixture of a flammable gas or vapour with a combustible dust or combustible flying, which can behave differently from the gas/vapour or dust individually (BS-EN 60079-10-2:2015).

1.2.5 Explosive Mists

Dispersed droplets of liquids which, in some situations, may form a flammable mist which may then give rise to an explosion hazard. It has been proved that aerosol sized droplets (sub-micron to 50 microns) will likely be the most easily ignitable portion of the mist cloud (BS-EN 60079-10-2:2015).

1.2.6 Supercritical Fluids

The defined state of a compound, mixture, or element above its critical pressure and critical temperature (IUPAC, 2014). Carbon dioxide in Carbon Capture and Storage (CCS) processing is frequently handled in its supercritical state.

1.3 Mass and Concentration Units in Process Safety

Mole

Amount of grams corresponding to the atomic or molecular weight of a substance

K mole

Amount of kilograms corresponding to the atomic or molecular weight of a substance

Molar fraction (Gas and liquid phase)

$$xi = \frac{Ni}{\Sigma N_i} \tag{1.24}$$

x_i = molar fraction of component i
N_i = number of moles or kmoles of component i

Partial pressures (Gas phase)

According to Dalton's law, within the limit of validity of ideal gas law:

$$p_i = y_i \cdot P \tag{1.25}$$

y_i = molar fraction of component i in the gas phase
p_i = pressure of component i
P = total pressure

1.3.1 Partial Volumes (Gas Phase)

According to Amagat's law, within the limit of validity of ideal gas law:

$$V_i = y_i \cdot V \tag{1.26}$$

y_i = molar fraction of component i in the gas phase
V_i = volume of component i
V = total volume

1.3.2 Mass Fraction (Gas and Liquid Phase)

$$x_{m_i} = \frac{m_i}{\sum m_i} \tag{1.27}$$

and:

$$x_{m_i} = \frac{w_i}{\sum w_i} \tag{1.28}$$

x_{m_i} = mass fraction of component i
m_i = partial pressure of component i
w_i = mass flow rate of component i

1.3.3 Mass to Volume Concentration (Gas and Liquid Phase)

$$c_i = \frac{m_i}{V(T)} \tag{1.29}$$

where:

c_i is the mass to volume concentration of component i.
$V(T)$ is the system volume at temperature T.

Particular mass to volume concentrations are (gas phase):

$$c_{Ni} = \frac{m_i}{V_N(T)} \tag{1.30}$$

$$c_{Si} = \frac{m_i}{V_S(T)} \tag{1.31}$$

where:

c_{Ni} is the mass concentration at normal conditions.
c_{Si} is the mass concentration at standard conditions.

Both states, normal and standard, are assumed in this book to be at 273.15 K and 1 atm, according to Hougen et al. (1954). Under these conditions the normal molar volumes are as follows:

$$\text{Volume of } 1\,\text{mol} = 22.414\,\text{L}$$
$$\text{Volume of } 1\,\text{kmol} = 22.414\,\text{m}^3$$

1.3.4 Parts per Million (Gas and Liquid Phase)

ppmw (weight)—typical in liquids

$$\text{ppmw} = \frac{m_i}{\sum m_i} \times 10^6 \tag{1.32}$$

If liquid is water, assuming water density as 1000 kg/m^3 or 1000 g/L:

$$\text{ppmw} = \frac{\text{g}}{\text{m}^3} = \frac{\text{mg}}{\text{L}} \tag{1.33}$$

1.3.5 Parts per Million (Gas Phase)

$$\text{ppmv} = y_i \times 10^6 = \frac{p_i}{P} \times 10^6 = \frac{v_i}{V} \times 10^6 \tag{1.34}$$

where all symbols are known.

1.3.6 Molar Concentration (Aqueous Solutions)

It is denoted as [X] and indicates moles/litre.

1.3.7 Concentration Units Conversion Summary

See Table 1.1.

Example 1.4

The IDLH (immediately dangerous to life and health) of sulphur sulphide is 100 ppmv. Calculate it as mg/Nm^3 and as molar fraction. (MW$\sim$34)

Solution

$$100\,\text{ppmv} = 100 \cdot \frac{MW}{22.414} = \frac{3400}{22.414} = 151.69\,\text{mg/Nm}^3 \tag{1.35}$$

$$100\,\text{ppmv} = \frac{100}{1{,}000{,}000} = 0.0001 \quad \text{molar fraction} \tag{1.36}$$

Table 1.1 Concentration units conversion summary

From	To	Multiply by
Molar fraction gas	ppmv	10^6
ppmv	Molar fraction gas	10^{-6}
Molar fraction gas	mg/Nm^3	$\frac{MW \cdot 1{,}000{,}000}{22.414}$
mg/Nm^3	Molar fraction gas	$\frac{22.414}{MW \cdot 1{,}000{,}000}$
mg/m^3	Molar fraction gas	$\frac{22.414 \cdot T}{273.15 \cdot 1{,}000{,}000 \cdot MW}$
Molar fraction gas	mg/m^3	$\frac{273.15}{22.414 \cdot T} \cdot 1{,}000{,}000 \cdot MW$
ppmv	mg/Nm^3	$\frac{MW}{22.414}$
mg/Nm^3	ppmv	$\frac{22.414}{MW}$
ppmv	mg/m^3	$\frac{273.15 \cdot MW}{22.414 \cdot T}$
mg/m^3	ppmv	$\frac{22.414 \cdot T}{MW \cdot 273.15}$

MW, molecular weight; *T*, temperature in K.

Example 1.5

A pressure vessel contains:

10 kg of molecular nitrogen	(MW~28)
20 kg of molecular oxygen	(MW~32)
50 kg of carbon dioxide	(MW~44)

Calculate molar and mass fractions at:

$T = 20°C$ and 1 atm

$T = 60°C$ and 3 atm

Solution

Molar and mass fractions don't depend on pressure and temperature.

$$N_{N_2} = \frac{10}{28} = 0.36\,\text{kmol} \tag{1.37}$$

$$N_{O_2} = \frac{20}{32} = 0.625\,\text{kmol} \tag{1.38}$$

$$N_{CO_2} = \frac{50}{44} = 1.14\text{kmol} \tag{1.39}$$

$$x_{N_2} = \frac{0.36}{0.36 + 0.625 + 1.14} = 0.17 \tag{1.40}$$

$$x_{O_2} = \frac{0.625}{0.36 + 0.625 + 1.14} = 0.3 \tag{1.41}$$

$$x_{CO_2} = \frac{1.14}{0.36 + 0.625 + 1.14} = 0.53 \tag{1.42}$$

$$x_{mN_2} = \frac{10}{10 + 20 + 50} = 0.125 \tag{1.43}$$

$$x_{mO_2} = \frac{20}{10 + 20 + 50} = 0.25 \tag{1.44}$$

$$x_{mCO_2} = \frac{50}{10 + 20 + 50} = 0.625 \tag{1.45}$$

1.4 Solutions and Chemical Equilibrium

A solution is defined as a homogeneous one-phase mixture of two or more substances. In process safety it is very frequent to deal with mixtures and solutions which may exists in any of the three states of matter, gaseous, liquid, and solid. Knowledge of solutions and mixtures chemistry is important to identify and calculate hazardous properties of the involved substances.

1.4.1 Gaseous Solutions

Examples of gaseous solutions are mixtures of gaseous hydrocarbons, air that is mainly a mixture of oxygen and nitrogen, off-gases from flaring, which typically contains carbon dioxide, sulphur dioxide, water, and nitrogen. Nonreactive gas mixtures present a high degree of homogeneity, so they can always be considered solutions. This is not always true for liquid and solid mixtures. Gaseous mixtures are governed by Dalton's law of partial pressures, that states that *the total pressure P of a mixture of n components i is equal to the sum of the partial pressures P_i of all the different gases*:

$$P = \sum_{1}^{n} p_i \tag{1.46}$$

$$p_i = P \cdot y_i \tag{1.47}$$

where y_i is the molar fraction of the component i.

Example 1.6

100 kg of solid sulphur are burnt in a combustor at atmospheric pressure. Knowing that 10% of air excess is used, find the partial pressure of nitrogen in the off gas. (Sulphur molecular weight: 32, nitrogen: 28, oxygen: 32).

Solution

The combustion reaction is:

$$S + O_2 + \frac{79}{21} \cdot N_2 \rightarrow SO_2 + \frac{79}{21} \cdot N_2 \tag{1.48}$$

100 kg of sulphur are equivalent to 3.125 kmol. From the reaction stoichiometry and considering 10% of air excess:

$$N_{SO_2} = 3.125\,\text{kmol} \tag{1.49}$$

$$N_{O_2} = 0.1 \cdot 3.125 = 0.3125\,\text{kmol} \tag{1.50}$$

$$N_{N_2} = 1.1 \cdot 3.125 \cdot \frac{79}{21} = 12.93\,\text{kmol} \tag{1.51}$$

$$P_{N_2} = \frac{12.93}{12.93 + 3.125 + 0.3125} = 0.79\,\text{atm} \tag{1.52}$$

This result is intuitive, due the equimolar S/O_2 ratio.

1.4.2 Kinetics and Equilibrium in Gas Reactive Mixtures

The gas phase reaction rate of the general chemical reaction presented in Eq. (1.2) may be written as:

$$r_f = k_f \cdot \prod_1^n p_{Ri}^{ri} \tag{1.53}$$

where r_f is the forward reaction rate, k_f is the kinetic constant, and p_{Ri} are the reactant's partial pressures. Some reactions may be reversible, therefore a similar equation may be written for the backward reaction:

$$r_b = k_b \cdot \prod_1^m p_{Qi}^{qi} \tag{1.54}$$

At equilibrium the two reaction rates are the same:

$$r_f = r_b \tag{1.55}$$

and

$$K_P = \frac{k_f}{k_b} = \frac{\prod_1^m p_{Qi}^{qi}}{\prod_1^n p_{Ri}^{ri}} \tag{1.56}$$

where K_P is the equilibrium constant.

Many important reactive mixtures in process safety reach the equilibrium.

1.4.3 Liquid Solutions

Liquid solutions are obtained by dissolving gaseous, liquid, or solid substances in liquids. Depending on the nature and the behaviour of the dissolved substances (solute), and of the liquid (solvent), a wide range of physical–chemical scenarios may be obtained, which have to be well understood in order for them to be properly analysed process safety wise.

Liquid–liquid solutions

Miscible liquids form homogeneous solutions, whereas immiscible liquids form two phase dispersed emulsions. A general criterion used to establish whether or not two or more liquids are miscible is comparing their polar features. The old saying *like dissolves like* is a very useful rule of thumb. Therefore, polar species, such as water, have the ability to engage in hydrogen bonding. Alcohols are less polar, but can form hydrogen bonding as well. Due to its strong polarity, water is an excellent solvent for many ionic species. Non-polar species do not have a permanent dipole, and therefore cannot form hydrogen bonding. Organic covalent liquids, such as many hydrocarbons, fall within this category.

The following general criteria can be adopted to predict solubility of chemicals:

- Symmetric structure molecules have a very low dipole moment and are not dissolved by water
- Molecules containing O—H and N—H can form hydrogen bonds
- Molecules containing fluorine and oxygen are expected to have a high dipole moment
- Pure hydrocarbons, oil and gasoline, are non-polar or weak molecules

Dipole moment gives just a very general indication of solubility of molecules. Table 1.2 includes the dipole moment for some organic and inorganic substances.

A common practice is to assume the following scale of polarity with respect to the dipole moment:

- Dipole moment < 0.4: Non polar molecule. Behaviour equivalent to homopolar covalent bond.
- $0.4 <$ Dipole moment < 1.7: Polar molecule. Behaviour equivalent to heteropolar covalent bond.
- Dipole moment > 1.7. Very polar (ionic) molecule.

Table 1.2 Dipole moment for some organic and inorganic substances

Substance	Dipole Moment, Debyes	Polarity
Ammonia	1.471[a]	Non-polar
Hydrogen sulphide	0.97[a]	Polar
Ethanol	1.69[a]	Polar
Methanol	1.70[a]	Polar
Ethane	0[a]	Non-polar
Propane	0.084[a]	Non-polar
Butane	0[a]	Non-polar
Pentane	0[a]	Non-polar
Hexane	0[a]	Non-polar
Ethylene	1.109[a]	Polar
Ethylene oxide	1.89[a]	Non-polar
Phenol	0[a]	Non-polar
Ethylbenzene	0.59[a]	Polar
Benzene	0[a]	Non-polar
Toluene	0.375[a]	Non-polar
o-Xylene	0.62[a]	Polar
m-Xylene	0.33[a]	Non-polar
p-Xylene	0[a]	Non-polar
Acetaldehyde	2.75[a]	Ionic
Acetic acid	1.7[a]	Polar
Diethanolamine	2.84[a]	Ionic
Aniline	1.3[a]	Polar
Pyridine	2.215[a]	Ionic
Sulphur dioxide (g)	1.63[a]	Polar
Sulphur trioxide	0[a]	Non-polar
Hydrogen chloride	1.109[a]	Polar
Carbon dioxide	0[a]	Non-polar
Water	1.84[b]	Ionic
Resins	2–3[c]	Ionic
Crude oils	<0.7[c]	Non-polar/polar
Asphaltenes	4–8[c]	Ionic

[a]Dean (1999).
[b]Poling et al. (2001).
[c]Riazi (2005).

Vapour–liquid equilibrium in liquid solutions

A liquid mixture may present toxic or flammable characteristics depending on concentrations of its components, both in the liquid and in the vapour phase. Therefore, it is important to understand the behaviour of solutions in both phases. The most general equilibrium relationship between vapour concentration and liquid concentration of a substance in a mixture is given by the equation:

$$y_A = K_A \cdot x_A \tag{1.57}$$

where y_A and x_A are the molar fraction in the vapour/gas phase and in the liquid phase respectively, and K_A is the distribution coefficient. K_A is an experimental datum, which depends on system temperature and pressure. A simple but approximate relation to describe the liquid–vapour equilibrium of a mixture is Raoult's, which states:

> *The partial pressure of solvent* p_A *over a solution equals the product of the vapour pressure of the pure solvent,* P_A^o *by the mole fraction of solvent,* x_A*, in the solution.*

$$p_A = P_A^o \cdot x_A \tag{1.58}$$

and, according to Eq. (1.47), if y_A is the vapour molar fraction and P the total pressure:

$$y_A = \frac{P_A^o \cdot x_A}{P} \tag{1.59}$$

The more dilute the solution (a high fraction of solvent typical of so called ideal solutions) the more accurate Raoult's assumption.

Example 1.7

A mixture of *n*-butane and *n*-pentane is in equilibrium at 2 atm e 30°C. Determine the liquid and vapour composition (Vapour pressures: *n*-butane = 3.2 atm, *n*-pentane = 0.78 atm)

Solution

Four equations are available with four unknown variables.

$$y_b = \frac{P_b^o \cdot x_b}{P} = \frac{3.2}{2} \cdot x_b \tag{1.60}$$

$$y_p = \frac{P_p^o \cdot x_p}{P} = \frac{0.78}{2} \cdot x_p \tag{1.61}$$

$$x_b + x_p = 1 \tag{1.62}$$

$$y_b + y_p = 1 \tag{1.63}$$

Solving:

$$x_b = 0.504, \quad x_p = 0.496, \quad y_b = 0.806, \quad y_p = 0.194 \tag{1.64}$$

An empirical correlation to calculate K_A for hydrocarbons and some inorganic gases is Hoffman's equation (Hoffman et al., 1953):

$$K_A = \frac{10^{(a + c \cdot F)}}{P} \tag{1.65}$$

where:

$$F = b \cdot \left(\frac{1}{T_B} - \frac{1}{T}\right) \tag{1.66}$$

Table 1.3 Parameters of Hoffman's equation

Substance	b (K)	T_B (K)
Nitrogen	261.1	60.6
Carbon dioxide	362.2	107.8
Hydrogen sulphide	631.1	183.9
Methane	166.7	52.2
Ethane	636.1	168.3
Propane	999.4	231.1
i-Butane	1131.7	261.7
n-Butane	1196.1	272.8
i-Pentane	1315.6	301.1
n-Pentane	1377.8	309.4
i-Hexane (all)	1497.8	335.0
n-Hexane	1544.4	342.2
n-Heptane	1704.4	371.7
n-Octane	1852.8	398.9
n-Nonane	1994.4	423.9
n-Decane	2126.7	447.2
C_6 (lumped)	1521.1	338.9

$$a = 0.0385 + 6.527 \cdot 10^{-3}P + 3.155 \cdot 10^{-5}P^2 \tag{1.67}$$

$$c = 0.89 - 2.4656 \cdot 10^{-3}P - 7.36261 \cdot 10^{-6}P^2 \tag{1.68}$$

and b and T_B (normal boiling temperature) are included in Table 1.3.

For $C7^+$ fractions the following equations can be used:

$$\gamma = 3.85 + 0.0135T + 0.02321P \tag{1.69}$$

$$b_{7^+} = 562.78 + 180\gamma - 2.364 \cdot \gamma^2 \tag{1.70}$$

$$T_{B7^+} = 167.22 + 33.25\gamma - 0.5394\gamma^2 \tag{1.71}$$

Pressure is given in bar.

1.4.4 Azeotropic Mixtures

Application of Raoult' law shows that mixtures of vapour composition are generally different from liquid composition, due to the different volatility (vapour pressures) of the components. This is not always true, because some mixtures behave as a single pure compound in correspondence to a specific composition and temperature at given pressures. Azeotropic composition is found at concentrations where volatility is reversed, as shown in Fig. 1.8. that represents the mixtures of two pure substances, A and B. In the left-hand zone, component A is more volatile than component B, whereas in the right-hand zone it is the opposite. Therefore

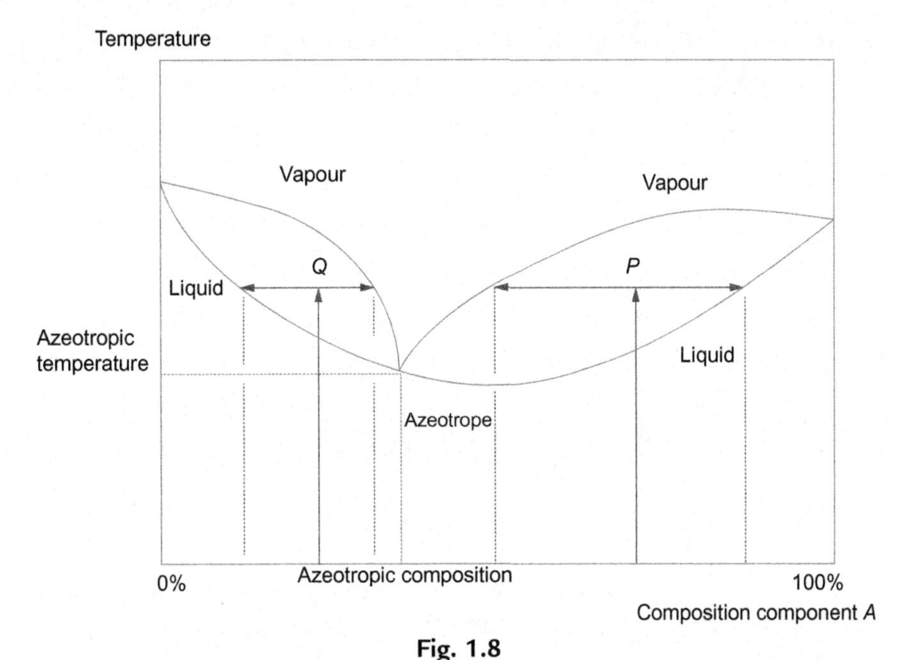

Fig. 1.8
Liquid boiling points and vapour condensation temperatures for minimum-boiling azeotrope mixtures of components *A* and *B*.

point *Q* in the liquid–vapour equilibrium zone corresponds to a liquid that is more rich in *B* and to a vapour that is more rich in *A* than the original composition. For point *P* it is the opposite. In correspondence to the azeotropic composition, and to the azeotropic temperature, vapour will have the same composition as liquid.

Table 1.4 includes some azeotropic mixtures with the indication of the azeotropic composition (first component) of the azeotropic temperature at 1 atm.

Table 1.4 Azeotropic mixtures at 1 atm (Dean, 1999)

1st Component	2nd Component	Azeotropic Composition (%)	Azeotropic Temperature (°C)
Hydrochloric acid	Water	20.24	110
Sulphuric acid	Water	98	338
Ethanol	Water	78.1	95.5
Ethyl acetate	Ethanol	71.8	69.2
Benzene	Ethanol	68.2	67.6
Toluene	Ethanol	76.7	32
n-Pentane	Ethanol	34.2	95
n-Exane	Ethanol	58.7	79

Table 1.5 Inverse of Henry's constants for some gases in water ($1/K_H \times 10^{-4}$, atm/molar fraction)

T (°C)	Air	CO_2	CO	C_2H_6	H_2	H_2S	CH_4	O_2
0	4.32	0.0728	3.52	1.26	5.79	0.0268	2.24	2.55
10	5.49	0.104	4.42	1.89	6.36	0.0367	2.97	3.27
20	6.64	0.142	5.36	2.63	6.83	0.0483	3.76	4.01
30	7.71	0.186	6.20	3.42	7.29	0.0609	4.49	4.75
40	8.70	0.233	6.96	4.23	7.51	0.0745	5.20	5.35
50	9.46	0.283	7.61	5.00	7.65	0.0844	5.77	5.88
60	10.1	0.341	8.21	5.65	7.65	0.1030	6.26	6.29
70	10.5		8.45	6.23	7.61	0.1190	6.66	6.63
80	10.7		8.45	6.61	7.55	0.1350	6.82	6.87
90	10.8		8.46	6.87	7.51	0.1440	6.92	6.99
100	10.7		8.46	6.92	7.45	0.1480	7.01	7.01

International Critical Tables of Numerical Data, Physics, Chemistry and Technology.

1.4.5 Gas–Liquid Equilibrium in Liquid Solutions

Solubility of gas in liquids increases with the gas's partial pressure over the liquid phase. For many gases the relationship between the concentration of the dissolved gas and the partial pressure over the liquid surface is expressed by Henry's law:

$$P_A = \frac{y_A}{K_H} \tag{1.72}$$

Henry's law is accurate at relative pressures and when the dissolved gas concentration is not very high. Table 1.5 shows inverse of Henry's constants for some gases in water.

Considering the frequent risk in the oil and gas and chemical industry, as well as in the wastewater management, and the very toxic effects of hydrogen sulphide, the following drawing is very effective to understand the big risk of this substance, even in relatively low concentrations in water (Fig. 1.9).

Example 1.8

A vessel with wastewater which contains 10 ppm (weight) of hydrogen sulphide at 30°C. The vessel is located below grade and is regularly accessed by a maintenance team through a ladder. Engineers are inquired about the necessity to install some gas detector in the confined space and possibly classify the zone as a restricted area. The very low sulphide concentration in the water convinces the team not to take into consideration the detection system (Fig. 1.10).

Solution

$$y_A = 10\,\text{ppm} = \frac{10\,\text{mg}\,H_2S}{1\,\text{kg water}} \cong \frac{10/(34,000)\,\text{mol}\,H_2S}{55\,\text{mol}\,H_2O} = 5.4 \cdot 10^{-6} \tag{1.73}$$

$$P_A = \frac{y_A}{K_H} = 609 \cdot 5.4 \cdot 10^{-6} = 0.0032886\,\text{atm} = 3288.6\,\text{ppmv} \tag{1.74}$$

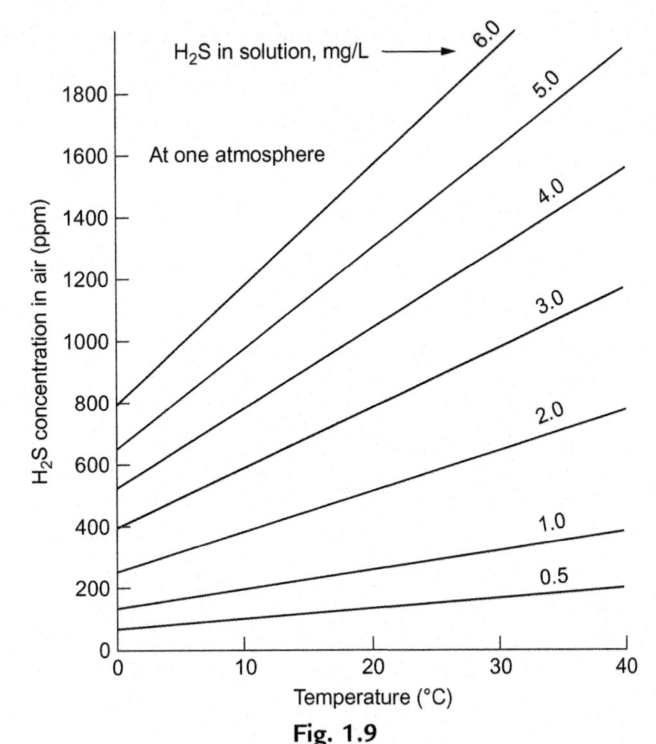

Fig. 1.9
Gas–liquid concentration of hydrogen sulphide in water (U.S. EPA, 1985).

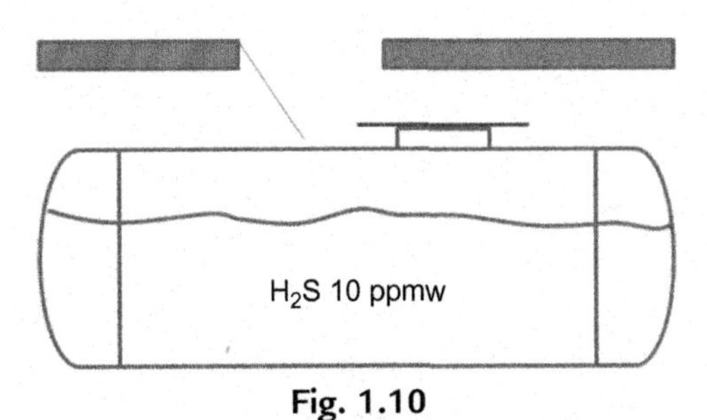

Fig. 1.10
Sour water vessel of Example 1.8.

The assumption made is wrong. This concentration kills immediately (OSHA, 2016).

Solubility data of hydrogen sulphide in hydrocarbon and derived organic solvents are included in Table 1.6 (Groysman, 2014)

Solubility data of H_2S in cyclohexane and *n*-decane at different molar fraction ranges are presented in Figs. 1.11 and 1.12 respectively (reproduced from Gerrard, 1980).

Table 1.6 Solubility of hydrogen sulphide in organic solvents

Solvent	Solubility	
	Molar Fraction	% Mass
n-Pentane	0.0507	2460
n-Hexane	0.0537	2.195
n-Heptane	0.0541	1.910
n-Octane	0.0556	1.726
n-Nonane	0.0575	1.595
n-Decane	0.0587	1.471
Ethanol	0.0177	1.314
Phenol	0.0200	0.773
Pyridine	0.0934	4.246
Benzene	0.0561	2.520
Toluene	0.0663	2.560
o-Xylene	0.0698	2.350

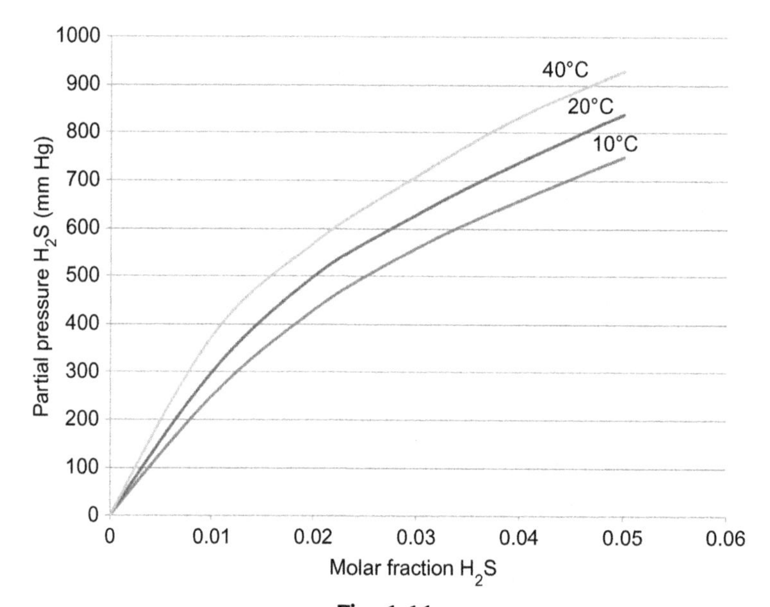

Fig. 1.11
Partial pressure vs molar fraction of H_2S in cyclohexane.

These data can be considered approximately as representative of hydrogen sulphide solubility in hydrocarbons, given the comparability of the molar fraction in Table 1.6.

For some very soluble gases, such as sulphur dioxide and ammonia, Henry's constant is very low, so the respective solutions are away from ideal, and more accurate data are necessary, as reported in Figs. 1.13 and 1.14.

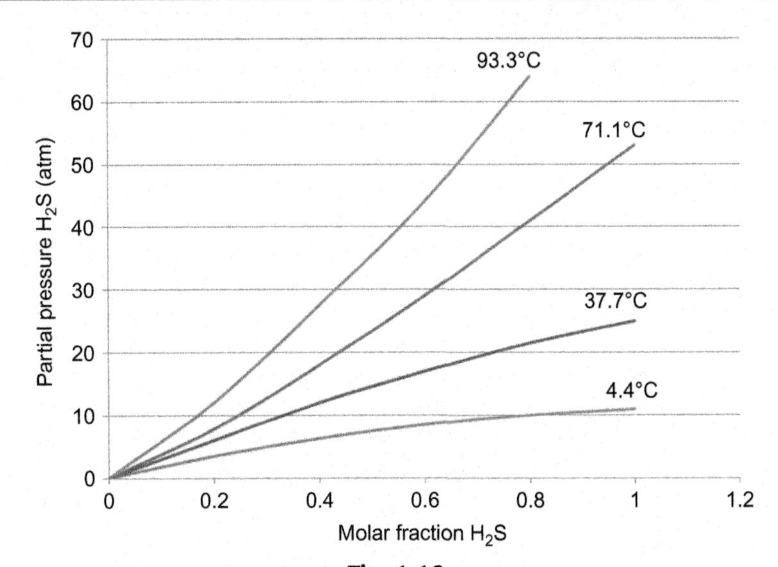

Fig. 1.12
Partial pressure vs molar fraction of H_2S in n-decane.

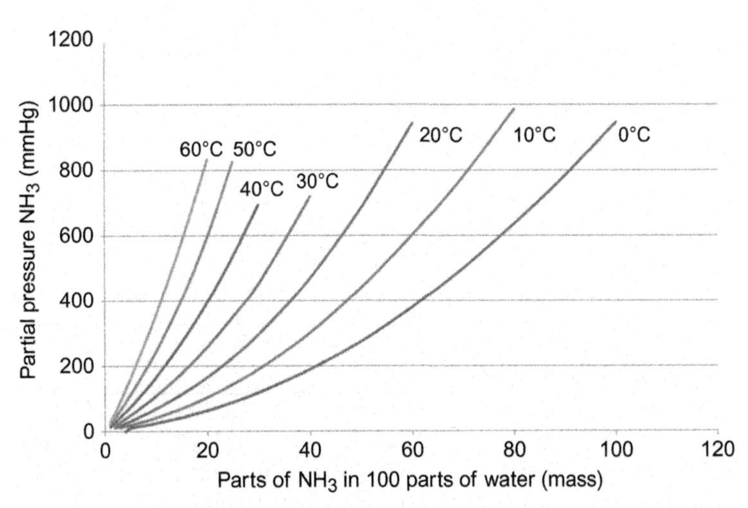

Fig. 1.13
Solubility values for NH_3 in water (Sherwood, 1925).

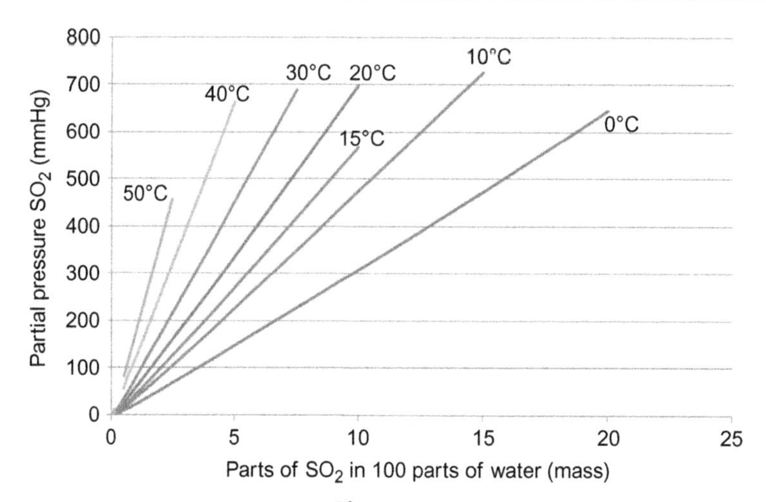

Fig. 1.14
Solubility values for SO_2 in water (Sherwood, 1925).

Inverse of Henry's constants (bars per mole fraction) for light hydrocarbons gases (methane, ethane, propane, *n*-butane and *i*-butane) in water at various temperatures (Kelvin) may be calculated according to the following equations (Riazi, 2005):

$$K_H = e^{\left(A_1 + A_2 \cdot T + \frac{A_3}{T} + A_4 \cdot \ln T\right)} \tag{1.75}$$

Coefficient A_1 to A_4 and applicable temperature and pressure ranges have been included in Table 1.7. The average absolute deviation percentage associated with Eq. (1.75) falls within the range 3.6–7.5.

Table 1.7 Henry's constants for light hydrocarbons in water (API Technical Data Book, 1997)

Hydrocarbon	*T* Range (K)	Pressure Range (bar)	A_1	A_2	A_3	A_4
Methane	274–444	1–31	569.29	0.107035	−19,537	−92.17
Ethane	279–444	1–28	109.42	−0.023090	−8006.3	−11.467
Propane	278–428	1–28	1114.68	0.205942	−39162.2	−181.505
n-Butane	277–444	1–28	182.41	−0.018160	−11418.06	−22.455
i-Butane	278–378	1–10	1731.13	0.429534	−52318.06	−293.567

In Fig. 1.15, Henry's constant of chlorine has been reproduced between 0°C and 100°C (Lange, 1949).

Hydrogen chloride solubility in water is high, as shown in Fig. 1.16.

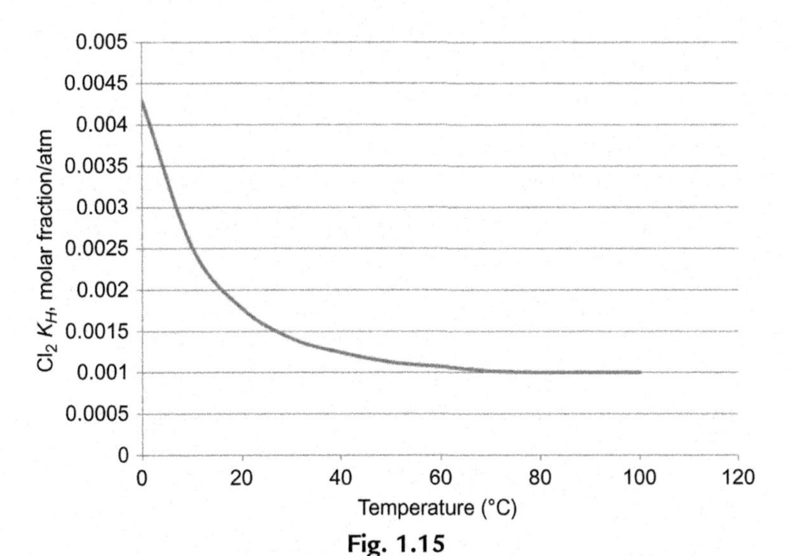

Fig. 1.15

Henry constant for chlorine. *Reproduced from data included in Lange, N.A., 1949. Lange's Handbook of chemistry, seventh ed. Handbook Publishers, Sandusky, OH.*

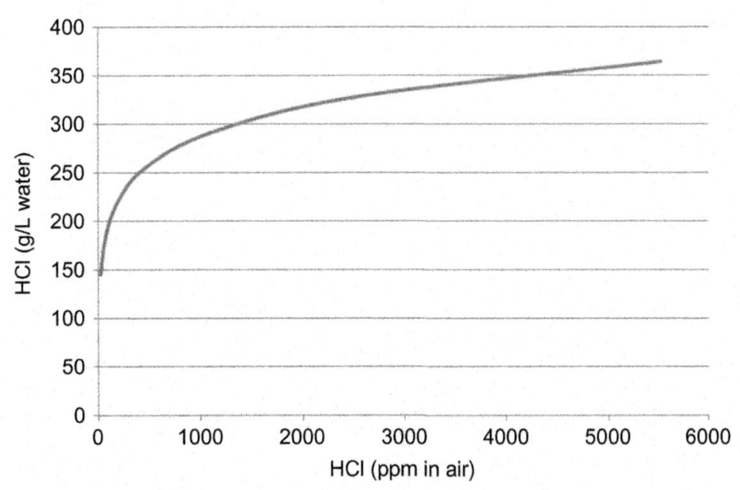

Fig. 1.16

Solubility of hydrogen chloride in water (Wight Washburn, 1929).

1.4.6 Gas–Liquid Equilibrium in Acid Gas Removal (AGR) Units

Sour gases such as hydrogen sulphide and carbon dioxide are removed via amine absorption with chemical reaction. Gas–liquid equilibrium in this case doesn't follow Henry's law due to the high absorption promoted by kinetics. Figs. 1.17 and 1.18 present a graphical plot of the experimental values provided by Barreau et al. (2006) relevant to the partial pressure of carbon dioxide and hydrogen sulphide against gas mole up to 25 wt % Diethanolamine (DEA) mole ratio at given temperatures.

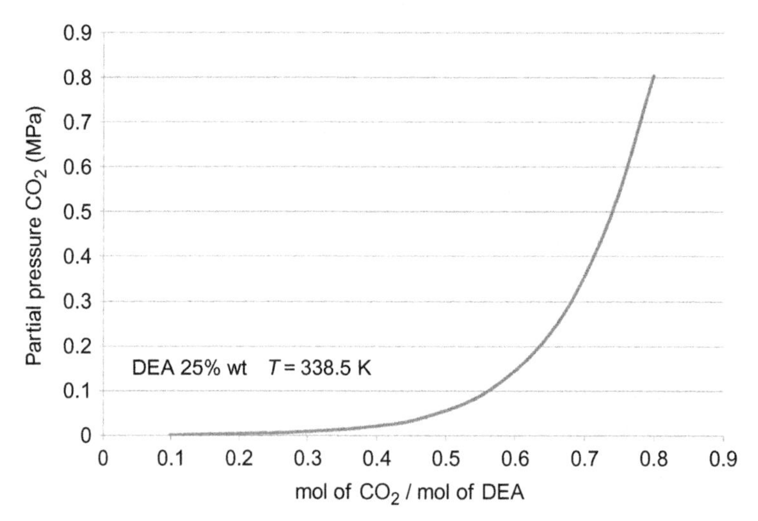

Fig. 1.17
Carbon dioxide solubility data in Diethanolamine (Barreau et al., 2006).

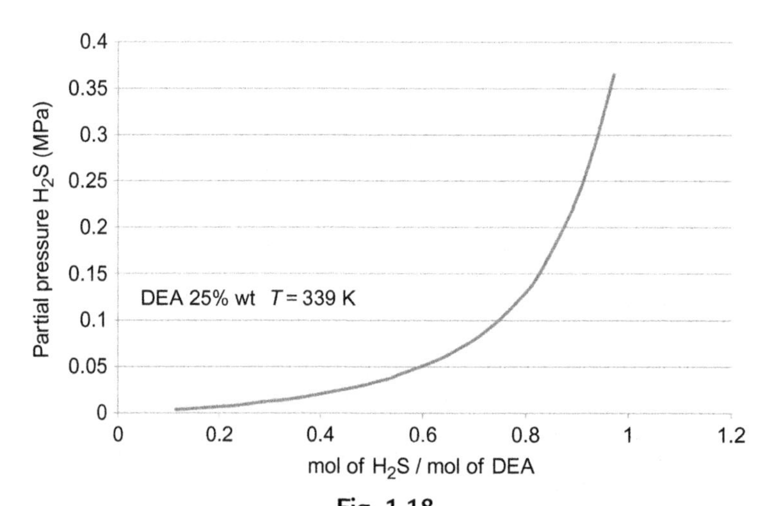

Fig. 1.18
Hydrogen sulphide solubility data in Diethanolamine (Barreau et al., 2006).

Mole fraction solubility of hydrogen sulphide in straight chain alkanes at 298.15 K, and a partial pressure of 1.013 bar can be approximated adopting the following equation, obtained from the experimental work of Makranczy et al. (1976):

$$100 \cdot x_{H_2S} = 0.129 \cdot n + 3.584 \quad (1.76)$$

where n is the no. of carbon atoms between 4 and 16.

1.4.7 Equilibria in Aqueous Solution

Pure water contains a very low concentrations of hydronium (H_3O^+) and hydroxide ions (OH^-) produced by water dissociation. At 24°C the ion-product constant for water is (moles^2/litre^2):

$$K_W = [H_3O^+] \times [OH^-] = 10^{-14} \quad (1.77)$$

$$pH = -\log_{10}[H^+] \quad (1.78)$$

where the symbols [ ] indicate the moles/litre concentration unit. Strong electrolytes such as sulphuric acid and sodium hydroxide completely dissociate in a water solution, whereas weak electrolytes, such as hydrogen sulphide and ammonia, partially ionise in water.

1.4.8 Hydrogen Sulphide

Hydrogen sulphide is a colourless gas easily soluble in water, where it dissociates as a diprotic acid according to the equilibrium reactions:

$$H_2S \rightleftarrows HS^- + H^+ \quad (1.79)$$

$$HS^- + H_2O \rightleftarrows S^- + H_3O^+ \quad (1.80)$$

and the equilibrium constants are (Treadwell, 1960):

$$K_{1-H_2S} = \frac{[HS^-] \times [H^+]}{[H_2S]} = 9.1 \times 10^{-8} \quad (1.81)$$

$$K_{2-H_2S} = \frac{[S^=] \times [H^+]}{[HS^-]} = 10^{-15} \quad (1.82)$$

Comparing the two constants, we see that the variation of HS^- and H^+ due to the second reaction is negligible, so their values in Eq. (1.81) can be considered equal and approximately:

$$[S^=] = K_{2-H_2S} = 10^{-15} \quad (1.83)$$

Multiplying Eqs. (1.81) and (1.82):

$$K_{1-H_2S} \cdot K_{2-H_2S} = 9.1 \cdot 10^{-23} \quad (1.84)$$

Example 1.9

Sour water contains 0.17 kg/L of non-ionised hydrogen sulphide. Air is to be blown through the liquid in order to strip out an amount equal to 1% of this acid. Calculate the initial and the final *pH*.

Solution

$$K_{1-H_2S} = \frac{[HS^-] \times [H^+]}{[H_2S]_o - [HS^-]} = \frac{x^2}{170/34 - x} = \frac{x^2}{5 - x} = 9.1 \times 10^{-8} \quad (1.85)$$

Since $[HS^-] \ll [H_2S]_o$:

$$\frac{x^2}{5 - x} \cong \frac{x^2}{5} = 9.1 \times 10^{-8} \quad (1.86)$$

$$x_o = [H^+]_o = \sqrt{5 \cdot 9.1 \times 10^{-8}} = 0.000675 \quad (1.87)$$

$$pH_o = -\log_{10} 0.000675 = 3.17 \quad (1.88)$$

$$x_1 = [H^+]_1 = \sqrt{0.05 \cdot 9.1 \times 10^{-8}} = 0.0000675 \quad (1.89)$$

$$pH_f = -\log_{10} 0.0000675 = 4.17 \quad (1.90)$$

The assumption to consider all dissociated H_2S as HS^- and H^+ can be generalised for other substances, provided the dissociation constant is smaller than 10^{-4}.

1.4.9 Sulphuric and Sulphurous Acid

Sulphur trioxide and dioxide are absorbed in water, producing, respectively, sulphuric and sulphurous acid. Sulphur dioxide is the main outcome of the combustion of sulphur compounds, even if a small amount of trioxide is formed, especially if a catalytic action is promoted by some metals, such as vanadium. Sulphuric acid may be considered totally dissociated into hydrogen and sulphate ions. Sulphurous acid is weaker and presents the following equilibria in water:

$$H_2SO_3 \rightleftarrows HSO_3^- + H^+ \quad (1.91)$$

$$HSO_3^- \rightleftarrows SO_3^= + H^+ \quad (1.92)$$

with the following equilibrium constant (Treadwell, 1960):

$$K_{1-H_2SO_3} = \frac{[HSO_3^-] \times [H^+]}{[H_2SO_3]} = 1.7 \times 10^{-2} \quad (1.93)$$

$$K_{2-H_2SO_3} = \frac{[SO_3^=] \times [H^+]}{[HSO_3^-]} = 5.0 \times 10^{-6} \quad (1.94)$$

1.4.10 Carbon Dioxide

In an aqueous solution, carbon dioxide is present as molecular CO_2 along with strong diprotic carbonic acid H_2CO_3 which is in equilibrium with its ionic species (*a*, Housecroft and Sharpe, 2008), (*b*, Treadwell, 1960)

$$CO_2 + H_2O \rightleftarrows H_2CO_3 \tag{1.95}$$

$$H_2CO_3 \rightleftarrows H^+ + HCO_3^- \tag{1.96}$$

$$HCO_3^- \rightleftarrows H^+ + CO_3^= \tag{1.97}$$

with:

$$K_{1-CO_2} = \frac{[H_2CO_3]}{[CO_2]} = 1.7 \times 10^{-3} \tag{1.98}$$

$$K_{2-CO_2} = \frac{[HCO_3^-][H^+]}{[H_2CO_3]} = 3 \times 10^{-7} \tag{1.99}$$

$$K_{3-CO_2} = \frac{[CO_3^=][H^+]}{[HCO_3^-]} = 5 \times 10^{-11} \tag{1.100}$$

Example 1.10

pH of carbonated water at 24°C is 5. Find the carbon dioxide pressure above the liquid surface.

Solution

Assuming all dissociated carbonic acid:

$$\frac{[HCO_3^-][H^+]}{[H_2CO_3]} \cong \frac{(10^{-5})^2}{[H_2CO_3]} = 3 \times 10^{-7} \tag{1.101}$$

$$[H_2CO_3] = 3.3 \times 10^{-4} \tag{1.102}$$

$$\frac{[H_2CO_3]}{[CO_2]} = \frac{0.00033}{[CO_2]} = 1.7 \times 10^{-3} \tag{1.103}$$

$$[CO_2] = 0.194\frac{\text{mol}}{\text{L}} \cong 0.0035 \text{ molar\_fraction} \tag{1.104}$$

Applying Henry's law:

$$P_A = \frac{y_A}{K_H} \tag{1.105}$$

And assuming from Table 1.5 $K_H{}^{-1} \cong 0.142 \times 10^4$ atm/molar faction:

$$P_A = 0.0035 \cdot 0.142 \cdot 10^4 = 4.96\,\text{atm} \tag{1.106}$$

This result is consistent with data reported in literature (Carroll et al., 1991) (Fig. 1.19).

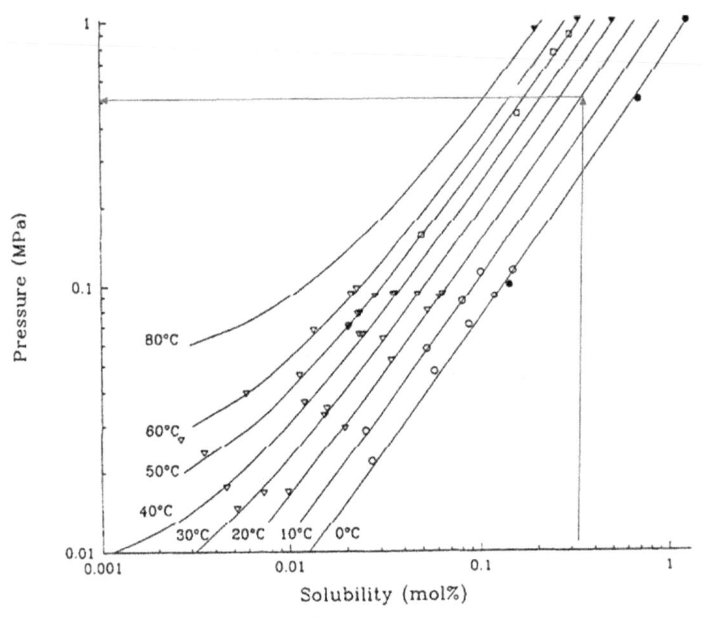

Fig. 1.19
Solubility of carbon dioxide with pressure in water at various temperatures (Carroll et al., 1991).

1.4.11 Ammonia

Ammonia is a colourless gas easy soluble in water where it dissociates according to the equilibrium reaction:

$$NH_3 + H_2O \rightleftarrows NH_4{}^+ + OH^- \tag{1.107}$$

and the equilibrium constant is (Treadwell, 1960):

$$K_{NH_3} = \frac{[NH_4{}^+] \times [OH^-]}{[NH_3]} = 1.8 \times 10^{-5} \tag{1.108}$$

1.4.12 Chlorine

Chlorine is a green gas with an irritating odour. It is easily absorbed by water and undergoes the following reaction:

$$Cl_2 + H_2O \rightleftarrows H^+ + OCl^- + HCl \tag{1.109}$$

with (Treadwell, 1960):

$$K_{Cl_2} = \frac{[H^+] \times [Cl^-] \times [HClO]}{[Cl_2]} = 3.0 \times 10^{-4} \tag{1.110}$$

Hydrogen chloride and molecular chlorine are in equilibrium with steam and oxygen according to Deacon's reaction:

$$2HCl + 0.5O_2 \rightleftarrows H_2O + Cl_2 \tag{1.111}$$

This equilibrium is described by the equilibrium constant:

$$K_P = \frac{p_{H_2O} \times p_{Cl_2}}{p_{HCl}^2 \times p_{O_2}^{0.5}} \tag{1.112}$$

where K_P (atm$^{-1/2}$) is defined as (Niessen, 2010):

$$K_P = \exp\left(-8.244 + 1.512 \times 10^{-4} T + \frac{7087}{T}\right) \tag{1.113}$$

Example 1.11

PENEX process is used to isomerise naphtha. The feedstock is sent to two reactors, which are charged with a catalyst activated by hydrogen chloride. A stabiliser column separates isomerate from overhead vapours, which are sent to a scrubber where hydrogen chloride is removed (Fig. 1.20). On the overhead line a pressure relief valve is connected to the flare. Analyse the operational scenario relevant to the PRV venting into the atmosphere and discuss the findings.

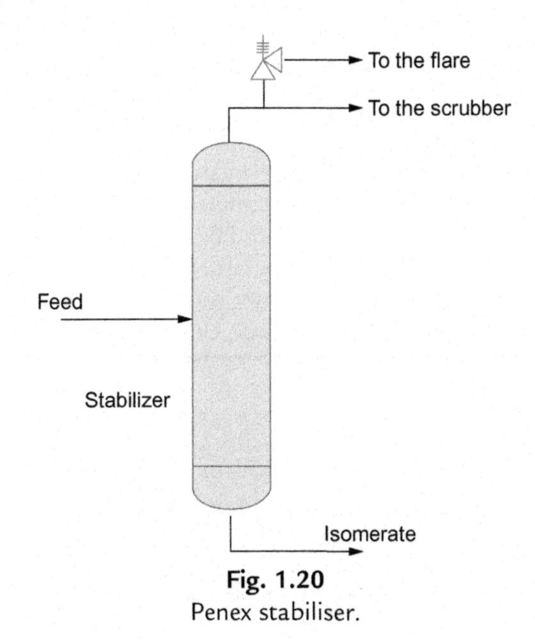

Fig. 1.20
Penex stabiliser.

Solution

It is assumed that flare combustion is not affected by air humidity, which is a realistic assumption. On this basis, steam partial pressure in the equilibrium equation may be considered equal to the molecular chlorine partial pressure (Fig. 1.21). Accordingly:

$$\frac{N_{Cl_2}}{N_{HCL}} = \frac{p_{Cl_2}}{P_{HCl}} = \sqrt{P_{O_2}^{0.5} \times K_P} = \sqrt{P_{O_2}^{0.5} \times \left(-8.244 + 1.512 \times 10^{-4} \times T + \frac{7087}{T}\right)} \quad (1.114)$$

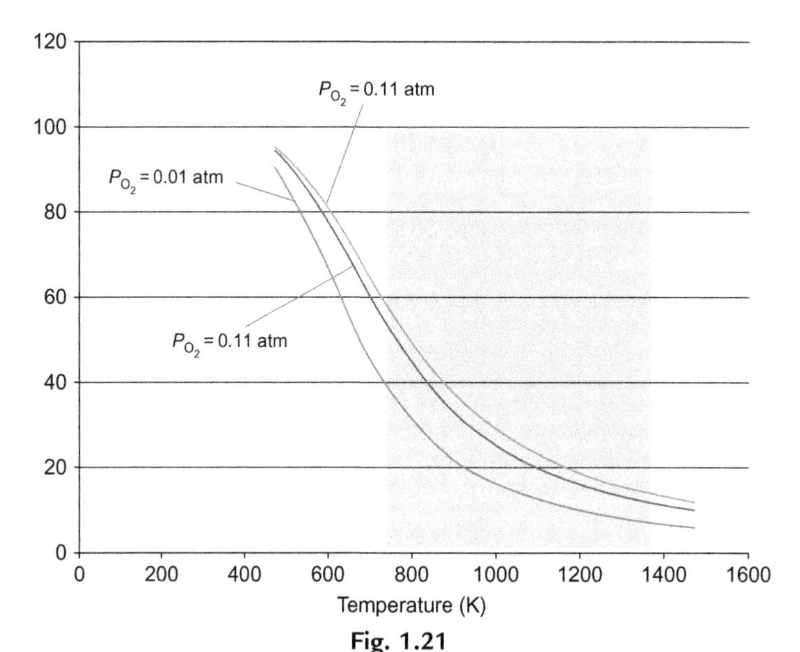

Fig. 1.21
Molecular chlorine fraction over chlorine content in Deacon equilibrium.

As per U.S. EPA (Fact Sheet EPA 452/F-03-019) the flare discharge temperature is between 500° C and 1100°C (773, 16 and 1373.16 K). It is determined that the chlorine-based molar fraction of molecular chlorine falls in the range 40%–57% and 6.5%–14%, depending on temperature and oxygen partial pressure. It is also seen that temperature affects equilibrium much more than oxygen partial pressure. This process segment is not safeguarded by a scrubber on the pressure relief valve line. Hydrogen chloride IDLH is 50 ppm, chlorine IDLH is 10 ppm.

1.4.13 Hydrolysis

Salts produced from weak acids, weak bases, or both undergo hydrolysis. Solutions of sulphides and ammonium salts hydrolyse when dissolved in water. Ammonium salts undergo the following hydrolysis reaction:

$$NH_4{}^+ + H_2O \rightleftarrows NH_3{}^- + H_3O^+ \quad (1.115)$$

and the equilibrium constant is the hydrolysis constant:

$$K_{i-NH_4{}^+} = \frac{[NH_3] \times [H_3O^+]}{[NH_4{}^+]} = 5.67 \times 10^{-10} \quad (1.116)$$

Table 1.8 Equilibrium constants for acids and bases in water solutions[a]

Substance	Formula	K_a	K_b
Acetic acid	CH_3COOH	$1.8\cdot10^{-5}$	
Ammonia	NH_3		$1.8\cdot10^{-5}$
Aniline	$C_6H_5NH_2$		$4.6\cdot10^{-10}$
Benzoic acid	C_6H_5OOH	$6.4\cdot10^{-5}$	
Boric acid	HBO_2	$6.5\cdot10^{-10}$	
Carbonic acid	$CO_2 \rightarrow H_2CO_3$[b]	$1.7\cdot10^{-3}$	
	$H_2CO_3 \rightarrow HCO3^-$	$3.5\cdot10^{-7}$	
	$HCO_3^- \rightarrow CO3^=$	$5.0\cdot10^{-11}$	
Chromic acid	$H_2CrO_4 \rightarrow HCrO_4^-$	$1.8\cdot10^{-1}$	
	$HCrO_4^- \rightarrow CrO_4^=$	$4.0\cdot10^{-7}$	
Diethylamine	$(C_2H_5)_2NH$		$1.3\cdot10^{-3}$
Dimethylamine	$(CH_3)_2NH$		$7.4\cdot10^{-4}$
Ethylamine	C_2H_5NH		$6.0\cdot10^{-4}$
Hydrazine	NH_2NH_2		$8.0\cdot10^{-7}$
Hydrofluoric acid	HF	$7.5\cdot10^{-4}$	
Hydroxylamine	NH_2OH		$1.0\cdot10^{-8}$
Hypochlorous acid	$HClO$	$5.0\cdot10^{-8}$	
Formic acid	$HCOOH$	$2.0\cdot10^{-4}$	
Methylamine	CH_3NH_2		$5.0\cdot10^{-4}$
Nitrous acid	HNO_2		$4.0\cdot10^{-4}$
Phosphoric acid	$H_3PO_4 \rightarrow H_2PO_4^-$	$7.5\cdot10^{-3}$	
	$H_2PO_4^- \rightarrow HPO_4^=$	$6.2\cdot10^{-8}$	
	$HPO_4^= \rightarrow PO_4^{\equiv}$	$1.0\cdot10^{-13}$	
Phenol	C_6H_5OH	$1.3\cdot10^{-10}$	
Pyridine	C_6H_5N		$1.4\cdot10^{-9}$
Sulphidric acid	$H_2S \rightarrow HS^-$	$9.1\cdot10^{-8}$	
	$HS^- \rightarrow S^=$	$1.2\cdot10^{-15}$	
Sulphurous acid	$H_2SO_3 \rightarrow HSO_3^-$	$1.7\cdot10^{-2}$	
	$HSO_3^- \rightarrow SO^=$	$5.0\cdot10^{-6}$	

[a]Treadwell (1960).
[b]Housecroft and Sharpe (2008).

Table 1.8 provides the equilibrium constants of some acids and bases (Treadwell, 1960).

1.5 Absorption and Adsorption

Toxic and flammable substances have to be removed from process streams either as part of the process or to reduce the health and environmental impact. For this purpose, they are absorbed by a liquid solution or adsorbed on porous media.

1.5.1 Absorption With Chemical Reaction

Absorption is the transfer of a gas or a vapour to a liquid. This process depends on the driving force existing between the liquid phase and the gas phase and is governed and limited by the gas–liquid equilibrium condition and, in turn, by the gas solubility. The lower the dissolved gas concentration, the greater is the mass transfer. This is the reason why absorption is promoted by chemical reactions occurring in the liquid body between gas A and reactant B, which reduces the concentration as gas is transferred to the liquid and is converted to D:

$$A(gas \rightarrow liq) + B(liq) \rightarrow D(liq) \tag{1.117}$$

In Fig. 1.22 p is partial pressure of A and C is concentration of reactant B in the liquid.

Applying Eq. (1.4) and assuming a steady state, the following overall and individual mass balance can be written (Table 1.9):

$$W_{in} - W_{out} = 0 \tag{1.118}$$

$$W_{A_{in}} - W_{A_{out}} - W_{A_{gen}} = 0 \tag{1.119}$$

$$W_{B_{in}} - W_{B_{out}} - W_{B_{gen}} = 0 \tag{1.120}$$

$$W_{D_{out}} = W_{D_{gen}} = W_{A_{gen}} + W_{B_{gen}} \tag{1.121}$$

1.5.2 Stripping

The operation of removing absorbed solutes from liquids is called stripping. This works according to the same principles of absorption, but just inverts the driving force with the aim to enhance the gas transfer from the liquid to the gas stream. Unlike absorption, stripping is promoted by high temperatures and low pressures. Stripping typically removes toxic or hazardous gases such as hydrogen sulphides and ammonia.

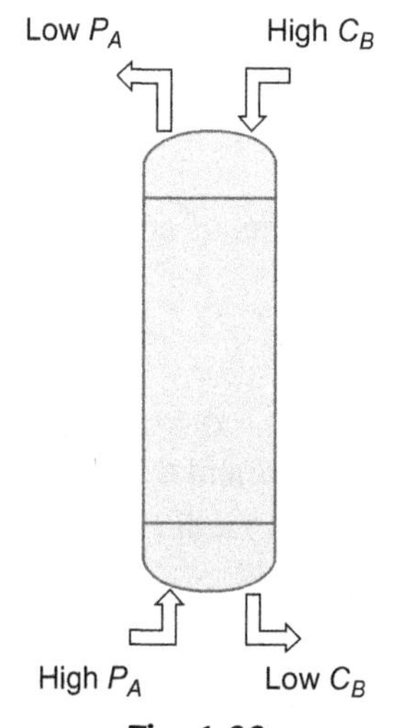

Fig. 1.22
Absorption with chemical reaction of gas or vapour A and liquid reactant B.

Table 1.9 Absorption of gases and vapours with selected reagents (Teller, 1960)

Gas	Reagent
Carbon dioxide	Carbonates
Carbon dioxide	Hydroxides
Carbon dioxide	Amines
Carbon monoxide	Cuprous amine complexes
Carbon monoxide	Cuprous ammonium chloride
Sulphur dioxide	Calcium hydroxide
Sulphur dioxide	Potassium hydroxide
Sulphur dioxide	Ozone–Water
Chlorine	Water
Chlorine	Ferrous chloride
Hydrogen sulphide	Amine
Hydrogen sulphide	Ferric hydroxide
Sulphur trioxide	Sulphuric acid
Ethylene	Potassium hydroxide
Ethylene	Trialkyl phosphates
Olephins	Cuprous amine complexes
Nitrogen monoxide	Ferrous sulphate
Nitrogen monoxide	Calcium hydroxide
Nitrogen monoxide	Sulphuric acid
Nitrogen monoxide	Water

1.5.3 Adsorption

Adsorption from gas phase is a mass transfer process of a gas or a vapour to a solid. When the molecules of the fluid come in contact with the adsorbent, equilibrium is established between the adsorbed gas or vapour and the fluid remaining in the gas phase. For a given adsorbate at a given temperature an isotherm can be constructed, relating the mass of adsorbate per unit of weight of adsorbent to the partial pressure of the adsorbate in the gas phase. The isotherm can be described by empirical equations such as Freundlich, Languimur or Brunauer, Emmett, and Teller's equations. Freundlich's equation is:

$$w = k \cdot p^m \tag{1.122}$$

where w is the mass of adsorbate per unit of weight of adsorbent, k and m are experimental constants, p is the partial pressure. A family of adsorption isotherms having the shape typical of adsorption on activated carbon is plotted in Fig. 1.23 (U.S. EPA, 2002). This, along with other isotherms whose shapes are convex upward throughout, are designated *Type I* isotherms. The Freundlich isotherm, which can be fit to a portion of a Type I curve, is commonly used in industrial design.

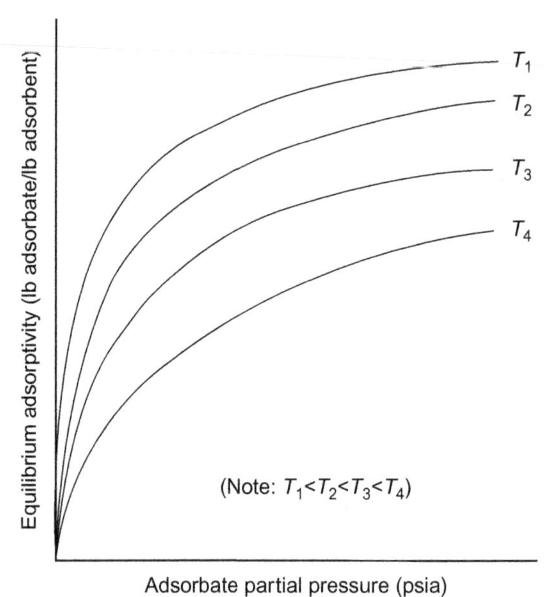

Fig. 1.23
Type I adsorption isotherms for hypothetical adsorbate (U.S. EPA, 2002).

Table 1.10 Parameters for selected adsorption isotherms (U.S. EPA, 2002)

Adsorbate	Adsorption Temperature (°F)	Isotherm Parameters		Range of Isotherm
		k	m	
Benzene	77	0.597	0.176	0.0001–0.005
Chlorobenzene	77	1.05	0.188	0.0001–0.01
Cychloexane	100	0.505	0.210	0.0001–0.05
Dichloroethane	77	0.976	0.281	0.0001–0.04
Phenol	104	0.855	0.153	0.0001–0.03
Trichloroethane	77	1.06	0.161	0.0001–0.04
Vinyl chloride	100	0.200	0.477	0.0001–0.05
m-Xylene	77	0.708	0.113	0.0001–0.001
	77	0.527	0.0703	0.0001–0.05
Acrylonitrile	100	0.935	0.424	0.0001–0.015
Acetone	100	0.412	0.389	0.0001–0.05
Toluene	77	0.551	0.110	0.0001–0.05

Table 1.10 presents Freundlich's parameters for some organic compounds (U.S. EPA, 2002).

Data adsorption of Calgon type BPL carbon not to be extrapolated outside these ranges.

Typical criteria apply in gas or vapour adsorption by solid media:

- The greater the molecular weight, the more the adsorption efficiency.
- Non-polar molecules are better adsorbed than soluble molecules.

- Non-soluble molecules are better adsorbed than polar molecules.
- Temperature increase may reduce the degree of adsorption.

1.6 Applications

1.6.1 Kinetics and Equilibrium of Sulphur Oxides

Sulphur dioxide and sulphur trioxide obey the following equilibrium:

$$SO_2 + 0.5O_2 \rightleftarrows SO_3 \tag{1.123}$$

And (Niessen, 2010) :

$$K_P = \frac{p_{SO_3}}{p_{SO_2} \times p_{O_2}^{0.5}} \tag{1.124}$$

$$\log_{10} K_P = \frac{5186.5}{T} + 0.611 \cdot \log_{10} T - 6.7497 \tag{1.125}$$

Real distribution of sulphur between dioxide and trioxide depends on many factors, but conversion to SO_3 is accelerated by metals working as catalysts (vanadium). Typically 2%–4% appears as trioxide. It has to be considered that conversion to SO_3 is promoted by high air excess, potentially resulting in much greater trioxide percentages.

1.6.2 Properties of Hydrogen Sulphide

See Tables 1.11–1.14.

1.6.3 Properties of Ammonia

See Tables 1.15 and 1.16.

1.6.4 Properties of Sulphur Dioxide

See Tables 1.17 and 1.18.

1.6.5 Properties of Sulphur Trioxide

See Tables 1.19 and 1.20.

Table 1.11 Properties of hydrogen sulphide

Name in Air	Hydrogen Sulphide
Name in water	Sulphydric acid
Formula	H_2S
CAS number	7783-06-4
EINECS number	231-977-3
UN number	1053
Molecular weight	34.081
$T_{BOIL\ (Norm)}$	212.8 K
T_{CRIT}	373.53 K
P_{CRIT}	9 MPa
Gas density	Gas density (15°C): 1.422 kg/m^3
State at 1 atm 25°C	Vapour
Specific gravity	1.17
Flash point	208.4 K (TNO)
AIT (air)	260°C (Glassman and Yetter, 2008)
MIE	0.77 (Glassman and Yetter, 2008)
LEL (1 atm 25°C)	4% (Glassman and Yetter, 2008)
UEL	44% (Glassman and Yetter, 2008)
Hazards	Inhalation–Fire–Explosion
Odour	Rotten eggs
Odour lower perception threshold	0.00047 (Parthum and Leffel, 1979)
Odour higher perception threshold	100 ppm (Knight and Presnell, 2005)

Table 1.12 Worker exposure limits of hydrogen sulphide

	Recommended Limits (ppm)			
Limit	NIOSH	OSHA	ACGIH	AIHA
IDLH	100	–	–	–
8-hour TWA	10	–	1	–
STEL	15	–	5	–
CEILING	50	10 (10 min)	–	–
ERPG 1	–	–	–	0.1
ERPG 2	–	–	–	30
ERPG 3	–	–	–	100

1.6.6 Properties of Carbon Monoxide

See Tables 1.21 and 1.22.

1.6.7 Properties of Carbon Dioxide

See Tables 1.23 and 1.24.

Table 1.13 Short-term symptoms and effects of hydrogen sulphide (OSHA, 2016)

Concentration (ppm)	Symptoms/Effects
0.00011–0.00033	Typical background concentrations
0.01–1.5	Odour threshold (when rotten egg smell is first noticeable to some). Odour becomes more offensive at 3–5 ppm. Above 30 ppm, odour described as sweet or sickeningly
2–5	Prolonged exposure may cause nausea, tearing of the eyes, headaches, or loss of sleep. Airway problems (bronchial constriction) in some asthma patients
20	Possible fatigue, loss of appetite, headache, irritability, poor memory, dizziness
50–100	Slight conjunctivitis (gas eye) and respiratory tract irritation after 1 h. May cause digestive upset and loss of appetite
100	Coughing, eye irritation, loss of smell after 2–15 min (olfactory fatigue). Altered breathing, drowsiness after 15–30 min. Throat irritation after 1 h. Gradual increase in severity of symptoms over several hours. Death may occur after 48 h
100–150	Loss of smell (olfactory fatigue or paralysis)
200–300	Marked conjunctivitis and respiratory tract irritation after 1 h. Pulmonary edema may occur from prolonged exposure
500–700	Staggering, collapse in 5 min. Serious damage to the eyes in 30 min. Death after 30–60 min
700–1000	Rapid unconsciousness, 'knockdown' or immediate collapse within 1–2 breaths, breathing stops, death within minutes
1000–2000	Nearly instant death

Table 1.14 Sectors where hydrogen sulphide may be present

Sector	Levels
Oil and gas	Low to very high
Petrochemical	Low to very high
Coke production	Medium
Sulphur production	Medium to high
Industrial wastewater treatment	Low to very high
Municipal wastewater treatment	Medium
Sewage management	Low to high
Waste treatment	Low to high
Landfilling	Low to high
Slaughterhouses	Low to medium
Tanneries	Low to medium
Iron smelting	Low to medium
Food processing	Low to medium
Pulp and paper industry	Low to high
Rubber industry	Low

Table 1.15 Properties of ammonia

Name	Ammonia
Formula	NH_3
CAS number	7664-41-7
EINECS number	215-647-6
UN number	1005
Molecular weight	17.03
$T_{BOIL\ (NORM)}$	239.72 K
T_{CRIT}	405.65 K
P_{CRIT}	11.3 MPa
Gas density	Gas density (15°C): 0.711 kg/m^3
State at 1 atm 25°C	Vapour
Specific gravity	0.58
Flash point	−82.4°C (TNO)
AIT (air)	651°C (Glassman and Yetter, 2008)
MIE	14 (Lees)
LEL	15% (Glassman and Yetter, 2008)
UEL	28% (Glassman and Yetter, 2008)
Hazards	Inhalation–Fire–Explosion
Odour	Pungent
Odour perception threshold	5 ppm (OSHA, 2016)

Table 1.16 Worker exposure limits of ammonia

	Recommended Limits (ppm)			
Limit	NIOSH	OSHA	ACGIH	AIHA
IDLH	300	–	–	–
8-h TWA	25	25	25	–
STEL	35	35	35	–
ERPG 1	–	–	–	25
ERPG 2	–	–	–	150
ERPG 3	–	–	–	1500

1.6.8 Properties of Chlorine

See Tables 1.25 and 1.26.

1.6.9 Properties of Benzene

See Tables 1.27 and 1.28.

Table 1.17 Properties of sulphur dioxide

Name	Sulphur Dioxide
Formula	SO_2
CAS number	7446-09-5
EINECS number	231-195-2
UN number	1079
Molecular weight	64.06
$T_{BOIL\ (NORM)}$	263.13 K
T_{CRIT}	430.75 K
P_{CRIT}	7.884 MPa
Gas density	Gas density (15°C): 2.674 kg/m^3
State at 1 atm 25°C	Vapour
Specific gravity	2.3
Flash point	NA
AIT	NA
MIE	NA
LEL	NA
UEL	NA
Hazards	Inhalation
Odour	Irritating

Table 1.18 Worker exposure limits of sulphur dioxide

Limit	Recommended Limits (ppm)			
	NIOSH	OSHA	ACGIH	AIHA
IDLH	100	–	–	–
8-h TWA	2	2	–	–
STEL	5	5	0.25	–
ERPG 1	–	–	–	0.3
ERPG 2	–	–	–	3
ERPG 3	–	–	–	25

Table 1.19 Properties of sulphur trioxide

Name	Sulphur Trioxide
Formula	SO_3
CAS number	7446-11-9
EINECS number	231-197-3
UN number	1079
Molecular weight	80.066
$T_{BOIL\ (NORM)}$	317.7 K
T_{CRIT}	490.85 K
P_{CRIT}	21.944 MPa
Gas density	Gas density (15°C): 3.342 kg/m^3
State at 1 atm 25°C	Liquid–vapour
Specific gravity	2.76
Flash point	NA
AIT	NA
MIE	NA
LEL	NA
UEL	NA
Hazards	Inhalation burning
Odour	Irritating

Table 1.20 Worker exposure limits of sulphur dioxide

Limit	Recommended Limits (ppm)			
	NIOSH	OSHA	ACGIH	AIHA
IDLH	–	–	–	–
8-h TWA	–	–	0.2	–
STEL	–	–	–	–
ERPG 1	–	–	–	02
ERPG 2	–	–	–	10
ERPG 3	–	–	–	120

Table 1.21 Properties of carbon monoxide

Name	Carbon Monoxide
Formula	CO
CAS number	211-128-3
EINECS number	211-128-3
UN number	1016
Molecular weight	28.01
$T_{BOIL\ (NORM)}$	81.07 K
T_{CRIT}	139.92 K
P_{CRIT}	3.49 MPa
Gas density	Gas density (15°C): 0.169 kg/m^3
State at 1 atm 25°C	Gas
Specific gravity	0.965
Flash point	NA
AIT	NA
MIE	NA
LEL	NA
UEL	NA
Hazards	Inhalation
Odour	Odourless

Table 1.22 Worker exposure limits of carbon monoxide

Limit	Recommended Limits (ppm)			
	NIOSH	OSHA	ACGIH	AIHA
IDLH	40,000	–	–	–
8-h TWA	25	35	25	–
Ceiling	200	200	–	–
ERPG 1	–	–	–	200
ERPG 2	–	–	–	350
ERPG 3	–	–	–	500

Table 1.23 Properties of carbon dioxide

Name	Carbon Dioxide
Formula	CO_2
CAS number	124-38-9
EINECS number	204-696-9
UN number	2187
Molecular weight	44.01
$T_{BOIL\ (SUBL)}$	194.66 K
T_{CRIT}	304.21 K
P_{CRIT}	7.39 MPa
Gas density	Gas density (15°C): 1.837 kg/m^3
State at 1 atm 25°C	Vapour
Specific gravity	2.3
Flash point	NA
AIT	NA
MIE	NA
LEL	NA
UEL	NA
Hazards	Inhalation
Odour	Odourless at low concentrations

Table 1.24 Worker exposure limits of carbon dioxide

Limit	Recommended Limits (ppm)			
	NIOSH	OSHA	ACGIH	AIHA
IDLH	40,000	–	–	–
8-h TWA	5000	5000	5000	–
STEL	30,000	30,000	30,000	–
ERPG 1	–	–	–	–
ERPG 2	–	–	–	–
ERPG 3	–	–	–	–

Table 1.25 Properties of chlorine

Name	Chlorine
Formula	Cl_2
CAS number	7782-50-5
EINECS number	231-959-5
UN number	1017
Molecular weight	70.91
$T_{BOIL\ (NORM)}$	239.12 K
T_{CRIT}	417.15 K
P_{CRIT}	7.79 MPa
Gas density	Gas density (15°C): 2.96 kg/m^3
State at 1 atm 25°C	Vapour
Specific gravity (gas)	2.41
Flash point	NA
AIT	NA
MIE	NA
LEL	NA
UEL	NA
Hazards	Inhalation
Odour	Pungent

Table 1.26 Worker exposure limits of chlorine

Limit	Recommended Limits (ppm)			
	NIOSH	OSHA	ACGIH	AIHA
IDLH	10	–	–	–
8-h TWA	0.5	–	0.5	–
STEL	1	–	1	–
Ceiling	–	0.5 (15 min)	–	–
ERPG 1	–	–	–	1
ERPG 2	–	–	–	3
ERPG 3	–	–	–	20

Table 1.27 Property limits of benzene

Name	Benzene
Formula	C_6H_6
CAS number	71-43-2
EINECS number	200-753-7
UN number	1114
Molecular weight	78.111
$T_{BOIL\ (NORM)}$	353.26 K
T_{CRIT}	562.16 K
P_{CRIT}	4.88 MPa
Gas density	Gas density (15°C): 3.26 kg/m^3
State at 1 atm 25°C	Vapour
Specific gravity (gas)	2.7
Flash point	262.15 K
AIT	580°C
MIE	0.2 mJ
LEL	1.2
UEL	7.8
Hazards	Inhalation contact
Odour	Sweet

Table 1.28 Worker exposure limits of benzene

Limit	Recommended Limits (ppm)			
	NIOSH	OSHA	ACGIH	AIHA
IDLH	10	–	–	–
8-h TWA	0.5	–	0.5	–
STEL	1	–	1	–
Ceiling	–	0.5 (15 min)	–	–
ERPG 1	–	–	–	1
ERPG 2	–	–	–	3
ERPG 3	–	–	–	20

References

API Technical Data Book, 1997. Petroleum Refining, sixth ed. American Petroleum Institute (API), Washington, DC.

Barreau, A., Blanchon le Bouhelec, E., Habchi Tounsi, P., Mougin, K.N., Lecomte, F., 2006. Absorption of H_2S and CO_2 in alkanolamine aqueous solution: experimental data and modelling with the electrolyte-NRTL model. Oil Gas Sci. Technol. - Rev. IFP 61 (3), 345–361.

BS-EN 60079-10-2, 2015. Explosive atmospheres - Part 10-2: Classification of Areas - Explosive dust atmospheres. CENELEC.

Carroll, J.J., Slupsky, J.D., Mather, A.E., 1991. The solubility of carbon dioxide in water at low pressure. J. Phys. Chem. Ref. Data 20, 1201.

Dean, J.A., 1999. Lange's Handbook of Chemistry, 15th ed. McGraw-Hill Inc.

Eckhoff, R.E., 2003. Dust Explosions, third ed. Gulf Professional Publishing.

Gerrard, W., 1980. Gas Solubilities. Widespread Applications, first ed. Pergamon Press.

Glassman, I., Yetter, R.A., 2008. Combustion, fourth ed. Elsevier.

Groysman, A., 2014. Corrosion in Systems for Storage and Transportation of Petroleum Products and Biofuels. Springer Science+Business Media, Dordrecht.

Hoffman, A.E., Crump, J.S., Hocott, C.R., 1953. Equilibrium Constants for a Gas Condensate System. Trans. AIME 198, 1–10.

Hougen, O.A., Watson, K.M., Ragatz, R.A., 1954. Chemical Process Principles, Part I: Material and Energy Balances, second ed. John Wiley and Sons Inc., New York.

Housecroft, C.E., Sharpe, A.G., 2008. Inorganic Chemsitry, Pearson Prentice Hall, third ed.

IUPAC, 2014. Compendium of Chemical Terminology, Golden Book, Version 2-3-3.

Knight, L., Presnell, E., 2005. Death by sewer gas: case report of a double fatality and review of the literature. Am. J. Forensic Med. Pathol. 26, 183.

Lange, N.A., 1949. Lange's Handbook of Chemistry, seventh ed. Handbook Publishers, Sandusky, OH.

Makranczy, J., Megyery-Balog, K., Rusz, L., Patyi, L., 1976. Solubility of gases in normal-alkanes. Hung. J. Ind. Chem. 4, 269–280.

Niessen, W.R., 2010. Combustion and Incineration Processes: Applications in Environmental Engineering, fourth ed. CRC Press Book.

OSHA, 2016. Occupational Safety and Health Administration (OSHA) Permissible Exposure Limits (PELS) from 29 CFR 1910.1000 (website).

Parthum, C.A., Leffel, R.E., 1979. Odor Control for Wastewater Facilities. Manual of Practice No. 22. Water Pollution Control Federation, Washington, DC.

Poling, B.E., Prausnitz, J.M., O'Connell, J.P., 2001. The Properties of Gases and Liquids, fifth ed. McGraw-Hill.

Riazi, M.R., 2005. Characterization and Properties of Petroleum Fractions. ASTM International.

Sherwood, T.K., 1925. Solubilities of sulfur dioxide and ammonia in water. Ind. Eng. Chem. 17 (7), 745–747.

Teller, A.J., 1960. Absorption With Chemical Reaction. Chem. Eng. 67, 111–124.

Treadwell, F.P., 1960. Chimica Analitica, Vol. I Analisi Qualitativa. Casa Editrice Francesco Vallardi.

U.S. EPA, 1985, Odor and Corrosion Control in Sanitary Sewerage Systems and Treatment Plants. Design Manual. EPA/62571-85-018.

U.S. EPA, 2002. EPA Air Pollution Control Cost Manual, 6 h ed., EPA/452/B-02-001.

U.S. EPA, Air Pollution Control Technology Fact Sheet EPA 452/F-03-019.

Wight Washburn, E., 1929. International Critical Tables of Numerical Data, Physics, Chemistry and Technology. National Research Council.

Further Reading

ACGIH, 2015. Book and documentation of the threshold limit values on chemical substances. In: American Conference of Governmental Industrial Hygienists, seventh ed.

AIHA, 2015. Emergency Response Planning Guidelines (ERPG) Levels. American Industrial Hygiene Association.
NIOSH, 2007. National Institute for Occupational Safety and Health. Recommended Exposure Limits (RELs) from the NIOSH Pocket Guide to Chemical Hazards. http://www.cdc.gov/niosh/npg.
Standing, M.B., 1979. A set of equations for computing equilibrium ratios of a crude oil/natural gas system at pressures below 1000 psia. J. Pet. Technol. 31, 1193–1195.

CHAPTER 2

Thermodynamics and Thermochemistry of Process Safety

The gods have not, of course, revealed all things to mortals from the beginning; but rather, seeking in the course of time, they discover what is better.

(Xenophanes of Colophon,c. 570–c. 475 BC)

2.1 Ideal Gases

The general ideal gas law is:

$$P \cdot V = n \cdot R \cdot T \tag{2.1}$$

where

- P is the absolute pressure.
- V is the volume.
- T is the absolute temperature.
- n is the number of moles, kmoles or lb-moles, depending on units.
- R is the universal gas constant.

Ideal gas law applies to partial pressure and partial volume:

$$P_A \cdot V = n_A \cdot R \cdot T \tag{2.2}$$

and

$$P \cdot V_B = n_B \cdot R \cdot T \tag{2.3}$$

where P_A and V_B are partial pressure and volume of component A and B, respectively.

Dalton's law

$$P_A = y_A \cdot P \rightarrow P = P_A + P_B + \ldots \tag{2.4}$$

Amagat's law

$$V_B = y_B \cdot V \rightarrow V = V_A + V_B + \ldots \tag{2.5}$$

where y_A and y_B are molar fractions of gaseous components A and B.

Process Safety Calculations. https://doi.org/10.1016/B978-0-08-101228-4.00002-2

Table 2.1 Values of gas law constants (Hougen et al., 1954)

Energy	Temperature Units	Pressure Units	Volume Units	Weight Units	R
–	K	atm	m^3	kmole	0.08206
–	K	atm	L	moles	0.08206
–	°R	atm	ft^3	lb-mole	0.7302
kcal	K			kmole	1.987
BTU	°R			lb-mole	1.987

2.1.1 Standard and Normal Conditions

Both states, normal and standard, are assumed in this book to be at 273.16 K and 1 atm, according to Hougen et al. (1954). Under these conditions the normal molar volumes are as follows:

$$\text{Volume of 1 mole} = 22.414\ \text{L}$$
$$\text{Volume of 1 kmole} = 22.414\ \text{m}^3$$

The ideal gas law tends to fail at lower temperatures or higher pressures, when intermolecular forces and molecular size become important (Table 2.1).

The ideal gas law is rigorously valid for pressures lower than 1 atm and at room temperature or more. A rule of thumb is that the ideal gas law can be used for pressures up to 10 atm, depending on the corresponding value of the temperature (Finlayson, 2006).

Example 2.1

A vessel contains a mixture of heavy hydrocarbons and heptane. This is the only volatile liquid and its molar fraction is 0.75. The free vessel head space is occupied by air. System temperature is 70°C, and pressure is 3 atm. Calculate the mass of air in a vessel head space of 10 m^3.

Problem data include

- Heptane molecular weight ~100
- Heptane vapour pressure (Pa) = $\exp\left[C_1 + \frac{C_2}{T} + C_3 \cdot \ln(T) + C_4 \cdot T^{C_5}\right]$
- $C_1 = 87.829$
- $C_2 = -6996.4$
- $C_3 = -9.8802$
- $C_4 = 7.2099E{-}06$
- $C_5 = 2$
- T (K)

Solution

$P°(343.15) = 40{,}538$ Pa ~ 0.40 atm

According to Raoult's law:

$$P_{C_7H_{16}} = P^{\circ}_{C_7H_{16}} \cdot x_{C_7H_{16}} = 0.4 \cdot 0.75 = 0.3\ \text{atm} \tag{2.6}$$

Applying Eq. (2.4)

$$P_{air} = P - P_{C_7H_{16}} = 3 - 0.3 = 2.7 \text{ atm} \tag{2.7}$$

and

$$W_{air} = \frac{P_{air} \cdot V}{R \cdot T} = \frac{2.7 \cdot 10}{0.08206 \cdot 343.15} = 0.96 \text{ kmoles} = 27.8 \text{ kg} \tag{2.8}$$

Example 2.2

1000 m^3/h of combustion gases (CG) at 0.98 atm with the following molar composition:

- Nitrogen: 79.2%
- Oxygen: 7.2%
- Carbon dioxide: 13.6%

are supplied to a coil heat exchanger inside a vessel where sour water (SW) at 15°C is fed. SW contains 100 ppmw of hydrogen sulphide. 10% of the entering SW is vaporised at approximately equilibrium condition (Fig. 2.1).

SW steam at an absolute pressure of 0.69 atm is mixed at the evaporator outlet with the cool CG. Both liquids leave the system at the same temperature (T_b) with the following composition:

- Nitrogen: 63.6%
- Oxygen: 5.7%
- Carbon dioxide: 10.7%
- Steam: 20.0%

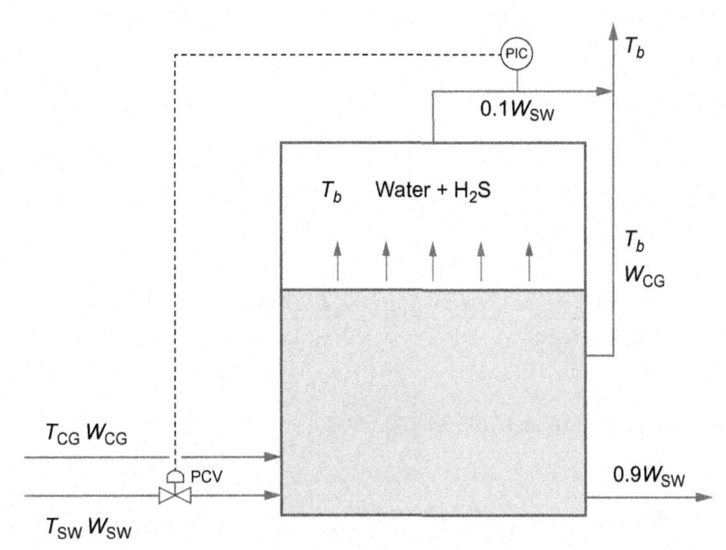

Fig. 2.1
Vapourisation of sour water.

Calculate:

a. the combustion gas temperature at the inlet
b. the vaporised mass flow rate
c. the overall volumetric flow rate at the outlet
d. the vapour sulphide concentration in the head space supposing that liquid to vapour volume ratio in the vessel is 1 to 3.

Solution

a. The calculation can be approached on the basis of 1 kmole of gas leaving the evaporator. The composition is:

- Steam: 0.2 kmoles/h
- Other gases: 0.8 kmoles/h

Vapour-liquid equilibrium temperature at 0.69 atm is 90°C. This is the temperature T_b of the gases leaving the evaporator. The thermal balance is:

$$W_{SW} \cdot c_{P_{SW}}(T_b - T_{SW}) + 0.1 \cdot W_{SW} \cdot \lambda(T_b) = W_{CG} \cdot c_{PCG} \cdot (T_{CG} - T_b) \tag{2.9}$$

where:

- the first term is the sensible heat transferred to all SW by combustion gas,
- the second term is the latent heat relating to the evaporation of 10% of the sour water,
- the third term is overall heat transferred by combustion gas,
- c_P are the average specific heats (SW~1 kcal/kg °C, CG~0.23 kcal/kg), and
- λ is the latent heat of vaporisation at 90°C ~545 kcal/kg.

According to molar ratios, assuming a molecular weight of CG ~ 30.5 (weighted average), Eq. (2.9) can be written as:

$$2 \cdot 18 \cdot c_{P_{SW}}(T_b - T_{SW}) + 0.2 \cdot 18 \cdot W_{SW} \cdot \lambda(T_b) = 0.8 \cdot 30.5 \cdot c_{P_{CG}}(T_{CG} - T_b) \tag{2.10}$$

$$T_{CG} = 920.72°C \tag{2.11}$$

b. Vaporised mass flow rate is:

$$W_{SW_{vap}} = \frac{0.2}{0.8} \cdot \frac{1000\,(mc/h)}{22.414\,(m^3/kmol)} \cdot \frac{273.16}{T_{CG}} \cdot 18 = \frac{0.2}{0.8} \cdot \frac{1000}{22.414} \cdot \frac{273.16}{273.16 + 920.72} \cdot 18 = 46\,kg/h \tag{2.12}$$

c. Overall volumetric flow rate at the outlet:

$$Q_{outlet} = Q_{SW_{vap}} + Q_{CG} = \left(\frac{1}{0.2} \cdot \frac{W_{SW_{vap}}}{18}\right) \cdot 22.414 \cdot \frac{T_b}{273.16} = \left(\frac{1}{0.2} \cdot \frac{46}{18}\right) \cdot 22.414 \cdot \frac{273.16 + 90}{273.16} = 380.8\,m^3/h \tag{2.13}$$

d. Vapour sulphide concentration in the head space
Hydrogen sulphide in the vapour phase (3 m^3):

$$P_{H_2S} = \frac{n_{H_2S_{vap}} \cdot R \cdot T_b}{V} \rightarrow n_{H_2S_{vap}} = \frac{P_{H_2S} \cdot V}{R \cdot T_b} \tag{2.14}$$

$$m_{H_2S_{vap}} = \frac{34 \cdot P_{H_2S} \cdot V}{R \cdot T_b} \tag{2.15}$$

$$m_{H_2S_{vap}} = \frac{34 \cdot P_{H_2S} \cdot V}{R \cdot T_b} \cdot 1000 = \frac{34 \cdot P_{H_2S} \cdot 3}{0.08206 \cdot 363} \cdot 1000 = 3424 \cdot P_{H_2S}\,(\text{atm})\,\text{g} \tag{2.16}$$

Hydrogen sulphide in the liquid phase (55.5 kmoles of water in 1 m^3):

$$x_{H_2S} = \frac{n_{H_2S_{liq}}}{55.5} \rightarrow n_{H_2S_{liq}} = 55.5 \cdot x_{H_2S} \tag{2.17}$$

$$m_{H_2S_{liq}} = 34 \cdot 1000 \cdot 55.5 \cdot x_{H_2S} = 1{,}887{,}000 \cdot x_{H_2S}\,\text{g} \tag{2.18}$$

Mass conservation requires that:

$$m_{H_2S_{liq}} + m_{H_2S_{vap}} = 1{,}887{,}000 \cdot x_{H_2S} + 3424 \cdot P_{H_2S} = 100\,\text{g} \tag{2.19}$$

Henry's equilibrium (Chapter 1):

$$\frac{P_{H_2S}}{x_{H_2S}} = 1030 \frac{\text{atm}}{\text{molar\_fraction}} \tag{2.20}$$

$$x_{H_2S} \cong 0.0000185 \tag{2.21}$$

$$P_{H_2S} \cong 0.019\,\text{atm} \tag{2.22}$$

Hydrogen sulphide vapour molar fraction is 0.019/0.69=0.0275 that corresponds to 27,500 ppmv.

2.2 Real Gases

2.2.1 Virial Equation of State

A generalised form of isotherms for gases and vapours is given by the following equation:

$$PV = a \cdot \left(1 + B \cdot P + C \cdot P^2 + D \cdot P^3 + \cdots\right) \tag{2.23}$$

Defining compressibility factor as *the product of pressure and molar volume divided by the gas constant and thermodynamic temperature* (IUPAC, 2014):

$$Z = \frac{P \cdot V}{R \cdot T} \tag{2.24}$$

and $a = R \cdot T$:

$$Z = 1 + B \cdot P + C \cdot P^2 + D \cdot P^3 + \cdots \tag{2.25}$$

this equation is known as virial expression, and coefficients B, C, D, etc. are the virial coefficients.

At relatively low pressures molecular interactions are weak, and, where no interactions exist, the virial coefficients would be zero, so the virial equation degenerates into the ideal gas equation. Therefore the value of compressibility factor is a measure of how much a gas or vapour approaches ideal state.

The expression:

$$P \cdot V = Z \cdot n \cdot R \cdot T \tag{2.26}$$

is the general state equation of gases and vapours. For an ideal gas Z is equal to 1.

2.2.2 Corresponding States

The following definitions are now introduced:

Reduced temperature
The ratio of the existing temperature of a substance to its critical temperature, both being expressed on an absolute scale.
Reduced pressure
The ratio of the existing pressure of a substance to its critical pressure, expressed on an absolute scale.
Reduced volume
The ratio of the existing molar volume to its critical molar volume.

Under conditions of equal reduced pressure and equal reduced temperature, substances are said to be in *corresponding states* (Hougen et al., 1954). The following is theorem of corresponding states:

All fluids, when compared at the same reduced temperature and reduced pressure, have approximately the same compressibility factor, and all deviate from ideal gas behaviour to about the same degree.

Many properties of different gases, in addition to the compressibility factor, are nearly the same at corresponding states. The similarity in the behaviour of substances at equal values of reduced temperature and pressure is referred to as the theory of corresponding states. It is only an approximation, but the discrepancies are small enough so that for many purposes a chart like Fig. 2.2 may be taken as applicable to all pure gases, until $T_r = 1$ to 2 and $Pr = 7$. For values greater than these, specific tabulated values have to be used (Table 2.2).

Example 2.3

A 5 L steel cylinder contains 400 g of nitrogen. Calculate the temperature at which the cylinder may be heated without exceeding 50 atm, knowing that the compressibility factor is equal to 0.945.

Solution

Nitrogen moles: 400/28 = 14.28 moles

Molar volume: 5/14.28 = 0.35 L/mole

Pressure: 50 atm

Critical pressure: 3.39 MPa = 33.5 atm

Critical temperature: 126.2 K

Reduced pressure: 1.49

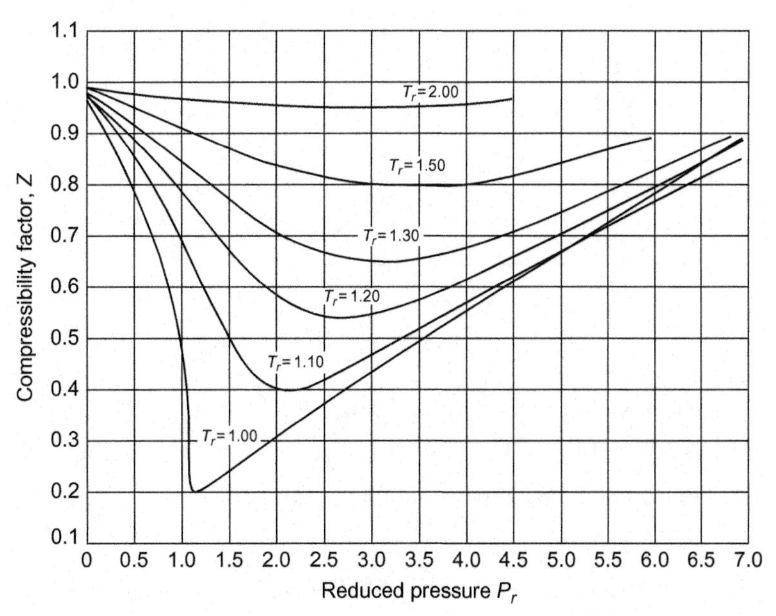

Fig. 2.2
Compressibility factor of corresponding states.

Table 2.2 Form and range of validity of state equations

Substance	Critical Temperature (K) T_C	Critical Pressure (MPa) P_C	Critical Volume (m^3/kmol) v_C	Acentric Factor ω
Methane	190.654	4.59	0.099	0.012
Ethane	305.32	4.85	0.146	0.100
Propane	369.83	4.21	0.200	0.152
n-Butane	425.12	3.77	0.255	0.200
n-Pentane	469.7	3.36	0.315	0.252
n-Hexane	507.6	3.04	0.373	0.301
n-Heptane	540.2	2.72	0.428	0.350
n-Octane	568.7	2.47	0.486	0.400
n- Nonane	594.6	2.31	0.54	0.444
n-Decane	617.7	2.09	0.602	0.492
Ethylene	282.34	5.03	0.132	0.087
Propylene	365.57	4.63	0.188	0.140
1-Butene	419.95	4.04	0.241	0.191
Acetylene	308.32	6.15	0.113	0.187
Benzene	562.16	4.88	0.261	0.210
Toluene	591.8	4.1	0.314	0.262
o-Xylene	630.33	3.74	0.374	0.310
m-Xylene	617.05	3.53	0.377	0.326
p-Xylene	616.23	3.5	0.381	0.322
Phenol	694.25	6.06	0.229	0.443
Air	132.45	3.79	0.092	0.035
Hydrogen	33.19	1.32	0.064	0.216
Oxygen	154.58	5.02	0.074	0.022
Nitrogen	126.2	3.39	0.089	0.038
Hydrogen sulphide	373.53	9.00	0.099	0.094
Ammonia	406.65	11.3	0.072	0.253
Carbon monoxide	132.92	3.49	0.095	0.048
Chlorine	417.20	7.71	7.71	0.069
Carbon dioxide	304.21	7.39	0.095	0.224
Sulphur dioxide	430.75	7.86	0.123	0.245
Water	647.13	21.94	0.056	0.345

Compressibility factor: 0.945

Entering Fig. 2.2 with $Pr \sim 1.5$ and $Z \sim 0.95$ it results
$T_r \sim 1.8 \rightarrow T_{max} = 1.8 \times T_C = 1.8 \times 126.2 = 227$ K

The same result is obtained applying the equation of real gas:

$$T = \frac{P \cdot V}{n \cdot Z \cdot R} = \frac{50 \cdot 5}{14.28 \cdot 0.945 \cdot 0.08206} = 225.7 \text{ K} \tag{2.27}$$

2.2.3 State Equations

A general equation of state represents the PVT behaviour of both liquids and vapours in a wide range of temperatures and pressures. The first practical equation of state was proposed by van der Waals:

$$P=\frac{R\cdot T}{v-b}-\frac{a}{v^2} \tag{2.28}$$

where v is the molar volume and a and b are specific gas constants. This equation gives an easy understanding of the reasons for departure of gases from ideal behaviour: constant b describes the effect of molecular size (volume reduction) and a the intermolecular cohesion. If v is large, the importance of b and a becomes proportionally negligible. Van der Waals'equation has a historical importance in gas thermodynamics, but its accuracy is no longer acceptable. More accurate state equations have been developed on the basis of the following generalised cubic equation of state (Smith et al., 2010):

$$P=\frac{R\cdot T}{v-b}-\frac{a(T)}{(v+\varepsilon\cdot b)\cdot(v+\sigma\cdot b)} \tag{2.29}$$

where ε and σ are pure numbers, the same for all substances, whereas $a(T)$ and b are substance dependent.

The Peng-Robinson (PR) (Peng and Robinson, 1976) and the Soave-Redlich-Kwong (SRK, Soave, 1972) equations of state (EOS) are the most used in the industry, especially for hydrocarbons' property prediction. The accuracy is comparable, although PR EOS is valid within a wider pressure and temperature range. Molar volumes calculated by SRK EOS are typically overestimated, i.e., densities are underestimated. Another major drawback of the SRK EOS is the poor liquid density prediction. The following table includes the equations along with the relevant parameters and constants (Ahmed, 2007). It must be noticed that temperature is given in °R, pressure in psi and constant gas R is equal to 10.73 psi ft^3/lb-mole °R (Table 2.3).

To use EOS with mixtures, the mixing rule has to be applied to determine constants α, a, and b.

$$(a\alpha)_m=\sum_i\sum_i\left[x_ix_j\sqrt{a_ia_j\alpha_i\alpha_j}(1-kij)\right] \tag{2.30}$$

$$b_m=\sum_i\left[x_ib_i\right] \tag{2.31}$$

where k_{ij} are empirical correction factors (called binary interaction coefficients). Soave (1972) suggested that no binary interaction coefficients are required for hydrocarbons.

Table 2.3 Form and range of validity of state equations

EOS	Form	Temperature Range	Pressure Range
Peng-Robinson	$P=\frac{R\cdot T}{v-b}-\frac{a\cdot\alpha}{v^2+2\cdot v\cdot b-b^2}$ Constant α $\left[1+(0.3796+1.54226\cdot\omega-0.26992\omega^2)\left(1-\sqrt{\frac{T}{Tc}}\right)\right]^2$	>−271°C Constant a $\frac{0.45724\cdot R^2\cdot T_c^2}{P_c}$	<100,000 kPa Constant b $\frac{0.07780\cdot R\cdot T_c}{P_c}$
Saove Redlich Kwong	$P=\frac{R\cdot T}{v-b}-\frac{a\cdot\alpha(T)}{v\cdot(v+b)}$ Constant α $\left[1+(0.48+1.57\cdot\omega-0.176\omega^2)\left(1-\sqrt{\frac{T}{Tc}}\right)\right]^2$	>−143°C Constant a $\frac{0.42747\cdot R^2\cdot T_c^2}{P_c}$	<35,000 kPa Constant b $\frac{0.08664\cdot R\cdot T_c}{P_c}$

Example 2.4

Resolve Example 2.3 using Peng-Robinson equation.

Solution

Nitrogen moles: 400/28 = 14.28 moles = 0.03 lb-mole

Molar volume: 5/14.28 = 0.35 L/mole = 5.57 ft^3/lb-mole

Pressure: 50 atm = 735 psi

Critical pressure: 3.39 MPa = 33.5 atm = 492.45 psi

Critical temperature = 126.2 K = 227.16°R

R = 10.73 psi ft^3/lb-mole °R

Acentric factor: 0.038

$$P=\frac{R\cdot T}{v-b}-\frac{a\cdot\alpha(T_r)^r}{v^2\cdot+2\cdot v\cdot b-b^2}=\frac{R\cdot T_c\cdot T_r}{v-b}-\frac{a\cdot\alpha(T_r)^r}{v^2\cdot+2\cdot v\cdot b-b^2}$$
$$=\frac{10.73\cdot 227.16\cdot T_r}{5.57-0.385}-\frac{5516.25\cdot\alpha(T_r)}{31.02+4.29-0.148} \tag{2.32}$$

$$735=\frac{10.73\cdot 227.16\cdot T_r}{5.57-0.385}-\frac{5516.25\cdot[1+(0.3796+1.54226\cdot 0.038-0.26992\cdot 0.038^2)(1-\sqrt{T_r})]^2}{31.02+4.29-0.148} \tag{2.33}$$

$$735=\frac{10.73\cdot 227.16\cdot T_r}{5.57-0.385}-\frac{5516.25\cdot[1+0.438\cdot(1-\sqrt{T_r})]^2}{31.02+4.29-0.148} \tag{2.34}$$

$$\frac{2437.42\cdot T_r}{5.185}-\frac{5516.25\cdot[1+0.438\cdot(1-\sqrt{T_r})]^2}{35.162}-735=0 \tag{2.35}$$

$$470.09\cdot T_r-156.88\cdot\left[1+0.438\cdot\left(1-\sqrt{T_r}\right)\right]^2-735=0 \tag{2.36}$$

Table 2.4 Peng-Robinson coefficients for Example 2.3

Parameter	Form	Value
α	$[1 + (0.3796 + 1.54226 \cdot 0.038 - 0.26992 \cdot 0.038^2)(1 - \sqrt{T_r})]^2$	$\alpha(T_r)$
a	$\frac{0.45724 \cdot 10.73^2 \cdot 227.16^2}{492.45}$	5516.25
b	$\frac{0.07780 \cdot 10.73 \cdot 227.16}{492.45}$	0.385

This equation can be easily solved assuming (Table 2.4):

$$\sqrt{T_r} = x \rightarrow T_r = x^2 \tag{2.37}$$

$$470.09 \cdot x^2 - 156.88 \cdot [1 + 0.438 \cdot (1 - x)]^2 - 735 = 0 \tag{2.38}$$

Two roots exist:

$x = -1.79$

$x = 1.34$

Taking the positive solution, it is:

$$\sqrt{T_r} = 1.34 \rightarrow Tr = 1.79 \rightarrow T = 407.89\ ^\circ\text{R} = 226.6\ \text{K} \tag{2.39}$$

2.3 Polytropic Transformations

Thermodynamic transformations are assumed to occur in such a way that the system under consideration stays infinitesimally close to a state of equilibrium. This scenario, known as the quasi-static transformation, is the foundation of classical thermodynamics. It will be shown that for some susbstances, such as carbon dioxide (Benintendi, 2014), this assumption is less accurate in some situations, taking significant departure from classic thermodynamics. Thermodynamic transformations of ideal gases assumed in process safety belong to the general family of *polytropic transformations*, where polytropic, from old Greek πολυτροπική, means many paths. This is referred to the different routes of a thermodynamic system migrating from state A to state B. The general formula describing a polytropic transformation is the following:

$$P \cdot V^n = \text{constant} \tag{2.40}$$

where n is a constant exponent.

For an ideal gas, it's easy to demonstrate that Eq. (2.40) can be expressed in terms of the other couples of thermodynamic variables:

$$T \cdot V^{n-1} = \text{constant} \tag{2.41}$$

Table 2.5 Exponents and thermodynamic pathways of polytropic transformations

Exponent n	Polytropic Form	Transformation	Law
0	$P=\text{constant}$	Isobaric	Charles
1	$P\cdot V=\text{constant}$	Isothermal	Boyle
γ	$P\cdot V^{\gamma}=\text{constant}$	Adiabatic	Poisson
$\pm\infty$	$P\cdot T^{-1}=\text{constant}$	Isochoric	Gay-Lussac

and

$$T\cdot P^{\frac{1-n}{n}}=\text{constant} \tag{2.42}$$

Depending on values of n, the transformation can represent specific thermodynamic conditions, as shown in Table 2.5:

2.4 State Functions

State functions are thermodynamic variables, the value of which does not depend on past history but only on present conditions. For a homogeneous pure substance, the definition of two thermodynamic coordinates, such as temperature and pressure, is sufficient to identify the value of a state function. It should be remembered that a state function is associated with exact differentials, whereas work and heat are not. Remarkable state functions in process safety are enthalpy, entropy, and internal energy.

2.4.1 Internal Energy

Internal energy refers to translational, rotational and vibrational energy of polyatomic molecules. Monoatomic molecules don't possess rotational and vibrational energy. The internal energy changes of a closed system are produced by thermal energy exchanged with its surroundings and by the work done by the system on the surroundings, or on the system by the surroundings. This is the first law of thermodynamics for closed systems, the mass of which is constant throughout the transformation:

$$dU=\delta Q+\delta W \tag{2.43}$$

$$\Delta U=Q+W \tag{2.44}$$

In this equation U is the internal energy, Q is the exchanged heat, W is the work done, and δ is used to mean that heat and work are not generally exact differentials. The sign convention has been recommended by the IUPAC>(Inczedy et al., 1997). Work is typically associated with a change in the volume of a fluid due to the force exerted on its surface. This is expressed by the formula:

$$W = -\int_{V_1}^{V_2} p \cdot dV \tag{2.45}$$

so

$$dU = \delta Q - \int_{V_1}^{V_2} p \cdot dV \tag{2.46}$$

From this equation we conclude that if the system volume does not change, then the change in internal energy coincides with heat exchanged at constant volume:

$$dU = \delta Q \tag{2.47}$$

This takes some important implications.

2.4.2 Enthalpy

Enthalpy is a state function defined as:

$$H = U + P \cdot V \tag{2.48}$$

Differentiating:

$$dH = U + P \cdot dV + V \cdot dP \tag{2.49}$$

For an isobaric process, Eq. (2.49) becomes:

$$dH = U + P \cdot dV \tag{2.50}$$

For closed systems:

$$\delta Q = U + P \cdot dV \tag{2.51}$$

We can conclude that, if system pressure does not change, the change in enthalpy coincides with heat exchanged at constant pressure. This takes some important implications in the calculations, because phase changes occur at constant pressures.

2.4.3 Entropy

A general definition of entropy is the following:

$$\Delta S = \int \frac{\delta Q}{T} \tag{2.52}$$

Entropy is a state function, whereas heat is not. The importance of entropy in thermodynamic and process safety calculations is due to two reasons:

- Many phenomena are substantially adiabatic, and if they are assumed as reversible, then they can be considered as isentropic.
- Being entropy a state function, its changes can be evaluated by the application of Eq. (2.52) to an arbitrary reversible process, being that the change is identical.

For an ideal gas, with constant specific heat ratio γ, undergoing a reversible adiabatic isentropic process, from Eq. (2.42) it can be obtained:

$$\frac{T_2}{T_1} = \left(\frac{P_2}{P_1}\right)^{\frac{\gamma-1}{\gamma}} \tag{2.53}$$

2.5 Thermodynamic Properties

2.5.1 Specific Heats

On the basis of the definitions of internal energy and enthalpy, specific heats at constant volume and pressure are respectively defined as:

$$c_V = \left(\frac{\partial U}{\partial t}\right)_V \tag{2.54}$$

$$c_P = \left(\frac{\partial H}{\partial t}\right)_P \tag{2.55}$$

These equations can be integrated:

$$\Delta U = \int c_V \cdot dT \tag{2.56}$$

$$\Delta H = \int c_P \cdot dT \tag{2.57}$$

These equations are strictly valid for constant-volume and constant-pressure reversible processes, whereas stirring work is not included.

For an ideal gas, Mayer's equation is valid:

$$c_P - c_V = R \tag{2.58}$$

For solids and for liquids at moderate and low pressures the effect of pressure on specific heat is negligible, so it varies only with temperature (Riazi, 2005) and $c_V = c_P = f(T)$ (Table 2.6).

Table 2.6 Specific heat at constant pressure of liquids, J/(kmol K) (Green and Perry, 2007)

Substance	Equation	C_1	C_2	C_3	C_4	C_5	T Range (K)
Methane	2	65.708	38,883	−257.95	614.07		90.69–190
Ethane	2	44.009	89,718	918.77	−1886		92–290
Propane	2	62.983	113,630	633.21	−873.46		85.47–360
Butane	1	191,030	−1675	12.5	−0.03874	4.6121E−05	134.86–400
Pentane	1	159,080	−270.5	0.99537			143.2–390
Hexane	1	172,120	−183.78	0.88734			177.83–460
Heptane	2	61.26	314,410	1824.6	−2547.9		182.57–520
Octane	1	224,830	−186.63	0.486	0.95891		216.38–460
Ethylene	1	247,390	−4428	40.936	−0.1697	0.00026816	104.00–252.70
Propylene	1	114,140	−343.72	1.0905			87.89–225.45
Methanol	1	105,800	−362.23	0.9379			175.47–400
Ethanol	1	102,640	−139.63	−0.030341	0.0020386		159.05–390
Benzene	1	162,940	−344.94	0.85562			278.68–500
Toluene	1	140,140	−152.3	0.695			178.18–500
Phenol	1	101,720	317.61				314.06–425
Air	1	−214,460	9185.1	−106.12	0.41616		75–115
Hydrogen	2	66.653	6765.9	−123.63	478.27		13.95–32
Oxygen	1	175,430	−6152.3	113.92	−0.92382	0.0027963	54.36–142
Nitrogen	1	281,970	−12,281	248	−2.2182	0.0074902	63.15–112
Ammonia	2	61.289	80,925	799.4	−2651		203.15–401.15
Chlorine	1	63,936	46.35	−0.1623			172.12–239.12
Carbon dioxide	1	−8,304,300	104,370	−433.33	0.60052		220–290
Water	1	276,370	−2090.1	8.125	−0.014116	9.3701E−06	273.16–533.15

Equation 1: $c_P = C_1 + C_2 \cdot T + C_3 \cdot T^2 + C_4 \cdot T^3 + C_5 \cdot T^4$.
Equation 2: $c_P = C_1^2/t + C_2 - 2 \cdot C_1 \cdot C_3 \cdot t - C_1 \cdot C_4 \cdot t^2 - C_3^2 \cdot t^3/3 - C_3 \cdot C_4 \cdot t^4/2 - C_4^2 \cdot t^5/5$ $(t = 1 - T/T_r)$

Example 2.5

Air in a 3 m^3 thermally insulated receiver at atmospheric pressure and 20°C is pressurised to 5 atm. Calculate the enthalpy change.

Solution

Ideal gas law can be used. Air undergoes an isochoric transformation, so Gay Lussac's law applies:

$$T_1 = T_o \cdot \frac{P_1}{P_o} = 293 \cdot \frac{5}{1} = 1465 \text{ K} \quad (2.59)$$

Moles in the receiver:

$$n = \frac{P \cdot V}{R \cdot T} = \frac{1 \cdot 3}{0.08206 \cdot 293} = 0.1248 \text{ kmoles} \quad (2.60)$$

Differentiating the equation of enthalpy in terms of molar enthalpy and molar volume, it is:

$$dh = du + p \cdot dv + v \cdot dp \quad (2.61)$$

that, for an isochoric process, reduces to:

$$dh = du + v \cdot dp \tag{2.62}$$

Being:

$$\Delta u = \int c_V \cdot dT \tag{2.63}$$

and, for in ideal gas:

$$c_P - c_V = R \tag{2.64}$$

Depending on the internal energy of temperature only, from Table 2.7 (Kobe et al., 1949–1954):

$$c_V = 6.713 \cdot + 0.00047 \cdot T + 0.115 \cdot 10^{-5} \cdot T^2 - 0.47 \cdot 10^{-9} \cdot T^3 - R \tag{2.65}$$

with $R = 1.987$ cal/mole K

Table 2.7 Specific heat at constant pressure of inorganic and organic compounds in the ideal gas cal/(mol K) (Kobe et al., 1949–1954)

Substance	a	$b \times 10^2$	$c \times 10^5$	$d \times 10^9$	T Range (K)
Methane	4.750	1.200	0.303	−2.630	273–1500
Ethane	1.648	4.124	−1.530	1.740	273–1500
Propane	−0.966	7.279	−3.755	7.580	273–1500
n-Butane	0.945	8.873	−4.380	8.360	273–1500
n-Pentane	1.618	10.850	−5.365	10.100	273–1500
n-Hexane	1.657	13.190	−6.844	13.780	273–1500
Ethylene	0.944	3.735	−1.993	4.220	273–1500
Propylene	0.753	5.691	−2.910	5.880	273–1500
Methanol	4.550	2.186	−0.291	−1.920	273–1500
Ethanol	4.750	5.006	−2.479	4.790	273–1500
Benzene	−8.650	11.578	−7.540	18.540	273–1500
Toluene	−8.213	13.357	−8.230	19.20	273–1500
Air	6.713	0.0470	0.115	0.470	273–1500
Hydrogen	6.952	−0.0457	0.096	−0.021	273–1500
Oxygen	6.085	0.363	−0.171	0.313	273–1800
Nitrogen	6.903	−0.037	0.193	−0.686	273–1800
Ammonia	6.584	0.612	0.237	−1.598	273–1500
Chlorine	6.821	0.571	−0.511	1.547	273–1500
Hydrogen sulphide	7.070	0.313	0.136	−0.787	273–1800
Carbon monoxide	6.726	0.156	−0.024	–	273–1800
Carbon dioxide	5.316	1.428	−0.836	1.784	273–1800
Sulphur dioxide	6.157	1.384	−0.910	2.057	273–1800
Sulphur trioxide	3.918	3.483	−2.675	7.744	273–1300
Steam	6.970	0.346	−0.048	–	273–3800

$c_P = a + b \cdot T + c \cdot T^2 + d \cdot T^3$.

$$\Delta u = \int_{T_o}^{T_1} c_V \cdot dT = 6.713 \cdot (T_1 - T_o) + 0.00047 \cdot \frac{(T_1 - T_o)^2}{2}$$
$$+ 0.115 \cdot 10^{-5} \cdot \frac{(T_1 - T_o)^3}{3} - 0.47 \cdot 10^{-9} \cdot \frac{(T_1 - T_o)^4}{4} - R \cdot (T_1 - T_o) \tag{2.66}$$

$$\int_{P_o}^{P_1} v \cdot dP = v_o \cdot \int_{P_o}^{P_1} dP = v_o \cdot (P_1 - P_o) = \frac{3 \text{ m}^3}{133.9 \text{ moles}} \cdot 202{,}650 \text{ Pa} = 4540.33 \text{ J/mole} \tag{2.67}$$

$$\Delta u = 6255.46 \text{ cal/mole} = 26{,}185.35 \text{ J/mole} \tag{2.68}$$

$$\Delta h = 26{,}185.35 + 4540.33 = 30{,}725.68 \text{ J/mole} \rightarrow \Delta H = 30.725.68 \times 124.8 = 3{,}834.57 \tag{2.69}$$

2.5.2 Vapour Pressure

Liquids vapour pressure only depends on temperature. A general relationship for pure substances is the following (Table 2.8):

$$P_V^{\circ}(\text{Pa}) = \exp\left[C_1 + \frac{C_2}{T} + C_3 \cdot \ln(T) + C_4 \cdot T^{C_5}\right] \tag{2.70}$$

Table 2.8 Vapour pressure of inorganic and organic liquids (*Pa*) (Green and Perry, 2007)

Substance	C_1	C_2	C_3	C_4	C_5	*T* Range (K)
Methane	39.205	−1324.4	−3.4366	3.1019E−05	2	90.69–190.56
Ethane	51.857	−2598.7	−5.1283	1.4913E−05	2	90.35–305.32
Propane	59.078	−3492.6	−6.0669	1.4913E−05	2	85.47–369.83
n-Butane	66.343	−4363.2	−7.046	9.4509E−06	2	134.86–425.12
n-Pentane	78.741	−5420.3	−8.8253	9.6171E−06	2	143.2–469.7
n-Hexane	104.65	−6995.5	−12.702	1.2381E−05	2	177.83–460
n-Heptane	87.829	−6996.4	−9.8802	7.2099E−06	2	182.57–540.2
n-Octane	96.084	−7900.2	−11.003	7.1802E−06	2	216.38–568.7
Ethylene	74.242	−2707.2	−9.8462	2.2457E−02	1	104.00–282.34
Propylene	57.263	−3382.4	−5.7707	1.0431E−05	2	87.89–225.45
Methanol	81.768	−6876	−8.7078	7.1926E−06	2	175.47–512.64
Ethanol	74.475	−7164.3	−7.327	3.1340E−06	2	159.05–513.92
Benzene	83.918	−6517.7	−9.3453	7.1182E−06	2	278.68–562.16
Toluene	80.877	6902.4	−8.7761	5.8034E−06	2	178.18–591.8
Phenol	95.444	−10,113	−10.09	6.7603E−18	6	314.06–694.25
Ammonia	90.843	−4669.7	−11.607	1.7194E−06	1	195.41–405.65
Chlorine	71.334	−3855	−8.5171	1.2378E−02	1	172.12–417.15
Carbon dioxide	140.54	−4735	−21.268	4.0909E−02	1	216.58–304.21
Water	73.649	−7258.2	−7.3037	4.1653E−06	2	273.16–647.13

Example 2.6

Benzene partial pressure in benzene-toluene vapour above a 50% molar solution of the same components is 2 bar. Calculate system temperature.

Solution

Applying Raoult's law:

$$p_B = P_B^o \cdot x_b \quad (2.71)$$

$$P_B^o = \frac{p_B}{x_B} = \frac{2}{0.5} = 4 \text{ bar} \quad (2.72)$$

Solving the equation for T:

$$4 = \exp\left[83.918 - \frac{6517.7}{T} - 9.3453 \cdot \ln(T) + 7.1182E - 06 \cdot T^2\right] \quad (2.73)$$

$T = 406\ K = 133°C$

2.5.3 *Latent Heat of Vaporisation*

Latent heat of vaporisation is the energy to be supplied in order for a liquid to be vaporised. *Latent* means that the heat transfer is not accompanied by a temperature change. Table 2.9 includes the latent heats of vaporisation for some organic and inorganic liquids according to the following equation:

$$\lambda_{vap} = C_1 \times \left(1 - \frac{T}{T_C}\right)^{\left[C_2 + C_3 \times \frac{T}{T_C} + C_4 \times \left(\frac{T}{T_C}\right)^2\right]} \quad (2.74)$$

Table 2.9 Heat of vaporisation of inorganic and organic liquids (J/kmole) (Green and Perry, 2007)

Substance	$C_1 \times 1E-07$	C_2	C_3	C_4	T Range (K)
Methane	1.0914	0.26087	−0.14694	0.22154	90.69–190.56
Ethane	2.1091	0.60646	−0.55492	0.32799	90.35–305.32
Propane	2.9209	0.78237	−0.77319	0.39246	85.47–369.83
n-Butane	3.6238	0.83370	−0.82274	0.39613	134.86–425.12
n-Pentane	3.9109	0.38681	0	0	143.2–469.7
n-Hexane	4.4544	0.39002	0	0	177.83–507.6
n-Heptane	5.0014	0.38795	0	0	182.57–540.2
n-Octane	5.5180	0.38467	0	0	216.38–568.7
Methanol	5.2390	0.36820	0	0	175.47–512.64
Ethanol	5.6900	0.33590	0	0	159.05–513.92
Benzene	4.7500	0.45238	0.05340	−0.11810	278.68–562.16
Toluene	5.0144	0.38590	0	0	178.18–591.8
Ammonia	3.1523	0.39140	−0.22890	0.23090	195.41–405.65
Chlorine	3.0680	0.84580	−0.90010	0.45300	172.12–417.15
Carbon dioxide	2.1730	0.38200	−0.43390	0.42213	216.58–304.21
Water	5.2053	0.31990	−0.21200	0.25795	273.16–647.13

2.5.4 Sound Speed of Liquids and Gases

The terms *sound speed or acoustic speed* are traditionally used in fluid-dynamics to indicate the velocity with which a differential pressure disturbance dp propagates in a fluid. Some authors prefer to use the term *adiabatic disturbance velocity* (Acton and Caputo, 1979). The general definition of the acoustic speed a of a fluid is given by the formula:

$$a = \sqrt{\left(\frac{\partial P}{\partial \rho}\right)_s} \tag{2.75}$$

This value is very large for incompressible fluids, in which very small pressure changes result in huge density variation. So that $a = 1000$ m/s. Sound speed is an important parameter in liquid pipelines in which a rapid pressure rise or pressure drop occurs, resulting in a negative or positive pressure wave, moving with a velocity a (Benintendi, 2016), with significant implications.

For an ideal gas, it can be easily shown that:

$$a = \sqrt{\frac{\gamma \cdot R \cdot T}{MW}} \tag{2.76}$$

where γ is the specific heats ratio.

The ratio of the fluid velocity to the acoustic speed is the Mach number, a dimensionless parameter which has great importance in high velocity phenomena occurring in process safety, such as explosions. According to Zucrow and Hoffman (1976), Mach number can be interpreted as the ratio of kinetic energy of directed fluid flow to the kinetic energy of random molecular motion (Table 2.10).

Table 2.10 Sound speed of various materials at 15.5°C and 1 atm (White, 2011)

Substance	*a* (m/s)
Gases	
Hydrogen	1294
Helium	1000
Air	340
Argon	317
Carbon dioxide	266
Methane	185
Liquids	
Glycerine	1860
Water	1490
Mercury	1450
Ethanol	1200

2.6 Heat Transfer Mechanisms

Heat transfer mechanisms are broadly discussed in the scientific literature (Kern, 1950; Incropera et al., 2011). In process safety, heat transfer scenarios are numerous and are often a combination of the individual mechanisms.

2.6.1 Thermal Conduction

Thermal conduction works at a molecular level. It is the governing mechanism in solids and is significant in liquids and gases at rest. The fundamental law is known as Fourier's law and is expressed by the three-dimensional vector equation:

$$q = -k \cdot \nabla T \tag{2.77}$$

where q is the heat flux (W/m^2) and k is the thermal conductivity (W/m K).

For each coordinate direction the scalar law becomes:

$$q_n = -k \cdot \frac{\partial T}{\partial n} \tag{2.78}$$

where n is the generic coordinate.

Thermal conductivity for some gases, liquids, and solids are included in Tables 2.11 and 2.12.

For steam, ammonia, and carbon dioxide vapours, the following formulas can be used (Acton and Caputo, 1979), kcal/m h °C (Table 2.13):

Table 2.11 Thermal conductivities of various materials (Bird et al., 2002)

Substance	T (K)	k (W/m K)
	Gases	
Hydrogen	100–200–300	0.06799–0.1282–0.1779
Oxygen	100–200–300	0.00904–0.01833–0.02657
Methane	100–200–300	0.01063–0.02184–0.03427
Nitrogen monoxide	200–300	0.01778–0.02590
Carbon dioxide	200–300	0.00950–0.01665
	Liquids	
1-Pentene	200–250–300	0.1461–0.1307–0.1153
Carbon tetrachloride	250–300–350	0.1092–0.09929–0.08935
Ethyl ether	250–300–350	0.1478–0.1274–0.1071
Ethyl alcohol	250–300–350	0.1808–0.1676–0.1544
Glycerol	300–350–400	0.2920–0.2977–0.3034
Water	300–350–400	0.6089–0.662–0.6848

Table 2.11 Thermal conductivities of various materials—cont'd

Substance	T (K)	k (W/m K)
	Solids	
Aluminium	373.2–573.2–873.2	205.9–268–423
Cadmium	273.2–373.2	93.0–90.4
Copper	291.2–373.2	384.1–379.9
Steel	291.2–373.2	46.9–44.8
Tin	273.2–373.2	63.93–59.8
Brick	–	0.63
Concrete	–	0.92
Earth's crust	–	1.7
Glass (soda)	473.2	0.71
Graphite	–	5.0
Sand (dry)	–	0.389
Wood, parallel to axis	–	0.126
Wood, normal to axis	–	0.038

Table 2.12 Thermal conductivities of some liquids (Acton and Caputo, 1979)

Substance	T (K)	k (kcal/m K)
Methyl alcohol	30	0.181
Aniline	30	0.150
Benzene	30	0.137
Gasoline	30	0.120
Kerosene	30	0.130
Mineral oil	30	0.120
Mercury	30	6.500
Crude oil	30	0.120

Table 2.13 Thermal conductivities of some vapours (Acton and Caputo, 1979)

Vapour	k (kcal/m h °C)
Water	$k(T) = 0.0145 + 5.5 \cdot 10^{-5} \cdot T(°C)$
Ammonia	$k(T) = 0.0180 + 8.0 \cdot 10^{-5} \cdot T(°C)$
Carbon dioxide	$k(T) = 0.0120 + 4.7 \cdot 10^{-5} \cdot T(°C)$

Example 2.7

A steel slab face 1 m thick, initially at 20°C, is put in contact with combustion gas at 600°C, while the opposite face is in contact with gasoline vapours in air at ambient temperature. Neglecting thermal resistances at both faces, calculate the time required for the cold face to reach the gasoline auto-ignition temperature of 250°C and start burning it.

According to Eurocodes (2005), average linear steel thermal conductivity and specific heat between 20°C and 600°C can be assumed as 660 J/kg K and 40 W/m K, respectively. Assume no heat transfers along the directions normal to the axis (Fig. 2.3).

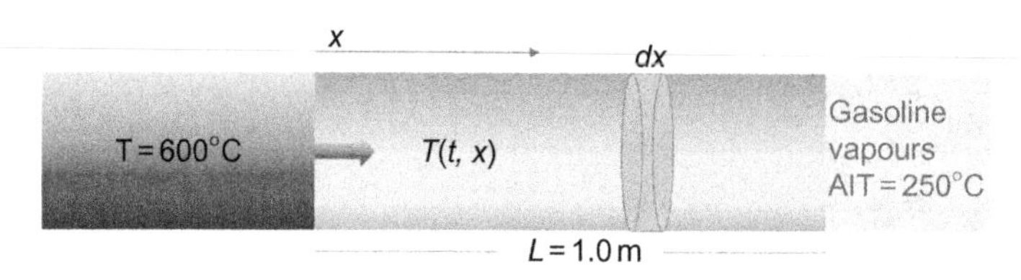

Fig. 2.3
Slab of Example 2.7.

A differential thermal balance gives (assuming a cross sectional area = 1 m^2):

$$\rho \cdot c \cdot dT \cdot dx = (q_x - q_{x+dx}) \cdot dt = k \cdot d\frac{\partial T}{\partial x} \cdot dt \tag{2.79}$$

where

- c is steel specific heat = 660 J/kg K
- ρ is steel density = 7850 kg/m^3
- k is steel thermal conductivity = 40 W/m K
- q is the heat flux, W/m^2

Rearranging:

$$\rho \cdot c \cdot dT \cdot dx = (q_x - q_{x+dx}) \cdot dt = k \cdot d\frac{\partial T}{\partial x} \cdot dt \tag{2.80}$$

$$\frac{\rho \cdot c}{k} \cdot \frac{\partial T}{\partial t} = \frac{\partial^2 T}{\partial x^2} \tag{2.81}$$

with the boundary conditions:

$x = 0$, $\forall t > 0$, $T = T_1 = 600°C$

$x = L$, $t = 0$, $T = T_o = 20°C$

The integral of differential equation (2.81) is given by the error function *erf* (Bird et al., 2002):

$$T = T_o + (T_1 - T_o) \cdot \left(1 - erf\frac{x}{\sqrt{\frac{4 \cdot k}{\rho \cdot c} \cdot t}}\right) \tag{2.82}$$

$$250 = 20 + 580 \cdot \left(1 - erf\frac{1}{0.00557 \cdot \sqrt{t}}\right) \tag{2.83}$$

Solving:

$$erf\frac{1}{0.00557 \cdot \sqrt{t}} = 0.6 \tag{2.84}$$

$$\frac{1}{0.00557 \cdot \sqrt{t}} = 0.6 \tag{2.85}$$

$$t = 17 \text{ s} \tag{2.86}$$

2.6.2 Thermal Convection

Thermal convection is important in fluids, especially when fluid bulk motion is so effective to govern heat transfer. It can be forced or natural, the former promoted by external fluid-dynamic agents, such as a fan or the wind, the latter by the natural density gradient and by the buoyancy around the object. Thermal convection describes interphase heat transfer in flow systems with fluid flowing either in a conduit or around/next to a solid object. A proportionality factor, h, has been introduced in the following equation:

$$Q = h \cdot A \cdot \Delta T \quad (2.87)$$

where Q is the interphase heat flow, A is the heat exchange area, ΔT is the temperature difference. Typical convective heat transfer scenarios are provided here (Fig. 2.4).

Typical heat transfer coefficients have been included in Table 2.14 (Bird et al., 2002).

Rigorous methods to calculate heat transfer coefficients in heat transfer scenarios can be found in sources such as Bird et al. (2002) and Foust et al. (1980). Configurations of natural and forced thermal convection have been described by means of empirical correlation using dimensionless numbers (Table 2.15).

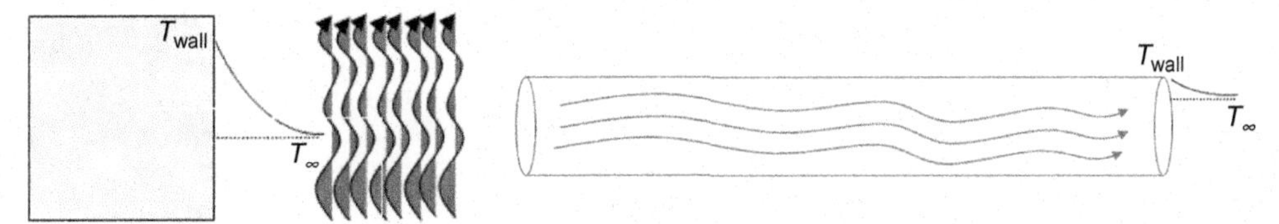

Fig. 2.4
Interphase heat transfer scenarios.

Table 2.14 Typical orders of magnitude for heat transfer coefficients (Bird et al., 2002)

System	h (W)
Free convection	
Hydrogen	1294
Helium	1000
Air	340
Argon	317
Carbon dioxide	266
Methane	185
Liquids	
Glycerine	1860
Water	1490
Mercury	1450
Ethanol	1200

Table 2.15 Definition of dimensionless numbers (D is a characteristic length)

Reynolds	$Re = \frac{\rho \cdot u \cdot D}{\mu}$	Prandtl	$Pr = \frac{c_P \cdot \mu}{k}$
Nusselt	$Nu = \frac{h \cdot D}{k}$	Grashof	$Gr = \frac{g \cdot \beta \cdot D^3 \cdot \Delta T}{\mu}$
Peclet	$Pe = \frac{u \cdot D \cdot \rho \cdot c_P}{k}$	Rayleigh	$Ra = \frac{g \cdot \beta \cdot D^3 \cdot \Delta T \cdot \rho^2 \cdot c_P}{\mu \cdot k}$
D = characteristic length β = thermal expansion coefficient	ρ = density k = thermal conductivity	μ = viscosity g = acceleration of gravity	u = velocity ΔT = temperature difference

Empirical correlations for natural thermal convective flow

Natural convection for vertical and inclined plates

Thermal convective flux for vertical flow can be generalised in many cases using the following general expression:

$$Nu = \frac{h}{k \cdot D} = C \cdot (Gr \cdot Pr)^n = C \cdot Ra^n \tag{2.88}$$

where for laminar flow ($10^4 \leq Ra \leq 10^9$), (McAdams, 1954):

$C = 0.59$

$n = 0.25$

and, for turbulent flow ($10^9 \leq Ra \leq 10^{13}$):

$C = 0.1$

$n = 0.33$

Example 2.8

A fire has broken out in a warehouse, which is separated from an adjacent room by an iron plate 2 m high, whose temperature is 480°C. Calculate the thermal flux from the warehouse to the room (Fig. 2.5).

Solution

All chemical-physical data are calculated at temperature $(T_s + T_\infty)/2 = 250$°C.

$D = 2$ m	$\rho = 0.675$ kg/m^3	$\mu = 2.8E-5$ kg/m s	$\Delta T = 460$°C
$\beta = 0.00191$	$k = 0.0241$ W/m K	$g = 9.81$ m/s^2	$c_P = 1034$ J/kg °C

$$Ra = \frac{g \cdot \beta \cdot L^3 \cdot (T_s - T_\infty)}{\frac{\mu}{\rho} \cdot \frac{k}{\rho \cdot c_P}} = \frac{9.81 \cdot 0.00191 \cdot 8 \cdot 460}{4.17E-5 \cdot \frac{0.0421}{0.675 \cdot 1034}} = 2.741298662 \cdot 10^{10} \tag{2.89}$$

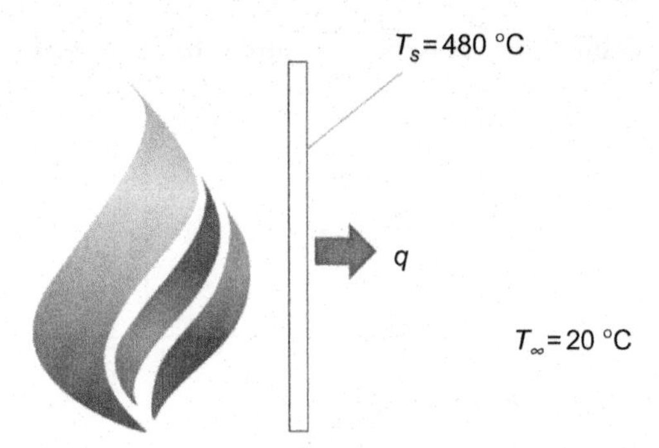

Fig. 2.5
Calculation scheme for Example 2.8.

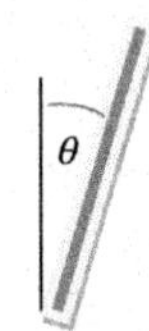

Fig. 2.6
Convective heat transfer from/to inclined plates.

$$h = k \cdot L \cdot C \cdot Ra^n = 0.0241 \cdot 2 \cdot 0.1 \cdot \left(2.741298662 \times 10^{10}\right)^{0.33} = 13.41\ \text{W/m}^2\ ^\circ\text{C} \quad (2.90)$$

$$q = 13.41 \cdot 460 = 6171\ \text{W/m}^2 \quad (2.91)$$

For inclined plates, it is recommended to replace g in Eq. (2.89) with $g \times \cos\theta$, even if this approach is strictly valid at the top and bottom surfaces (Fig. 2.6).

Natural convection for horizontal plates

For horizontal plates, the following correlation can be used, provided the characteristic length D is replaced by the plate surface to perimeter ratio:

Upper Surface of Hot Plate or Lower Surface of Cold Plate (Lloyd and Moran, 1974)

$$Nu = 0.54 \cdot Ra^{0.25} \ \left(10^4 \leq Ra \leq 10^7,\ Pr \geq 0.7\right) \quad (2.92)$$

$$Nu = 0.15 \cdot Ra^{0.33} \left(10^7 \leq Ra \leq 10^{11},\ \text{all} Pr\right) \quad (2.93)$$

Lower Surface of Hot Plate or Upper Surface of Cold Plate (Radziemska and Lewandowski, 2001)

$$Nu = 0.52 \cdot Ra^{0.20} \left(10^4 \leq Ra \leq 10^9,\ Pr \geq 0.7\right) \quad (2.94)$$

Table 2.16 Constants C and n for equation

Ra	C	n
10^{-10}–10^{-2}	0.675	0.058
10^{-2}–10^{2}	1.020	0.148
10^{2}–10^{4}	0.850	0.188
10^{4}–10^{7}	0.480	0.250
10^{7}–10^{12}	0.125	0.333

Natural convection for long horizontal cylinders

The following relationship can be used with C and n given in Table 2.16 (Morgan, 1975):

$$Nu = C \cdot Ra^n \tag{2.95}$$

Natural convection for spheres in fluids

The following relationship can be used for spheres in fluids with $Pr \geq 0.7$ and $Ra \leq 10^{11}$ (Churchill, 2002):

$$Nu = 2 + \frac{0.589 \cdot Ra^{0.25}}{\left[1 + \left(\frac{0.469}{Pr}\right)^{9/16}\right]^{4/9}} \tag{2.96}$$

Empirical correlations for forced thermal convective flow

Turbulent flow in pipes

Turbulent flow inside pipes longer than 50 diameters can be dealt with according the following simplified correlations:

$$Nu = 0.023 \cdot Pr^{0.4} \cdot Re^{0.8} (\text{heating}) \tag{2.97}$$

$$Nu = 0.024 \cdot Pr^{0.3} \cdot Re^{0.8} (\text{cooling}) \tag{2.98}$$

For gases this simple correlation may be used:

$$Nu = 0.025 \cdot Re^{0.8} \tag{2.98}$$

Cylinders in cross flow

For circular cylinders with $Re \cdot Pr \geq 0.2$ (Churchill and Bernstein, 1977):

$$Nu = 0.3 + \frac{0.62 \cdot Re^{0.5} \cdot Pr^{0.33}}{\left[1 + \left(\frac{0.4}{Pr}\right)^{2/3}\right]^{1/4}} \cdot \left[1 + \left(\frac{Re}{282{,}000}\right)^{5/8}\right]^{4/5} \tag{2.99}$$

where all properties are evaluated at the film temperature (Fig. 2.7).

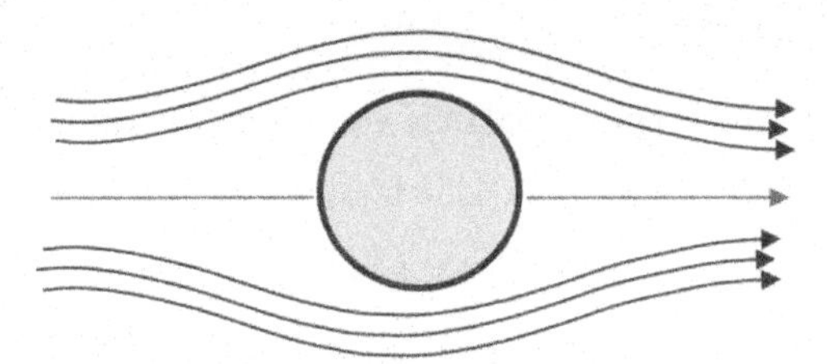

Fig. 2.7
Forced convective heat transfer for cylinders in cross flow.

Flow around spheres

For spheres with constant surface temperature in a flowing fluid approaching a uniform velocity, the mean Nusselt number is given by the following correlation, the accuracy of which is high for low Reynolds number:

$$Nu = 0.60 \cdot Re^{0.5} \cdot Pr^{0.33} \tag{2.100}$$

Whitaker (1972) recommends the following correlation for $3.5 \leq Re \leq 7.6 \cdot 10^4$, $0.71 \leq Pr \leq 380$, $1 \leq \mu/\mu_s \leq 3.2$:

$$Nu = 2 + \left(0.4 \cdot Re^{1/2} + 0.06 \cdot Re^{2/3}\right) \cdot Pr^{0.4} \cdot \left(\frac{\mu}{\mu_s}\right)^{1/4} \tag{2.101}$$

where all properties except μ_s are evaluated at T_∞.

Example 2.9

An LPG sphere ($D = 10$ m) is radiated by an external fire and its external mean wall temperature has risen up to 400°C. Wind is blowing at 20 m/s with a temperature of 0°C. Calculate the heat removal rate due to wind around the sphere in this scenario, as well as the final mean wall temperature if wind velocity drops to 10 m/s, assuming that heat radiation from the sphere is negligible (all transport data to be calculated at $(T_s + T_\infty)/2$).

Solution

	Scenario 1	
$T_s = 400°C$		$T_\infty = 0°C$
$D = 10$ m	$u = 20$ m/s	$\nu = \mu/\rho = 3.463E-5$ m²/s
$c_P = 1026$ J/kg °C	$k = 0.0386$ W/m K	$\mu = 2.58E-5$ Pa s
	Scenario 2	
$T_s =$ unknown		$T_\infty = 0°C$
$D = 10$ m	$u = 10$ m/s	$\nu = 4.4E-5$ m²/s
	$Pr = 0.685$	

Scenario 1

At steady state, according to the energy balance equation (Chapter 1):

$$Q_{Hacc} = Q_{Hin} - Q_{Hout} + Q_{Hgen} \tag{2.102}$$

$$Q_{Hin} = Q_{Hout} \tag{2.103}$$

$$Re = \frac{20 \cdot 10}{3.463 \cdot 10^{-5}} = 5,775,339 \tag{2.104}$$

$$Pr = \frac{1026 \cdot 2.58 \cdot 10^{-5}}{0.0386} = 0.685 \tag{2.105}$$

$$Nu = 0.6 \cdot 5,775,339^{0.5} \cdot 0.685^{0.33} = 1273 \tag{2.106}$$

$$h = \frac{1273 \cdot 0.0386}{10} = 4.9 \text{ W/m}^2 \text{ °C} \tag{2.107}$$

$$Q_{Hout} = 4.9 \cdot 4 \cdot \pi \cdot 25 \cdot 400 = 617,014 \text{ W} \tag{2.108}$$

Scenario 2

$$Q_{Hin} = Q_{Hout} = 617,014 \text{ W} \tag{2.109}$$

Assuming $T_s = 567$°C:

$$Re = \frac{10 \cdot 10}{4.4 \cdot 10^{-5}} = 2,272,727 \tag{2.110}$$

$$Nu = 0.6 \cdot 2,272,727^{0.5} \cdot 0.685^{0.33} = 798 \tag{2.111}$$

$$h = \frac{798 \cdot 0.0435}{10} = 3.47 \text{ W/m}^2 \text{ °C} \tag{2.112}$$

$$Q_{Hout} = 3.47 \cdot 4 \cdot \pi \cdot 25 \cdot 567 = 617,792 \text{ W} \tag{2.113}$$

T_s is accepted.

2.6.3 Thermal Radiation

Thermal radiation is an electromagnetic mechanism, which consists of heat being transported at the speed of light through regions of space that are devoid of matter. Due to its nature, heat radiation is the governing mechanism of many fire scenarios, especially in the medium and far-field. For real non-black surfaces A_1 and A_2 (Fig. 2.8), the net radiation rate is given by the adapted Stefan-Boltzmann's law:

$$Q_H = \sigma \cdot A_1 \cdot F_{12} \cdot \left(\varepsilon_1 \cdot T_1{}^4 - \alpha_1 \cdot T_2{}^4\right) \tag{2.114}$$

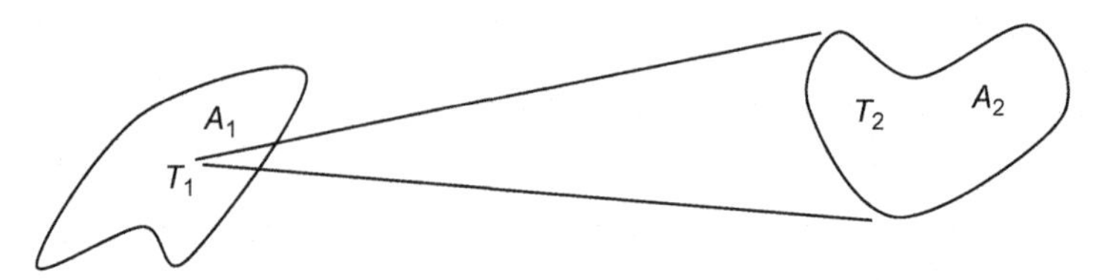

Fig. 2.8
Radiation exchange between surfaces A_1 and A_2.

where

- σ is the Stefan-Boltzmann constant (5.670×10^{-8} W/m$^2 \cdot$ K^4)
- A_1 is the area of body 1
- F_{12} is the view factor, representing the fraction of radiation leaving body 1 that its directly intercepted by body 2
- ε_1 is the emissivity of surface 1
- α_1 is the emissivity of surface 1 estimated as the value of ε at T_1

The reciprocity relation states:

$$A_1 \cdot F_{12} = A_2 \cdot F_{21} \tag{2.115}$$

Emissivities of solid surfaces

See Tables 2.17 and 2.18.

Table 2.17 Emissivity of selected materials (Modest, 2013)

Material	Temperature (°C)	ε
Aluminium foil	20	0.04
Black paint	25	0.95
Black epoxy paint	25	0.89
Black enamel paint	95	0.81
Brick (red)	20	0.93
Brick (building)	1000	0.45
Cast iron, polished	200	0.21
Chromium plate	95	0.12
Chromium plate	400 (after heating)	1.25
Chromium plate	35 (after heating)	0.15
Concrete, tiles	1000	0.63
Concrete, rough	38	0.94
Copper, electroplated	20	0.03
Glass, smooth	20	0.94
Glass (Pyrex, lead, soda)	260–540	0.95–0.85
Graphite	25	0.91
Iron, polished	425–1025	0.14–0.38
Lead	125–225	0.057–0.075
Mild steel	25	0.1–0.15
Mild steel	230–1065	0.20–0.35
Nickel, electroplated	20	0.03
Rubber	23–25	0.94–0.86
Silica (average)	20–35	0.83–0.84
Steel, polished	100	0.066
Stainless steels	35–425	0.13–0.32
Stainless steel, 304	215–490	0.44–0.36
Steel (after 42 h heating at 525°C)	215–525	0.62–0.73
Stainless steel, 316	25	0.1
Stainless steel, 316	230–1050	0.26–0.66
White acrylic r. paint	95	0.92
White epoxy paint	25	0.88

Table 2.18 Emissivity of iron and steels at different temperatures (Modest, 2013)

Material	100°C ε	500°C ε	1000°C ε
Iron	0.07	0.14	0.24
Steel, polished	0.13–0.21	0.18–0.26	0.55–0.80
Steel, cleaned	0.21–0.38	0.25–0.42	0.50–0.77

View factors

Infinitely long parallel cylinders

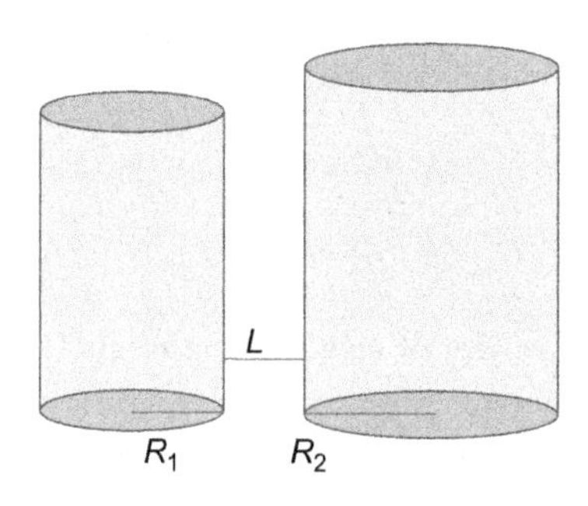

$$F_{12}=\frac{1}{2\pi}\left\{\pi+\left[C^2-(R+1)^2\right]^{1/2}-\left[C^2-(R-1)^2\right]^{1/2}\right.$$
$$\left.+(R-1)\cdot\cos^{-1}\left[\frac{R}{C}-\frac{1}{C}\right]-(R+1)\cdot\cos^{-1}\left[\frac{R}{C}+\frac{1}{C}\right]\right\} \qquad (2.116)$$

$$R=\frac{R_2}{R_1},\quad B=\frac{L}{R_1},\quad C=1+R+B \qquad (2.117)$$

Infinitely long parallel cylinders of the same diameter

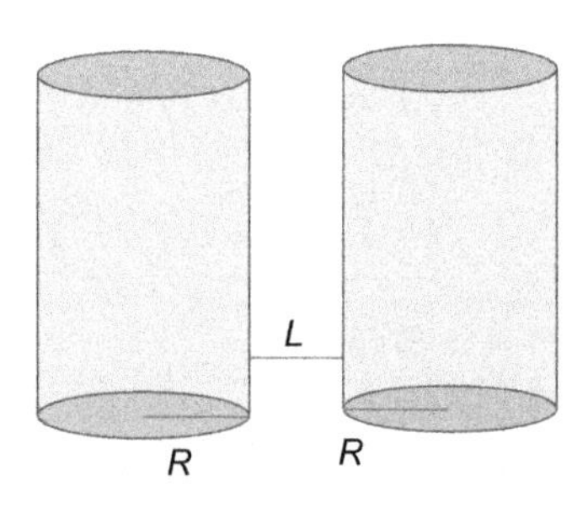

$$F_{12}=\frac{1}{\pi}\left(\sin^{-1}\frac{1}{B}+\sqrt{B^2-1}-B\right) \qquad (2.118)$$

$$B=1+\frac{L}{2\cdot R} \qquad (2.119)$$

According to Juul (1982), for a L to R ratio $= 3.0$, the application of infinite length view factors to a case of finite length does not cause any significant errors ($\cong 5\%$). Therefore, it is reasonable to use the infinite length view case to calculate view factors for cylinders of finite length.

Perpendicular surfaces with a common edge

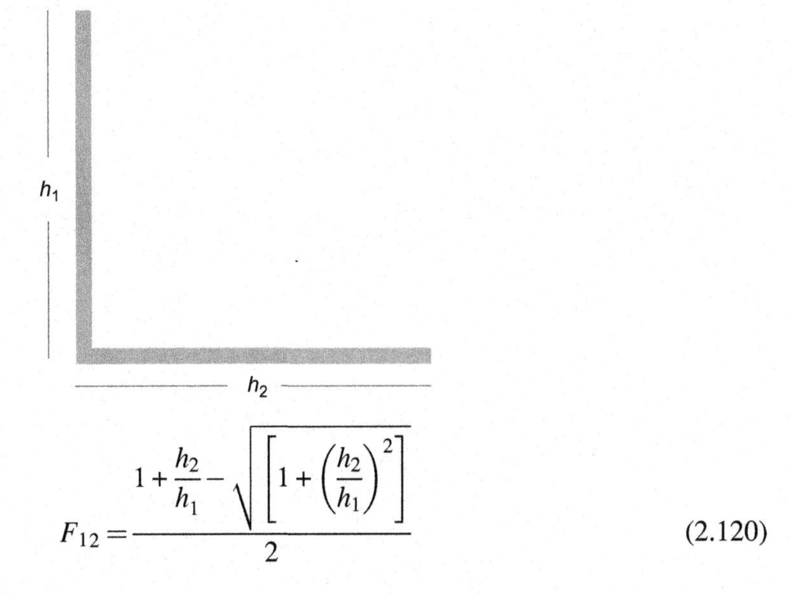

$$F_{12} = \frac{1 + \frac{h_2}{h_1} - \sqrt{\left[1 + \left(\frac{h_2}{h_1}\right)^2\right]}}{2} \tag{2.120}$$

Linear and circular surfaces

$$F_{12} = \frac{r}{l_1 - l_2} \cdot \left[\tan^{-1}\frac{l_1}{l} - \tan^{-1}\frac{l_2}{l}\right] \tag{2.121}$$

Coaxial parallel disks

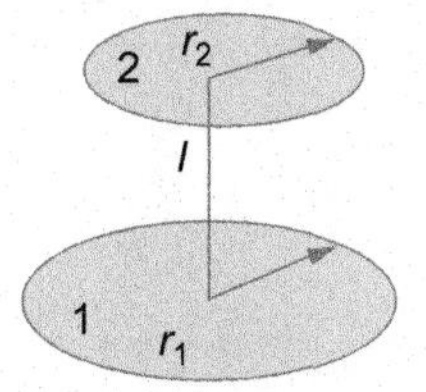

$$F_{12}=\frac{1}{2}\cdot\left\{\left[1+\frac{1+\left(\frac{r_2}{l}\right)^2}{\left(\frac{r_1}{l}\right)^2}\right]-\sqrt{\left[1+\frac{1+\left(\frac{r_2}{l}\right)^2}{\left(\frac{r_1}{l}\right)^2}\right]^2-4\cdot\left(\frac{r_2}{r_1}\right)^2}\right\} \tag{2.122}$$

Parallel rectangles

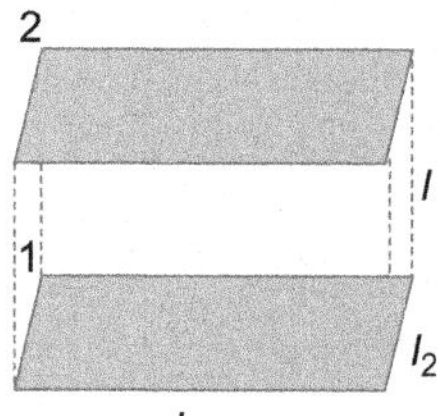

$$F_{12}=\frac{2}{\pi\cdot\frac{l_1\cdot l_2}{l^2}}\cdot\left\{\begin{array}{l}\ln\left[\frac{\left(1+\left(\frac{l_1}{l}\right)^2\right)\cdot\left(1+\left(\frac{l_2}{l}\right)^2\right)}{1+\left(\frac{l_1}{l}\right)^2+\left(\frac{l_2}{l}\right)^2}\right]^{1/2}+\left(\frac{l_1}{l}\right)\cdot\left(1+\left(\frac{l_2}{l}\right)^2\right)^{1/2}\cdot\tan^{-1}\frac{\left(\frac{l_1}{l}\right)}{\left(1+\left(\frac{l_2}{l}\right)^2\right)^{1/2}}\\+\left(\frac{l_2}{l}\right)\cdot\left(1+\left(\frac{l_1}{l}\right)^2\right)^{1/2}\cdot\tan^{-1}\frac{\left(\frac{l_2}{l}\right)}{\left(1+\left(\frac{l_1}{l}\right)^2\right)^{1/2}}-\left(\frac{l_1}{l}\right)\cdot\tan^{-1}\left(\frac{l_1}{l}\right)-\left(\frac{l_2}{l}\right)\cdot\tan^{-1}\left(\frac{l_2}{l}\right)\end{array}\right\} \tag{2.123}$$

Perpendicular rectangles with a common edge

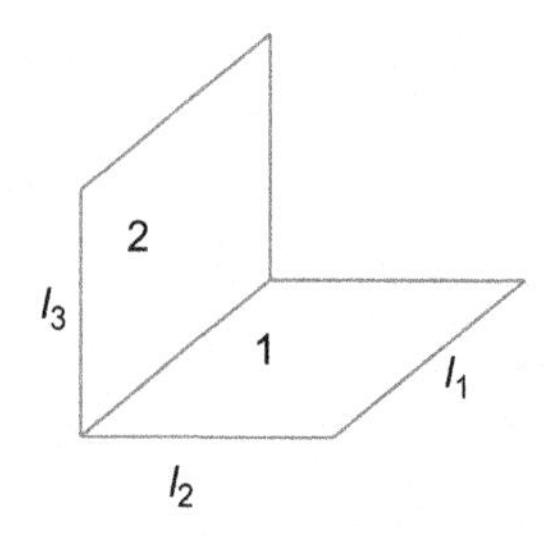

$$F_{12}=\frac{l_1}{\pi\cdot l_2}\cdot\left\{\begin{array}{l}\frac{l_2}{l_1}\cdot\tan^{-1}\frac{l_1}{l_2}+\frac{l_3}{l_1}\cdot\tan^{-1}\frac{l_1}{l_3}-\left[\left(\frac{l_2}{l_1}\right)^2+\left(\frac{l_3}{l_1}\right)^2\right]^{1/2}\cdot\tan^{-1}\frac{1}{\left[\left(\frac{l_2}{l_1}\right)^2+\left(\frac{l_3}{l_1}\right)^2\right]}\\ +\frac{1}{4}\cdot\ln\left[\frac{\left(1+\left(\frac{l_2}{l_1}\right)^2\right)\cdot\left(1+\left(\frac{l_3}{l_1}\right)^2\right)}{1+\left(\frac{l_2}{l_1}\right)^2+\left(\frac{l_3}{l_1}\right)^2}\cdot\left(\frac{\left(\frac{l_2}{l_1}\right)^2\cdot\left(1+\left(\frac{l_2}{l_1}\right)^2+\left(\frac{l_3}{l_1}\right)^2\right)}{\left(1+\left(\frac{l_2}{l_1}\right)^2\right)\cdot\left(\left(\frac{l_2}{l_1}\right)^2+\left(\frac{l_3}{l_1}\right)^2\right)}\right)^{\left(\frac{l_2}{l_1}\right)^2}\times\left(\frac{\left(\frac{l_3}{l_1}\right)^2\cdot\left(1+\left(\frac{l_2}{l_1}\right)^2+\left(\frac{l_3}{l_1}\right)^2\right)}{\left(1+\left(\frac{l_3}{l_1}\right)^2\right)\cdot\left(\left(\frac{l_2}{l_1}\right)^2+\left(\frac{l_3}{l_1}\right)^2\right)}\right)^{\left(\frac{l_3}{l_1}\right)^2}\right]\end{array}\right\} \quad (2.124)$$

Sphere to rectangle with $r < d$

$$D_1=\frac{d}{l_1},D_2=\frac{d}{l_2} \quad (2.125)$$

$$F_{1-2}=\frac{1}{4\cdot\pi}\cdot\sqrt{\frac{1}{D_1{}^2+D_2{}^2+D_1{}^2\cdot D_2{}^2}} \quad (2.126)$$

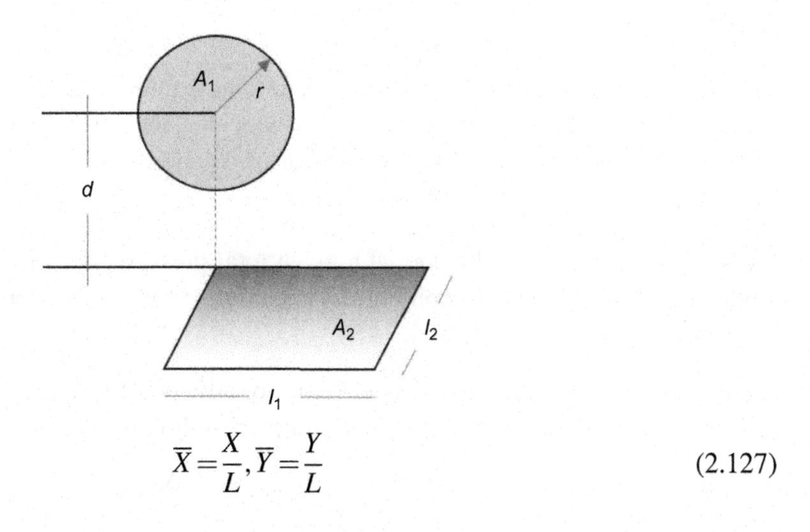

$$\overline{X}=\frac{X}{L},\overline{Y}=\frac{Y}{L} \quad (2.127)$$

$$F_{BA}=\left\{\frac{2}{\overline{Y}\cdot\pi\cdot\overline{X}}\cdot\left[\ln\sqrt{\frac{\left(1+\overline{X}^{2}\right)\cdot\left(1+\overline{Y}^{2}\right)}{1+\overline{X}^{2}+\overline{Y}^{2}}}\right]+\overline{X}\cdot\sqrt{1+\overline{Y}^{2}}\cdot\tan^{-1}\frac{\overline{X}}{\sqrt{1+\overline{Y}^{2}}}\right.$$
$$\left.+\overline{Y}\cdot\sqrt{1+\overline{X}^{2}}\cdot\tan^{-1}\frac{\overline{Y}}{\sqrt{1+\overline{X}^{2}}}-\overline{X}\cdot\tan^{-1}\overline{X}-\overline{Y}\right\} \tag{2.128}$$

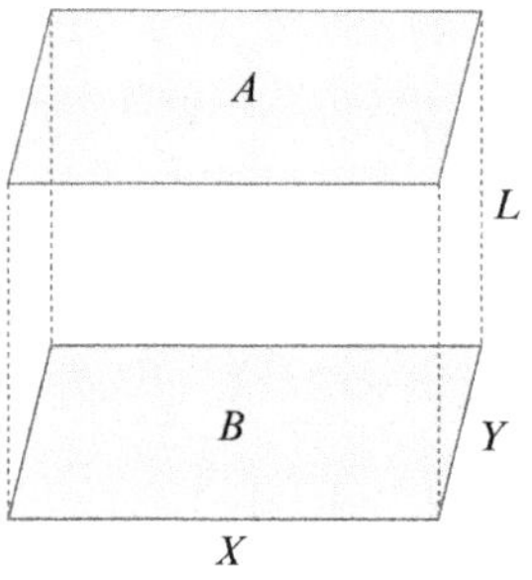

Thermal radiation and emissivities of gases

Gases such as carbon dioxide and water absorb heat radiated by a hot surface in specific zones of the infrared spectrum and can radiate heat to a target in the same wavelength regions. This behaviour is typical for asymmetrical molecules, also including carbon monoxide, hydrocarbons, water vapour, sulphur dioxide, ammonia, hydrogen chloride, and sulphide. Hottel (1954) has provided the fundamental equation of heat radiated transmission from a gas at temperature T_G to a black element of a surface at temperature T_S:

$$Q_G=\varepsilon_G\cdot\sigma\cdot{T_G}^4 \tag{2.129}$$

Q_G is specifically related to a hemispherical gas mass of radius L and radiating gas partial pressure P_j. ε_G is the gas emissivity, depending on L, P_j, and total gas mass pressure.

Water vapour

The emissivity of water vapour depends on gas temperature T_G and on the product $L\times P_{w.}$, as shown on Fig. 2.9 (Hottel, 1954).The emissivities calculated on this diagram have to be multiplied by the factor C_w to account for the contribution of P_W and P_T, as shown in Fig. 2.10 (Hottel, 1954).

Carbon dioxide

The emissivity of carbon dioxide, at a given total pressure, depends on T_G and on the product $L\times P_w$ only, as shown on Fig. 2.11 (Hottel, 1954). Correction factors can be calculated on the basis of Fig. 2.12. (Hottel, 1954).

Hottel has developed the following specific formula and the associated diagrams for gas mixtures containing both water vapour and carbon dioxide (Fig. 2.13):

$$\varepsilon_G=\varepsilon_W+\varepsilon_C-\Delta\varepsilon \tag{2.130}$$

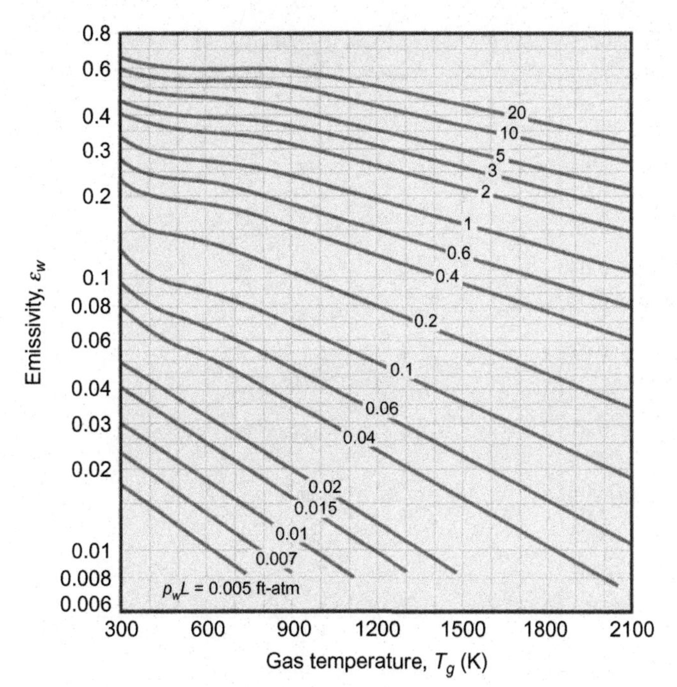

Fig. 2.9
Emissivity of water vapour (Hottel, 1954).

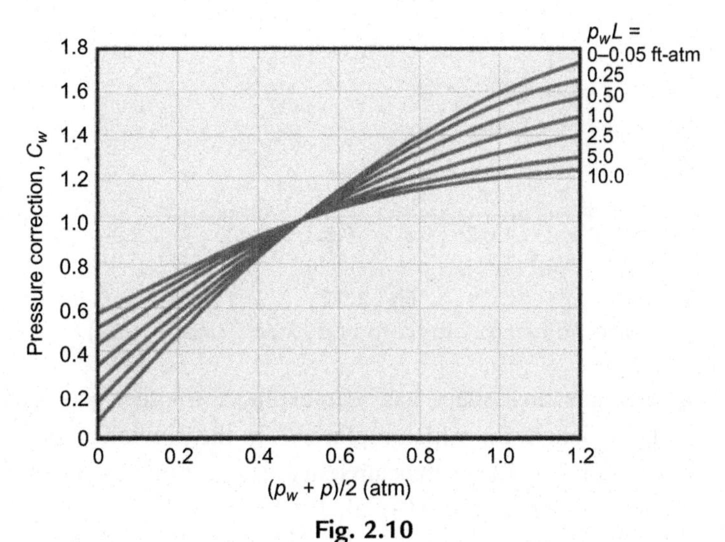

Fig. 2.10
Correction factors for water vapour (Hottel, 1954).

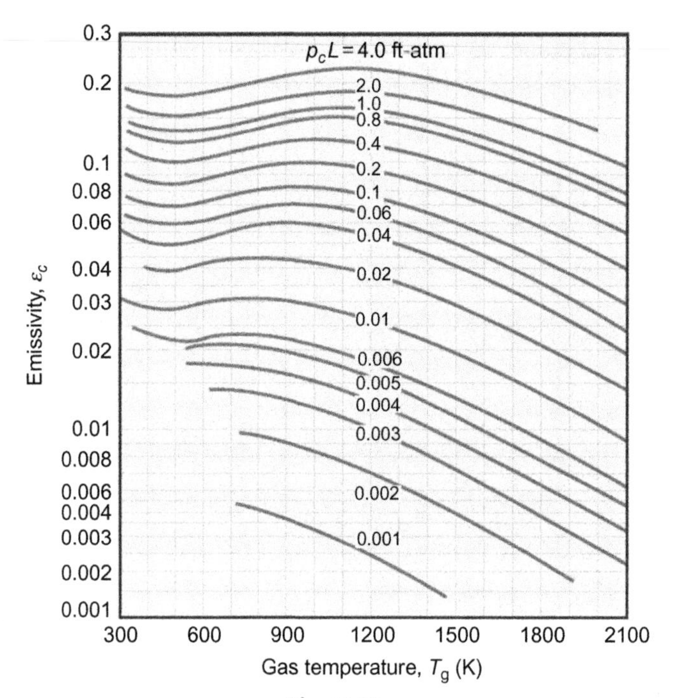

Fig. 2.11
Emissivity of carbon dioxide (Hottel, 1954).

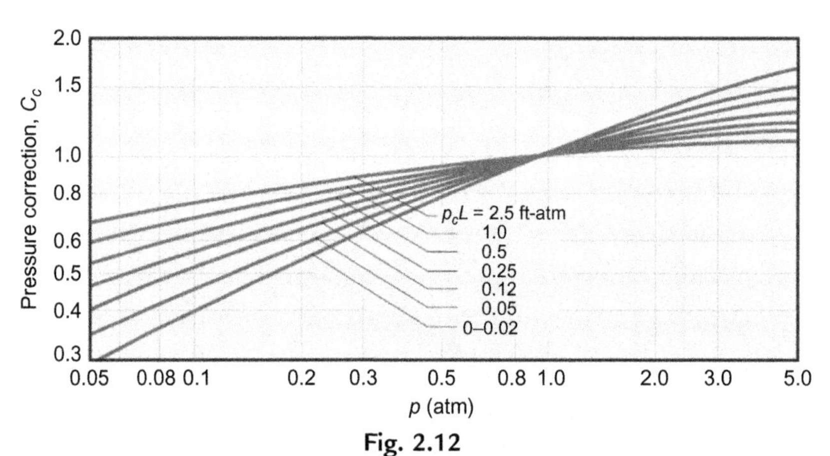

Fig. 2.12
Correction factors for carbon dioxide (Hottel, 1954).

Finally the hemispherical gas mass shape has been extended to other geometries, introducing the concept of mean beam length L_b, which replaces L in the previous diagrams. It is worth noting that, for low values of P_GL, the bean beam length is four times the mean hydraulic radius of the enclosure, as shown in Table 2.19 (Hottel, 1954).

The net heat radiation rate between the gas mass and a black body surface of area A_s is calculated with the following formula (Table 2.20):

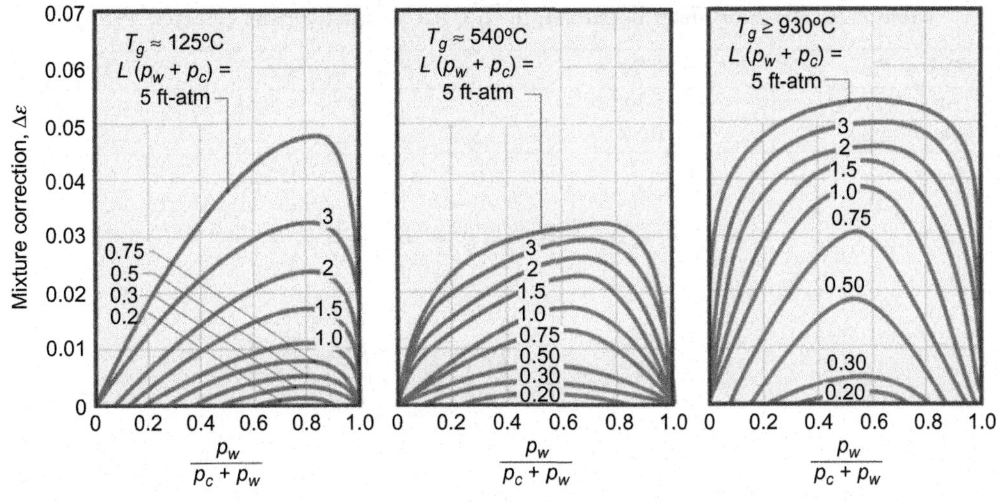

Fig. 2.13
Correction factors for gas mixtures containing water vapour and carbon dioxide (Hottel, 1954).

Table 2.19 Mean beam lengths for gas radiation (Hottel, 1954)

Shape	Characterising Dimension X	Factor by which X has to be multiplied to obtain L_b	
		$P_G L \asymp 0$	$P_G L > 0$
Sphere	Diameter	2/3	0.60
Infinite cylinder	Diameter	1	0.90
Semi-infinite cylinder, radiating to centre of base	Diameter	–	0.90
Right-circular cylinder, height = diameter radiating to centre of base	Diameter	–	0.77
Right-circular cylinder, height = diameter radiating to whole surface	Diameter	2/3	0.60
Infinite cylinder of half-circular cross section radiating to spot on middle flat side	Radius	–	1.26
Rectangular parallelepipeds:		2/3	Table 2.20
– 1:1:1 (cube)	Edge	0.9	Table 2.20
– 1:1:4 radiating to 1 × 4 face		0.86	Table 2.20
radiating to 1 × 1 face	Shortest edge	0.89	Table 2.20
radiating to all faces		1.18	Table 2.20
– 1:2:6 radiating to 2 × 6 face	Shortest edge	1.24	Table 2.20
radiating to 1 × 6 face		1.18	Table 2.20
radiating to 1 × 2 face		1.20	Table 2.20
radiating to all faces		2	Table 2.20
– 1:∞:∞ (infinite parallel planes)	Distance between planes		

Table 2.20 Ratio of mean beam length to L for parallelepipeds (Hottel, 1954)

P_CL or P_WL	0.01	0.1	1
Carbon dioxide	0.85	0.80	0.77
Water vapour	0.97	0.93	0.85

$$Q_G = A_s \cdot \sigma \cdot \left(\varepsilon_G \cdot T_G{}^4 - \alpha_G \cdot T_s{}^4\right) \tag{2.131}$$

where

$$\alpha_w = \varepsilon_w \left(T_s, p_w L_b \cdot \frac{T_s}{T_G}\right) \cdot C_w \cdot \left(\frac{T_G}{T_s}\right)^{0.45} \tag{2.132}$$

and

$$\alpha_c = \varepsilon_c \left(T_s, p_c L_b \cdot \frac{T_s}{T_G}\right) \cdot C_c \cdot \left(\frac{T_G}{T_s}\right)^{0.65} \tag{2.133}$$

ε and C can be found on the previous diagrams, where the emissivities are evaluated considering the variables included in brackets. Once again, in case of simultaneous presence of water and carbon dioxide it will be:

$$\alpha_G = \alpha_w + \alpha_c - \Delta\alpha \tag{2.134}$$

were $\Delta\alpha = \Delta\varepsilon$.

Example 2.10

Methane is burnt with 10% air excess and fed at 1 atm to a 150 mm diameter pipe. Gas and pipe inlet and outlet temperatures are 1094°C/427°C and 538°C/316°C respectively. What is the heat rate radiated from the gas to the pipe, if its length is 10 m and it can be considered a black body?

Solution

From Table 2.19 $L_b = 0.9 \times 0.15 = 0.135$ m $= 0.44$ ft

The combustion reaction is:

$$CH_4 + 1.1O_2 + 1.1 \cdot 3.76N_2 \rightarrow CO_2 + 2H_2O + 0.1O_2 + 1.1 \cdot 3.76N_2$$

The molar fractions of the combustion products and the related values are calculated and included in the following table. Mean values have been assumed for the gas and the pipe wall temperature T_G and T_s, and the corrective curves for $T > 1700$°F have been used for the determination of $\Delta\varepsilon$ (Table 2.21).

$$\begin{aligned} Q_G &= A_s \cdot \sigma \cdot \left(\varepsilon_G \cdot T_G{}^4 - \alpha_G \cdot T_s{}^4\right) \\ &= \pi \cdot 0.15 \cdot 10 \cdot 5.670 \times 10^{-8} \cdot \left(0.137 \cdot 1089\ (\mathrm{K})^4 - 0.17 \cdot 644\ (\mathrm{K})^4\right) = 43{,}647\ \mathrm{W} \end{aligned} \tag{2.135}$$

Table 2.21 Molar fractions and radiative coefficients for Example 2.10

	Carbon Dioxide	Oxygen	Water Vapour	Nitrogen	Overall
Moles	1	0.1	2	4.13	7.23
Molar fractions	0.14	0.014	0.28	0.57	1
Partial pressure (atm)	0.14	0.014	0.28	0.57	1
Partial pressure × L_b (ft atm)	0.06		0.12		
C_c	1				
C_w			1.2		
Emissivity (T_G=1960°R)	0.063 × 1		0.062 × 1.2		
$\Delta\varepsilon$ $[P_w/(P_c+P_w)]=0.66)$			≅0		
$\varepsilon_G=\varepsilon_W+\varepsilon_C-\Delta\varepsilon$			0.137		
T_s (°R)			1159		
T_G/T_s			1.69		
T_s/T_G			0.59		
$p_c \cdot L_b \cdot \frac{T_s}{T_G}$			0.036		
$p_w \cdot L_b \cdot \frac{T_s}{T_G}$			0.072		
ε_C $(T_s, L_bT_s/T_G)$			0.050		
ε_w $(T_s, L_bT_s/T_G)$			0.068		
α_C $(T_s, L_bT_s/T_G,C, T_G/T_s)$			0.07		
α_w $(T_s, L_bT_s/T_G, C,T_G/T_s)$			0.1		
$\alpha_G=\alpha_W+\alpha_C\text{-}\Delta\alpha$			0.17		

This calculation does not consider the heat exchanged by conduction and convection.

2.7 Applications

2.7.1 Isothermal Processes

Many processes take place at constant temperature. Phase changes are isothermal processes. Other scenarios may be isothermal either because no thermal effect is associated with the process, or because heat is removed so that temperature is constant. In some cases, temperature constancy is essential for the system to maintain its stable configuration in order to avoid serious hazardous scenarios such as thermal runaway and overheating. An interesting case study is free expansion.

Free expansion

Example 2.11

A thermally insulated 10 m^3 pressure vessel contains nitrogen at 20°C and 5 atm. This vessel has an internal partition which divides it into two equal parts. The partition is removed and the gas is allowed to expand with no appreciable friction effects. Calculate the final temperature for both an ideal and a real gas, assuming van der Waal's state equation (Fig. 2.14).

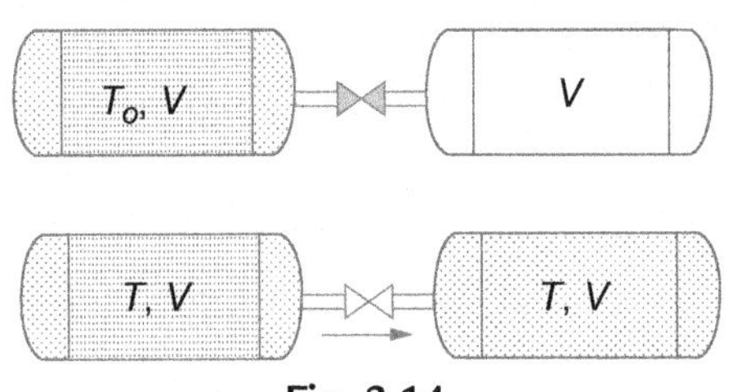

Fig. 2.14
Free expansion.

Solution
In both cases, first law of thermodynamics applies (specific units):

$$\Delta u = Q + w \tag{2.136}$$

During the expansion no work is done on or by the vessel, nor is heat transferred to or from the vessel. Consequently:

$$\Delta u = 0 \tag{2.137}$$

For an ideal gas, according to the kinetic theory, internal energy is a function of temperature only. For a diatomic gas:

$$u = \frac{5}{2}RT \tag{2.138}$$

We can conclude that no temperature change will occur.

The more the gas departs from the ideal conditions (high temperature and low pressure), the less the final temperature will be close to the initial temperature, as shown by Joule. Internal energy is an exact differential, which can be written as:

$$du = \left(\frac{\partial u}{\partial v}\right)_T dv + \left(\frac{\partial u}{\partial T}\right)_v dT \tag{2.139}$$

According to the first law of thermodynamics:

$$du = \left(\frac{\partial u}{\partial v}\right)_T dv + \left(\frac{\partial u}{\partial T}\right)_v dT = 0 \tag{2.140}$$

$$\left(\frac{\partial T}{\partial v}\right)_u = -\frac{\left(\frac{\partial u}{\partial v}\right)_T}{\left(\frac{\partial u}{\partial T}\right)_v} \tag{2.141}$$

By definition:

$$\left(\frac{\partial u}{\partial T}\right)_v = c_V \tag{2.142}$$

and, according to Maxwell relation:

$$\left(\frac{\partial u}{\partial v}\right)_T = \left[T \cdot \left(\frac{\partial P}{\partial T}\right)_V - P\right] \tag{2.143}$$

And the Joule coefficient, μ_u:

$$\mu_u = \left(\frac{\partial T}{\partial v}\right)_u = -\frac{1}{c_V} \cdot \left[T \cdot \left(\frac{\partial P}{\partial T}\right)_V - P\right] \tag{2.144}$$

For an ideal gas:

$$\left(\frac{\partial P}{\partial T}\right)_V = \frac{R}{v} = \frac{P}{T} \tag{2.145}$$

and, replacing in the equation:

$$\mu_u = \left(\frac{\partial T}{\partial v}\right)_u = -\frac{1}{c_v} \cdot \left[T \cdot \frac{P}{T} - P\right] = 0 \tag{2.146}$$

For a real gas, according to van der Waals equation:

$$P = \frac{R \cdot T}{v - b} - \frac{a}{v^2} \tag{2.147}$$

$$\mu_u = \left(\frac{\partial T}{\partial v}\right)_u = -\frac{1}{c_v} \cdot \left[T \cdot \left(\frac{R}{v - b}\right) - \left(\frac{R \cdot T}{v - b} - \frac{a}{v^2}\right)\right] = -\frac{a}{c_v \cdot v^2} \tag{2.148}$$

Integrating, considering that specific heat at constant volume is substantially constant (Table 2.22):

Table 2.22 Calculation data for Example 2.11

Initial volume	v_o	m^3	$v_o = \frac{V_o}{n} = \frac{R \cdot T_o}{P_o} = \frac{0.00008206 \cdot 293.16}{5} = 0.0048\ m^3/mole$	
Final volume	v	m^3	$0.0096\ m^3/mole$	
Van der Waals constant	a	$J\ m^3/mole^2$	$a \cong 014\ J\ m^3/mole^2$	Hougen et al. (1954)
Specific heat	c_v	J/mole	20.8 J/mole K	

$$\Delta T = -\int_{v_o}^{v} \frac{a}{c_v \cdot v^2} \cdot dv = \frac{a}{c_v} \cdot \left(\frac{1}{v} - \frac{1}{v_o}\right) \tag{2.149}$$

$\Delta T = -0.7°C$

Temperature drop is lower than 1°C.

2.7.2 Isochoric processes

Isochoric processes are very frequent in process safety. Chemical reactions taking place in closed reactors are isochoric processes. Explosive events occurring inside containments (confined explosions) are an important isochoric scenario in process safety engineering.

Example 2.12

1 m^3 iron cubic vessel with 1-m side contains a stoichiometric mixture of air and methane at 1 atm and at ambient temperature of 18°C. The mixture is ignited. Calculate the time required for the system to return to the ambient temperature, assuming:

- ideal gas law
- adiabatic combustion (very fast)
- uniform and identical gas and wall temperature
- methane heating value: 890,739 J/mole
- no wind
- a convective heat transfer coefficient equal to 20 W/m^2 K
- an emissivity factor for iron equal to 0.3

Solution

According to Hess' law, we can imagine the explosion process to occur in accordance with the following physical-chemical path, consisting of three steps (Fig. 2.15):

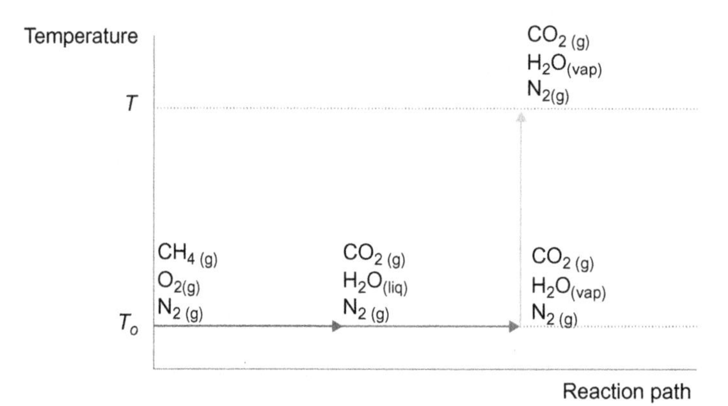

Fig. 2.15
Reaction path for Example 2.12, where T_o is the initial temperature and T is the final (maximum) temperature.

Basis: 1 mole CH_4

First step— Isothermal combustion with final liquid water (Hougen et al., 1954):

$$CH_{4(g)} + 2O_{2(g)} + 2 \cdot 3.76\,N_{2(g)} \rightarrow CO_{2(g)} + 2H_2O_{(liq)} + 2 \cdot 3.76N_{2(g)}$$

$H_1 = 890{,}802$ J/mole CH_4

Second step—Liquid water vaporisation:

$$2H_2O_{(liq)} \rightarrow 2H_2O_{(vap)}$$

Latent heat of water at 100°C: 40,726 J/mole H_2O

$H_2 = 2 \times 40{,}726$ J/mole $H_2O = 81{,}451$ J

Third Step—Reaction products heating to final temperature at constant volume:

Reactants (moles|)			
Methane 1 3.95	Oxygen 2 7.9	Nitrogen 7.52 29.72	Total 10.52 41.57
Reactants (molar fractions)			
0.095	0.19	0.715	1
Products (moles)			
Carbon dioxide 1 3.95	Water 2 7.9	Nitrogen 0.715 29.72	Total 10.52 41.57
Products (molar fractions)			
0.095	0.19	0.715	1
Average specific heats at constant pressure (J/mole K), c_P			
49.15	43.2	32.76	–
Average specific heats at constant volume (J/mole K), $c_P - R$ (8.31 J K^{-1} $mole^{-1}$)			
40.9	34.9	24.45	–
Average specific heats at constant volume × molar fraction (J/mole K)			
3.89	6.63	17.48	–
Average mixture specific heat at constant volume (J/mole K)			
$c_v = 3.89 + 6.63 + 17.48 = 28$			

$H_3 = 41.57 \times 28 \times (T - 291.16)$

Energy balance:

$$H_1 + H_2 + H_3 = 890{,}802 + 81{,}451 + 41.57 \times 28 \times (T - 291.16) = 890{,}739 \times 3.95 \quad (2.150)$$

Solving: $T = 2187$°C

It is worth noting that the adiabatic temperature at constant volume is significantly greater than the adiabatic temperature at constant pressure, due to the lower specific heat.

Neglecting thermal conduction, the system will exchange heat with the environment according to the following equation:

$$N\,(\text{moles})\cdot c_v\frac{dT}{dt}=-\left[A\cdot h\cdot(T-T_{amb})+A\cdot\varepsilon\cdot\sigma\cdot\left(T^4-T_{amb}\right)\right] \tag{2.151}$$

$$41.57\cdot 28\cdot\frac{dT}{dt}=-\left[6\cdot 20\cdot(T-T_{amb})+6\cdot 0.3\cdot 5.67\times 10^{-8}\cdot\left(T^4-T_{amb}\right)\right] \tag{2.152}$$

Integrating, it is shown that it will take around 86 h for the system to go back to the ambient temperature, according to Fig. 2.16.

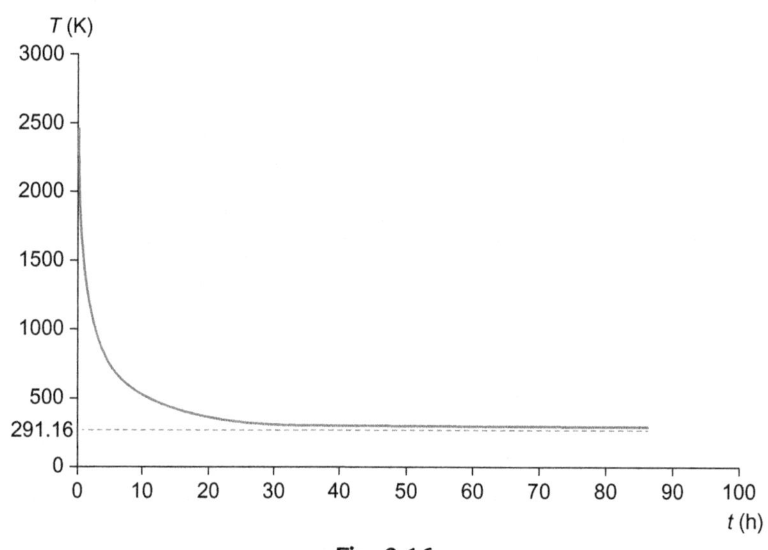

Fig. 2.16
Temperature decay of the Example 2.12.

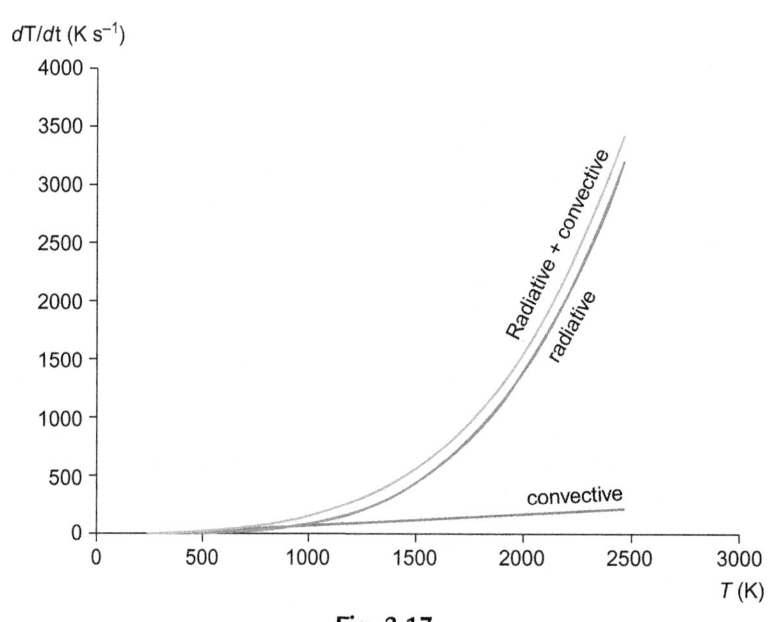

Fig. 2.17
Temperature rise rate contributors for the Example 2.12.

It is interesting to differentiate the two contributors, convective and radiative, to heat transfer (Fig. 2.17).

2.7.3 *Isobaric Processes*

Example 2.13

An atmospheric tank blanketed with nitrogen contains hexane at 25°C. The head space volume is 300 m^3. Temperature rises to 40°C, so that the PRV opens and vents out the overpressure. Neglecting liquid expansion, calculate the amount of gas vented out to the atmosphere and the composition of the gas in the head space at both temperatures (Fig. 2.18).

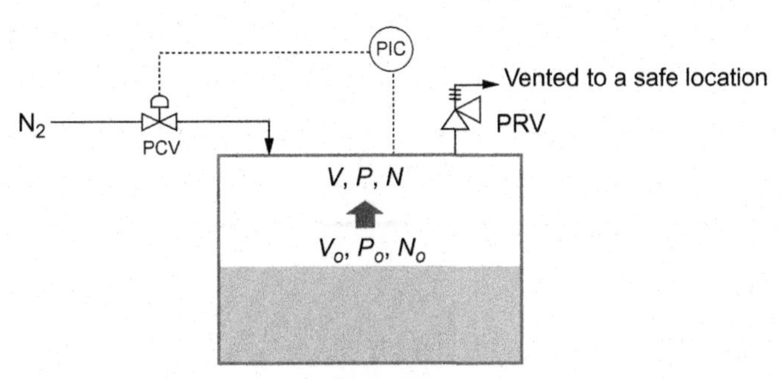

Fig. 2.18
Atmospheric tank of Example 2.13.

System at 25°C			
Temperature 298.16 K Head space volume 300 m^3	Pressure 1 atm Hexane composition 20% (molar)	Hexane vapour pressure 0.2 atm Nitrogen composition 80% (molar)	Nitrogen partial pressure 0.8 atm (abs) Total gas amount (kmoles) $N=\frac{1\cdot 300}{0.08206\cdot 298.16}=12.26$ kmoles
Total gas amount (kg)			
$M_o=0.2\cdot 12.26\cdot 86+0.8\cdot 12.26\cdot 28=485.5$ kg			
System at 40°C			
Temperature 313.16 K Head space volume 300 m^3	Pressure 1 atm Hexane composition 36% (molar)	Hexane vapour pressure 0.36 atm Nitrogen composition 64% (molar)	Nitrogen partial pressure 0.64 atm Total gas amount (kmoles) $N=\frac{1\cdot 300}{0.08206\cdot 313.16}=11.67$ kmoles
Total gas amount (kg)			
$M=0.36\cdot 11.67\cdot 86+0.64\cdot 11.67\cdot 28=570.5$ kg			

Solution

To calculate the hexane amount released to the atmosphere, we note that, soon before the PRV opening the head-space nitrogen would have been the same, $0.8 \times 12.26 = 9.808$ kmoles, whereas hexane concentration would have been increased in accordance with the vapour pressure at the new temperature. Consistently, 0.36 atm would have been the new hexane partial pressure, so that:

$$\frac{N_{hex}}{N_{hex} + 9.808} = 0.36 \quad (2.153)$$

Solving, $N_{hex} = 5.517$ kmoles, and the total kmoles would be $5.517 + 9.808 = 15.325$ kmoles. Finally, the following molar and mass amounts have been released:

Molar release:

$$N_{rel} = 15.325 - 11.67 = 3.655 \text{ kmoles} \quad (2.154)$$

Mass release is calculated assuming complete mixing of nitrogen and hexane:

$$M_{rel} = 3.655 \times (0.36 \times 86 + 0.64 \times 28) = 113.159\,(hexane) + 65.5\,(nitrogen) = 178.66kg \quad (2.155)$$

2.7.4 Adiabatic Processes

Adiabatic transformations occur very frequently in process safety. Theory, experience, and case history have shown that most of the fluid release, loss of containment, and jet phenomena are so rapid that an adiabatic model is often the most adequate to describe them. In some instances, speeds are so high not to allow the system to reach a final equilibrium state, so that equilibrium thermodynamics is not adequate to describe the phenomenon. Specific cases have been examined in the next chapters.

Example 2.14

High pressure methane at 10 atm and 20°C undergoes a very rapid expansion through a PRV within a flare header. If the downstream pressure is 1 atm absolute, calculate the associated temperature.

Solution

Assuming ideal gas law and adiabatic transformation, it results:

$$T_1 \times P_1^{(1-k)/k} = const \quad (2.156)$$

where k is the specific heat ratio, that for methane is 1.32.

$$293.16 \times 10^{-0.24242} = 167.7 \text{ K atm} \quad (2.157)$$

$$T_2 = \frac{167.7}{1^{-0.24242}} = 167.7 \text{ K} \quad (2.158)$$

2.7.5 Thermodynamics of LNG

Definition of LNG

Liquefied natural gas (LNG) is a colourless fluid in the liquid state composed predominantly of methane, which may contain minor quantities of ethane, propane, nitrogen or other components normally found in natural gas (BS-EN, 1160). Pressure-enthalpy diagrams for methane and ethane are shown in Figs 2.19 and 2.20 (Aartun, 2002).

Physical-chemical data of LNG

See Tables 2.23 and 2.24.

2.7.6 Thermodynamics of Pressurised Liquids

If a saturated liquid stored at a pressure higher than the atmospheric pressure is released to the atmosphere, some of the liquid vaporises. This phenomenon is known as a *flash stage* in chemical engineering (Fig. 2.21), where a liquid L_o, at initial saturation pressure P_o and temperature T_o, is throttled to a lower pressure P_1.

No heat is assumed to be transferred to the liquid, so it has to accommodate its state to the new pressure, vaporising a part of it at the saturation temperature corresponding to P_1. This is expressed by the following mass and heat balances:

$$L_o = V_1 + L_1 \tag{2.159}$$

$$x_1 = \frac{L_1}{L} \tag{2.160}$$

$$y_1 = \frac{V_1}{L} \tag{2.161}$$

$$-\int_{T_o}^{T_1} c_{pL}(T) \cdot dT = y_1 \cdot \lambda(T_1) \tag{2.162}$$

We have five unknown variables, which are V_1, L_1, x_1, y_1, and T_1. A fifth equation is the relation between temperature T_1 and vapour pressure at this temperature:

$$P_1(T_1) = P_{V1}^o(T_1) = f(T_1) \tag{2.163}$$

Table 2.23 Physical-chemical data (BS-EN, 1160)

Composition	Density	Temperature
Methane (>75%) Nitrogen (<5%)	430–470 kg/m^3 max 520 kg/m^3 1.35 kg · m^3/°C	−166°C to 157°C 1.25 × 10–4°C/Pa (*Pa* is LNG vapour pressure in pascal)

Table 2.24 Physical-chemical data (Benintendi and Rega, 2014)

	Units	LNG 1[a]	LNG 2[a]	LNG 3[a]	LNG4[b]	Methane[c]	LNG[d]	LNG[e]	LNG[f]	LNG[g]	LNG[h]	LNG[i]	LNG[j]	LNG[k]	LNG[l]	LNG[m]
Methane	mole%	99.8	93.9	87.2	–	–	96.9	87.93	91.692	87.876	99.72	86.98	93.32	98	92	
Ethane	mole%	0.07	3.26	8.61	–	–	2.7	7.73	4.605	7.515	0.06	9.35	4.65	1.40	6.00	
Propane	mole%	–	0.69	2.74	–	–	0.3	2.51	2.402	3.006	0.0005	2.33	0.84	0.4	1	
i-Butane	mole%	–	0.12	0.42	–	–	0.1	0.50	1.301	1.603	0.0005	0.63	0.18	0.1	–	
n-Butane	mole%	–	0.15	0.65	–	–	–	0.72	–	–						
Pentane	mole%	–	0.09	0.02	–	–	–	0.61	–	–	–	–	–	–	–	
Nitrogen	mole%	0.17	1.79	0.36	–	–	–	–	–	–	0.2	0.71	101	0.1	1	
Molecular weight	kg/ kmole	16.07			16.043	16.043	16.55	17.91	18.615	18.77	–	–	–	–	–	
Bubble point temperature	°C	−161.4	−165.3	−161.3	–	−161.5	−161.05	−160.04	−159.9		–	–	–	–	–	
Liquid density at boiling point	kg/m^3	422	448.8	468.7	422.6 at bp (methane) 450 at bp (LNG)	422.5	430.9	452.8	463.6	452.9	–	–	–	–	–	
Liquid density approximate gradient with temperature	$kg/m^3/K$	1.4	–	–	–	–	–	–	–	–	–	–				
Gas density at NTP	kg/m^3		0.651	0.6685	0.6894	0.7459	0.7751	0.7829								
Vapour density at boiling at boiling point	kg/m^3		1.82	1.81	1.799	1.776	1.763	1783								
Latent heat of vaporisation	kj/kg	525.6	679.5	675.5	510											
Approximate rate of evaporation on water (rate per unit	kg/m^2h	600														

Table 2.24 Physical-chemical data—cont'd

	Units	LNG 1	LNG 2	LNG 3	LNG4	Methane	LNG	LNG	LNG	LNG	LNG	LNG	LNG	LNG	LNG	LNG
area after 60s)																
LNG viscosity at 160°C	Pa s	$1 \cdot 10^{-4}$–$2 \cdot 10^{-4}$														
Critical temperature	K		190.6													
Critical pressure	Pa				$4.64 \cdot 10^6$											
Vapour specific heat at constant pressure	kj/kg K				2.2											
Ratio of specific heats					1.30815											
Surface tension at −161°C	N/m															1014

[a]ISO 16903.
[b]MKOPSC (Lees's 2012).
[c]Methane (Woodward and Pitbaldo, 2010).
[d]Trinidad (Woodward and Pitbaldo, 2010).
[e]Algeria (Woodward and Pitbaldo, 2010).
[f]Nigeria (Woodward and Pitbaldo, 2010).
[g]Oman (Woodward and Pitbaldo, 2010).
[h]Alaska (DOT).
[i]Algeria (DOT).
[j]Baltimore Gas Si Electric (DOT).
[k]New York City (DOT).
[l]San Diego Gas & Electric (DOT) http://www.beg.utexas.edu/energyecon/INTRODUCTION%20TO%20LNG%20Update%202012.pdf.
[m]National Oceanic and Atmospheric Administration, Database of Hazardous Materials, available at: http://cairieochemicals.noaa.gov/chris/LNC.pdf.

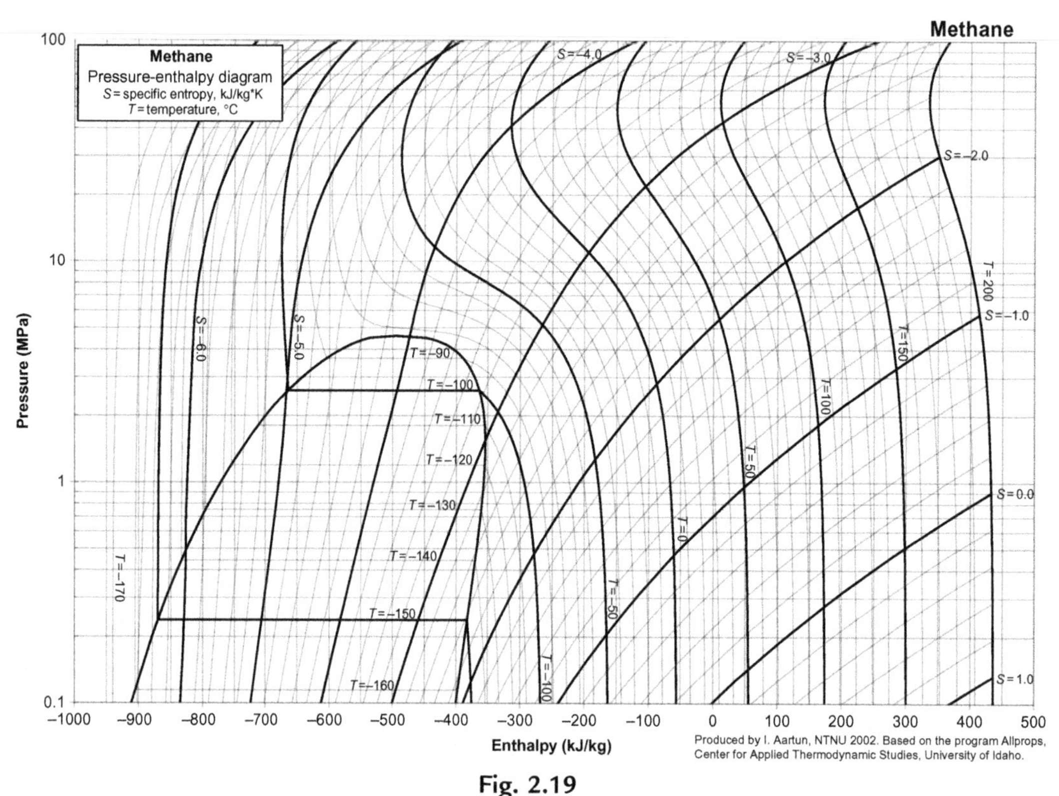

Fig. 2.19
Pressure enthalpy diagram for methane (Aartun, 2002).

The flash scenario often met in chemical engineering occurs in process safety when a loss of containment takes place, forcing a liquid in equilibrium with its vapour at a pressure significantly greater than its normal boiling point, to adjust its thermodynamic state to the atmospheric condition. In this case T_1 is the normal boiling point. This simple approach has been proposed by Crowl and Louvar (2011). The model has been discussed further in this book. Pressurised flashing hydrocarbons are typically propane and butane, their mixtures (LPG), and pentane. LNG (predominantly methane) is stored by refrigeration at atmospheric pressure. Non-hydrocarbon remarkable pressurised liquids are chlorine, ammonia, and carbon dioxide. Vapour pressure curves of the mentioned liquids are in Fig. 2.22 (broad vapour pressure range) and Fig. 2.23 (low vapour pressure range).

With reference to these liquids, the following reference scenarios have been considered, and the corresponding vaporised fraction has been calculated, adopting an average value of the specific heat:

- for propane, butane, and their mixture (LPG), and chlorine, a storage temperature of 20°C has been assumed; this value is significantly far from the normal boiling point, so a flash evaporation is likely.

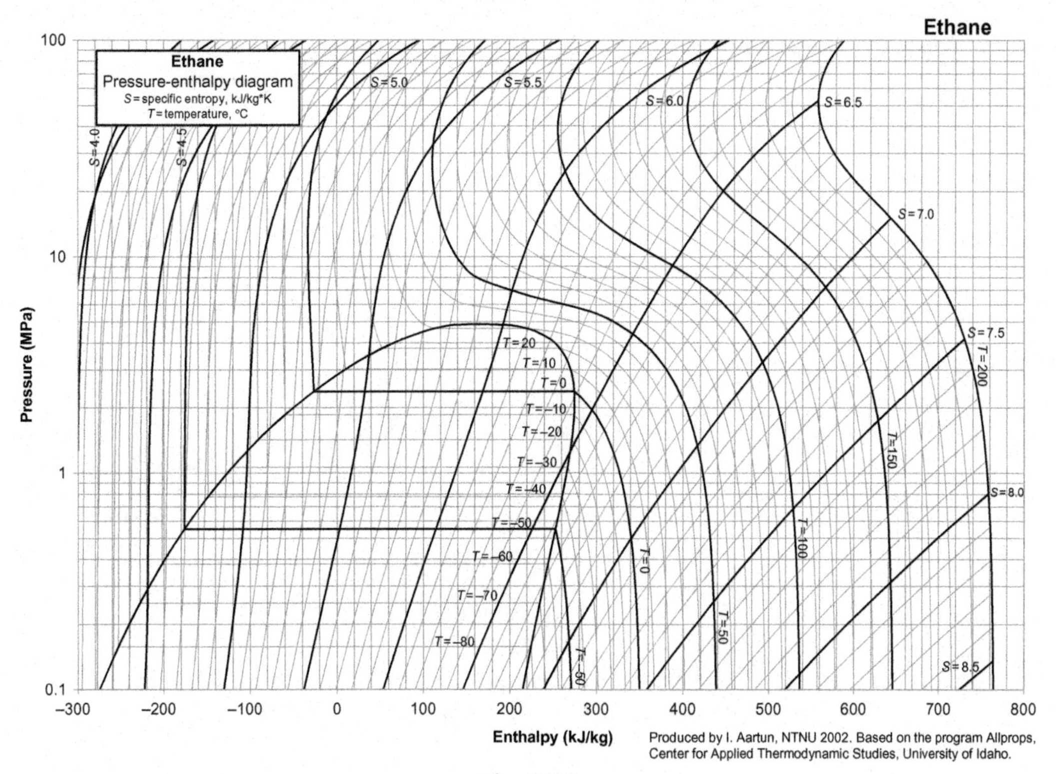

Fig. 2.20
Pressure enthalpy diagram for ethane (Aartun, 2002).

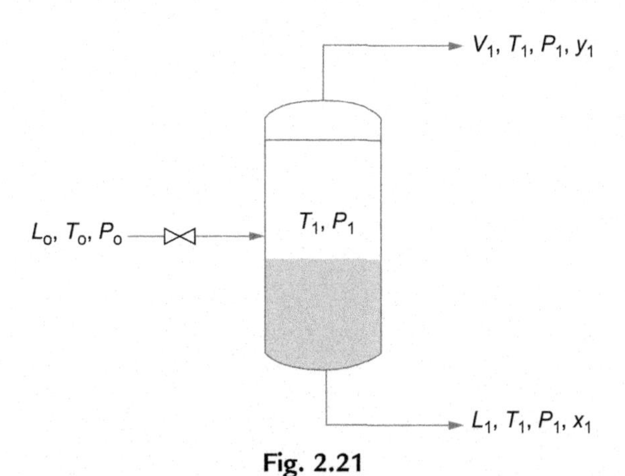

Fig. 2.21
Flash drum.

– pentane has a relatively high normal boiling point, so a storage temperature of 50°C has been assumed; it is important to note that at 20°C, pentane vapour pressure is 0.56 atm, so, in principle, pentane atmospheric flash is expectedly limited. The situation would be different above 40°C.

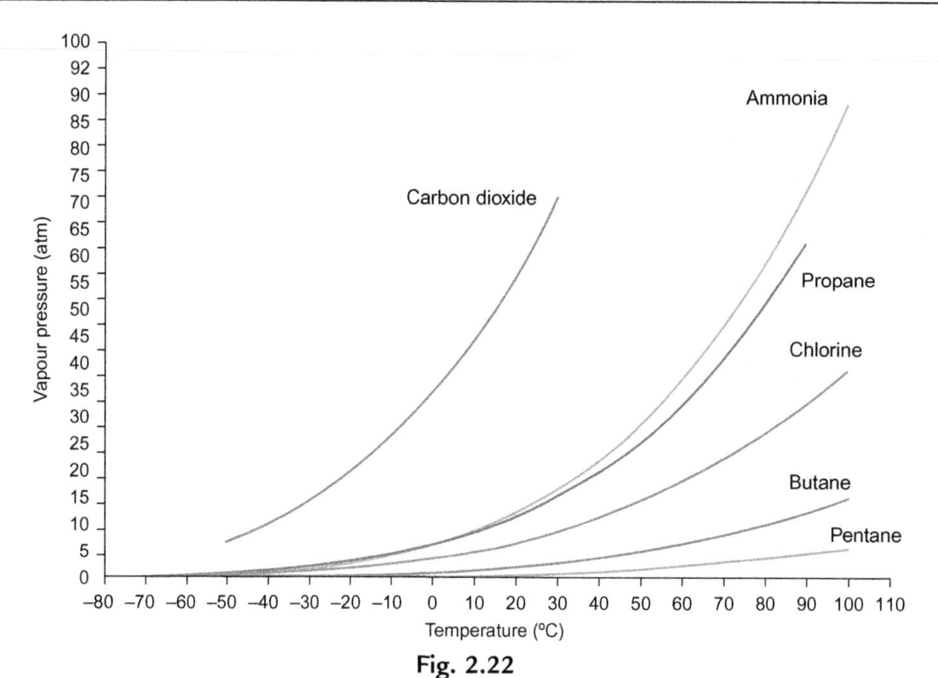

Fig. 2.22

High vapour pressure range of liquefied gas. *Elaborated from Green, D., Perry R., 2007. Perry's Chemical Engineers' Handbook, eighth ed. McGraw-Hill Professional Publishing.*

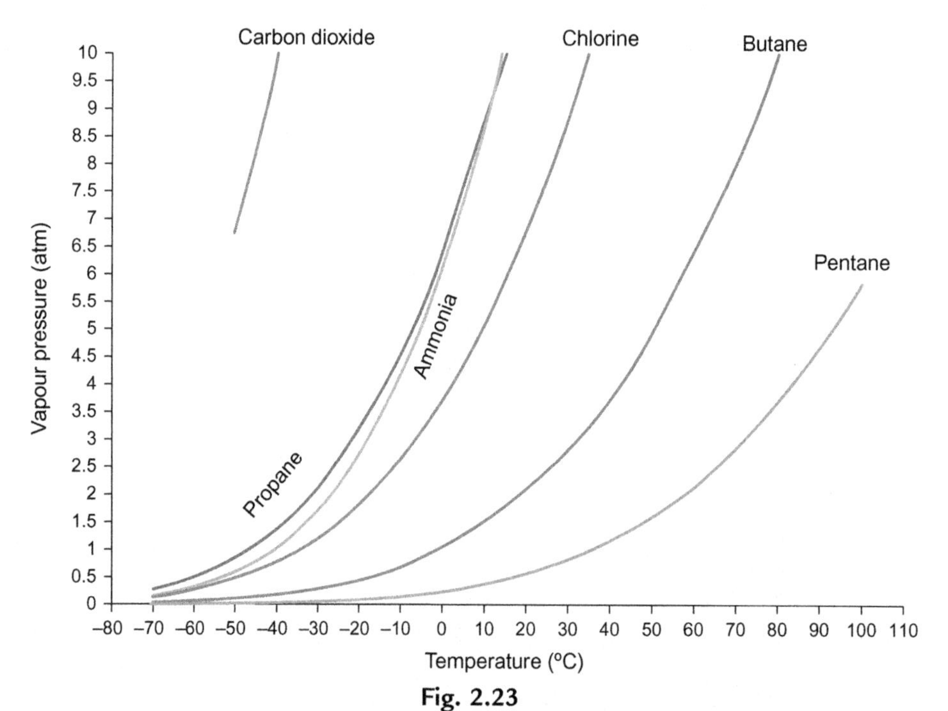

Fig. 2.23

Low vapour pressure range of liquefied gas. *Elaborated from Green, D., Perry R., 2007. Perry's Chemical Engineers' Handbook, eighth ed. McGraw-Hill Professional Publishing.*

- typical *flash tank* process data have been assumed for refrigerated ammonia; even if the corresponding temperature is much lower than ambient temperature, the flash process is considered very fast and adiabatic.
- refrigerated carbon dioxide has been considered; at 1 atm liquid carbon dioxide may not exist, and solid formation would occur, along with subsequent sublimation; the vaporised fraction in Table 2.25 has been calculated accounting for this particularity.

The following table shows the vaporised fractions for pressurised liquids, except carbon dioxide, calculated with the approximated formula (average specific heat):

$$y = -\frac{c_{P_L} \cdot (T_b - T_o)}{\lambda} \tag{2.164}$$

Carbon dioxide liquid is in equilibrium with its vapour until the triple point (5.2 bar and −56.4°C), after which liquid may not exist, and vapour will be in equilibrium with solid. The easiest way is to carry out an energy balance equating liquid enthalpy to the final vapour-solid:

$$H_o = y \cdot H_{1v} + (1 - y) \cdot H_{1s} \tag{2.165}$$

where (Plank and Kuprianoff, 1929)

- H_o is the liquid enthalpy: 370,000 J/kg
- H_{1v} is the equilibrium vapour enthalpy at 1 atm: 643,000 J/kg
- H_{1s} is the equilibrium solid enthalpy at 1 atm: 70,000 J/kg

Solving:

$$y = 0.52 \tag{2.166}$$

This approach does not take into account carbon dioxide non-equilibrium behaviour and any significant changes of kinetic terms (Benintendi, 2014), but it is a simple and useful approximation.

Table 2.25 Vaporised fractions of pressurised liquids in assumed storage conditions

Liquefied Gas	Storage Temperature, T_o (°C)	Storage Pressure (atm) (Vapour Pressure)	Normal Boiling Point, T_b (°C)	Specific Heat of Liquid (J/kg K)	Latent Heat of Vaporisation, λ (J/kg)	Vaporised Fraction
Propane	20	8.29	−42	2500	340,000	0.46
Butane	20	2.07	−1	2380	370,000	0.13
Pentane	50	1.6	36	2287	357,000	0.09
Carbon dioxide	−20	20	−78.45 (sublimation)			0.52
Chlorine	20	9.02	−34	987.9	287,790	0.185
Ammonia	−5	3.4	−33.4	2400	1,370,000	0.05

2.7.7 Thermodynamics of LPG

LPG stands for Liquefied Petroleum Gas. It is defined as *any material having a vapour pressure not exceeding that allowed for commercial propane that is composed predominately of the following hydrocarbons, either by themselves (except propylene) or as mixtures: propane, propylene, butane (normal butane or isobutene), and butylenes* (NFPA 58, 2014). High vapour pressure liquefied petroleum gases also include pentane, hexane, and heptane. Higher hydrocarbons are liquefied by pressurisation, whereas methane and ethane are liquefied by refrigeration. The typical process and plant configuration met in process safety studies consists of the storage spheres, unless mounded bullet pressure vessels are adopted. Loss of containment of storage spheres can results in various incident scenarios, which will be specifically discussed in detail in the following chapters. Pressure-enthalpy diagrams for propane and n-butane are shown in Figs 2.24 and 2.25.

2.7.8 Thermodynamics of Carbon Dioxide

Properties of carbon dioxide have been provided in Chapter 1. At the triple point (−56.6°C, 5.1 atm), this substance exists simultaneously as a solid, liquid, and vapour. Below the triple point, only vapour and solid may coexist. Carbon dioxide can be stored in liquid and gaseous conditions. Liquid carbon dioxide in uninsulated cylinders is stored at about

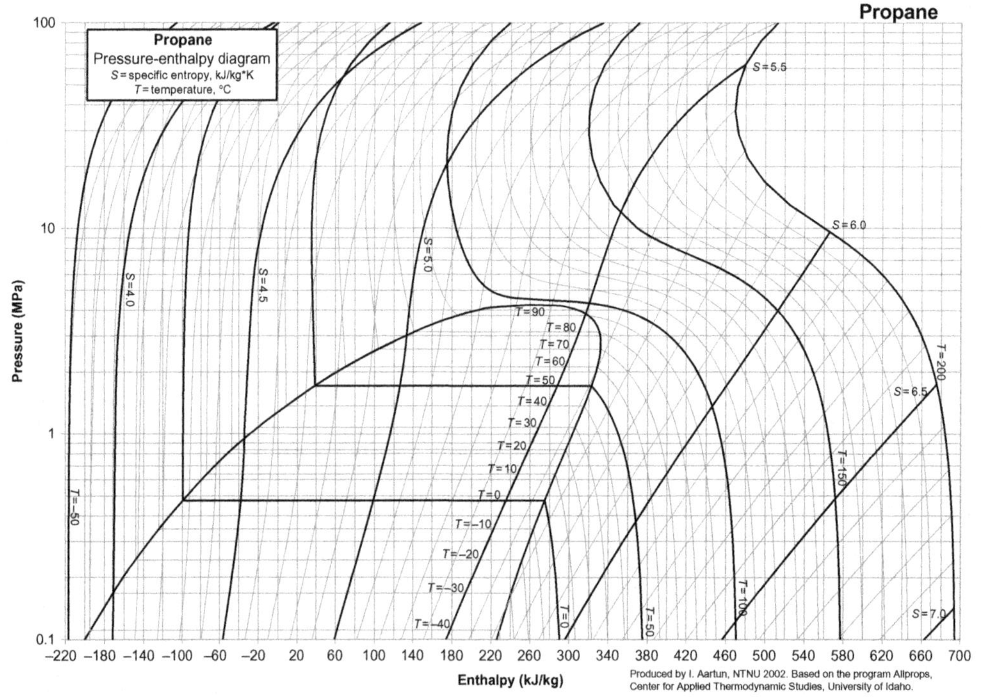

Fig. 2.24
Pressure-enthalpy diagram for propane (Aartun, 2002).

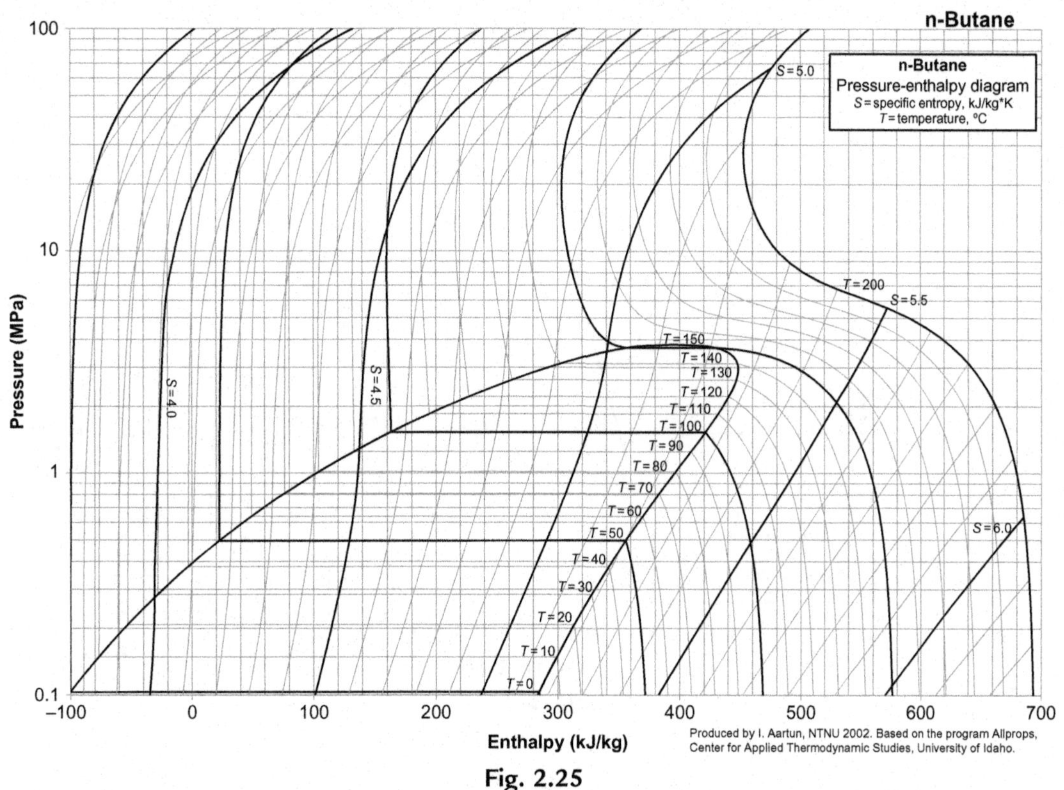

Fig. 2.25
Pressure-enthalpy diagram for n-butane (Aartun, 2002).

850 psig (AIGA, 2009). Storage in insulated DOT-approved (minimum) 454 L cylinders is at least at 40 psia, and at no more than 500 psig (AIGA, 2009). Carbon dioxide is transported either in insulated or uninsulated cylinders, and containers commonly have a working pressure between 200 and 500 psig (AIGA, 2009).

Carbon dioxide in CCS (Carbon Capture and Storage) is generally transported at temperature and pressure ranges between 12°C and 44°C, and 85 and 149.6 atm, respectively (Mohitpour et al., 2007; Kinder Morgan, 2006).

Pressure-enthalpy diagram has been included in Fig. 2.26.

2.7.9 Thermodynamics of Ammonia

Properties of ammonia have been provided in Chapter 1. Anhydrous ammonia is stored in large atmospheric tanks at about −33°C. In refrigeration systems, ammonia is typically compressed up to 30–60 psig and then up to a condensing pressure of 150–225 psig. Smaller volumes are stored under pressure at 16–25 bar, depending on the volume (Appl, 1999) (Fig. 2.27).

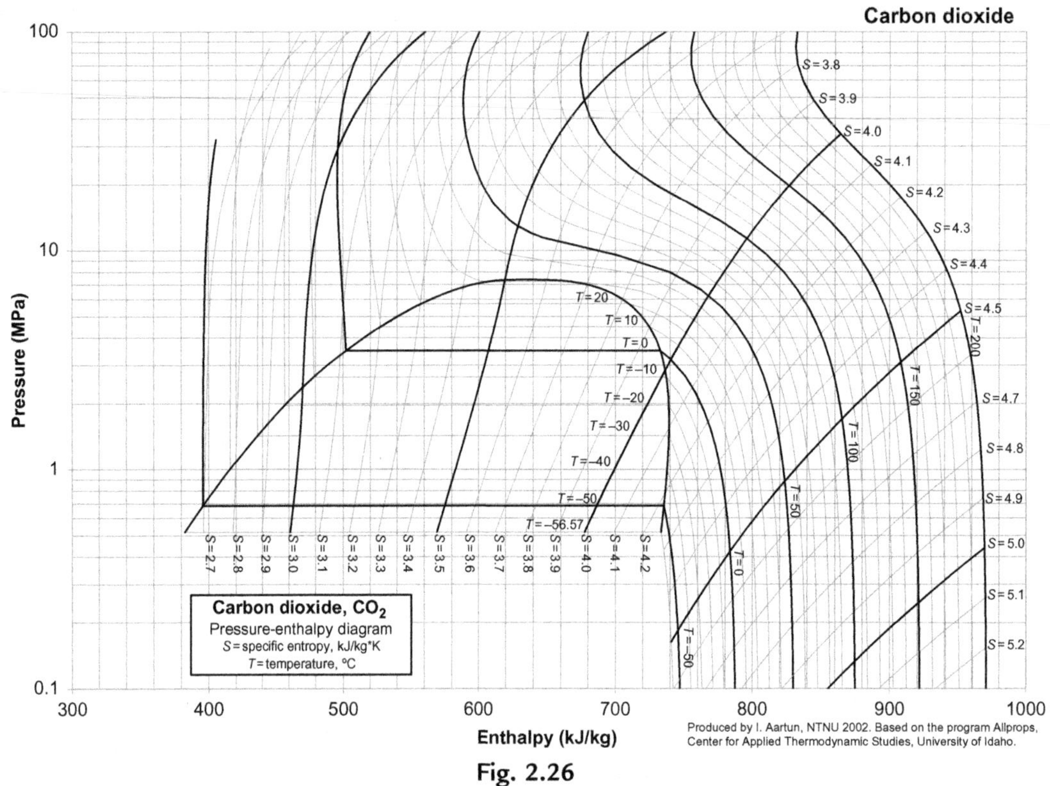

Fig. 2.26
Pressure-enthalpy diagram for carbon dioxide (Aartun, 2002).

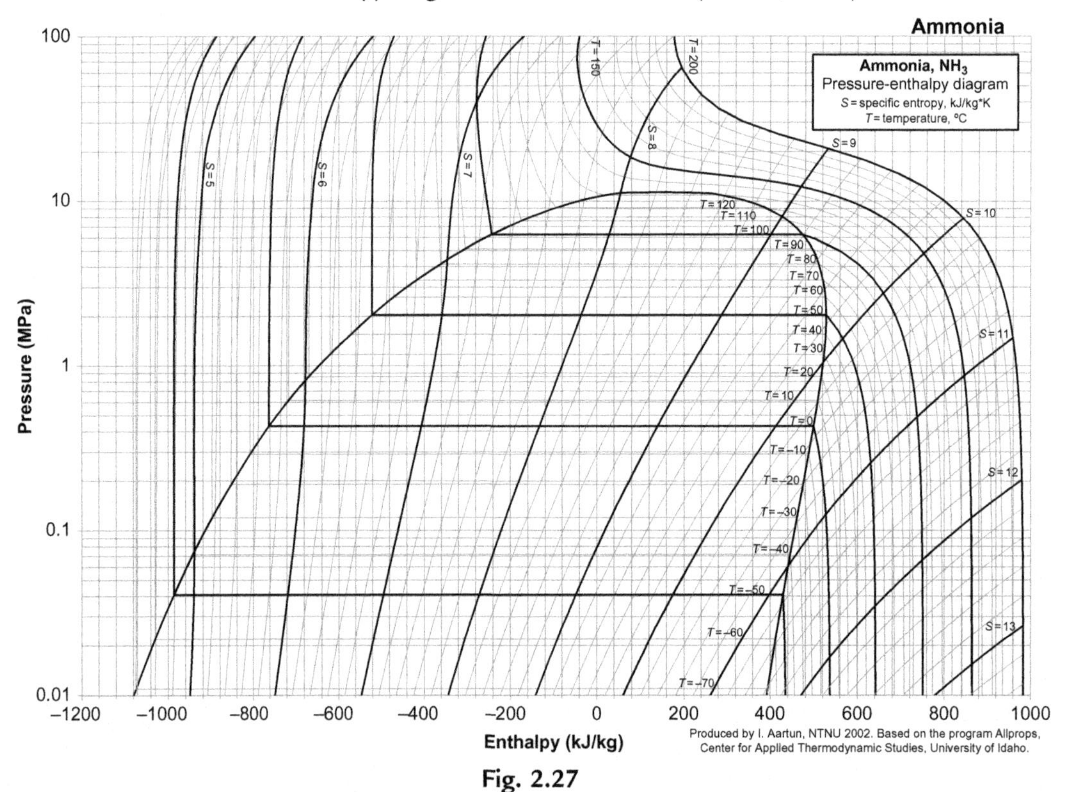

Fig. 2.27
Pressure-enthalpy diagram for ammonia (Aartun, 2002).

2.7.10 Thermodynamics of Chlorine

Properties of chlorine have been provided in Chapter 1. Liquid chlorine is typically supplied in tanker lorries, in cylinders containing 1 tonne, and in standard cylinders containing approximately 45 or 70 kg. Smaller cylinders containing 5 or 9 kg are also seen. A pressure-enthalpy diagram is included in Fig. 2.28 (Green and Perry, 2007).

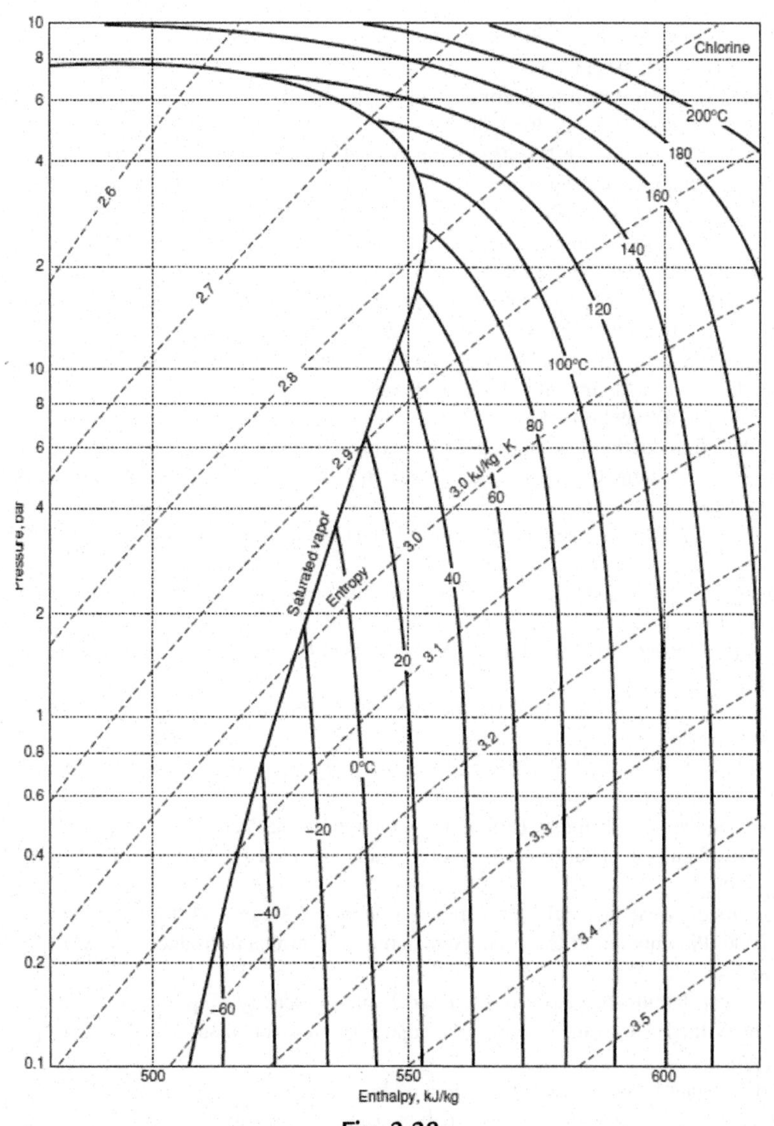

Fig. 2.28
Pressure-enthalpy diagram for chlorine (Green and Perry, 2007).

References

Aartun, I., 2002. Pressure Enthalpy Diagrams. Centre for Applied Thermodynamics Studies, University of Idaho, Idaho.

Acton, O., Caputo, A., 1979. Introduzione allo Studio delle Macchine. UTET, Turin.

Ahmed, T., 2007. Equations of State and PVT Analysis. Applications for Improved Reservoir Modeling. Gulf Publishing Company.

Appl, M., 1999. Ammonia. Principles and Industrial Practice. Wiley-VCH.

Asia Industrial Gases Association (AIGA), 2009. Carbon Dioxide, AIGA 068/10.

Benintendi, R., 2014. Non-equilibrium phenomena in carbon dioxide expansion. Process Saf. Environ. Prot. 92 (1), 47–59.

Benintendi, R., 2016. Influence of the elastic behaviour of liquids on transient flow from pressurised containments. J. Loss Prev. Process Ind. 40, 537–545.

Benintendi, R., Rega, S., 2014. A unified thermodynamic framework for LNG rapid phase transition on water. IChemE Chem. Eng. Res. Des. 92, 3055–3071.

Bird, R.B., Stewart, W.E., Lightfoot, E.N., 2002. Transport Phenomena, second ed. Wiley & Sons, Inc.

BS-EN 1160, 1997. Installations and equipment for liquefied natural gas. General Characteristics of Liquefied Natural Gas. ISBN: 0-580-26446-7.

Churchill, S.W., 2002. Free convection around immersed bodies. In: Hewitt, G.F. (Ed.), Heat Exchanger Design Handbook. Begell House, New York (Section 2.5.7).

Churchill, S.W., Bernstein, M., 1977. A correlating equation for forced convection from gases and liquids to a circular cylinder in crossflow. J. Heat Transfer 99, 300.

Crowl, D.A., Louvar, J.F., 2011. Chemical process safety: fundamentals with applications. In: International Series in Physical and Chemical Engineering, third ed. Prentice Hall.

Eurocode 3, 2005. Design of Steel Structures – Part 1–2: General Rules – Structural Fire Design.

Finlayson, B.A., 2006. Introduction to Chemical Engineering Computing. John Wiley & Sons Inc.

Foust, A.S., Wenzel, L.A., Clump, C.W., Maus, L., Andersen, L.B., 1980. Principles of Unit Operations, second ed. John Wiley & Sons.

Green, D., Perry, R., 2007. Perry's Chemical Engineers' Handbook, eighth ed. McGraw-Hill Professional Publishing.

Hougen, O.A., Watson, K.M., Ragatz, R.A., 1954. Chemical Process Principles, Part I: Material and Energy Balances, second ed. John Wiley and Sons Inc., New York.

Hottel, H.C., 1954. Radiant-heat transmission. In: McAdams, W.H. (Ed.), Heat Transmission. third ed. McGraw-Hill, New York.

Incropera, F.P., DeWitt, D.P., Bergman, T.L., Lavine, A.S., 2011. Fundamentals of Heat and Mass Transfer, seventh ed. John Wiley & Sons, Inc.

Inczedy, J., Lengyel, T., Ure, A.M., 1997. Compendium of analytical nomenclature. The Orange Book, third ed. IUPAC.

IUPAC, 2014. Compendium of Chemical Terminology, Golden Book, Version 2-3-3. .

Juul, N.H., 1982. View factors in radiation between two parallel oriented cylinders of finite lengths. J. Heat Transfer 104 (2), 384–388.

Kern, D.Q., 1950. Process Heat Transfer, International Student ed. Mcgraw-Hill International Book Company.

Kinder Morgan, 2006. Presentation to Canadian Association of Petroleum Producers (CAPP) Investment Symposium, June 13.

Kobe, K.A., 1949–1954.Thermochemistry for the Petrochemical Industry. Reprinted from Petroleum Refinery.

Lloyd, J.R., Moran, W.R., 1974. Natural convection adjacent to horizontal surface of various planforms. J. Heat Transfer 96, 443.

Modest, M.F., 2013. Radiative Heat Transfer. Academic Press.

Mohitpour, M., Golshan, H., Murray, A., 2007. Pipeline Design & Construction: A Practical Approach, third ed. ASME Press, New York.

Morgan, V.T., 1975. The overall convective heat transfer from smooth circular cylinders. In: Irvine, T.F., Hartnett, J.P. (Eds.), Advances in Heat Transfer, vol. 11. Academic Press, New York, pp. 199–264.
NFPA 58, 2014. Liquefied Petroleum Gas Code. National Fire.
Peng, D., Robinson, D.B., 1976. A new two-constant equation of state. Ind. Eng. Chem. Fundam. 15, 59–64.
Plank, R., Kuprianoff, J., 1929. Die Thermischen Eigenschaften der Kohlensaure im Gasformigen, Flussigen und Festen Zustand, Beihefte zur Zeitschrift fur die gesamte Kalte-Industrie, Reihe 1, Heft 1.
Radziemska, E., Lewandowski, W.M., 2001. Heat transfer by free convection from an isothermal vertical round plate in unlimited space. Appl. Energy 68, 347.
Riazi, M.R., 2005. Characterization and Properties of Petroleum Fractions. ASTM International.
Smith, J.M., Van Ness, H.C., Abbott, M.M., 2010. Introduction to Chemical Engineering Thermodynamics, seventh ed. The Mc Graw-Hill Companies Inc.
Soave, G., 1972. Equilibrium constants from a modified Redlich-Kwong equation of state. Chem. Eng. Sci. 27, 1197–1203.
Whitaker, S., 1972. Forced convection heat transfer correlations for flow in pipes, past flat plates, single cylinders, single spheres, and for flow in packed beds and tube bundles. AIChE J. 18, 361.
White, F.M., 2011. Fluid Mechanics, seventh ed. McGraw Hill.
Woodward, J.L., Pitbaldo, R., 2010. LNG Risk Based Safety: Modeling and Consequence Analysis. Wiley.
Zucrow, M.J., Hoffman, J.D., 1976. Gas Dynamics. John Wiley & Sons Inc.

Further Reading

Domalski, E.S., 1972. Selected values of heats of combustion and heats of formation of organic compounds. containing the elements C, H, N, O, P and S. J. Phys. Chem. Ref. Data. 1 (2).

CHAPTER 3

Reaction Engineering for Process Safety

πάντα ῥεῖ (Eraclitus, c. 535 – c. 475 BC)

3.1 Background

Chemical kinetics is a fundamental branch of chemistry and chemical engineering. Process safety is often involved in dealing with wanted chemical reactions, the outcome of which may pose some potential hazards to people and assets, and with unwanted chemical reactions, which are caused or promoted by the specific nature of process and substances. Reactive hazard assessment is an important aspect of process engineering. This chapter analyses the reactive scenario from the point of view of the process safety engineer.

3.2 Reactive Hazards

Reactive hazards are identified by the international regulation and standards (EPA, 2005; REACH regulation, 2006; CLP regulation, 2008; OSHA, 1910.119; NFPA 49; HSE, 2008) and by the process safety literature (Johnson et al., CCPS, 2003; Bretherick and Urben, 2006; Sax, 1999-2014). Fig. 3.1 (Benintendi and Alfonzo, 2013) shows a flow chart useful for the identification and assessment of the main hazardous reactive scenarios. Classification of inherent hazard severity (IHS) of substances based on NFPA 400 rating has been shown in Fig. 3.2. Further configuration hazard severity (CHS) and process hazard severity (PHS) codes have been defined in Figs. 3.3 and 3.4, which account for spacing/logistics and process severity features respectively (Benintendi and Alfonzo, 2013). This approach could be effective in providing a quick process and logistic ranking of reactive hazard.

3.3 Homogeneous Reactions

Reaction rate for homogeneous reactions is generally assumed to follow the general Arrhenius law:

$$r_k = A(T) \cdot \prod_{1}^{n} c_i^{n_i} \tag{3.1}$$

Process Safety Calculations. https://doi.org/10.1016/B978-0-08-101228-4.00003-4

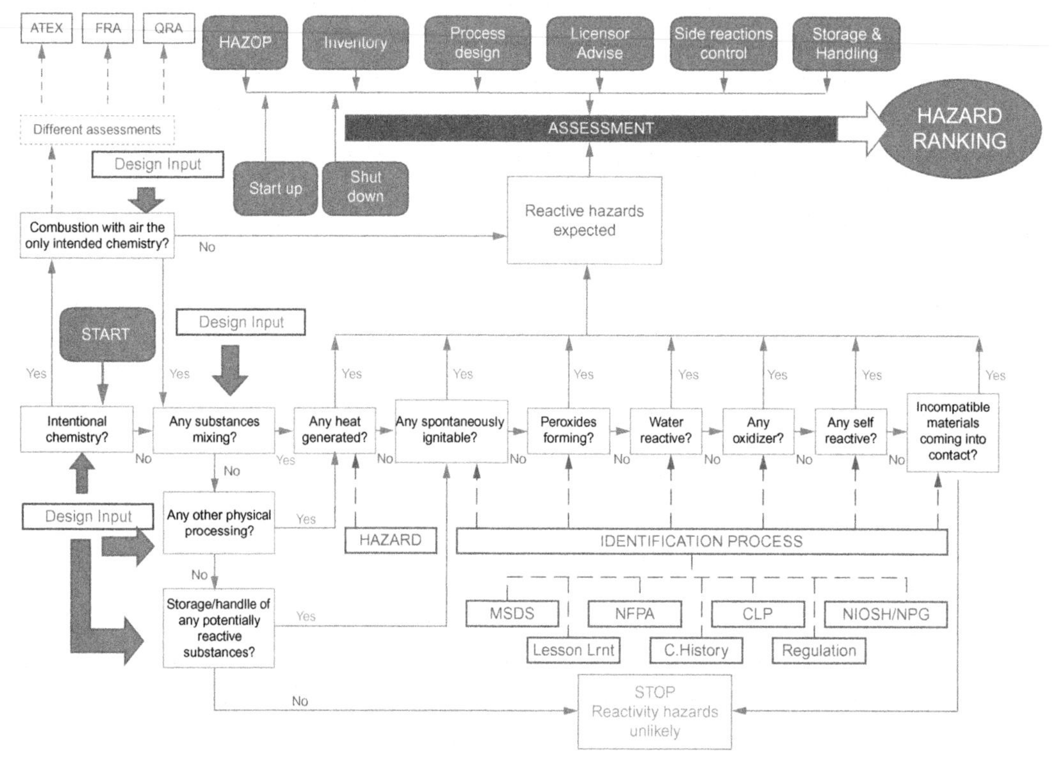

Fig. 3.1
Reactive hazard flow chart (Benintendi and Alfonzo, 2013).

Inherent hazard severity (IHS)		
4	NFPA 4	Extremely reactive
3	NFPA 2	Highly reactive
2	NFPA 1	Moderately reactive
1	NFPA 1	Weakly reactive
0	NFPA 0	Unreactive

Fig. 3.2
Inherent hazard severity (IHS).

where:

- r_k is the reaction rate.
- c_i is the concentration of reactant i.
- n_i is the reaction order with respect to component i.
- $A(T)$ is the kinetic constant defined as:

Configuration hazard severity (CHS)	
4	IHS is 4 and the plant configuration hazard is high
3	IHS is 4 and the plant configuration factor is medium OR IHS is 3 and the plant configuration factor is high
2	IHS is 4 and the plant configuration factor is low OR IHS is 3 and the plant configuration factor is medium OR IHS is 2 and the plant configuration factor is high
1	IHS is 4 and the plant configuration factor is very low OR IHS is 3 and the plant configuration factor is low OR IHS is 2 and the plant configuration factor is medium OR IHS is 1 and the plant configuration factor is high

Fig. 3.3
Configuration hazard severity (CHS).

Process hazard severity (PHS)	
4	CHS is 4 and the process hazard is high
3	CHS is 4 and the plant configuration factor is medium OR CHS is 3 and the plant configuration factor is high
2	CHS is 4 and the plant configuration factor is low OR CHS is 3 and the plant configuration factor is medium OR CHS is 2 and the plant configuration factor is high
1	CHS is 4 and the plant configuration factor is very low OR CHS is 3 and the plant configuration factor is low OR CHS is 2 and the plant configuration factor is medium OR CHS is 1 and the plant configuration factor is high

Fig. 3.4
Process hazard severity (PHS).

$$A(T) = A^o \cdot e^{-\frac{E_a}{R \cdot T}} \tag{3.2}$$

A^o is the frequency factor, and E_a is the energy of activation.

3.3.1 Hydrocarbons

Table 3.1 includes frequency factors and activation energies for the simplified single-step kinetic mechanism relevant to hydrocarbons, carbon monoxide, and hydrogen.

Table 3.1 Rate parameters for hydrocarbons single-step simplified kinetic mechanism (Westbrook and Dryer, 1981)

Hydrocarbon	$A \times 10^{-6}$	(E_a/R) (K)	a	b
Methane	130	24.4	−0.30	1.30
Ethane	34	15.0	0.10	1.65
Propane	27	15.0	0.10	1.65
Butane	23	15.0	0.15	1.60
Pentane	20	15.0	0.25	1.50
Hexane	18	15.0	0.25	1.50
Eptane	16	15.0	0.25	1.50
Octane	14	15.0	0.25	1.50
Nonane	13	15.0	0.25	1.50
Decane	12	15.0	0.25	1.50
Methyl alcohol	101	15.0	0.25	1.50
Ethyl alcohol	47	15.0	0.25	1.60
Benzene	6	15.0	−0.10	1.85
Toluene	5	15.0	−0.10	1.85
Ethylene	63	15.0	0.10	1.65
Propylene	13	15.0	−0.10	1.85
Acetylene	205	15.0	0.50	1.25

Units: cm, s, mole, kcal, K.

$$C_nH_m + \left(\frac{n}{2} + \frac{m}{4}\right) O_2 \rightarrow nCO_2 + \frac{m}{2} H_2O \tag{3.3}$$

$$r_{C_nH_m} = A \cdot \exp\left(-\frac{E_a}{R \cdot T}\right) \cdot [C_nH_m]^a \cdot [O_2]^b \tag{3.4}$$

3.3.2 Carbon Monoxide

Carbon monoxide single-step reaction rate can be simplified according to the following equation (Dryer and Glassman, 1973):

$$CO + 0.5O_2 \rightarrow CO_2 \tag{3.5}$$

$$r_{CO} = 1.3 \cdot 10^{10} \exp\left(-\frac{20{,}130}{T}\right) \cdot [CO] \cdot [H_2O]^{0.5} [O_2]^{0.25} \tag{3.6}$$

where the dependence on $[H_2O]$ is due to the significant CO consumption by its reaction with OH.

3.3.3 Hydrogen Sulphide

The single-step reaction rate of hydrogen sulphide combustion is not widely dealt with in the literature. An approximate but validated equation has been developed by Vervisch et al. (2004):

$$2H_2S + 3O_2 \rightarrow 2SO_2 + 2H_2O \tag{3.7}$$

$$r_{H_2S} = -\rho \cdot K_o \cdot K_\phi \cdot \left(\frac{S_L(\phi)}{S_L(1)}\right)^2 \cdot Y_{H_2S} \cdot Y_{O_2}{}^{3/2} \cdot \exp\left(-\frac{T_{AC}}{T}\right) \tag{3.8}$$

with:

- ρ is the density.
- s is the stoichiometric mass fraction ratio (oxygen/hydrogen sulphide) = 1.4117.
- ϕ is the equivalence ratio $= s \cdot \left(\frac{Y_{H_2S}}{Y_{O_2}}\right)$
- K_o is the preexponential factor.
- K_ϕ is a function of the equivalence ratio.
- $S_L(1)$ is the stoichiometric flame speed.
- $S_L(\phi)$ is the flame speed corresponding to the equivalence ratio.
- Y is the mass fraction.
- X is the molar fraction.
- T_{AC} is the activation temperature = 11,000 K.

valid for combustion in pure oxygen and flame time greater than 2.5 (Tables 3.2–3.4).

Table 3.2 Kinetic parameters of Eq. (3.8)

X_{O_2}	$S_L(\phi = 1)$ (m/s)	K_o (s^{-1})
0.21	0.421	2.70×10^{10}
0.40	2.372	5.78×10^{10}
0.60	4.199	8.41×10^{10}
0.80	5.657	9.74×10^{10}
1.00	7.040	1.11×10^{11}

Table 3.3 Kinetic parameters of Eq. (3.8)

Equivalence ratio ϕ	Approximate $SL(\phi)/SL(1)$ for $X_{O_2} > 0.4$
0.5	0.80
1.0	1.00
1.5	0.80
2.0	0.60
2.5	0.50
3.0	0.40
3.5	0.30
4.0	0.27

Table 3.4 Kinetic parameters of Eq. (3.8)

Equivalence ratio ϕ	$K_o \cdot K_\phi$
0.5	1.70×10^{11}
1.0	1.05×10^{11}
1.5	9.0×10^{10}

3.3.4 Nitrogen Oxides (NO_x)

Nitrogen oxides are produced by nitrogen and oxygen at high temperatures. Thermal NO_x is significantly produced by oxygen and nitrogen reaction at temperatures higher than 1200°C. The reaction rate has been proposed by Bowman (1992), based on Zeldovitch kinetics (Zeldovitch et al., 1947):

$$\frac{d[NO_x]}{dt} = 1.45 \cdot 10^{17} \cdot T^{-1/2} \cdot \exp\left(\frac{-69,460}{T}\right) \cdot [N_2] \cdot [O_2]^{1/2} \tag{3.9}$$

where [X] denotes moles/cm^3, time is in seconds and temperature is in Kelvin. De Soete's equation provides the fraction of initial fuel bound nitrogen converted to NO with the same notations.

$$Y = \left[\frac{2}{1/Y - \{([2.5 \times 10^3 [N_f]_o])([T\exp(-3150T)[O_2]])\}}\right] - 1. \tag{3.10}$$

3.4 Heterogeneous Reactions

3.4.1 Solid-Catalysed Reactions

Heterogeneous reactions take place at an interface across which heat and mass transfer occurs.

Many catalytic gas-solid and liquid-solid reactions generally follow this kinetics. The Langmuir-Hinshelwood Hougen-Watson (LHHW) reaction rate for the surface-reaction-control catalytic gas reaction:

$$A + B \rightarrow R + S \tag{3.11}$$

can be written as (Albright, 2009):

$$r_A = -\frac{k \cdot K_A \cdot K_B \cdot p_A \cdot p_B}{(1 + K_A \cdot p_A + K_B \cdot p_B + K_R \cdot p_R + K_s \cdot p_s)^2} \tag{3.12}$$

where:

- k is the kinetic constant
- K_i are the adsorption equilibrium constants
- p_i are the partial pressures

3.4.2 Noncatalytic Gas Solid Reactions

For noncatalytic gas solid reactions, such as combustion reaction of particle matter, a general consumption rate can be written, which includes both controlling mechanism factors, kinetic and diffusional (Field et al., 1967):

$$w = \frac{p_{O_2}}{\frac{1}{k_s} + \frac{1}{k_d}} \tag{3.13}$$

where:

- w is the rate of solid consumption [g/(cm^2 s)]
- p_{O_2} is oxygen partial pressure (atm)
- k_s is the kinetic rate constant
- k_d is the diffusional rate constant

Eq. (3.13) is strictly valid for soot, but the physical-chemical mechanism is reasonably the same for any combustible solids. It is worth noting that this approach can explain organic dust explosion evolution, when a very high diffusional rate promotes kinetics to be controlling and very fast.

For small carbonaceous particles of diameter d (centimetres), diffusional and kinetic constants are (Niessen, 2010):

$$k_s = 0.13 \cdot \exp\left[\left(\frac{-35{,}000}{R}\right) \cdot \left(\frac{1}{T} - \frac{1}{1600}\right)\right] \tag{3.14}$$

$$k_d = \frac{4.35 \times 10^{-6} \times T^{0.75}}{d} \tag{3.15}$$

with:

- R (1.986 cal/(mole K))
- k_s [g/(cm^2 s atm)]
- k_d [g/(cm^2 s atm)]

An important noncatalytic gas solid reaction, such as combustion of coal particles in a fluidised bed under kinetic control, is given by Levenspiel (1999) (Fig. 3.5):

$$-\rho_C \cdot \frac{dr_C}{dt} = k_s \cdot C_{O_2} \tag{3.16}$$

where:

- ρ_C is the unreacted coal bulk density
- r_C is the radius of the unreacted spherical particle at time t
- k_s is the kinetic constant of first degree reaction

3.5 Reactor Schemes

Ideal reactors are broadly used in process engineering due to their simplicity and satisfactory approach. The same schemes can frequently be adopted to resolve various configurations occurring in process safety. When chemical reactions take place, the risk analyst should be able to select the most appropriate reactive scheme with the aim to identify the hazardous scenarios (Figs. 3.6–3.8). A full coverage of reactor theory and applications may be found in Froment et al. (2011) and Satterfield (1970).

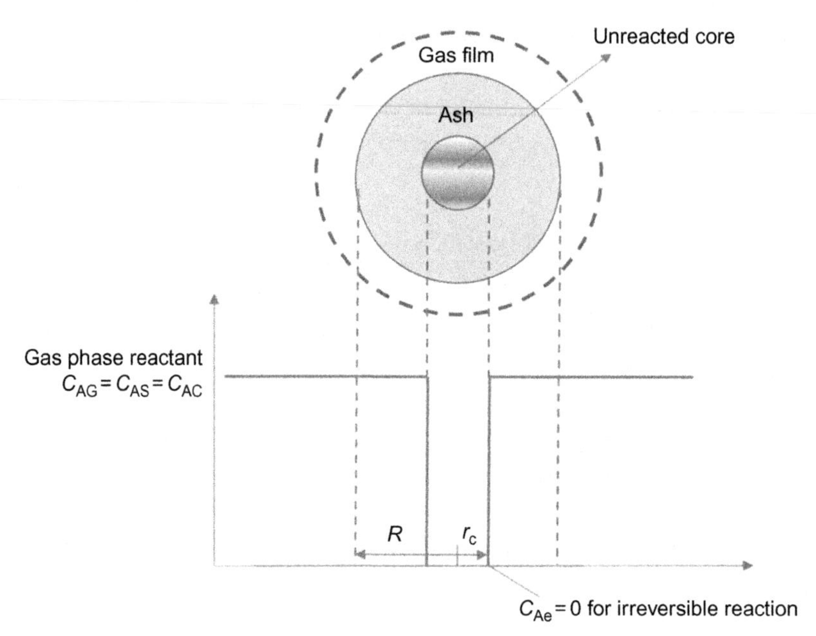

Fig. 3.5
Coal particle combustion under kinetic control.

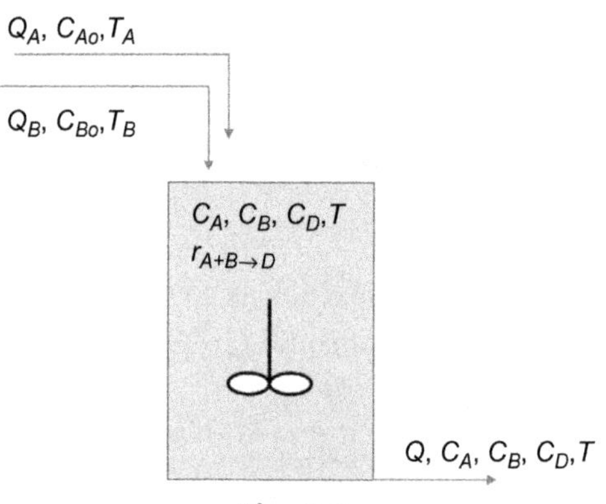

Fig. 3.6
CSTR.

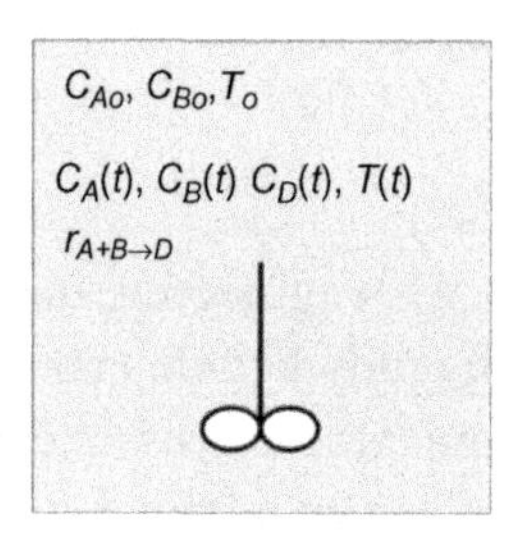

Fig. 3.7
Batch reactor.

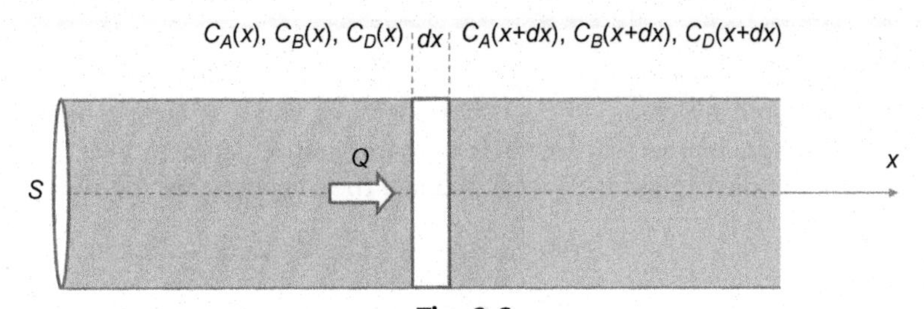

Fig. 3.8
Plug flow reactor.

3.5.1 CSTR (Continuous Stirred Tank Reactor)

$$A + B \rightarrow D$$

Mass balance

$$\text{IN} - \text{OUT} + \text{GEN} = 0 \tag{3.17}$$

$$Q \cdot C_D = k \cdot C_A^a \cdot C_B^b \cdot V \tag{3.18}$$

Thermal balance

$$\rho \cdot Q \cdot c_p \cdot (T - T_o) = k \cdot C_A^a \cdot C_B^b \cdot V \cdot \Delta H_R \tag{3.19}$$

3.5.2 BATCH-DSTR (Discontinuous Stirred Tank Reactor)

$$A + B \rightarrow D$$

Mass balance

$$\text{GEN} = \text{ACC} \tag{3.20}$$

$$\frac{dC_D}{dt} = k \cdot C_A^a \cdot C_B^b \tag{3.21}$$

Thermal balance

$$\rho \cdot c_V \cdot \frac{dT(t)}{dt} = k \cdot C_A(t)^a \cdot C_B(t)^b \cdot \Delta H_R \tag{3.22}$$

3.5.3 PFR (Plug Flow Reactor)

Mass balance

$$Q \cdot C_D(x) = Q \cdot C_D(x + dx) + k \cdot C_A(x)^a \cdot C_B(x)^b \cdot S \cdot dx \tag{3.23}$$

$$Q \cdot \frac{dC_D(x)}{dx} = k \cdot C_A(x)^a \cdot C_B(x)^b \cdot S \tag{3.24}$$

Thermal balance

$$\rho \cdot Q \cdot c_p \cdot \frac{dT}{dx} = k \cdot C_A^a \cdot C_B^b \cdot S \cdot \Delta H_R \tag{3.25}$$

Example 3.1

A mixture of methane and stoichiometric air contained in a vessel at atmospheric pressure is uniformly heated, until gas temperature reaches methane ignition temperature (537°C). Identify the most adequate reactor configuration and calculate the time required for all the gas amount to be ignited.

Solution

The most suitable configuration is the Batch-DSTR reactor.

$$CH_4 + 2O_2 \rightarrow CO_2 + 2H_2O$$

Reference	Nitrogen fraction	Oxygen fraction	Methane fraction
Air	0.79	0.21	0.105
1 mole	0.715	0.19	0.095

Across the reaction the overall number of moles does not change, being the number of moles of reactants and products the same at any time. With reference to 1 cm^3, the total number of moles is:

$$N = \frac{P \cdot V}{R \cdot T} = \frac{1 \times 1}{82.05784 \times 293.16} = 4.16 \times 10^{-5} \text{ moles/cm}^3 \tag{3.26}$$

From the previous table:

$$y_{O_2} = 2 \times y_{CH_4} \tag{3.27}$$

So:

$$N_{O_2} = 2N_{CH_4} \tag{3.28}$$

$$\frac{d[CH_4]}{dt} = -A \cdot \exp\left(-\frac{E_a}{R \cdot T}\right) \cdot [CH_4]^a \cdot 2^b \cdot [CH_4]^b \tag{3.29}$$

$$\frac{d[CH_4]}{dt} = -A \cdot \exp\left(-\frac{E_a}{R \cdot T}\right) \cdot 2^b \cdot [CH_4]^{a+b} \tag{3.30}$$

$$\rho \cdot c_V \cdot \frac{dT}{dt} = A \cdot \exp\left(-\frac{E_a}{R \cdot T}\right) \cdot 2^b \cdot [CH_4]^{a+b} \cdot \Delta H_r \tag{3.31}$$

$$\rho \cdot c_V \cdot \frac{dT}{d[CH_4]} = -\Delta H_r \tag{3.32}$$

$$T = T_o + \frac{\Delta H_r}{\rho \cdot c_V} \cdot \left([CH_4]_0 - [CH_4]\right) \tag{3.33}$$

$$\frac{d[CH_4]}{dt} = -A \cdot \exp\left(-\frac{E_a}{R \cdot \left[T_o + \frac{\Delta H_r}{\rho \cdot c_V}\left([CH_4]_0 - [CH_4]\right)\right]}\right) \cdot 2^b \cdot [CH_4]^{a+b} \tag{3.34}$$

T_o (K)	E_a/R (K)	ΔH_r (J/mole)	ρ (moles/cm^3)	c_V (joule/mole K)	$[CH_4]_0$ (mole/cm^3)
810.16	24.4	890,310	4.16×10^{-5}	20.8	3.9×10^{-6}

$$t = -\int_{[CH_4]_0}^{0} \frac{d[CH_4]}{A \cdot \exp\left(-\frac{E_a}{R \cdot \left[T_o + \frac{\Delta H_r}{\rho \cdot c_V}\left([CH_4]_0 - [CH_4]\right)\right]}\right) \cdot 2^b \cdot [CH_4]^{a+b}} \tag{3.35}$$

$$= 3.9 \times 10^{-8}\ s$$

It is worth noting that the very short combustion time is due to the absence of any mass transfer resistances, according to the assumption that temperature is the same everywhere in the gas volume (Fig. 3.9).

Definite integral:

$$\int_0^{3.952\times 10^{-6}} \frac{1}{1.3\times 10^8 \exp\left(-\frac{24.4}{810.16+1028927053\left(3.952\times 10^{-6}-x\right)}\right) 2.46\, x}\, dx = \tag{3.36}$$

$$-3.90701\times 10^{-8}$$

Visual representation of the integral:

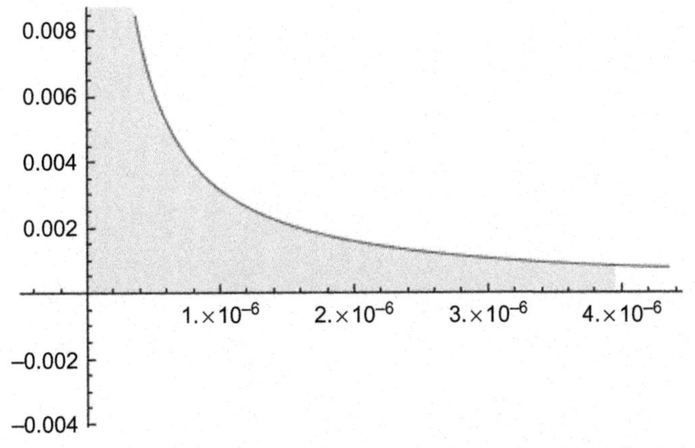

Fig. 3.9
Solution of Example 3.1.

Example 3.2

Air at atmospheric pressure and 1600°C, coming from a combustion chamber, where a very small amount of VOCs is burnt, is fed at a velocity of 7 m/s to a quenching tower through a 25 m long duct. Assuming that products of combustion can be neglected, and that no nitrogen oxides are generated in the combustion chamber, identify the most suitable reactor configuration and calculate the NO_x generated in the duct as NO.

Solution

The most suitable configuration is a plug flow reactor.

$$\frac{d[NO_x]}{dt} = 1.45 \cdot 10^{17} \cdot T^{-1/2} \cdot \exp\left(\frac{-69,460}{T}\right) \cdot [N_2] \cdot [O_2]^{1/2} \tag{3.37}$$

Assuming 1 initial mole of air only and the following reaction as representative of NO_x formation:

$$N_2 + O_2 \rightarrow 2NO \tag{3.38}$$

it will result:

$$n_{N_2 0} + n_{O_2 0} = 1 \tag{3.39}$$

$$n_{N_2}(t) + n_{O_2}(t) + n_{NO}(t) = 1 \tag{3.40}$$

$$n_{N_2}(t) + 0.5 n_{NO}(t) = 0.79 \tag{3.41}$$

$$n_{O_2}(t) + 0.5 n_{NO}(t) = 0.21 \tag{3.42}$$

where n_i indicates the number of moles of species -*i*- which coincides with the corresponding molar fraction y_i (Fig. 3.10).

For this specific case:

$$\frac{dy_{NO_x}}{dt} = 1.45 \cdot 10^{17} \cdot T^{-1/2} \cdot \exp\left(\frac{-69,460}{T}\right) \cdot \frac{P}{R \cdot T} \cdot \left(0.71 - \frac{1}{2} y_{NO_x}\right) \cdot \left(0.29 - \frac{1}{2} y_{NO_x}\right)^{1/2} \tag{3.43}$$

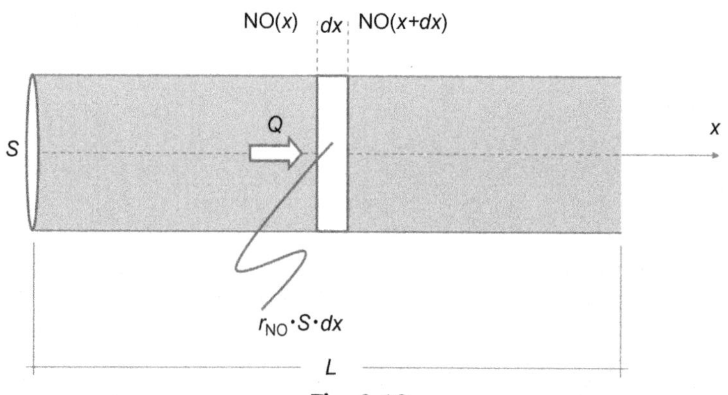

Fig. 3.10
PFR of Example 3.2.

With:

$$\frac{P}{R \cdot T} = \frac{1}{82.05784 \cdot 1873.16} = 6.5 \cdot 10^{-6}\ \text{moles/cm}^3 \tag{3.44}$$

$$Q \cdot \left([\text{NO}]_{x+dx} - [\text{NO}]_x\right) = r_{\text{NO}} \cdot S \cdot dx \tag{3.45}$$

Being $Q/S = 7$ m/s $= 700$ cm/s:

$$\frac{dy_{\text{NO}_x}}{dt} = \frac{1.45 \cdot 10^{17} \cdot T^{-1/2} \cdot \exp\left(\frac{-69,460}{T}\right) \cdot \left(6.5 \cdot 10^{-6}\right)^{1/2} \cdot \left(0.79 - \frac{1}{2} y_{\text{NO}_x}\right) \cdot \left(0.21 - \frac{1}{2} y_{\text{NO}_x}\right)^{1/2}}{700} \tag{3.46}$$

Solving:

$$-5.47 \cdot 10^6 \cdot \left[\tan^{-1}\left(0.928477 \cdot \sqrt{0.42 - y_{\text{NO}_L}}\right) - \tan^{-1}\left(0.928477 \cdot \sqrt{0.42 - y_{\text{NO}_o}}\right)\right] = L - 0 \tag{3.47}$$

$$y_{\text{NO}_o} = 0; \quad L = 2500\ \text{cm} \tag{3.48}$$

So:

$$y_{\text{NO}_L} = 0.00087 \text{ corresponding to } 0.0056\ \text{moles/m}^3 = 168\ \text{mg/m}^3 \tag{3.49}$$

3.6 Combustion Reactions

Combustion is an oxidation-reduction reaction of a fuel (reductant) and oxygen (oxidant) with heat development. It is worth noting that oxidants other than oxygen may participate in combustion reactions with explosive behaviour. For example, chlorine is much more reactive than oxygen and participates in the reaction as a pure gas, unlike oxygen of air. Like oxygen, chlorine, when is mixed with hydrocarbons in the gas phase, can explode.

3.6.1 Definitions

Lower Flammable Limit (*LFL*): the concentration of flammable gas, vapour, mist, or dust in a given oxidant below which a flammable or explosive atmosphere will not be formed (BS EN 60079-10-1, 2015).

Upper Flammable Limit (*LFL*): the concentration of flammable gas, vapour, mist, or dust in a given oxidant above which a flammable or explosive atmosphere will not be formed (BS EN 60079-10-1, 2015).

Flash point: lowest liquid temperature at which, under certain standardised conditions, a liquid gives off vapours in quantity such as to be capable of forming an ignitable vapour/air mixture (BS EN 60079-10-1, 2015).

Auto Ignition Temperature: minimum temperature required to initiate self-sustained combustion independtently of any heating (Benedetti, 2005) NFPA Pocket Guide to Inspecting Flammable Liquids.

Hybrid mixture: mixture of flammable gas or vapour with dust (BS EN 60079-10-1, 2015).

Limiting oxidant concentration: The concentration of oxidant in a fuel-oxidant-diluent mixture below which a deflagration cannot occur under specified conditions (NFPA 69, 2014). This definition is very important in combustion prevention through inertisation.
Minimum ignition energy (*MIE*): the minimum amount of energy required to ignite a combustible vapour, gas, or dust cloud. This parameter is very important, especially in the European ATEX frame.
Burning rate: specific combustion rate of a liquid fuel in a pool flame configuration.
Adiabatic flame temperature: temperature that results from a complete combustion process that occurs without any work, heat transfer, or changes in kinetic or potential energy.
Combustible dust: finely divided solid particles, 500 μm or less in nominal size, which may form an explosive mixture with air at atmospheric pressure and normal temperatures (IEC 60079-10-2). This definition is very important, because it provides a threshold for potentially explosive dust sizes. It also entails specific implication in the hazardous classification frame.
Ignition sources:
Quenching distance: minimum critical size of a channel (distance between plate walls or tube diameter) through which a flame can stably propagate. The quenching distance is a thermodynamical property and is a very important parameter. Upon it flame arrestors have been based and sized. In correspondence to the quenching distance the ignition energy coincides with the MIE.

3.6.2 Combustion of Hydrocarbons

The general hydrocarbon combustion reaction can be written as:

$$C_xH_y + \left(x+\frac{1}{4}y\right)O_2 + \frac{79}{21}\left(x+\frac{1}{4}y\right)N_2 \rightarrow xCO_2 + \frac{1}{2}yH_2O + \frac{79}{21}\left(x+\frac{1}{4}y\right)N_2 \tag{3.50}$$

For an oxygenated hydrocarbon the combustion reaction can be written:

$$C_xH_yO_z + \left(x+\frac{1}{4}y-\frac{1}{2}z\right)O_2 + \frac{79}{21}\left(x+\frac{1}{4}y-\frac{1}{2}z\right)N_2 \rightarrow xCO_2 + \frac{1}{2}yH_2O + \frac{79}{21}\left(x+\frac{1}{4}y-\frac{1}{2}z\right)N_2 \tag{3.51}$$

Lower Flammability Limit (LFL)

Jones (1938), as reported by Bodurtha (1980), has determined that the Lower Flammability Limit (LFL) of organic compounds at around 20°C and 1 atm is about 55% of the stoichiometric concentration in air.

Example 3.3

Calculate the LFL for hexane in accordance with Jones' rule.

Solution

The stoichiometric concentration of any oxygenated hydrocarbons can be parameterised according to the following equation:

$$C_{st} = 100 \times \frac{1}{\left[\frac{100}{21} \cdot \left(x + \frac{1}{4}y - \frac{1}{2}z\right)\right] + 1} \% \tag{3.52}$$

Where:

- Number 1 at the numerator and at the denominator (outside the square brackets) indicates 1 mole of hexane
- The term in square brackets indicates the molecules of air with respect to stoichiometric oxygen concentration (round brackets)
- The ratio represents the molar fraction

The specific reaction is:

$$C_6H_{14} + 19O_2 \rightarrow 6CO_2 + 7H_2O \tag{3.53}$$

where:

x	*y*	*z*
6	14	0

$$C_{st} = 100 \times \frac{1}{\left[\frac{100}{21} \cdot \left(6 + \frac{1}{4} \cdot 14\right)\right] + 1} = 100 \times \frac{1}{\left[\frac{100}{21} \cdot 9.5\right] + 1} = 2.16\% \tag{3.54}$$

$$LFL = 0.55 \times C_{st} = 1.1\% \tag{3.55}$$

The dependence of LFL on temperature is described by the following formula (Zabetakis et al., 1951), which shows a decrease of 8% by a temperature increase of 100°C.

$$LFL_t = LFL_{25°C} + \left(0.8 \cdot LFL_{25°C} \times 10^{-3})\right) \cdot (t - 25) \tag{3.56}$$

The effect of pressure on LFL is slight. Reduction of LFL for ethane from 1 atm to 35 and to 69 atm is 10.5% and 22.8%, respectively (Lewis and Von Elbe, 1987).

Upper Flammability Limit (UFL)

According to Bodurtha (1980), for many compounds the UFL at around 20°C and 1 atm is about 3.5 times of the stoichiometric concentration in air.

Example 3.4

Calculate the UFL for hexane.

Solution

The specific reaction is:

$$C_6H_{14} + \frac{19}{2}O_2 \rightarrow 6CO_2 + 7H_2O \tag{3.57}$$

where (Eq. 3.5.2):

x	*y*	*z*
6	14	0

$$C_{st} = 100 \times \frac{1}{\left[\frac{100}{21} \cdot \left(6 + \frac{1}{4} \cdot 14\right)\right] + 1} = 100 \times \frac{1}{\left[\frac{100}{21} \cdot 9.5\right] + 1} = 2.16\% \tag{3.58}$$

$$UFL = 3.5 \times C_{st} = 7.5\% \tag{3.59}$$

The dependence of UFL on temperature is described by the following formula (Zabetakis et al., 1951), which shows an increase of 8% by a temperature increase of 100°C.

$$UFL_t = UFL_{25°C} + \left(0.8 \cdot UFL_{25°C} \times 10^{-3}\right)) \cdot (t - 25) \tag{3.60}$$

The effect of pressure on UFL is significant for several saturated hydrocarbons from 0.1 to 20.7 MPa (Bodurtha, 1980):

$$UFL_P \approx UFL_{0.101\ \mathrm{Pa}} + 20.6 \times (\ \ln P + 1) \tag{3.61}$$

Limits of flammability in pure oxygen

Limits of flammability for many gases in pure oxygen have been included in Table 3.5. For those gases that are not included in this table the following equation can be used (Chen, 2011):

$$LFL_{ox} = \frac{LFL_{air} \cdot \overline{c}_{P_{O_2}}}{LFL_{air} \cdot \overline{c}_{P_{O_2}} + (1 - LFL_{air}) \cdot \left(0.21 \cdot \overline{c}_{P_{O_2}} + 0.79 \cdot \overline{c}_{P_{N_2}}\right)} \tag{3.62}$$

$$UFL_{ox} = \frac{UFL_{air} \cdot \overline{c}_{P_f} + 0.79 \cdot (1 - UFL_{air}) \cdot \overline{c}_{P_{N_2}}}{0.79 \cdot LFL_{air} \cdot \left(\overline{c}_{P_f} - \overline{c}_{P_{N_2}} + \left(0.21 \cdot \overline{c}_{P_f} + 0.79 \cdot \overline{c}_{P_{N_2}}\right)\right.} \tag{3.63}$$

According to Chen, average predictive errors are from 4.94% to 6.67% for some listed hydrocarbons, hydrogen, and carbon monoxide.

Quenching distance

The quenching distance can be easily predicted, on the basis of the *MIE*, according to Calcote et al. (1952). The following equation:

$$d_Q = 10^{(10.22 \cdot \log(MIE \cdot 10) + 1)/25.21} \tag{3.64}$$

fits well the d_Q of many combustible vapours and gases whose values fall very close to the straight line of the *MIE* vs. d_Q diagram provided by Glassman and Yetter (2008). In Eq. (3.64), *MIE* is in mJ and d_Q in mm. The following diagram depicts the Eq. (3.64) (Fig. 3.11 and Table 3.5).

3.6.3 Combustion of Nitrogenated Compounds

Combustion of compounds containing nitrogen (nitro-alkanes, pyridine) at high temperature leads predominantly to NO formation (De Soete, 1975). Lower temperatures limit NO formation and further oxidation NO_2 takes place, promoted by ambient temperature and oxygen concentration.

3.6.4 Combustion of Sulphur Compounds

Sulphur compounds are predominantly converted to sulphur dioxide. This is in equilibrium with sulphur trioxide according to the reaction:

$$SO_2 + 0.5O_2 \rightleftarrows SO_3 \tag{3.65}$$

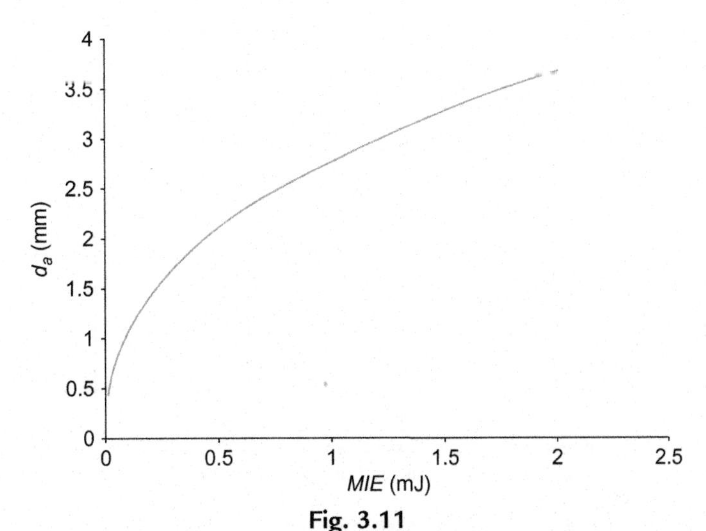

Fig. 3.11
Quenching distance as a function of MIE for combustible vapours and gases.

Table 3.5 Flammability and combustion properties of substances

Substance	LFL (%) Air	UFL (%) Air	LFL (%) Oxygen	UFL (%) Oxygen	AIT Air (°C)	MIE Air (mJ)
Methane	5.00	15.0	5.10	61.0	536.85	0.21
Ethane	2.90	13.0	3.00	66.0	471.85	0.23
Propane	2.00	9.50	2.30	55.0	449.85	0.25
n-Butane	1.50	9.00	1.80	49.0	287.85	0.25
n-Pentane	1.30	8.00			242.85	0.28
n-Hexane	1.05	7.68			224.85	0.24
n-Heptane	1.00	7.00			203.85	0.24
n-Octane	0.80	6.50			205.85	0.25(i)
n-Nonane	0.70	5.60			204.85	
n-Decane	0.70	5.40			200.85	
Ethylene	2.70	36.0	3.00	80.0	450.00	0.084
Propylene	2.00	11.0	2.10	53.0	455.00	0.18
1-Butene	1.60	9.30	1.80	58.0	383.85	
2-Butene	1.70	9.70	1.80	58.0	325.00	
Acetylene	2.50	80.0			305	0.017
Benzene	1.40	7.10			561.85	0.20
Toluene	1.20	7.10			535.85	0.24
o-Xylene	1.00	6.00			463.85	0.20
Phenol	1.52	8.76			714.85	
Methanol	7.30	36.0			463.85	0.14
Ethanol	3.30	19.0			356.00	0.23
Ethylbenzene	1.0	6.70			431.85	
Ethylacetate	2.20	11.40			426.85	0.23
Ethylene glycol	3.20	33.00			39,585	
Ethylene oxide	3.00	100			428.85	0.065
Acetaldehyde	4.00	60.0	4.00	93.0	175.00	0.130
Formaldehyde	7.00	73.0			430.00	
Cyclopropane	2.40	10.4	2.50	60.0	497.85	0.17
Cyclohexane	1.30	7.80			245.00	0.22
Acetic acid	4.05	16.0			426.85	
Acetone	2.60	13.0			465.00	0.19
Aniline	1.30	11.0			616.85	
Vinyl chloride	3.60	33.0	4.00	70.0	471.85	<0.3
Crude oil						
Jet fuel (JP 4)	1.30	8.00			240	
Gasoline	1.20	7.10			440	
Kerosene	0.70	5.00			210	
Diesel fuel	0.60	7.50			225	
Naphtha	0.80	5.00			290.00	
Ethylamine	3.50	14.0			383.85	0.19
Diethylamine	1.70	10.0			312	
Mono-ethanolamine	3.02	24.06			410.00	
Diethanolamine	1.69	16.25			661.85	
Hydrogen	4.00	75.0	4.00	94.0	400.00	0.016
Hydrogen sulphide	4.30	45.0			260.00	0.068
Ammonia	16.0	25.0			650.85	680
Carbon monoxide	12.50	74.0			608.85	<0.3

1. Flammibility limits in air have been collected from Green and Perry (2007) and Zabetakis et al. (1951).
2. Flammability limits in pure oxygen have been collected from Bodurtha (1980) and Coward and Jones (1952).
3. AIT have been collected from Glassman and Yetter (2008) and Zabetakis et al. (1951).
4. Standard enthalpy of combustion have been collected from Green and Perry (2007).
5. Burning rates have been collected from Babrauskas (1995) and NFPA (1997).
6. Flame speeds in pure oxygen have been collected from Bradley (1969).
7. Flame speeds in pure oxygen have been collected from Glassman and Yetter (2008).
8. Limiting Oxidant Concentrations have been collected from NFPA 69 (2014).
9. Limiting Oxidant Concentration for hydrogen has been collected from Kuchta (1985).

MIE Oxygen (mJ)	Flash Point (°C)	Standard Net Enthalpy of Combustion (kJ/kmole × 1E−06)	Burning Rate (kg/m^2 s)	Flame Speed (m/s) at Stoichiometric Condition	Adiabatic Flame Temperature (°C)	Adjusted LOC in Air (%)
0.0027	−187.15	0.80262		3.3–4 (oxygen) 0.434 (air)	2780 (oxygen) 1953 (air)	10
0.0019	−130.15	1.42864		0.445	1987	9.0
0.0021	−104.00	2.04311		0.456	1984	9.5
0.0090	−69.00	2.65732		0.448	1987	10
	−40.00	3.24494	0.126	0.427	1989	10
0.0060	−21.65	3.85510	0.074		1991	10
	−4.15	4.46473	0.101	0.422	1992	9.5
	12.85	5.07415			1993	
	30.85	5.68455			1994	
	45.85	6.29422			1994	
0.00094	−146.85	1.32300		0.680	2097	8.0
	−108.15	1.92620		0.702	2052	9.5
	−79.81	2.54080			2039	9.5
	<−34	2.53000			2033	
0.0002	−18.15	1.25700		0.1.44	2268	
	−11.15	3.13600	0.048	0.476	2058	10.1
	4.85	3.73400	0.112		2044	9.5
	16.85	4.33300	0.09		2035	
	79.85	2.92100				
	10.85	0.63820	0.017	0.712		8.0
	12.85	1.23500	0.015			8.5
	15.00	4.34480	0.121		2038	9.0
	11.00	2.06100	0.064	0.370		
	110.85	1.05270				
	−19.81	1.21800				
	−38.15	1.10450		0.414		
	−52.96	0.52680				
	92.00	1.95930			2071	9.5
	−18.15	3.65600	0.122		2033	
	11.00	0.78660	0.033			
0.0024	−18.15	1.65900	0.041	0.426		9.5
	70.00	3.3930		0.466		
	−78.15	1.1780				13.4
		4.26E+04 kJ/kg	0.045			
		4.31E+04 kJ/kg	0.051			9.5
		4.44E+04 kJ/kg	0.055			10
			0.039			
		4.30E+04 kJ/kg				8.0
		4.34E+04 kJ/kg	0.035			
	−6.00	4.49E+04 kJ/kg				
	−39.00	1.58740				
		2.80030				
	85.00	1.52089				
	151.85	2.41050				
0.0012	−260.15	0.06904		1.7 (air) 9–11 (oxygen)	2045 (air) 2500 (oxygen)	4.6 (Kuchta)
	−107.15	0.51800				7.5
	–	0.31683				
	−206.15	0.28300		0.18 (air + water) 0.72–1.08 (oxygen + water)	2925 (air + water) 2000 (oxygen + water)	3.5

and (Niessen, 2010):

$$K_P = \frac{p_{SO_3}}{p_{SO_2} \times p_{O_2}^{0.5}} \tag{3.66}$$

$$\log_{10} K_P = \frac{5186.5}{T} + 0.611 \cdot \log_{10} T - 6.7497 \tag{3.67}$$

Partial pressures are expressed in atmospheres and temperature in Kelvin degrees. Reaction (3.65) is catalysed by vanadium oxides.

3.6.5 Combustion of Chlorinated Compounds

Combustion of chlorinated compounds follows the Deacon's equilibrium:

$$2HCl + 0.5O_2 \rightleftarrows H_2O + Cl_2 \tag{3.68}$$

that is described by the ratio:

$$K_P = \frac{p_{H_2O} \times p_{Cl_2}}{p_{HCl}^2 \times p_{O_2}^{0.5}} \tag{3.69}$$

where K_P ($atm^{-1/2}$) is defined as (Niessen, 2010):

$$K_P = \exp\left(-8.244 + 1.512 \times 10^{-4} T + \frac{7087}{T}\right) \tag{3.70}$$

It is worth noting that chlorine formed according to Deacon equilibrium seems to be a promoter of dioxin formation through the precursor chlorophenol, according to the mechanism (U.S. EPA, 2003):

$$Cl_2 + \text{phenol} \rightarrow \text{chlorophenol} \tag{3.71}$$

$$\text{chlorophenol} + 0.5O_2 \rightarrow \text{dioxin} + Cl_2 \tag{3.72}$$

where the second reaction is promoted by $CuCl_2$ as catalyst.

3.6.6 Combustion with Halogens Like Oxidant

Hydrogen burns in a chlorine atmosphere according to the reaction (Treadwell, 1960):

$$Cl_2 + H_2 \rightarrow 2HCl \tag{3.73}$$

Combustion heat of this reaction is 44 kcal/mole, according to Lewis and von Elbe (1987).

Chlorine, and generally halogens, behave like oxygen when organic compounds and flammable mixtures are formed (Mannan, Lees, 2012). Flammability limits of many gaseous fuels at given temperatures are provided by Gustin and Fines (1995).

Photochemical mechanisms activate highly reactive radicals and dissociated atoms. For example chlorine dissociates according to the photochemical reaction (Lewis and von Elbe, 1987):

$$Cl_2 + h\nu \rightarrow 2Cl \tag{3.74}$$

Reactions between alkanes and chlorine or bromine will take place in the presence of a flame, whereas fluorine will react also in the cold and dark, producing carbon and the corresponding halide:

$$(n+1)X_2 + C_nH_{2n+2} \rightarrow nC + (2n+2)HX \tag{3.75}$$

X being the generic halogen. For example, for methane:

$$2X_2 + CH_4 \rightarrow C + 4HX \tag{3.76}$$

Reactions with iodine will not take place.

Chlorine also reacts with acetylene, and the reaction is exothermic:

$$Cl_{2(gas)} + H_2C_{2(gas)} \rightarrow HCl_{(liq)} + HC_2Cl_{(gas)} \tag{3.77}$$

Halogens react exothermically with iron at relatively low temperatures, around 100°C for chlorine (Gustin and Fines, 1995), according to the reactivity sequence:

$$\text{Chlorine} > \text{Bromine} > \text{Iodine} \tag{3.78}$$

and to the following general combustion reaction:

$$2Fe + 3Cl_2 \rightarrow 2FeCl_3 \tag{3.79}$$

Chlorine can explode if mixed with many other oxidisable gases, such as solvents and ammonia. Due to the very high reactivity, flammability limits and corresponding ignition energy levels are more favourable to promote explosions than oxygen (Tables 3.6–3.8).

Table 3.6 Limits of flammability of methane in chlorine (Bartkowiak and Zabetakis, 1960)

		Temperature (°C)		
Pressure (Psig)	**Limits (% v/v)**	**25**	**100**	**200**
0	Lower	5.6	3.6	0.6
	Upper	70	66	–
100	Lower	–	2.4	6
	Upper	72	76	77
200	Lower	–	–	–
	Upper	73	72	75

Table 3.7 Limits of flammability of ethane in chlorine (Bartkowiak and Zabetakis, 1960)

		Temperature (°C)		
Pressure (Psig)	**Limits (% v/v)**	**25**	**100**	**200**
0	Lower	6.1	32.5	2.5
	Upper	58	48	–
100	Lower	3.5	1.0	1.0
	Upper	63	75	82
200	Lower	–	–	–
	Upper	66	73	76

Table 3.8 Limits of flammability of hydrogen in chlorine (Umland, 1954)

Pressure (atm abs)	Limits (% v/v)	$T=25°C$
1	Lower	4.1
	Upper	89

3.6.7 Reaction of Combustion With Oxides of Nitrogen

Mixtures of nitrogen oxides and oxidisable substances can propagate flame. The lower limits in oxygen, and in a wide variety of oxygen-nitrogen mixtures, are essentially the same as those in air at the same temperature and pressure (Tables 3.9–3.12).

Table 3.9 Limits of flammability of butane in oxides of nitrogen (Bodurtha, 1980)

Oxidant	Pressure (atm abs)	Limits (% v/v)	$T=25°C$
Nitrous oxide	1	Lower	2.5
		Upper	20
Nitric oxide		Lower	7.6
		Upper	12.5

Table 3.10 Limits of flammability of hydrogen in oxides of nitrogen (Scott et al., 1957)

Oxidant	Pressure (atm abs)	Limits (% v/v)	$T=25°C$
N_2O	1	Lower	3
		Upper	84
NO		Lower	6.6
		Upper	66

Table 3.11 Limits of flammability of olephynes in nitrous oxide (Scott et al., 1957)

Combustible	Oxidant	Limits	(% v/v)
Ethylene	Nitrous oxide	Lower	1.9
		Upper	40
Propylene		Lower	1.4
		Upper	29

Table 3.12 Limits of flammability in nitrous oxide (Scott et al., 1957)

Combustible	Nitrous Oxide Limits (%v/v)	Nitric Oxide Limits (% v/v)
Ammonia	2.20	
Carbon monoxide	72.0	31.0
Hydrogen sulphide	19.0	48.0
	84.0	20.0
		55.0
Cyclopropane	1.60	
	30.0	

3.6.8 Combustion of Phosphorated Compounds

Combustion of phosphorated compounds produces deca-oxide of phosphorous, according to the following reaction:

$$4P + 5O_2 \rightarrow P_4O_{10}$$

Reference	Nitrogen fraction	Oxygen fraction	Methane fraction
Air	0.79	0.21	0.105
1 mole	0.715	0.19	0.095

3.7 Reaction Heat

Chemical reactions can be accompanied by significant thermal effects. These can cause serious and sometimes catastrophic consequences. Incident case history has shown that quite often even simple phenomena, such as dilution and neutralisation, have been ignored, resulting in dramatic outcomes.

3.8 Combustion Heat

The standard net enthalpy of combustion is enthalpy rise occurring when a substance in its standard state at 298.15 K and 101,325 Pa undergoes oxidation to products obtained in complete combustion. Unless otherwise stated the value is for the combustion of the gaseous substance to produce H_2O(g), CO_2(g), F_2(g), Cl_2(g), Br_2(g), I_2(g), SO_2(g), N_2(g), H_3PO_4(s), and SiO_2(crystobalite). It is given in joules per kilomole (J/kmol) at 298.15 K. Heat of combustion values in the Project 911 database are defined as negative values for purposes of sign convention. The standard net enthalpy heat of combustion is also known as the Net Heating Value, or the lower heating value (LHV).

The quantity known as higher heating value (HHV) (or *gross energy,* or *upper heating value* or *gross calorific value* (GCV), or *higher calorific value* (HCV)) is determined by bringing all the products of combustion back to the original precombustion temperature, and in particular condensing any vapour produced.

3.9 Heat of Solution

Heat of solution is the amount of energy associated with the dissolution of a solute in a solvent at constant pressure. This can be an exothermic or an endothermic process. The exothermic energy is often non negligible, and planned or unwanted mixing processes may result in catastrophic failures in storage tanks and vessels. The following table includes heat developed in water at different temperatures for sodium hydroxide and sulphuric acid (Table 3.13).

For sulphuric acid, on the basis of the enthalpy of formation provided at different concentrations in water by the U.S. NIST and presented by Bhatt and Vora (2004), an interpolation curve has been obtained, which can be used for quick estimation of the thermal effects of dilution. Table 3.14 includes the corresponding values for some moles of water mixed with sulphuric acid (Fig. 3.12).

Table 3.13 Heat of solution of sodium hydroxide and sulphuric acid

Substance	Dilution in water	Heat (kJ/mole)	References
Sodium hydroxide	Infinite at 18°C	42.6	Green and Perry (2007)
Sulphuric acid	2.65% at 25°C	74.63	Bhatt and Vora (2004)

Table 3.14 Heat of solution of sodium hydroxide and sulphuric acid

Moles of Water per Mole of Sulphuric Acid	Sulphuric Acid (% mass)	Heat of Solution (kJ/mole)
200	2.65	74.63
150	3.5	74.19
100	5.16	73.64
75	6.77	73.29
50	9.82	72.77
40	12	72.46
30	15.4	71.98
25	17.9	71.59
20	21.4	70.92
15	26.6	69.62
12	31.2	68.13
10	35.3	66.53
9	37.7	65.43
8	40.5	64.08
7	43.8	62.37
6	47.6	60.22
5	52.1	57.48
4	57.6	53.88
3	64.5	48.91
2	73.1	41.44
1	84.5	27.79
–	93	16
0	100	0

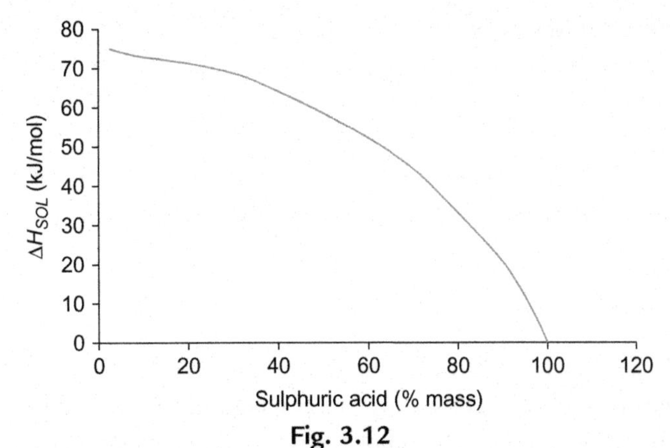

Fig. 3.12
Solution heat for sulphuric acid.

Example 3.5

30 L of concentrated sulphuric acid are inadvertently added to 100 L of pure water initially at 25°C. Calculate the final temperature of the solution.

Solution

	Sulphuric acid	Water	Solution
Density (g/l)	1830	1000	–
Mass (g)	54,900	100,000	154,900
Mass (%)	35.5	64.5	100
Moles	560.2	5555.5	–
Specific heat (cal/g °C)	–	–	0.72

From Table 3.14 $\Delta H_{sol} \cong 66$ kJ/mole = 15,770 cal/mole. The overall dilution heat is:

$$Q_{sol} = 15,770 \times 560.2 = 8,834,354 \text{ cal} \quad (3.80)$$

$$\Delta T = \frac{8,834,354}{154,900 \times 0.72} = 79.2°\text{C} \quad (3.81)$$

The final temperature would be 104°C, which is very close to the normal boiling point of the solution. This example shows how potentially dangerous can be the incautious action to mix water and chemicals, or to clean up a spillage by just pouring water.

3.10 Heat of Neutralisation

Reaction of neutralisation is the combination of an acid with a base which, if occurs at stoichiometric concentrations, theoretically leads water pH to 7. The effects are comparable to those analysed for the exothermic dissolution of electrolytes in water. The peculiarity of the reaction of neutralisation is that for each mole of water produced, 13.7 kcal will be released, independently of the identity of the strong electrolytes.

3.11 Endothermic Processes

Endothermic processes generate a negative amount of energy, so the system temperature decreases. Solving sodium chloride in water is a typical example of an endothermic process. In closed equipment such as pressure vessels and piping, if not properly vented and accurately foreseen, an endothermic reaction could cause serious structural damages.

3.12 Pyrophoricity

Pyrophoric materials are substances that ignite instantly upon simple exposure to oxygen. They do not require an ignition source. Thus, they are considered to have zero minimum ignition energy. The definition of pyrophoricity is not restricted to finely divided powders.

3.12.1 Pyrophoric Substances

Pure metals, alloys, carbides, hydrides, nonmetallic materials, and even liquids may show pyrophoric behaviour. According to Housecroft and Sharpe (2012), transition (d-block) metals present a general pyrophoric behaviour.

3.12.2 Pyrophoricity Scenarios

Iron may form pyrophoric iron sulphide (pyrite) in presence of H_2S:

$$Fe_2O_3 + 3H_2S \rightarrow 2FeS + 3H_2O + S \quad (3.82)$$

Pyrite oxidation produces heat:

$$4FeS + 3O_2 \rightarrow 2Fe_2O_3 + 4S + Heat_{react} \quad (3.83)$$

$$4FeS + 7O_2 \rightarrow 2Fe_2O_3 + 4SO_2 + Heat_{react} \quad (3.84)$$

Organometallic compounds like $Li(CH_3)$, $Zn(CH_3)_2$, $B(CH_3$, and $A(CH_3)_6$ are spontaneously flammable in air. According to Glassman et al. (1992), the general size of particles that are pyrophoric is of the order 0.01 μm.

3.13 Reactivity of Remarkable Substances

3.13.1 Ammonium Nitrate (AN)

Ammonium nitrate is a salt formed through the reaction of ammonia and nitric acid:

$$HNO_3 + NH_3 \rightarrow NH_4NO_3 \quad (3.85)$$

Under specific conditions, ammonium nitrate decomposes, producing ammonia and nitric acid as well as nitrogen dioxide when exposed to external heat above 200°C. Above 250°C other oxides of nitrogen can be formed, along with hydroxyl radicals, which propagate the reaction. It is deemed that the final products are molecular nitrogen, oxygen, and water (Cagnina et al., 2013), accompanied by a heat reaction of 350 kcal/kg according to Mannan (2012). This is the key parameter to understand the explosive behaviour of ammonium nitrate. High temperature and confinement can trigger explosive phenomena, even if sensitivity to other substances has been statistically the most frequent triggering factor. Some main incompatible chemicals are listed by Bretherick and Urben (2006):

- Ammonia
- Chloride salts
- Powdered metals
- Acids
- Organic fuels
- Ammonium sulphate
- Sulphides

3.13.2 Chlorates

Metallic chlorates (sodium, potassium, calcium, barium) don't burn when heated in a fire, but decompose, providing oxygen, which sustains a fire and significantly increases the flame temperature and speed of burning. For example, potassium chlorate decomposes according to the following reaction:

$$2KClO_3(s) \rightarrow 2KCl(s) + 3O_2(g) \tag{3.86}$$

Many incidents have occurred in warehouses where sodium chlorate has been involved in fires, resulting in violent explosions (Benintendi and Round, 2014).

3.13.3 Organic Peroxides and Hydrogen Peroxide

These substances contain the unstable 0-0 peroxy bond in their molecular structure, which is a source of instability, causing accelerating decomposition. Furthermore, they can oxidise organic compounds and reducing agents.

Hydrogen peroxide shows a similar behaviour in addition to promoting and increasing fire hazard as chlorates.

3.14 Self Heating

Kohlbrand (1990) has defined *self heating* as the process whereby a system increases in temperature because the contents are reacting or decomposing exothermically, and the resulting heat is not capable of being transferred to the surroundings. The traditional theoretical frames utilised to describe the self heating cases are identified as the Semenov model, the Frank-Kamenetskii model, and the Thomas model.

3.14.1 Semenov Model

This model assumes that the heat conductance in the mass is significantly greater than the heat rate between system and surroundings, so that the temperature profile is flat. This model is applied in well mixed systems (Fig. 3.13).

3.14.2 Frank-Kamenetskii Model

Frank-Kamenetskii model assumes that the heat conduction in the mass is low, so that a temperature gradient exists, with a maximum in the centre. This is typically the case of large, unmixed, or reacting exothermically systems (Fig. 3.14).

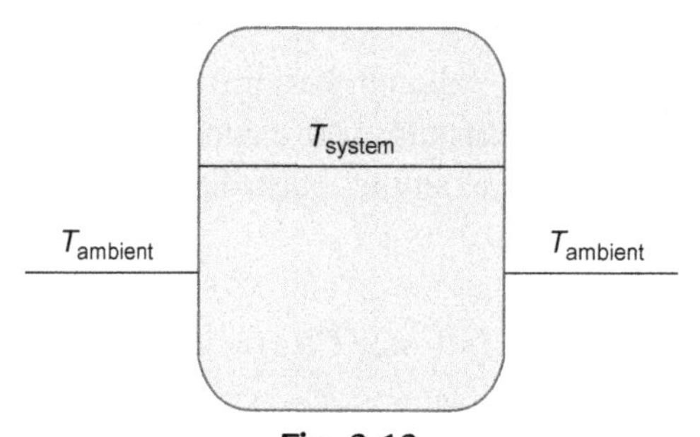

Fig. 3.13
Semenov model.

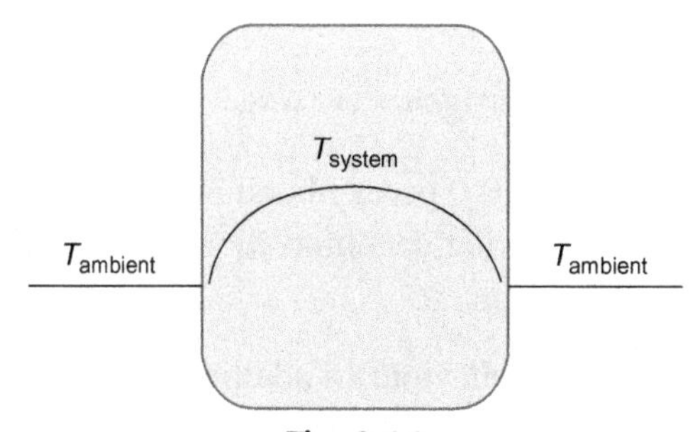

Fig. 3.14
Frank-Kamenetskii model.

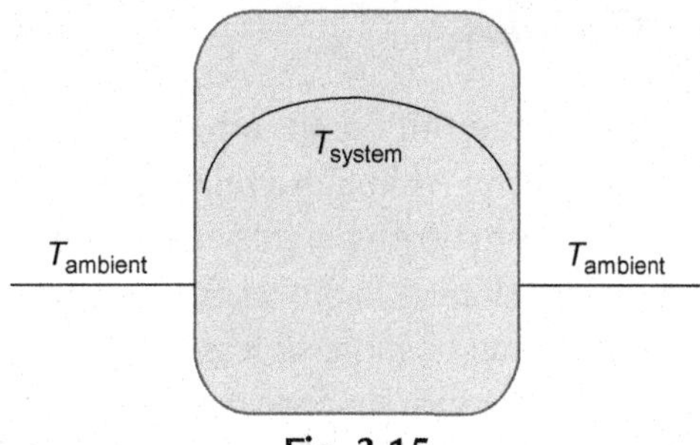

Fig. 3.15
Thomas model.

3.14.3 Thomas Model

Thomas model applies when both heat conduction in the substance and heat transfer to the ambient or cooling system are important (Fig. 3.15).

For screening evaluation the Biot number can be used to determine which model to adopt:

$$Bi = \frac{h \cdot R}{k}$$

with:

- *Bi* is Biot number
- *h* is the coefficient of heat transfer to the cooling system composed of the radiative part and of the convective part, being the latter more important for liquids
- *k* is the susbstance conductivity
- *R* is the radius of the vessel

13.14.4 Choice of a Model

Typically:

Process safety implications of self heating along with thermal stability calculations procedures have been discussed in Chapter 10. Table 3.15 relates BIOT number to model choice.

Table 3.15 Parameters for choice of model

Bi	Model
$Bi < 0.1$	Frank-Kamenetskii
$0.1 < Bi < 10$	Thomas
$Bi < 0.1$	Semenov

3.15 Water and Spray Curtains

Water and spray curtains may be successfully used to reduce the toxic impact of hazardous substances dispersed into the atmosphere. Although a relatively limited number of applications have been implemented, these systems may offer significant potential for the reduction of toxic or flammable cloud concentration developing in process areas. Curtains work according to two different mechanisms, each of which can be more or less important depending on factors such as the chemical nature of the hazardous gas, its concentration, and its fluid-dynamic characteristic (momentum, density, temperature). These mechanisms are

- simple dilution of cloud concentration;
- absorption, and eventually chemical reaction;

and can take place simultaneously to a certain extent.

Absorption is particularly important when a significant reduction is required, either because of high hazardous gas content or its great potential for harm. In this respect, the effectiveness of absorption is strongly affected by the solubility properties of the substances. Typically, ammonia and hydrochloric acid are easily absorbed in pure water, whereas hydrogen sulphide has a very low solubility. Affecting parameters are the following:

- Henry constant
- mass transfer properties such as transfer coefficients in the gas and liquid film
- surface to volume ratio of the absorbent spray
- droplet size
- reactivity towards specific electrolytes

The theoretical frame for absorption is the Lewis and Whitman two-film model, depicted in Fig. 3.16. This model typically assumes that the interface gas–liquid concentrations are in Henry equilibrium, if applicable.

Absorption of substances, like hydrogen sulphide, which show a very low water solubility, may be promoted by solutes, which increase the overall absorption power by one or more orders of magnitude, depending on their concentration in water. Effective modelling and experimental results have been obtained by means of Na_2CO_3 water solutions in the range 0.1%–40% (weight), against H_2S concentrations in air ranging from a few ppm to percentages. The main chemical reaction is the following:

$$H_2S + Na_2CO_3 \rightarrow NaHS + NaHCO_3$$

Unlike pure absorption in water, hydrogen sulphide absorption with a chemical reaction of sodium or potassium carbonate is promoted by high temperatures.

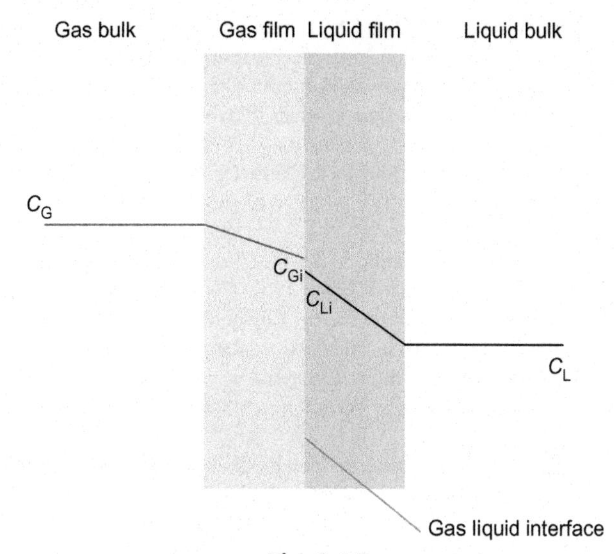

Fig. 3.16
Lewis and Whitman two-film model.

References

Albright, L.F., 2009. Albright's Chemical Engineering Handbook. CRC Press Taylor & Francis Group.

Babrauskas, V., 1995. SFPE Handbook, 2nd ed. National Fire Protection Association, Quincy, MA.

Bartkowiak, A., Zabetakis, M.G., 1960. Flammability limits of methane and ethane in chlorine at ambient and elevated temperatures and pressure. BuMines Rept. of Inv. 5610.

Benedetti, R.P., 2005. NFPA Pocket Guide to Inspecting Flammable Liquids. National Fire Protection Association Inc.

Benintendi, R., Alfonzo, J., 2013. Identification and analysis of the key drivers for a systemic and process-specific reactive hazard assessment (RHA) methodology. 2013 International Symposium. Mary Kay O'Connor Process Safety Center, College Station, TX.

Benintendi, R., Round, S., 2014. Design a Safe Hazardous Materials Warehouse. Hydrocarbon Processing. Gulf Publishing.

Bhatt, B.I., Vora, S.M., 2004. Stoichiometry, 4th ed. Tata McGraw-Hill Education.

Bodurtha, F.T., 1980. Industrial Explosion Prevention and Protection. McGrawHill Book Company.

Bowman, C.T., 1992. Control of combustion generated nitrogen oxides emissions: technology driven by regulation. In: Proceedings of the Twenty Fourth Symposium (International) on Combustion. Combustion Institute, Pittsburgh, PA, USA, pp. 859–878.

Bradley, J.N., 1969. Flame and Combustion Phenomena. Barnes & Noble.

Bretherick, L., Urben, P., 2006. Bretherick's Handbook of Reactive Chemical Hazards, 7th ed. Elsevier.

Cagnina, S., Rotureau, P., Adamo, C., 2013. Study of incompatibility of ammonium nitrate and its mechanism of decomposition by theoretical approach. In: 14 International Symposium on Loss Prevention and Safety Promotion in the Process Industry, May, Florence, Italy.

Calcote, H.F., Gregory Jr., C.A., Barnett, C.M., Gilmer, R.B., 1952. Spark ignition: effect of molecular structure. Ind. Eng. Chem. 44, 2656.

Chen, C.-C., 2011. A study on estimating flammability limit in oxygen. Ind. Eng. Chem. Res. 50 (17), 10283–10291.

Coward, H.F., Jones, G.W., 1952. Limit of Flammability of Gases and Vapors, Bureau of Mines Bulletin 503.

De Soete, G., 1975. Overall reaction rates of NO and N_2 formation from fuel nitrogen. In: 15th International Symposium on Combustion. The Combustion Institute, Pittsburgh, USA, pp. 1093–1102.
Dryer, F.L., Glassman, I., 1973. High Temperature Oxidation of CO and CH_4. In: Fourteenth Symposium (International) on Combustion. The Combustion Institute, Pittsburgh, PA.
EPA, U.S., 2003. Exposure and Human Health Reassessment of 2,3,7,8-Tetrachlorodibenzo-p-Dioxin (TCDD) and Related Compounds. Exposure Assessment and Risk Characterization Group National Center for Environmental Assessment—Washington Office—Office of Research and Development U.S. Environmental Protection Agency, Washington, DC.
EPA, 2005. Managing Chemical Reactivity Hazards, 550-F-04-005.
Field, M.A., Gill, D.W., Morgan, B.B., Hawksley, P.E.W., 1967. Combustion of Pulverized Coal. British Coal Utilization Research Association, Leatherhead, Surrey, England.
Froment, G.F., Bischoff, K.B., De Wilde, J., 2011. Chemical Reactor Analysis and Design. Wiley.
Glassman, I., Yetter, R.A., 2008. Combustion, 4th ed. Elsevier.
Glassman, I., Papas, P., Brezinsky, K., 1992. A new definition and theory of metal pyrophoricity. J. Combust. Sci. Technol. 83(1–3).
Green, D., Perry, R., 2007. Perry's Chemical Engineers' Handbook, 8th ed. McGraw-Hill Professional Publishing.
Gustin, J.L., Fines, A., 1995. Safety of chlorination reactions. In: 8th International Symposium on Loss Prevention and Safety Promotion in the Process Industries, Antwerpenpp. 157–169.
Housecroft, C., Sharpe, A.G., 2012. Inorganic Chemistry, 4th ed. Pearson, New York.
HSE, 2008. Designing and Operating Safe Chemical Reaction Processes.
Johnson, R.W., Rudy, S.W., Unwin, S.D., 2003. Essential Practises for Managing Chemical Reactivity Hazards, CCPS. American Institute of Chemical Engineers.
Jones, G.W., 1938. Inflammation limits and their practical application in hazardous industrial operations. Chem. Rev. 22 (1), 1–26.
Kohlbrand, H., 1990. The Dow Chemical Co., Personal Communication.
Kuchta, J.M., 1985, Investigation of Fire and Explosion Accidents in the Chemical, Mining, and Fuel-Related Industries: A Manual U.S. Bureau of Mines Bulletin 680, 84 pp.
Levenspiel, O., 1999. Chemical Reaction Engineering, 3rd ed. John Wiley & Sons.
Lewis, B., von Elbe, G., 1987. Combustion, Flames and Explosions of Gases, 3rd ed. Academic Press, Inc./Harcourt Brace Jovanovich Publishers.
Mannan, S., 2012. Lees' Loss Prevention in the Process Industries. 4th ed. Butterworth-Heinemann.
National Fire Protection Association, 1997. Fire Protection Handbook, 18th ed. NFPA, Quincy, MA.
NFPA 49. Table of Common Hazardous Chemicals.
NFPA 69. Standard on Explosion Prevention Systems 2014 Edition.
Niessen, W.R., 2010. Combustion and Incineration Processes: Applications in Environmental Engineering, 4th ed. CRC Press.
Official Journal of the European Union, L 353/1, 2006. Regulation (EC) No. 1907/2006 of the European Parliament and of the Council of 18 December 2006 concerning the Registration, Evaluation, Authorisation and Restriction of Chemicals (REACH), Establishing a European Chemicals Agency, Amending Directive 1999/45/EC and Repealing Council Regulation (EEC) No. 793/93 and Commission Regulation (EC) No. 1488/94 as well as Council Directive 76/769/EEC and Commission Directives 91/155/EEC, 93/67/EEC, 93/105/EC and 2000/21/EC.
Official Journal of the European Union, L 396/1, 2008. Regulation (EC) No. 1272/2008 of the European Parliament and of the Council of 16 December 2008 on Classification, Labelling and Packaging (CLP) of Substances and Mixtures, Amending and Repealing Directives 67/548/EEC and 1999/45/EC, and Amending Regulation (EC) No. 1907/2006.
OSHA, Process Safety Management of Highly Hazardous Chemicals, 1910.119, Regulations (Standard – 29 CFR).
Satterfield, C.N., 1970. Mass transfer in Heterogeneous Catalysis. Massachusetts Institute of Technology Press, Cambridge, MA.
Sax's Dangerous Properties of Industrial Materials, 1999–2014. John Wiley and Sons, Inc.

Scott, F.E., Van Dolah, R.W., Zabetakis, M.G., 1957. The flammability characteristics of the system H_2-NO-N_2O-Air. In: 6th Symposium on Combustion. Reinhold Pub. Co, New York, pp. 540–545.
Treadwell, F.P., 1960. Chimica Analitica, Vol. I Analisi Qualitativa, Casa Editrice Francesco Vallardi.
Umland, A.W., 1954. Explosive limits of hydrogen-chlorine mixtures. J. Electrochem. Soc. 101, 626–631.
Vervisch, L., Labégorre, B., Réveillon, J., 2004. Hydrogen-sulphur oxy-flame analysis and single-step flame tabulated chemistry. Fuel 83 (4–5), 605–614.
Westbrook, C.K., Dryer, F.L., 1981. Simplified reaction mechanisms for the oxidation of hydrocarbon fuels in flames. Combust. Sci. Technol. 27, 31–43.
Zabetakis, M.G., Scott, G.S., Jones, G.W., 1951. Limits of flammability of paraffin hydrocarbons in air. Ind. Eng. Chem. 43 (9), 2120–2124.
Zeldovitch, B., Sadovnikov, P., Frank-Kamenetski, D., 1947. Oxidation of Nitrogen in Combustion, Academy of Sciences (USSR), Inst. of Chem. Physics, Moscow-Leningrad.

Further Reading

NFPA 400, 2016. Hazardous Materials Code.
NIST Chemistry WebBook, http://webbook.nist.gov/chemistry/.

CHAPTER 4

Fluid-Dynamics for Process Safety

If you are to deal with water ask experience first, then the theory.

(Leonardo da Vinci, 1452–1519)

4.1 Equations of Conservation

Classical mechanics is based on equations of conservations, which are a very powerful tool to approach and resolve many process safety scenarios. They are:

1. Equation of conservation of momentum
2. Equation of conservation of momentum
3. Equation of conservation of energy

This chapter will familiarise the reader with these equation, providing several calculation cases.

4.1.1 Equation of Mass Conservation

A fluid of density ρ moves with vectoral velocity $\boldsymbol{u}$ through a deformable control volume V at rest. Any elemental area dA of the surface is identified by the unit normal $\boldsymbol{n}$. No chemical reaction is assumed to take place (Fig. 4.1).

$$\frac{dm}{dt} = \frac{d}{dt}\left(\int_V \rho \cdot dV\right) + \int_A \rho \cdot \boldsymbol{u} \times \boldsymbol{n} \cdot dA \tag{4.1}$$

where $\frac{dm}{dt}$ refers to generated mass, which is equal to zero by definition:

$$\frac{dm}{dt} = \frac{d}{dt}\left(\int_V \rho \cdot dV\right) + \int_A \rho \cdot \boldsymbol{u} \times \boldsymbol{n} \cdot dA = 0 \tag{4.2}$$

$\boldsymbol{u} \times \boldsymbol{n}$ is assumed to indicate *the scalar product* of fluid velocity $\boldsymbol{u}$ and unit normal $\boldsymbol{n}$, i.e.:

$$u \times \cos\theta \tag{4.3}$$

where u is the module of the velocity vector, and θ is the angle between this one and the normal unit vector of the differential area dA.

Process Safety Calculations. https://doi.org/10.1016/B978-0-08-101228-4.00004-6

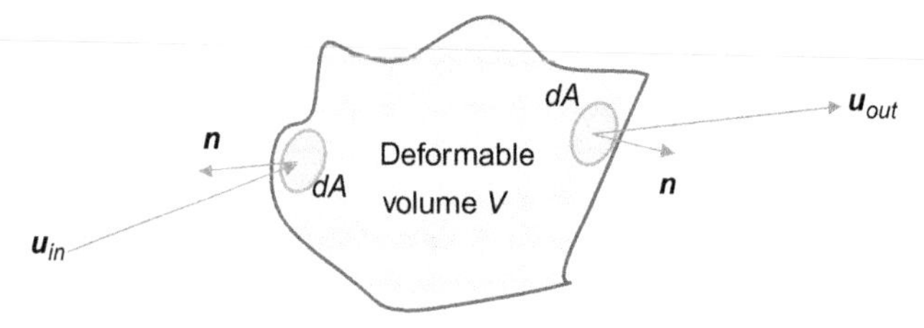

Fig. 4.1
Motion of a fluid through a control volume V.

For a fixed volume, equation becomes:

$$\left(\int_V \frac{d\rho}{dt} \cdot dV\right) + \int_A \rho \cdot \boldsymbol{u} \times \boldsymbol{n} \cdot dA = 0 \tag{4.4}$$

And if density does not depend on time, the equation of conservation of mass reduces to:

$$\int_A \rho \cdot \boldsymbol{u} \times \boldsymbol{n} \cdot dA = 0 \tag{4.5}$$

A specific application of conservation mass is the one relevant to incompressible fluids. In this case:

$\frac{d\rho}{dt} = 0$ and ρ is constant, so:

$$\int_A \boldsymbol{u} \cdot \boldsymbol{n} \times dA = 0 \tag{4.6}$$

Typical applications of the equation of conservation of mass are related to the motion of compressible and incompressible fluids in ducts, where in every section, respectively (W is the mass flow rate and Q is the volumetric flow rate) (Figs. 4.2 and 4.3):

$$W = \rho \cdot u \cdot A = \text{const} \tag{4.7}$$

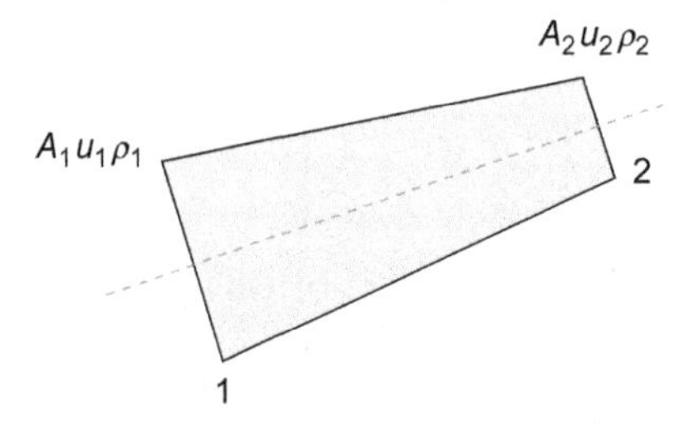

Fig. 4.2
Motion of compressible fluid in a duct.

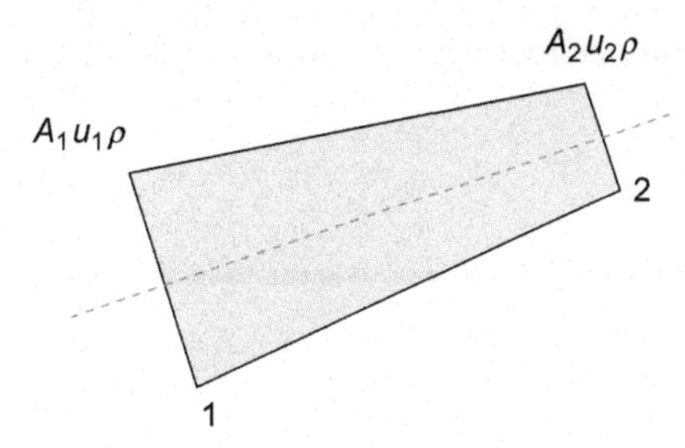

Fig. 4.3
Motion of incompressible fluid in a duct.

$$Q = u \cdot A = \text{const} \tag{4.8}$$

Velocity u has to be considered as the average velocity in the section. In some cases, this has to be calculated on the basis of the motion regime in the duct. An important study case is the steady state laminar flow in circular tubes of fluids of constant density and viscosity driven by a pressure difference and gravity. The application of momentum transfer in Newton's law gives the following formula for the parabolic velocity profile in the tube (Bird et al., 2002) (Fig. 4.4):

$$u(r) = \frac{(P_1 - P_2)}{4 \cdot \mu \cdot L} \cdot R^2 \cdot \left[1 - \left(\frac{r}{R}\right)^2\right] \tag{4.9}$$

where:

- P is the the sum of the pressure and gravitational terms in the cross section.
- R is the radius.
- r is the generic radius.
- μ is the dynamic viscosity.
- L is the distance between the sections.

The average velocity is calculated dividing the flow rate by the cross sectional area:

$$u(r) = \frac{(P_1 - P_2)}{8 \cdot \mu \cdot L} \cdot R^2 \tag{4.10}$$

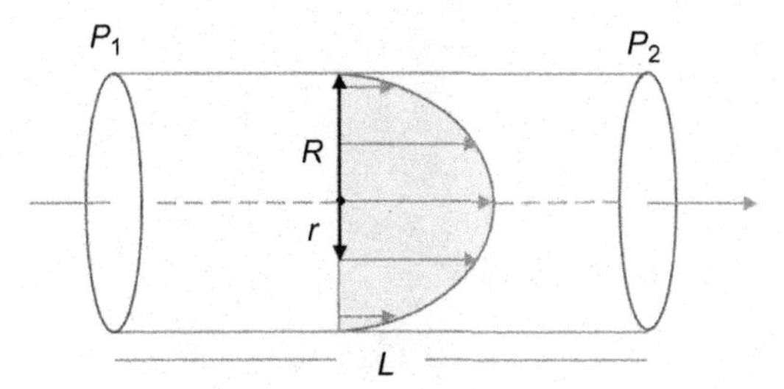

Fig. 4.4
Velocity parabolic profile of laminar flow of a viscous incompressible fluid.

So, mass conservation equation for incompressible fluids becomes Hagen Poiseuille formula:

$$W = \rho \cdot \frac{(P_1 - P_2)}{8 \cdot \mu \cdot L} \cdot \pi \cdot R^4 \tag{4.11}$$

Example 4.1

A thermally insulated tank, partially filled with water initially at atmospheric pressure, is fed with additional water at a constant flow rate of 1 m^3/h. The air contained in the headspace is compressed as long as new water is fed. The vessel can withstand a maximum pressure of 2 atm absolute. Find the variation law in water level and in air pressure, the time needed for the vessel to collapse, and the air temperature when this happens (Fig. 4.5). Assume that air is not absorbed by water.

Q	m^3/h	1
V_o	m^3	3
T_o	K	298.16
P_o	Pa	1
P_1	Pa	2
D	m	1
γ	–	1.4

Solution

Following the mass conservation principle, the liquid volume in the vessel v will increase according to the law:

$$\frac{dv}{dt} = \pi \cdot \frac{D^2}{4} \cdot \frac{dh}{dt} = Q \tag{4.12}$$

Integrating:

$$h(t) = h_o + Q \cdot \frac{4 \cdot t}{\pi \cdot D^2} \tag{4.13}$$

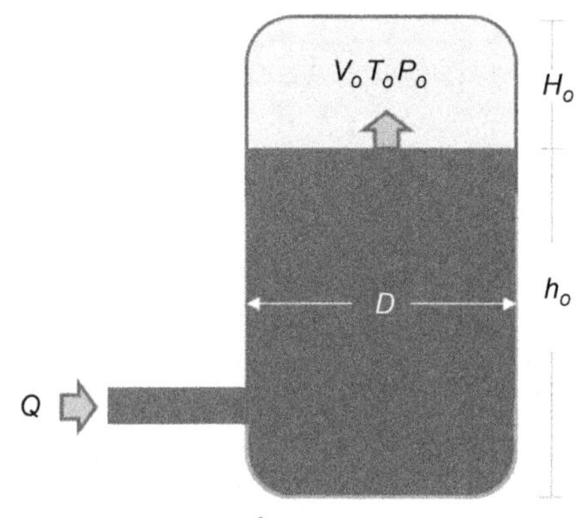

Fig. 4.5
Thermally insulated tank of Example 4.1.

The air volume in the headspace will decrease at the same rate:

$$\frac{dV}{dt} = Q \tag{4.14}$$

An adiabatic compression can be assumed, so that:

$$T_1 \times P_1^{(1-\gamma)/\gamma} = T_o \times P_o^{(1-\gamma)/\gamma} = 298.16 \times 1^{(1-1.4)/1.4} = 298.16 = const \tag{4.15}$$

$$T_1 = \frac{298.16}{2^{\frac{(1-1.4)}{1.4}}} = 363.5\,\mathrm{K} \tag{4.16}$$

For an adiabiatic transformation it is also:

$$P \times V^{\gamma} = const \tag{4.17}$$

$$1 \times 3^{1.4} = 4.65 \tag{4.18}$$

$$V_1 = \left(\frac{4.65}{2}\right)^{1/1.4} = 1.83\,\mathrm{m}^3 \tag{4.19}$$

The time required to reach a pressure of 2 atm can be easily calculated, equating the volume variation to the liquid fed at constant flow rate Q.

$$t_1 = \frac{V_o - V_1}{Q} = \frac{3 - 1.83}{1} = 1.17\,\mathrm{h} \tag{4.20}$$

Example 4.2

Air is supplied to a compressor through a 0.5×0.6 m^2 duct at 1 atm absolute, and 20°C at an average velocity of 20 m/s, and leaves the compressor at 40°C and 5 atm absolute. If the delivery duct has a size of 0.25×0.3 m^2, the ideal gas constant is 0.082 atm m^3/kmoles K, and the air molecular weight is 29, find the downstream air velocity.

Solution

$$V_2 = V_1 \frac{P_1}{P_2} \times \frac{T_2}{T_1} = u_1 \times S_1 \times \frac{P_1}{P_2} \times \frac{T_2}{T_1} = 20 \times 0.5 \times 0.6 \times \frac{1 \times 313}{5 \times 293} = 1.29\,\mathrm{m}^3/\mathrm{s} \tag{4.21}$$

$$u_2 = \frac{V_2}{S_2} = \frac{1.29}{0.25 \times 0.3} = 17.1\,\mathrm{m/s} \tag{4.22}$$

4.1.2 Equation of Conservation of Momentum

Momentum of a body of mass m is a vector $\boldsymbol{p}$ defined as:

$$\boldsymbol{p} = m \cdot \boldsymbol{u} \tag{4.23}$$

The second Newton's law defines the entity force $\boldsymbol{F}$ as:

$$\boldsymbol{F} = \frac{d(m \cdot \boldsymbol{u})}{dt} = \frac{d\boldsymbol{p}}{dt} \tag{4.24}$$

The first law states that, unless acted by an unbalanced force, a body stays at rest, or stays in motion with the same initial speed, i.e. if the overall external forces component is zero, body momentum does not change. So:

$$\boldsymbol{p} = m \cdot \boldsymbol{u} = \text{constant} \tag{4.25}$$

Also, any momentum change is due to the result of all the external forces acting on the body:

$$\sum \boldsymbol{F} = \frac{d(m \cdot \boldsymbol{u})}{dt} = \frac{d\boldsymbol{p}}{dt} \tag{4.26}$$

With reference to the control volume depicted in Fig. 4.1, adopting the substantial derivative for a fluid stream, the momentum flux is defined as:

$$\sum \boldsymbol{F} = \frac{\partial}{\partial t} \int_V \boldsymbol{u} \cdot \rho \cdot dV + \int_A \boldsymbol{u} \cdot \rho \cdot (\boldsymbol{u} \times \boldsymbol{n}) \cdot dA \tag{4.27}$$

For a steady state stream, this becomes:

$$\sum \boldsymbol{F} = \int_A \boldsymbol{u} \cdot \rho \cdot (\boldsymbol{u} \times \boldsymbol{n}) \cdot dA \tag{4.28}$$

Typical process safety applications of momentum equations deal with fluids flowing through ducts or pipes, and jet flows from holes and orifices (Fig. 4.6):

$$\sum F = \rho_1 \cdot u_1^2 \cdot A_1 - \rho_2 \cdot u_2^2 \cdot A_2 \tag{4.29}$$

This reduces to the conservation equation in cases of no active forces acting on the system.

$$0 = \rho_1 \cdot u_1^2 \cdot A_1 - \rho_2 \cdot u_2^2 \cdot A_2 \tag{4.30}$$

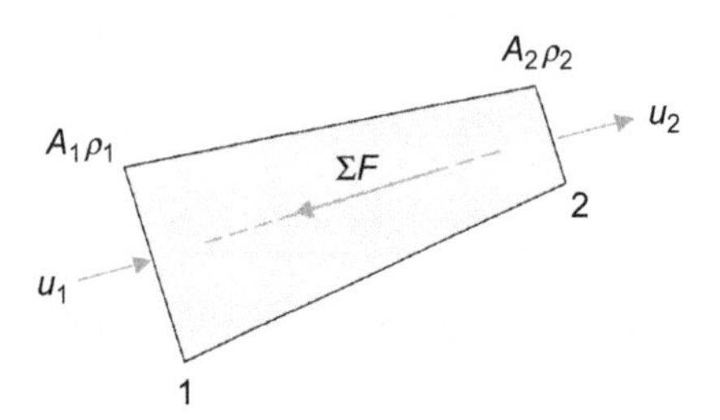

Fig. 4.6
Balance of forces in momentum equation.

Example 4.3

Discuss the scenario included in Fig. 4.6. Resolve the balance to find whether any active force is acting on the system, assuming no friction effects.

Solution

Eq. (4.29) can be written as:

$$\sum F = u_1 \cdot (\rho_1 \cdot u_1 \cdot A_1) - u_2 \cdot (\rho_2 \cdot u_2 \cdot A_2) \tag{4.31}$$

According to the mass conservation law, terms in brackets are identical and coincide with the invariant mass flow rate W:

$$\sum F = W \cdot (u_1 - u_2) \tag{4.32}$$

It is found that a force is acting opposite to the stream direction, being that the downstream velocity is greater than the upstream one. In a duct with no cross section variation along the axis, this force would not exist.

Example 4.4

A water tank has a discharge pipe located h metres below the free surface level, which is kept constant by feeding water at a negligible velocity. Applying the momentum equation, find a formula for the discharge velocity (Fig. 4.7).

Solution

The acting force can be described in terms of the hydrostatic load as described by Stevin's law:

$$P = \rho \cdot g \cdot h \tag{4.33}$$

Hydrostatic pressure varies with water level, so a differential balance is needed for the calculation (Fig. 4.8).

General equation:

$$\sum F = W \cdot (u_1 - u_2) \tag{4.34}$$

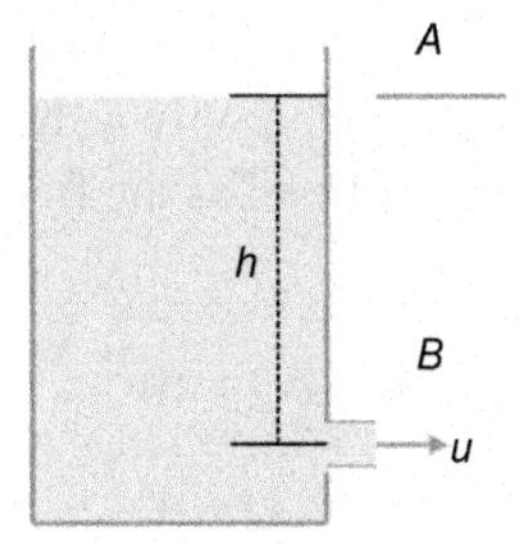

Fig. 4.7
Water tank of Example 4.4.

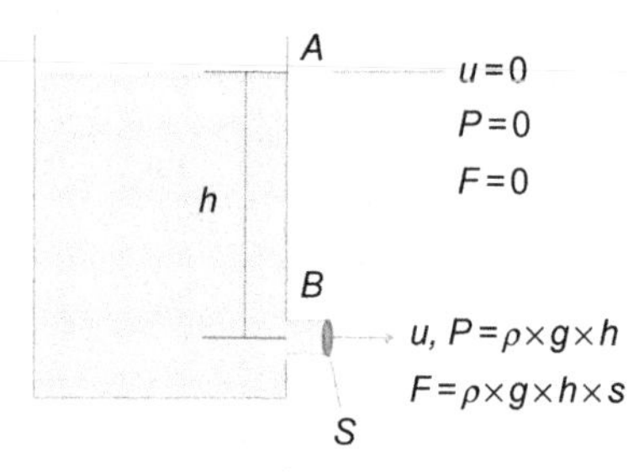

Fig. 4.8
Balance of forces on tank from Example 4.4.

becomes:

$$F(h) = \rho \cdot g \cdot h \cdot A \tag{4.35}$$

$$F(h) = [\rho \cdot A \cdot u(h)] \cdot u(h) = W \cdot u(h) \tag{4.36}$$

With:

$$[\rho \cdot A \cdot u(h)] = W = \text{const} \tag{4.37}$$

Differentiating and equating:

$$\rho \cdot g \cdot A \cdot dh = [\rho \cdot A \cdot u(h)] \cdot du(h) \tag{4.38}$$

Integrating:

$$g \cdot \int_0^h dh = \int_0^u u \cdot du \tag{4.39}$$

$$u(h) = \sqrt{2 \cdot g \cdot h} \tag{4.40}$$

that is Torricelli's velocity.

4.1.3 Equations of Conservation of Energy

Equations of conservation of energy are defined for closed and open systems. In Chapter 2 closed systems have been identified as those systems in which the mass is constant throughout the transformation, so energy can be exchanged only as heat and work with the surroundings. The law of energy conservation has been defined as the first law of thermodynamics for closed systems:

$$dU = \delta Q + \delta W \tag{4.41}$$

$$\Delta U = Q + W \tag{4.42}$$

In this equation U is the internal energy, Q is the exchanged heat, and W is the work done.

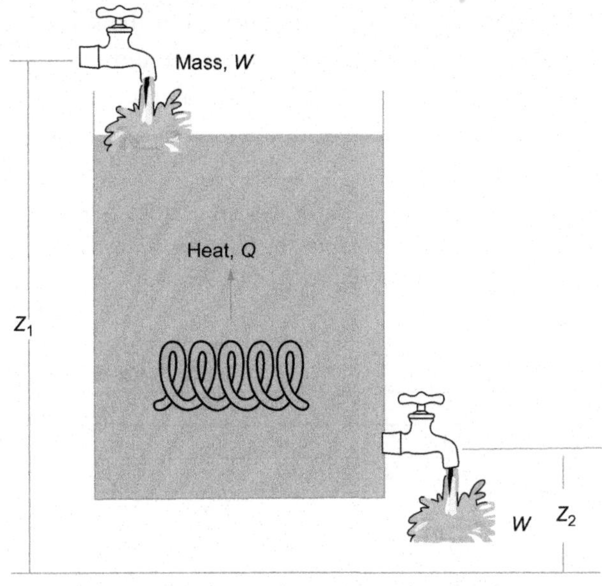

Fig. 4.9
Open system.

In addition to energy, an open system can exchange its mass with other systems. A typical open system is the tank depicted in Fig. 4.9.

The law of conservation of energy for this system is:

$$h_2 + \frac{u_2^2}{2} + g \cdot z_2 = h_1 + \frac{u_1^2}{2} + g \cdot z_1 + Q + W \tag{4.43}$$

In this equation, the following units should be used:

- All energies are specific energies per mass unit [J/kg].
- Velocities are in [m/s].
- Heights z are in [m].
- g (acceleration of gravity) is in [m/s^2].

Q is heat transferred to the system. W is work done on the system

If no significant heat and work are exchanged, and enthalpy variation is negligible with respect to gravitational and kinetic terms, as it happens for water flow at constant temperature, the equation of conservation of energy reduces to the Bernoulli equation, due to the constancy of the internal energy fraction of the enthalpy:

$$\frac{P_2}{\rho_2} + \frac{u_2^2}{2} + g \cdot z_2 = \frac{P_1}{\rho_1} + \frac{u_1^2}{2} + g \cdot z_1 \tag{4.44}$$

where pressure is in Pascal, and the density is in kg/m^3. For the sake of correctness, it must be remembered that Bernoulli's law is strictly valid for a liquid motion along a streamline. Nevertheless, for practical calculations, it is applied by taking average values.

Example 4.5

Saturated steam at 2.8 bar is mixed with a water stream at 20°C with the aim to produce 15,000 kg/h of water at 80°C. The two streams are fed to a 2″ horizontal static mixer with negligible internal friction. Find the required amount of steam.

Solution

Process Data	Units	Vapour		Cold Water		Hot water	
Mass flow rate	kg/h	w_1		w_2		w_3	15,000
Temperature	°C	131.2		20		80	
Enthalpy	J/kg	h_1	2, 721,500	h_2	83,720	h_3	334,880
Density	kg/m^3	–		–		1030	

Water velocity at the mixer outlet is:

$$u_2 \cong \frac{15,000}{3600 \cdot 1030 \cdot \pi \cdot 0.025^2} = 2.06\,\text{m/s} \tag{4.45}$$

and specific kinetic energy:

$$\frac{u_2^2}{2} \cong \frac{4.25}{2} = 2.125\text{J} \tag{4.46}$$

Steam inlet velocity is, expectedly, much greater than water velocity, but the corresponding kinetic energy is reasonsably much smaller than enthalpies. Therefore, we decide to neglect the kinetic terms.

Eq. (4.43) becomes:

$$w_1 \cdot h_1 + w_2 \cdot h_2 = w_3 \cdot h_3 \tag{4.47}$$

$$w_1 \cdot 2,721,500 + w_2 \cdot 83,720 = 15,000 \cdot 334,380 \tag{4.48}$$

and conservation of mass:

$$w_1 + w_2 = w_3 \tag{4.49}$$

$$w_1 + w_2 = 15,000 \tag{4.50}$$

Solving:

$$w_1 = 1425.4\text{kg} \tag{4.51}$$

Example 4.6

Resolve Example 4.4 applying the equation of conservation of energy

Solution

The process is isothermal, and the fluid is incompressible. Bernoulli's law can be applied:

$$g \cdot h = \frac{u(h)^2}{2} \tag{4.52}$$

$$u(h) = \sqrt{2 \cdot g \cdot h} \tag{4.53}$$

4.2 Joule–Thomson Expansion in Process Safety

If a gas undergoes a continuous throttling, free to expand through a valve, an orifice, or a porous medium with no appreciable friction and heat transfer, gravitational terms are negligible, so Eq. (4.43) reduces to:

$$h_2 + \frac{{u_2}^2}{2} = h_1 + \frac{{u_1}^2}{2} \tag{4.54}$$

Provided kinetic terms are comparable or significantly lower than enthalpies:

$$h_2 = h_1 \tag{4.55}$$

This isenthalpic expansion is called Joule–Thomson expansion. Plotting the final pressure vs. temperature downstream values, starting from a specific upstream temperature, a curve is generated that is the locus of all the points representing initial and final states of same enthalpy. Changing the initial temperature at the same pressure, different isenthalpic curves are obtained. The slope of the isenthalpic curve is the Joule–Thomson coefficient, defined as:

$$\mu_{JT} = \left(\frac{\partial T}{\partial P}\right)_h \tag{4.56}$$

The curve fitting all points where the Joule–Thomson coefficient is zero is called the *inversion curve*.

An approximate plot of the inversion and the throttling curves for nitrogen has been included in Fig. 4.10 (interpolated from TS data of Din, 1962). The zone inside the inversion curve, where $\mu_{JT} > 0$, is the cooling region, i.e. expansion will cool the gas. The external zone is the heating region. Almost all gases show positive coefficients at ordinary temperature and below reasonably high expansion pressures, so throttling will cool them. Hydrogen, helium, and neon will be heated at ordinary temperature.

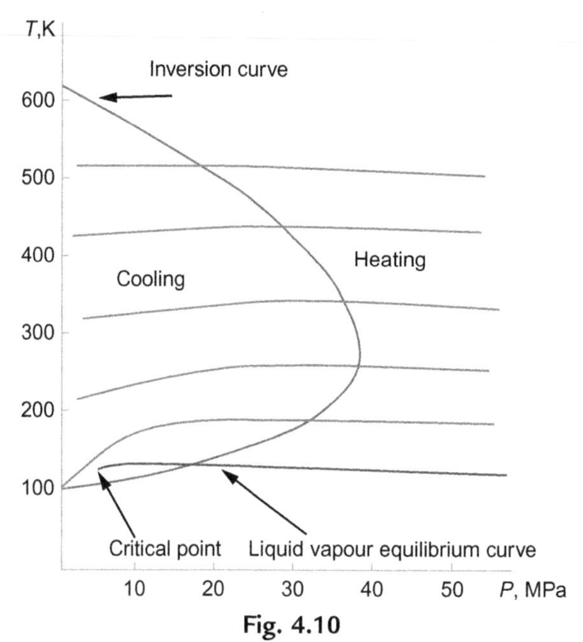

Fig. 4.10

Throttling and inversion curves for nitrogen. *Elaboration of Din, F., 1962. Thermodynamic Functions of Gases, Volume 1: Ammonia, Carbon Dioxide, and Carbon Monoxide; Volume 2: Air, Acetylene, Ethylene, Propane and Argon; Volume 3: Ethane, Methane, and Nitrogen. Butterworths.*

Throttling phenomena are very important in process safety. Many accidental gas releases occur from pressurised containments and follow Joule–Thomson like expansion processes. It is very important to remember the different behaviour of hydrogen, and to keep in mind that the isenthalpic assumption is a simplification of the equation of conservation of energy justified by the neglectability or simple deletion of the other terms. The general assumption of an isenthalpic expansion can cause serious mistakes in the calculations if not careful assessed.

Example 4.7

Emergency depressurisation of carbon dioxide from a compression station is implemented by means of a pressure relief valve. Carbon dioxide is vented to the atmosphere from stagnant conditions at 40 barg and 20°C. At the valve outlet, choked condition can be assumed, so that 280 m/s sonic velocity is attained by the gas. Analyse this expansion case (Fig. 4.11).

Solution

The analysis can be carried out with reference to the following pressure–enthalpy diagram (Fig. 4.12).

The equation of conservation of energy can be written:

$$h_2 + \frac{u_2^{\,2}}{2} = h_1 \tag{4.57}$$

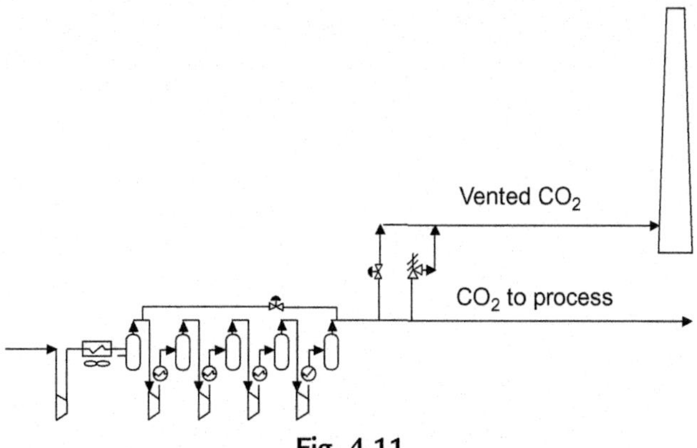

Fig. 4.11
Emergency depressurisation of CO_2.

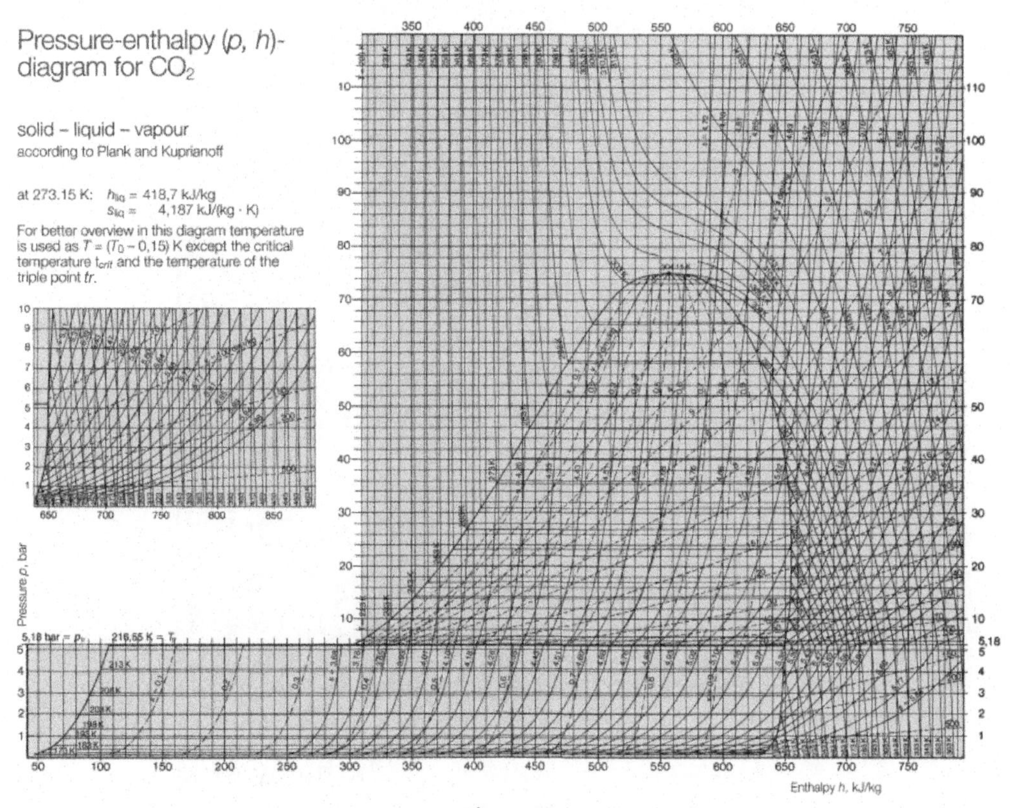

Fig. 4.12
Pressure–enthalpy of CO_2 (Plank and Kuprianoff, 1929).

where the subscript 1 refers to the upstream condition, and the subscript 2 refers to the downstream condition. Specific enthalpy h_1 can be read on the diagram (Fig. 4.13):

$$h_1 = 673.5\,\text{kJ/kg} \tag{4.58}$$

And:

$$h_2 = 673{,}500 - \frac{280^2}{2} = 634{,}300\,\text{J/kg} = 634.3\,\text{kJ/kg} \tag{4.59}$$

The transition from state 1 to state 2 can be followed on the diagram (Fig. 4.14):

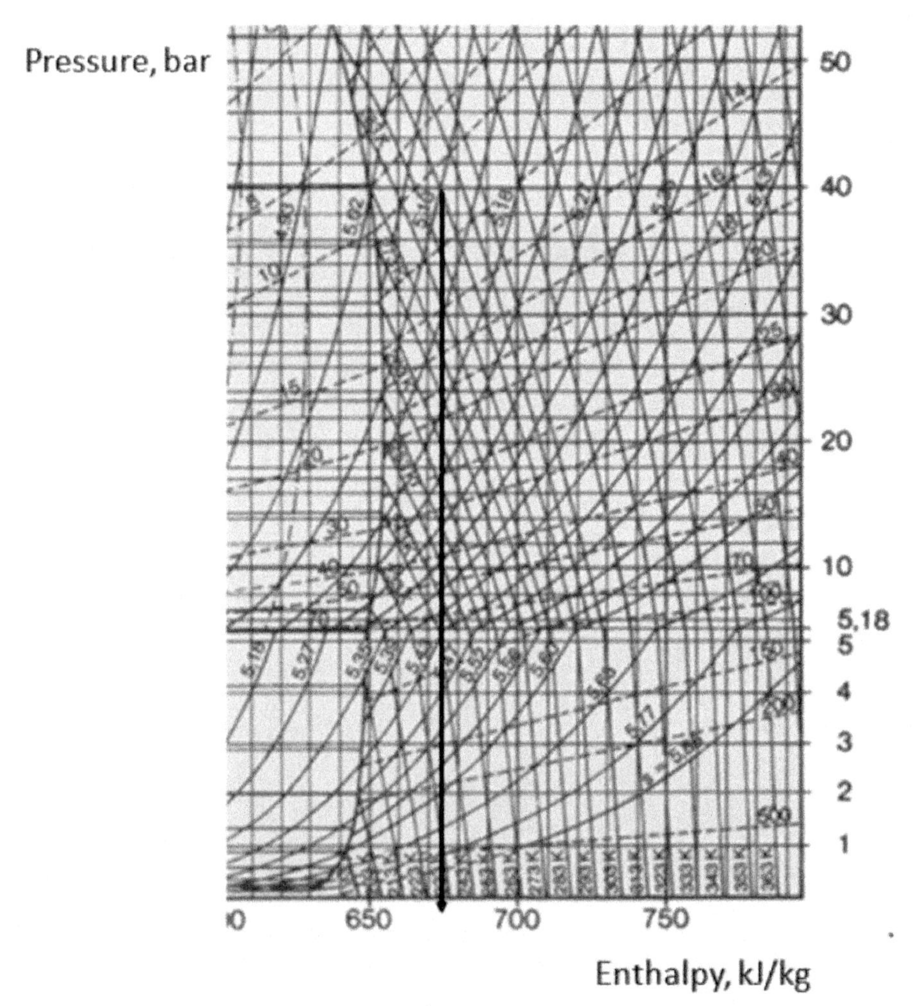

Fig. 4.13
Enthalpy of CO_2 for upstream condition 1 (Plank and Kuprianoff, 1929).

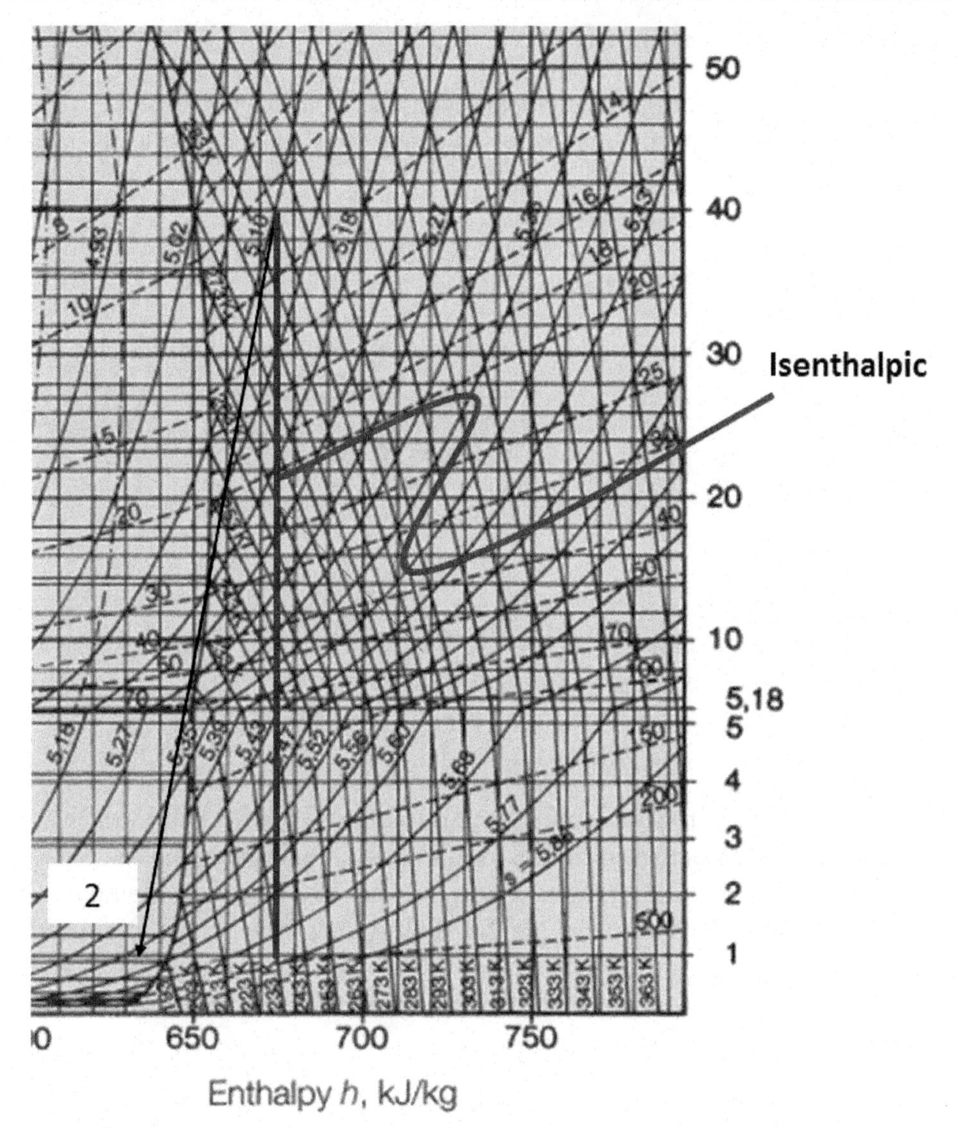

Fig. 4.14
Enthalpy of CO_2 for downstreamm condition 2 (Plank and Kuprianoff, 1929).

Conclusions:

- Expansion cannot be assumed as isenthalpic (Joule–Thomson).
- The final state can fall within the vapour–solid equilibrium zone.
- Consequence of this inaccuracy is the underestimation of solid formation, plugging of valve and piping, and snow-out with potentially very serious process safety implications.

4.3 Turbulent and Laminar Flows

The identification of the flow regime in process safety is a fundamental aspect. Successful selection of models and calculation approaches is strictly dependant on the understanding of the characteristics of flow.

4.3.1 Laminar–Turbulent Transition

The Reynolds number is the dimensionless parameter to be considered for the flow regime definition.

$$Re = \frac{\rho \cdot u \cdot D}{\mu} = \frac{u \cdot D}{\nu} \tag{4.60}$$

where:

- ρ is the fluid density.
- μ is the fluid dynamic viscosity.
- ν is the fluid kinematic viscosity.
- u is fluid velocity.
- D is the pipe diameter or a significant flow entity.

Laminar regimes are related to moderate Reynolds numbers and to the prevalence of viscous and diffusive transport mechanisms on the inertial and convective actions. Guidance values for the transition ranges, relating to various process configurations, which can be found in Table 4.1 (Bird et al., 2002). These values have to be considered for a screening approach, especially for flow around spheres, whose transition zone is affected by a certain complexity. Flow regimes of jets have been discussed in paragraphs further in this chapter.

Table 4.1 Guidance values for flow regime transitions (Bird et al., 2002)

Fluid-Dynamic Scenario	Laminar Range	Transition Range	Turbulent Range
Annular flow	<2000	2000	>2000
Tangential laminar flow along a flat plate	<100,000	100,000–500,000	>500,000
Flow in circular pipes	<2100	2100	>2100
Flow around spheres	<200,000	200,000	>200,000
Transverse flow around circular cylinders	<10	10–10,000	>10,000

4.4 Liquid Elasticity (Bulk Modulus)

The influence of liquid elasticity is a significant aspect of fluid-dynamics, which can also entail important implications in process safety. Specifically, for compressible fluids, this property can play an important role in release scenarios and can impact the structural load of vessels and piping. Pascal's principles state that if an external pressure is applied to a liquid, this will be transmitted uniformly to all parts of the liquid, regardless of their location. Liquids are characterised by a very high pressure increase, also corresponding to a relatively small density increase. This may result in the potential accumulation of a very high elastic energy, which can drive the flow regime until it persists and can expose containment to significant stresses. The key parameter representing liquid elasticity is bulk modulus B_n, defined as:

$$B_n = -v \cdot \left(\frac{\partial P}{\partial v}\right)_n \tag{4.61}$$

where:

- v is the specific volume.
- P is liquid pressure.
- n is the thermodynamic pathway.

Many processes are thermodynamically possible, but only the isothermal bulk modulus and the adiabatic bulk modulus are practically used:

$$B_T = -v \cdot \left(\frac{\partial P}{\partial v}\right)_T \tag{4.62}$$

$$B_K = -v \cdot \left(\frac{\partial P}{\partial v}\right)_K \tag{4.63}$$

B_T and B_K of a liquid of given API gravity, and at pressure P (psig), and temperature T (°R) can be calculated through empirical ARCO formulas as given by Shashi Menon (2004):

$$B_T = a_T + b_T \cdot P - c_T \cdot T^{1/2} + d_T \cdot T^{3/2} - e_K \cdot API^{3/2} \tag{4.64}$$

$$B_K = a_K + b_K \cdot P - c_K \cdot T^{1/2} - d_K \cdot API - e_K \cdot API^2 + f_K \cdot T \cdot API \tag{4.65}$$

where:

- a, b, c, d, e, f are constants.
- P is the pressure, psig.
- T is the temperature, °R.

Table 4.2 shows the values of constants (Shashi Menon, 2004), and Table 4.3 includes B_T and B_K values for some process liquids at 20°C and 1 atm.

Table 4.2 ARCO equation for bulk modulus calculation (Shashi Menon, 2004)

Isothermal	Value	Adiabatic	Value
a_T	$2.619 \cdot 10^6$	a_T	$1.286 \cdot 10^6$
b_T	9.203	b_T	13.55
c_T	$1.417 \cdot 10^5$	c_T	$4.122 \cdot 10^4$
d_T	73.05	d_T	$4.53 \cdot 10^3$
e_T	341	e_T	10.59
–	–	f_K	3.228

Table 4.3 Isothermal and adiabatic bulk moduli for some liquids at 1 atm—20°C (Shashi Menon, 2004)

Liquid	Isothermal BM (N/mm^2)	Reference	Adiabatic BM (N/mm^2)	Reference
Ammonia	–	–	1.82×10^9	White (2011)
Benzene	1.06×10^9	Avallone et al. (2007)	1.53×10^9	Avallone et al. (2007)
Ethanol	8.9×10^8	Avallone et al. (2007)	1.06×10^9	Avallone et al. (2007)
Ethylene glycol	–	–	3.05×10^9	White (2011)
Freon 12	–	–	7.95×10^8	White (2011)
Kerosene	1.29×10^9	Avallone et al. (2007)	1.44×10^9	Avallone et al. (2007)
Methanol	–	–	1.03×10^9	White (2011)
SAE 10W oil	–	–	1.31×10^9	White (2011)
Water	2.17×10^9	Avallone et al. (2007)	2.17×10^9	Avallone et al. (2007)
Seawater	2.33×10^9	Avallone et al. (2007)	2.37×10^9	Avallone et al. (2007)

ARCO equations have no specific theoretical basis, but are very useful for a general identification of the variables affecting bulk moduli value. Their general correctness can be proved according to Bair (2014), which, on the basis of Murnaghan EoS, has proposed the following equation for the isothermal bulk modulus of some diesel fuels:

$$B_T = K_o + K_o' \cdot P \tag{4.66}$$

$$K_o = K_{oo} \cdot \exp(-\beta_k \cdot T) \tag{4.67}$$

Values of β_K, K_o', and K_{oo} for four investigated diesel fuels have been included in Table 4.4 (Bair, 2014). Finally, bulk moduli relevant to water (Totten and De Negri, 2011), propane (Younglove and Ely, 1987), and benzene (Forsythe, 2003) at different pressures have been included in Table 4.5.

Table 4.4 Values of Murnaghan EoS (Bair, 2014)

Constant	Sample 16A ELV	Sample ED95	Sample B20	Sample B100
K'_o	9.677	8.353	10.700	9.437
K_{oo} (GPa)	6.267	3.838	7.438	5.643
β_K (K^{-1})	0.006894	0.004680	0.005971	0.004829

Table 4.5 Bulk moduli for some liquids at different pressure (Totten and De Negri, 2011; Younglove and Ely, 1987; Forsythe, 2003)

Liquid	Pressure (bar)	Temperature (°C)	Bulk Modulus (bar)	Reference
Water	1	20	22,000	Totten and De Negri (2011)
	100		22,200	
	1000		22,700	
Propane	1	20	2450	Bulk modulus has been calculated through sound velocity using data provided by Younglove and Ely (1987)
	100		3700	
	1000		13,000	
Benzene	5	17	11,236	Forsythe (2003)
	200	20	12,987	
	400	20	14,925	

4.4.1 Bulk Modulus and Sound Velocity

Sound velocity can be expressed in terms of the bulk modulus according to the Newton–Laplace equation:

$$a = \sqrt{\frac{B}{\rho}} \tag{4.68}$$

Due to the sound velocity definition:

$$a = \sqrt{\left(\frac{\partial P}{\partial \rho}\right)_S} \tag{4.69}$$

we recognise that B in Eq. (4.68) is the adiabatic bulk modulus.

4.5 Fluid Hammer (Surge)

Fluid hammer, also known as *surge*, is a transient phenomenon caused by the fast forced stop of fluid flow, resulting velocity change, and pressure wave that propagates in the liquid. Surge can

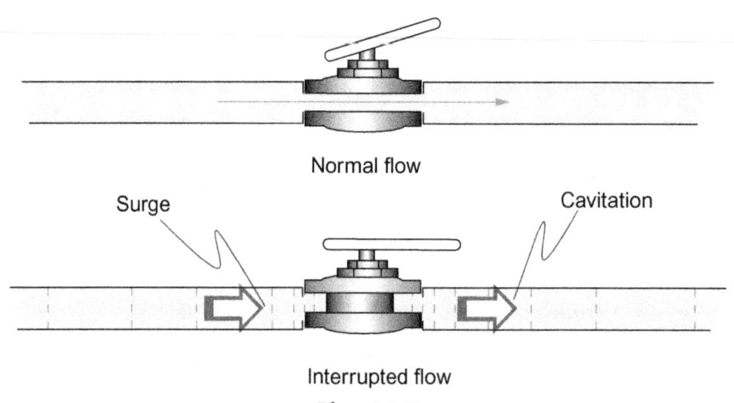

Fig. 4.15
Fluid hammer.

affect both liquid and gaseous equipment (Fig. 4.15). The most significant, but not limited to, *surge* scenarios in process safety are:

- Accidental or planned, sudden or progressive, closure of a valve in a pipeline. This can cause a pressure rise in the upstream section and cavitation (buckling) in the downstream section:
- Sudden change in flow, pressure, and density in compressors caused by various events, such as variations in downstream demand, or fluctuations, which can result in a downstream pressure greater than the delivery pressure. This is the typical reversal flow for compressors in HAZOP studies.

4.5.1 Water Hammer in Pipelines

As the valve closes, at time dt the amount of liquid $r \cdot A \cdot dx$ has stopped, and, according to equation of momentum, this one has to equate the impulse of the forces acting on the differential volume (Fig. 4.16).

$$\rho \cdot A \cdot u_o \cdot dx = A \cdot \Delta P \cdot dt \tag{4.70}$$

where u_o is the steady state velocity prior to the valve closure. Rearranging and considering that:

$$a = \frac{dx}{dt} \tag{4.71}$$

is liquid sound velocity, it results:

$$\Delta P = \rho \cdot a \cdot u_o \tag{4.72}$$

that is Joukowsky' s formula (1898). At time t the pressure wave has travelled into the pipeline section far L_t from the valve:

$$L_t = a \cdot t \tag{4.73}$$

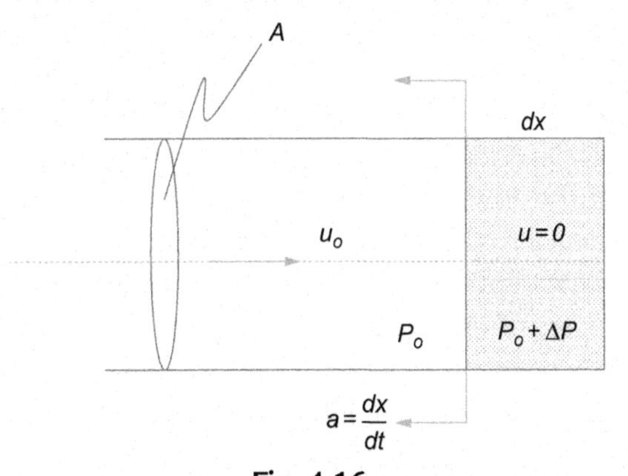

Fig. 4.16
Differential balance for water hammer.

Other important parameters are the overpressure caused by a partial closure of valve reducing flow velocity from u_o to u_1:

$$\Delta P = \rho \cdot a \cdot (u_o - u_1) \tag{4.74}$$

and the force F_V acting on the valve resulting from the upstream and downstream pressure difference:

$$F_V = A \cdot (P_U - P_D) \tag{4.75}$$

Example 4.8

Water flow rate in a pipeline 500 m long is 1500 m^3/h. Pipeline diameter is 0.5 m. A valve is suddenly closed. Find:

- the resulting overpressure
- the time taken for the overpressure to propagate to all the pipeline
- the force acting on the valve, considering that downstream of the valve water is supplied to the free surface of a basin

Water sound velocity is 1490 m/s.

Solution

In order to apply Joukowsky' s equation, water velocity is needed:

$$u_o = \frac{1,500}{3600 \cdot \pi \cdot 0.25^2} = 2.12\,\text{m/s} \tag{4.76}$$

$$\Delta P = 1000 \cdot 1490 \cdot 2.12 = 3,158,800\,\text{Pa} = 32.21\,\text{kg/cm}^2 \tag{4.77}$$

Downstream relative pressure is zero, so:

$$F_V = 3,158,800 \cdot \pi \cdot 0.25^2 = 619,914.5\,\text{N} = 63,212.125\,\text{kg} \tag{4.78}$$

4.6 Theory of Jets

Release of liquids, vapours, and gases from pressurised containments is one the most important scenarios in process safety. Not only do these events represent in themselves a hazardous configuration, but most of the downstream dispersion phenomena are triggered by them and strictly depend on their characteristics. A broad literature has been developed about jet physics; however, due to the wide range of scenarios, and to the complex nature of governing parameters, a significant level of uncertainty is still found in practical calculations. This paragraph aims to present a practical approach to jets in consequence assessment, along with many applicative cases and numerical examples, selected on the basis of the most frequent hazardous scenarios met in consequence assessment.

4.6.1 Definitions

Jet

For the purpose of this book, a jet can be defined as a *vapour, gaseous, liquid or multi-phase flow released from pressurised containments with significant momentum.*

Turbulent gas jet

According to Oosthuizen (1983), a wide range of Reynolds numbers exist where transitional fluid dynamic instabilities are shown for circular gas jet flows until $Re = 10{,}000$. On the other hand, API 521 states that the jet-momentum forces of release are usually dominant, provided the gas velocity is greater than 12 m/s, or the jet to wind velocity ratio is less than 50, and gas Reynolds number is:

$$Re \geq 15{,}400 \cdot \frac{\rho_g}{\rho_{air}} \tag{4.79}$$

where:

- ρ_g is gas density.
- ρ_{air} is air density.

if these conditions are not met, the released material could be wind dominated. In this book, a fully developed turbulent jet will be assumed at Reynolds greater than 10,000.

Transitional gas jet

According to Oosthuizen's findings (1983), a gas jet showing significant instabilities is expected in the Reynolds range between 5000 and 10,000.

Laminar gas jet

Gas jet flow can possibly be expected as laminar if the Reynolds number is significantly smaller than 5000, and, according to Rankin et al. (1983), laminar steady solutions have been found at Reynolds numbers up to 1000.

4.6.2 Choked and Unchoked Jet

For a gas jet, with reference to a stagnant pressure P_o, the reduction of the downstream pressure will result in increasing the mass flow rate (unchoked flow) up to a specific pressure (choked pressure), below which no further increase in the flow rate will be observed. The exit velocity will coincide with the sound velocity at the prevailing condition, and the flow rate will be choked. Choked pressure is therefore a process property of the system, depending only on the stagnant pressure and the specific substance. The flow rate for an adiabatic expansion through an orifice can be calculated by the formula:

$$W = P_o \cdot A \cdot C_D \cdot \sqrt{2 \cdot \frac{[g \cdot MW \cdot \gamma]}{R \cdot T_o \cdot (\gamma - 1)} \cdot \left[\left(\frac{P}{P_o} \right)^{2/\gamma} - \left(\frac{P}{P_o} \right) \right]^{(\gamma-1)/\gamma}} \tag{4.80}$$

where:

- P_o: upstream pressure
- P: downstream pressure
- MW: molecular weight
- T_o: downstream temperature, K
- R: ideal gas constant
- γ: specific heats ratio
- g: acceleration of gravity
- C_D: coefficient of discharge

and:

$$W_c = P_o \cdot A \cdot C_D \cdot \sqrt{\frac{g \cdot MW \cdot \gamma}{R \cdot T_o} \cdot \left[\frac{2}{\gamma + 1} \right]^{(\gamma+1)/(\gamma-1)}} \tag{4.81}$$

In order to decide which formula is to be used, P_{choked} will be preliminarily calculated and compared to the downstream pressure. If the latter is equal or lower, then the choked flow rate will be adopted. According to CCPS (1999), a threshold pressure for real gas can be assumed at 1.4 barg (20 psig). A value of 0.61 is suggested for C_D (CCPS), a value for 1 for screening purposes. For the sake of simplicity, in the calculations C_D has often been omitted.

It is worth noting that as the gas velocity is choked, if the upstream pressure increases, the mass flow rate increases as well due to the density rise, not due to the velocity rise.

4.6.3 Isothermal Turbulent (Choked) Gas Steady Jet

According to CCPS, an upstream stagnation pressure of 20 psig can be assumed as the pressure threshold for a choked flow. A turbulent gas jet propagates from the stagnant condition inside the containment to the zone where initial momentum is lost, and the outcome of the flow becomes predominantly governed by other agents, such as buoyancy and wind. In this section, jet transition through the various stages are presented with the aim to calculate all affecting variables.

4.6.3.1 Choked Jet

1. Stagnant condition
   The gas is assumed to be stagnant and at steady state.
   - $T = T_o$
   - $P = P_o$
   - $\rho = \rho_o$
   - $u = 0$
2. Outlet plane
   At the outlet, flow is sonic, and the expansion is assumed to take place according to an adiabatic transformation (Fig. 4.17). The fluid dynamic data at the outlet can be calculated as follows (Hess et al., 1973):

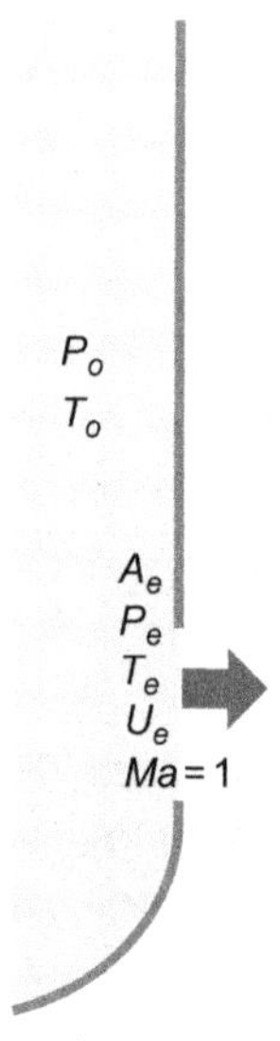

Fig. 4.17
Choked jet—Outlet section.

- $T = T_e$
- $P = P_e$
- $\rho = \rho_o$
- $u = u_e = a_e$

$$P_e = P_o \cdot \left(\frac{2}{\gamma+1}\right)^{\frac{\gamma}{\gamma-1}} \tag{4.82}$$

$$T_e = T_o \cdot \left(\frac{2}{\gamma+1}\right) \tag{4.83}$$

$$\rho_e = \rho_o \cdot \left(\frac{2}{\gamma+1}\right)^{\frac{\gamma}{\gamma-1}} \tag{4.84}$$

$$u_e = a_e = \sqrt{\left(\frac{2 \cdot \gamma}{\gamma+1} \cdot R \cdot T_o\right)} \tag{4.85}$$

3. Shock plane
   Jets with a high pressure ratio are classified as underexpanded. The originated expansion waves meet the jet boundary and are reflected as compression waves. The overall result is the well-known barrel-shaped shock, which covers a jet zone substantially unaffected by any entrainment contribution from the surrounding calm gas. Jet gas recompression creates a normal shock, the Mach disk, immediately downstream of which the flow becomes sonic and then subsonic (Fig. 4.18). The barrel-shaped zone shows very high Mach numbers, it is called the silence zone, and no air entrainment.
   The knowledge of Mach disk distance from the exit plane is essential to identify the position of the first jet section where entrainment begins, as it is the door of the transition zone of the jet. If a combustible or toxic gas is emitted to calm air, this is the first section where mixing with air takes place. The mentioned distance may be estimated according to the following empirical formula of Ashkenas and Sherman (1966):

$$x_m = 0.67 \cdot D_e \cdot \left(\frac{P_o}{P_e}\right) \tag{4.86}$$

   This can be assumed as the distance from the outlet where the jet speed is sonic and air entrainment begins. The related zone is very complex and can be simplified for calculation purposes according to the following equivalent scheme (Fig. 4.19).
   A pseudo-diameter D_s is assumed next to Mach disk, within the supersonic zone at the so called shock-plane, where pressure is assumed to be atmospheric and conservation equations can be applied (Ramskill, 1985):

$$D_s = D_e \cdot \left(\frac{P_o}{P_a}\right)^{\frac{\gamma-1}{2 \cdot \gamma}} \tag{4.87}$$

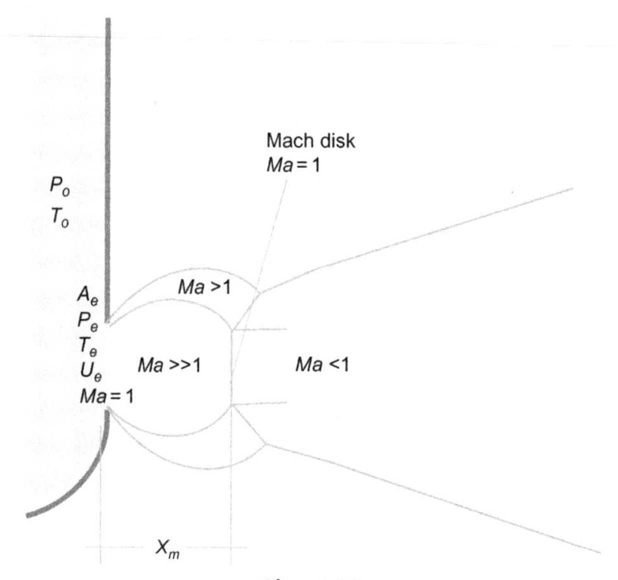

Fig. 4.18
Choked jet—Mach disk and shock plane.

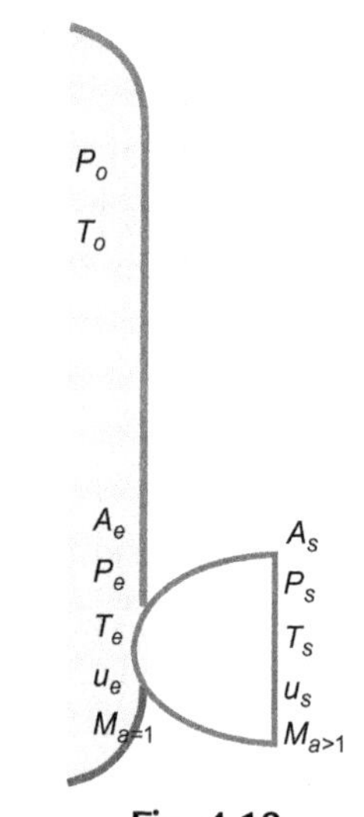

Fig. 4.19
Choked jet—Simplified scheme.

where $P_a = P_s$ is the atmospheric pressure. The condition at the shock plane is well represented by an adiabatic expansion. Temperature and density can be calculated:

$$T_s = T_e \cdot \left(\frac{P_e}{P_a}\right)^{\frac{1-\gamma}{\gamma}} \tag{4.88}$$

$$\rho_s = \rho_e \cdot \left(\frac{P_s}{P_e}\right)^{\frac{1}{\gamma}} \tag{4.89}$$

Supersonic velocity of the shock plane is recommended to be calculated through the application of the equation of conservation of energy, which is less affected by the dimensional uncertainty of the diameter size:

$$h_e + \frac{u_e^2}{2} = h_s + \frac{u_s^2}{2} \tag{4.90}$$

$$u_s = \left[2 \cdot \left(h_e + \frac{u_e{}^2}{2} - h_s\right)\right]^{0.5} \tag{4.91}$$

Example 4.9

Air at 20°C and 10 atm absolute (1,013,000 Pa) is released from a vessel through a 20 mm hole. Analyse the jet flow until the beginning of the transition (entrainment) zone.

Parameter	Vessel	Outlet	Shock Plane
P (Pa)	1,013,000	Unknown	1,013,000
T (K)	293.16	Unknown	Unknown
ρ (kg/m^3)	12.10	Unknown	Unknown
D (m)	—	0.02	Unknown
u (m/s)	—	Unknown	Unknown
γ	1.4	1.4	1.4
R (m^3 Pa K^{-1} kmol^{-1})	8314.472		

Solution

Exit Plane

Flow is choked, so the following equations apply:

$$P_e = 1{,}013{,}000 \cdot \left(\frac{2}{2.4}\right)^{\frac{1.4}{0.4}} = 535{,}149.5\,\text{Pa} = 5.28\,\text{atm} \tag{4.92}$$

$$T_e = 293.16 \cdot \left(\frac{2}{2.4}\right) = 244\,\text{K} \tag{4.93}$$

$$\rho_e = 12.1 \cdot \left(\frac{2}{2.4}\right)^{\frac{1.4}{0.4}} = 6.4\,\text{kg/m}^3 \tag{4.94}$$

$$u_e = a_e = \sqrt{\left(\frac{2 \cdot 1.4}{1.4+1} \cdot \frac{8.314}{0.029} \cdot 293.16\right)} = 313\,\text{m/s} \tag{4.95}$$

Mass flow rate

$$W_c = 6.4 \cdot \pi \cdot 0.01^2 \cdot 313 = 0.63\,\text{kg/s} \tag{4.96}$$

Shock Plane

$$D_s = 0.02 \cdot \left(\frac{10}{1}\right)^{\frac{0.4}{2.8}} = 0.027\,\text{m} \tag{4.97}$$

$$T_s = 244 \cdot \left(\frac{535,149}{101,300}\right)^{-\frac{0.4}{1.4}} = 151.7\,\mathrm{K} \quad (4.98)$$

$$\rho_s = 6.4 \cdot \left(\frac{101,300}{535,149}\right)^{\frac{1}{1.4}} = 1.95\,\mathrm{kg/m^3} \quad (4.99)$$

$$u_s = \left[2 \cdot \left(242,000 + \frac{313^2}{2} - 151,000\right)\right]^{0.5} = 529\,\mathrm{m/s} \quad (4.100)$$

The jet will enter the transition zone, where air will start being entrained at 151.7 K with a velocity of 529 m/s.

The Mach disk is located at:

$$x_m = 0.67 \cdot 0.02 \cdot \left(\frac{1,013,000}{535,149}\right) = 0.025\,\mathrm{m} \quad (4.101)$$

from the outlet. It is evident that the entrainment can be assumed as started at the outlet itself, given the very small distance from Mach disk, but the full calculation provides the very important information of the jet temperature prior to air entrainment. Thermal effects can be particularly important for two phase flows, or for substances such as carbon dioxide, which can show solid formation and snow-out.

Transition zone

The transition zone typically has the following characteristics (Fig. 4.20):

- the turbulent intensity is not completely developed, to a variable extent, also depending on the Reynolds number.
- the velocity profile follows a decay law approximately proportional to $x^{-0.5}$ according to Yue (1999).
- entrainment takes place, as stated by Boguslawski and Popiel (1979) and Hill (1972), with a lower coefficient with respect to the following fully developed jet.

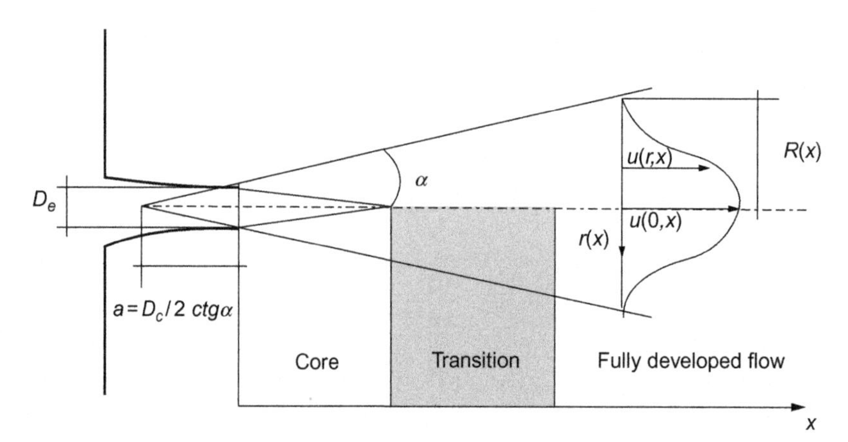

Fig. 4.20
Transition zone of jet flow.

- The transition zone begins at two to six diameters from the outlet (Goodfellow and Tähti, 2001) and terminates at a distance that depends on the Reynolds number. For a momentum driven jet with moderate velocity, this distance extends to 8–10 diameters (Goodfellow and Tähti, 2001) and for high velocity jets, it may be assumed to be at $x > 25 \div 30\ D_e$ according to Bogey and Bailly (2006).

Fully developed (self-similarity) turbulent zone

Beyond the transition zone, the jet forms a conical shape volume with a virtual origin placed at a distance from the opening which, for simplicity's sake, has been assumed equal to zero in the next calculations, with an axial similar velocity profile following a Gauss error function (Fig. 4.21).

The jet entrains air downstream of the Mach disk, so that the gas density at the boundary approaches the air density ρ_a not far from the nozzle exit. Thring and Newby (1953) introduced the concept of equivalent nozzle, which has the same momentum and velocity as the operating nozzle, W_e and u_e, but the density of the entrained fluid, typically air. The application of the equivalence criterion gives:

$$D_{eq} = D_e \cdot \sqrt{\frac{\rho_e}{\rho_a}} \tag{4.102}$$

where ρ_e is the density of the fluid at the outlet, and ρ_a is the air density at ambient conditions. The adoption of the equivalent diameter equivalence allows one to consider the density as constant. The fully developed jet ($x > 30D$) is properly described by the Gaussian relationship (self-similarity zone):

$$u(r,x) = u(0,x) \cdot \exp\left(-\frac{r^2}{2 \cdot \sigma^2}\right) \tag{4.103}$$

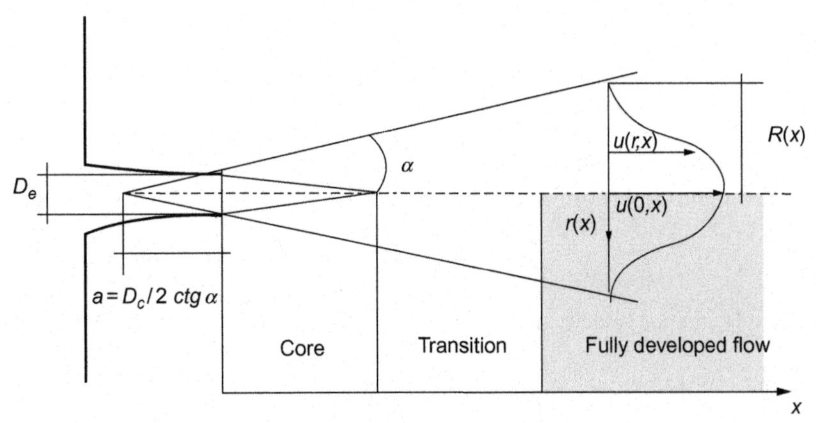

Fig. 4.21
Self similarity zone of jet flow.

Pope (2000) found the following general relations for self-similarity zone:

$$u(0,x) = u_e \cdot B \cdot \frac{D_{eq}}{x} \tag{4.104}$$

where B falls in a range around 6.

$$u(0,x) = 6 \cdot u_e \cdot \frac{D_{eq}}{x} \tag{4.105}$$

The same author found that the radial distance at which jet velocity is half of centreline velocity is given by the formula:

$$R_{1/2}(x) = S \cdot x \tag{4.106}$$

where S is very close to 0.1.

$$R_{1/2}(x) = 0.1 \cdot x \tag{4.107}$$

On the basis of these evidences we can write:

$$u(R_{1/2}, x) = u(0,x) \cdot \exp\left(-\frac{{R_{1/2}}^2}{2 \cdot \sigma^2}\right) = \frac{u(0,x)}{2} \tag{4.108}$$

$$\exp\left(-\frac{{R_{1/2}}^2}{2 \cdot \sigma^2}\right) = \frac{1}{2} \tag{4.109}$$

$$\exp\left(-\frac{0.01 \cdot x^2}{2 \cdot \sigma^2}\right) = \frac{1}{2} \tag{4.110}$$

$$\left(-\frac{0.01 \cdot x^2}{2 \cdot \sigma^2}\right) = \ln\frac{1}{2} = -0.693 \tag{4.111}$$

$$\sigma = \sqrt{\left(\frac{0.01 \cdot}{2 \cdot 0.693}\right)} \cdot x = 0.085 \cdot x \tag{4.112}$$

which is very close to the value provided by Shepelev (1961),and:

$$u(r,x) = u(0,x) \cdot \exp\left[-\frac{1}{2} \cdot \left(-\frac{r}{0.082 \cdot x}\right)^2\right] \tag{4.113}$$

The typical range of the jet half-angle α is 9.4°–11.8°. The latter value is generally accepted and is independent of the Reynolds number being tan 11.8° ≈ 0.2, so:

$$R(x) = \frac{x}{5} \tag{4.114}$$

Knowing the angle, the jet velocity averaged over the cross section can be calculated.

$$u_{av}(x) = \frac{\int_0^{+\infty} u(0,x) \cdot \exp\left[-\frac{1}{2} \cdot \left(-\frac{r}{0.082 \cdot x}\right)^2\right] \cdot 2 \cdot \pi \cdot r \cdot dr}{\pi \cdot R(x)^2} \tag{4.115}$$

Table 4.6 Jet velocity formulas

Velocity	Formula	Reference
Centreline, $u(0,x)$	$6\cdot\frac{D_{eq}}{x}\cdot u_e$	Pope (2000) Benintendi (2010)
At any (r, x) point of the jet	$6\cdot\frac{D_{eq}}{x}\cdot u_e\cdot \exp\left[-\frac{1}{2}\left(\frac{r}{0.082\cdot x}\right)^2\right]$	Calculated on the basis of Pope's data (2000)
Average velocity, $u_{av}(x)$	$0.36\cdot u(0,x)=2.16\cdot\frac{D_{eq}}{x}\cdot u_e$	Calculated
Mass flow, $W(x)$	$\pi\cdot\rho_a\cdot 2.16\cdot\frac{D_{eq}}{25}\cdot x$	Calculated
Fully developed jet momentum	$u_{av}(x)\cdot\pi\cdot\rho_a\cdot R(x)^2=2.16^2\cdot\pi\cdot\rho_a\cdot\frac{D_{eq}^2}{25}\cdot u_e$	Calculated

$$u_{av}(x)=\frac{u(0,x)\cdot 2\cdot\pi\cdot 0.085^2\cdot x^2\cdot \exp\left[\frac{1}{2}\cdot\left(-\frac{r}{0.082\cdot x}\right)^2\right]_{+\infty}^{0}}{\pi\cdot\frac{x^2}{25}}=50\cdot 0.085^2\cdot u(0,x)$$
$$=0.36\cdot u(0,x) \tag{4.116}$$

Table 4.6 includes the calculation formula for remarkable values of jet velocity and mass flow rate.

The accepted entrainment formula was developed by Ricou and Spalding in their original work (1961):

$$\frac{W(x)}{W_e}=0.32\cdot\frac{x}{D_{eq}} \tag{4.117}$$

where W_e is the mass flow rate at the outlet. The same formula is recommended by the standard API 521 (2014). Reciprocal of previous formula is the mass fraction of the substance averaged through the jet cross-section:

$$y_{mass}=3.1\cdot\frac{D_{eq}}{x} \tag{4.118}$$

The distance at which a given average molar fraction y is attained has been developed by Benintendi (2010), on the basis of the Ricou and Spalding equation:

$$x_y=3.1\cdot\left[\left(\frac{1}{y\cdot MW}-\frac{1}{MW}\right)\cdot MW_a+1\right]\cdot D_{eq} \tag{4.119}$$

where MW and MW_a are the molecular weights of the substance and of air, respectively.

Chen and Rodi (1980) have proposed this equation for the mass fraction on the jet axis (centreline):

$$y(0,r)=5.4\cdot\frac{D_{eq}}{x} \tag{4.120}$$

Accordingly, the molar fraction at any point of the jet can be assumed as:

$$y(x,r)=5.4\cdot\frac{D_{eq}}{x}\cdot \exp\left[-\frac{1}{2}\cdot\left(-\frac{r}{0.085\cdot x}\right)^2\right] \tag{4.121}$$

Table 4.7 Jet concentration formulas

Molar fraction	Formula	Reference
Centreline, $u(0,x)$	$5.4 \cdot \frac{D_{eq}}{x}$	Chen and Rodi (1980)
At any (r, x) point of the jet	$5.4 \cdot \frac{D_{eq}}{x} \cdot u_e \cdot \exp\left[-\frac{1}{2}\left(\frac{r}{0.085 \cdot x}\right)^2\right]$	Calculated on the basis of Chen and Rodi (1980) and Pope's data (2000)
Average molar fraction, $y_{av}(x)$	$y_{av}(x)^* = \frac{1}{\frac{MW}{MW_a} \cdot \left(0.32 \cdot \frac{x}{D_{eq}} - 1\right) + 1}$	Calculated

*Valid for $x \geq D_{eq}/0.32$.

Table 4.7 includes the calculation formula for remarkable values of the jet concentrations:

According to CCPS (1999), the equation of the entrainment should be adopted for distances from outlet to diameter ratios ranging from 7 to 100 only.

Example 4.10

Resolve Example 4.9, assuming that the jet is composed by hydrogen sulphide, and find (Fig. 4.22):

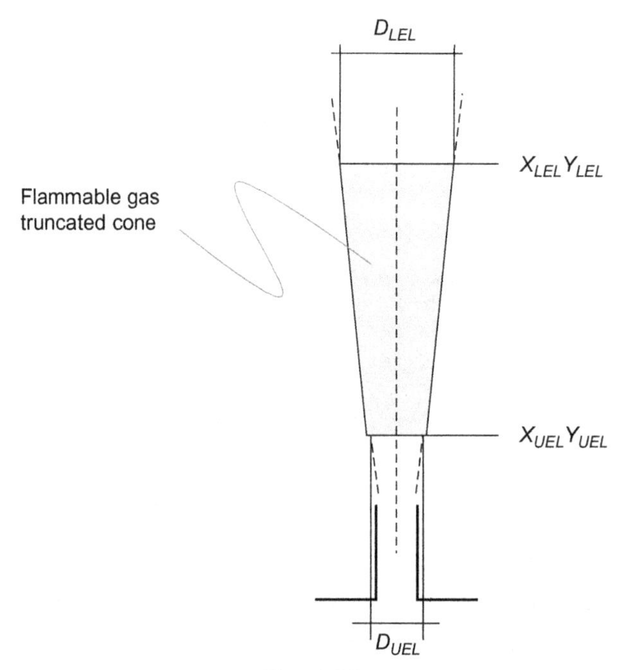

Fig. 4.22
Truncated conical model for turbulent jet.

1. The distance at which the average concentration is equal to the upper explosion limit (UEL) and the related cross section diameter
2. The distance at which the average concentration is equal to the lower explosion limit (LEL) and the related cross section diameter
3. The distance at which the average concentration is equal to 1000 ppm
4. The behaviour of the average velocity along the axis

Parameter	Vessel	Outlet	Far Field
P (Pa)	1,013,000	Unknown	
T (K)	293.16	Unknown	
ρ (kg/m^3)	14.1	Unknown	
D (m)	–	0.02	
u (m/s)	–	Unknown	
γ	1.4	1.4	1.4
UEL (molar fraction)			0.44
LEL (molar fraction)			0.04
R (m^3 Pa K^{-1} kmol^{-1})	8314.472		

Solution

Exit Plane

Flow is choked, so the following equations apply:

$$P_e = 1,013,000 \cdot \left(\frac{2}{2.3}\right)^{\frac{1.3}{0.3}} = 552,822.2\,\text{Pa} \tag{4.122}$$

$$T_e = 293.16 \cdot \left(\frac{2}{2.3}\right) = 255\,\text{K} \tag{4.123}$$

$$\rho_e = 14.1 \cdot \left(\frac{2}{2.3}\right)^{\frac{1.3}{0.3}} = 7.69\,\text{kg/m}^3 \tag{4.124}$$

$$u_e = a_e = \sqrt{\left(\frac{2 \cdot 1.3}{2.3} \cdot \frac{8.314}{0.034} \cdot 293.16\right)} = 285/\text{s} \tag{4.125}$$

Mass flow rate

$$W_c = 7.69 \cdot \pi \cdot 0.01^2 \cdot 285 = 0.69\,\text{kg/s} \tag{4.126}$$

Flammable gas truncated cone

$$x_{UEL} = 3.1 \cdot \left[\left(\frac{1}{0.44 \cdot 34} - \frac{1}{34}\right) \cdot 29 + 1\right] \cdot \sqrt{\frac{7.69}{1.2}} \cdot 0.02 = 0.33\,\text{m} \tag{4.127}$$

$$x_{LEL} = 3.1 \cdot \left[\left(\frac{1}{0.04 \cdot 34} - \frac{1}{34}\right) \cdot 29 + 1\right] \cdot \sqrt{\frac{7.69}{1.2}} \cdot 0.02 = 3.37\,\text{m} \tag{4.128}$$

Endpoint distance (1000 ppm)

$$x_{1000} = 3.1 \cdot \left[\left(\frac{1}{0.001 \cdot 34} - \frac{1}{34}\right) \cdot 29 + 1\right] \cdot \sqrt{\frac{7.69}{1.2}} \cdot 0.02 = 133.9\,\text{m} \tag{4.129}$$

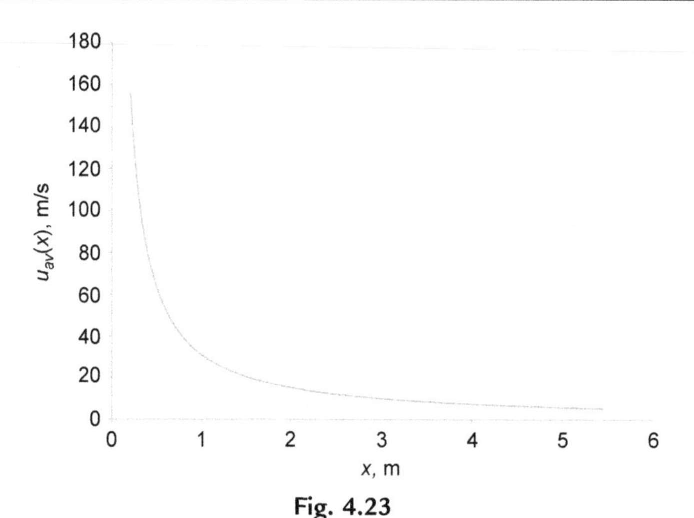

Fig. 4.23
Decay of jet velocity with distance from 10 diameters plane from the outlet.

The following diagram shows the behaviour of the average velocity with time from a distance from the outlet equal to around ten diameters. We can assume that at a distance from the outlet where velocity is 10 m/s wind velocity may potentially have an effect on the jet, i.e. momentum can no longer be considered as predominant. Consequently, the endpoint concentration is affected by a significant uncertainty (Fig. 4.23).

4.6.4 Unsteady Gas Jet

It is beneficial to notice that at a distance equal to 100 diameters most of hydrocarbon mixtures leave the flammable range. It must be considered that the calculated scenario can be considered valid only for the first moments of the release or if the release is stopped soon. In fact, the more the jet lasts, the more the adjacent area around the jet accumulates the emitted gas, resulting in a progressive concentration rise. In this case, the steady state approach of the Ricou entrainment equation should be replaced by a transient approach, accounting for the concentration change into the entrained air.

A simple, non-validated method to predict the growth of a jet concentration truncated cone with time could be adopted according to the following assumption:

- the mass flow is constant.
- the concentration gradient rises only in the jet transversal plane, due to the lateral entrainment; accordingly, no concentration change is expected along the axis.
- the average concentration formula does not change.
- the angle of the steady state truncated cone, that is assumed to be consistent with the formula:

$$R(x) = \frac{x}{5} \tag{4.130}$$

has the same mathematical behaviour, i.e.:

$$R(x) = \frac{x}{k(t)} \tag{4.131}$$

where $k(t)$ is the varying angle cotangent.

The transient balance on the truncated cone can be written as (Figs. 4.24 and 4.25):

$$\frac{MW \cdot y_{av} \cdot V(t) \cdot 273.16}{22.414 \cdot T} = W_o \cdot t \tag{4.132}$$

where:

- $V(t)$ is the varying volume.
- 22,414 is the molar volume at normal conditions (273.16 K, 1 atm).

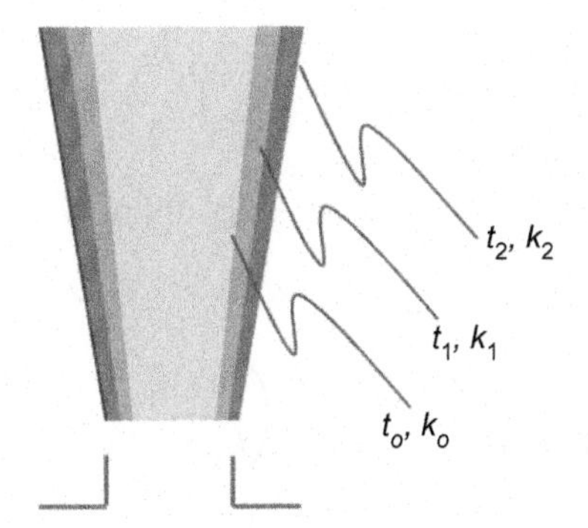

Fig. 4.24
Transient truncated cone jet flow.

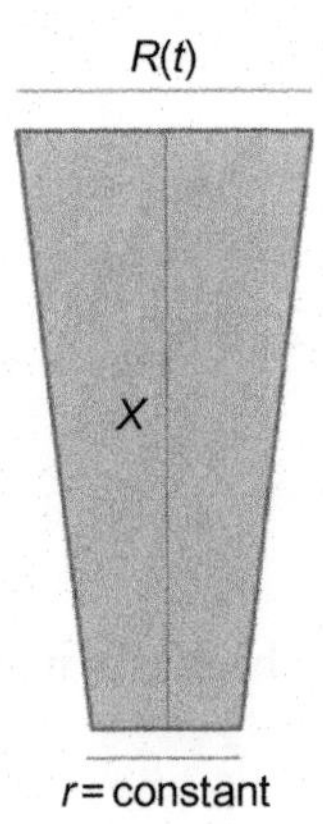

Fig. 4.25
Transient truncated cone jet flow.

- T is jet temperature, K.
- W_o is the mass flow.
- t is the release duration.

with:

$$V(t) = \frac{1}{3} \cdot \pi \cdot x \cdot \left[r^2 + R(t)^2 + r \cdot R(t) \right] \tag{4.133}$$

Being:

$$R(x) = \frac{x}{k(t)} \tag{4.134}$$

$$V(t) = \frac{1}{3} \cdot \pi \cdot x \cdot \left[r^2 + \left(\frac{x}{k(t)} \right)^2 + r \cdot \frac{x}{k(t)} \right] \tag{4.135}$$

In this equation x and r are constant, so $k(t)$ is the only unknown variable. The mass balance can be written as:

$$\frac{MW \cdot y_{av} \cdot 273.16 \cdot \pi \cdot x}{3 \cdot 22.414 \cdot T} \cdot \left[r^2 + \left(\frac{x}{k(t)} \right)^2 + r \cdot \frac{x}{k(t)} \right] = W_o \cdot t \tag{4.136}$$

and, being:

$$y_{av}(x) = \frac{1}{\frac{MW}{MW_a} \cdot \left(0.32 \cdot \frac{x}{D_{eq}} - 1 \right) + 1} \tag{4.137}$$

$$\left[\frac{1}{\frac{MW}{MW_a} \cdot \left(0.32 \cdot \frac{x}{D_{eq}} - 1 \right) + 1} \cdot \frac{MW \cdot 273.16 \cdot \pi \cdot x}{3 \cdot 22.414 \cdot T} \right] \cdot \left[r^2 + \left(\frac{x}{k(t)} \right)^2 + r \cdot \frac{x}{k(t)} \right] = W_o \cdot t \tag{4.138}$$

This is a second degree equation which can be easily resolved and the evolution of $k(t)$ with time predicted.

This method should be considered as a screening method to assess the influence of the jet on the surrounding atmosphere.

4.6.5 Transient Release of Turbulent Jets

In the previous approach, it has been assumed that the jet's driving force was related to a constant pressure. In real cases, pressure will decay. A general formula useful for taking into account the pressure and inventory decay is provided by Spouge (1999):

$$W(t) = W_o \cdot \exp\left(-\frac{W_o}{M_o} \cdot t\right) \tag{4.139}$$

where:

- $W(t)$ is the mass flow rate.
- W_o is the initial flow rate.
- M_o is the inventory.

4.6.6 Effect of Wind on Turbulent Jet

The deflection of the jet by the wind crossflow results in a particular change of the jet cross sectional shape. Fig. 4.26 shows the kidney-like profile of the originally round shape of the jet. Internally, the shear stress distribution creates a recirculation of the streamlines.

The axis of the deflected jet, the locus where concentrations and velocities are maximum, can be identified according to Patrick (1967) (Fig. 4.27):

$$\frac{y}{D_e} = \alpha^{0.85} \cdot \left(\frac{x}{D_e}\right)^n \tag{4.140}$$

where:

- α is the jet to wind velocity ratio falling in the range from 6 to 50.
- exponent n is 0.38 from velocity and 0.34 from concentration measurements.

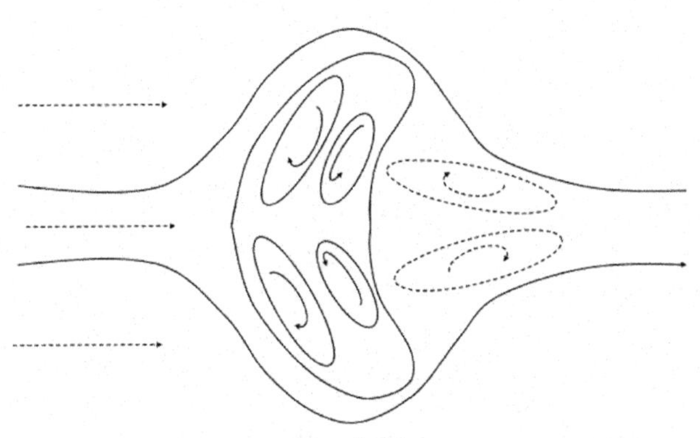

Fig. 4.26
Jet kidney profile.

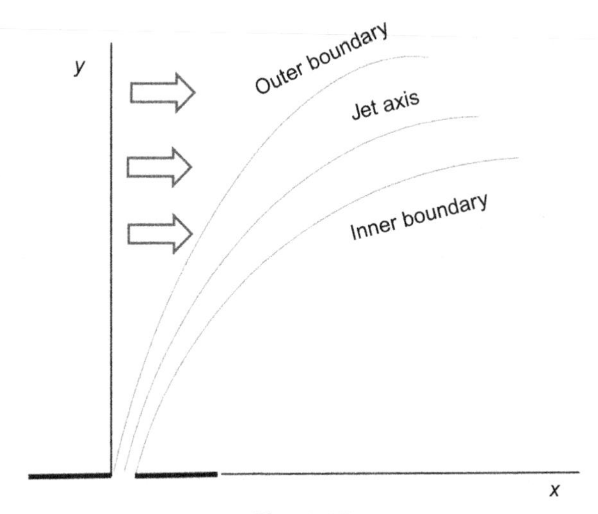

Fig. 4.27
Axis and boundaries of deflected jet.

Example 4.11

For Example 4.10, find the deflection at 2.5 m from the outlet if wind velocity is 7 m/s (Fig. 4.28).

Solution

y	D_e	α	n
M	m	–	–
2.5	0.02	44.7	0.36

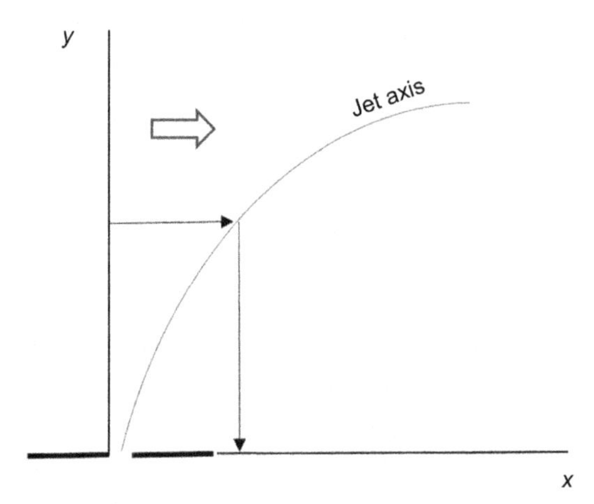

Fig. 4.28
Axis of deflected jet.

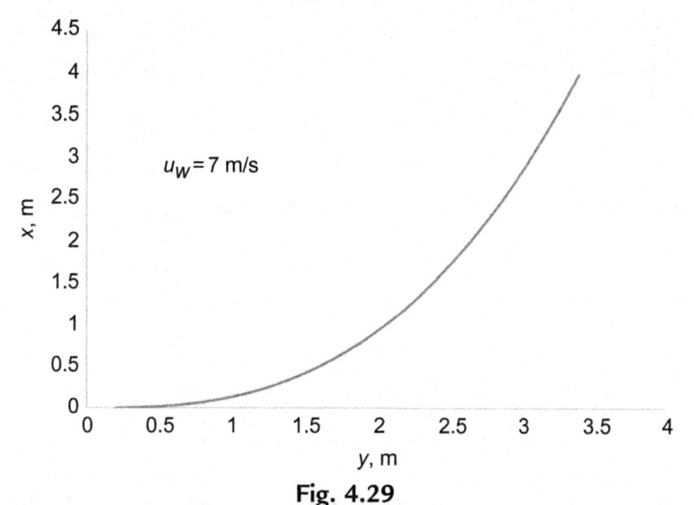

Fig. 4.29
Penetration distance for Example 4.11.

$$x = D_e \cdot \left(\frac{y}{D_e \cdot \alpha^{0.85}}\right)^{\frac{1}{n}} \tag{4.141}$$

Solving:

$$x = 1.7\text{m}$$

In the next diagram, the penetration distance has been plotted (Fig. 4.29).

4.6.7 Non Isothermal Jets

Temperature distribution in non-isothermal jets follows the Gaussian behaviour of velocity and concentration fields. According to Shepelev (1961), the following practical formula can be used (Fig. 4.30):

$$T(r,x) = T_a + [T(0,x) - T_a] \cdot \exp\left(-\frac{1}{4} \cdot \frac{r}{0.082 \cdot x}\right)^2 \tag{4.142}$$

Solution of Eq. (4.142) requires the knowledge of centreline temperature decay $T(0,x)$. This can be estimated according to the formula provided by (Goodfellow and Tähti, 2001):

$$T(0,x) = T_a + (T_o - T_a) \cdot K_2 \cdot \frac{\sqrt{A_o}}{x} \tag{4.143}$$

In this formula, T_o is the initial centreline temperature, A_o is the hole area size, and K_2 can be approximately assumed equal to 6 for horizontal jets (Li et al., 1993).

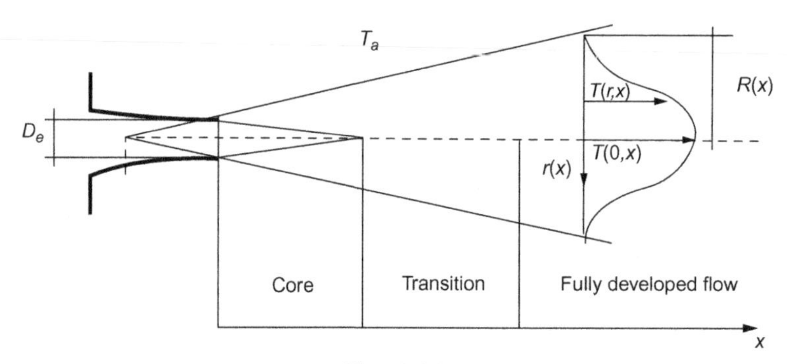

Fig. 4.30
Self-similarity temperature profile.

Example 4.12

Find the distance from the outlet at which centreline temperature of air jet of Example 4.9 will be 20°C. Assume an ambient temperature of 30°C.

Solution

Ambient Temperature (K)	Outlet Plane Fluid Temperature (K)	*T*(0,*x*) (K)
303.16	244	293.16

$$x=\frac{(244-303.16)}{(293.16-303.16)}\cdot 6\cdot 0.02\sqrt{\frac{\pi}{4}}=0\cdot 62\,\mathrm{m} \tag{4.144}$$

4.6.8 Unchoked (Subsonic) Flow

Exit subsonic flow condition, as reported by Hess et al. (1973), consists of the following properties:

$$\frac{P_e}{P_v}<\left(\frac{\gamma+1}{2}\right)^{\frac{\gamma}{\gamma+1}} \tag{4.145}$$

$$P_e=P_a \tag{4.146}$$

$$\frac{T_e}{T_v}=\left(\frac{P_a}{P_v}\right)^{\frac{\gamma-1}{\gamma}} \tag{4.147}$$

$$\frac{\rho_e}{\rho_v}=\left(\frac{P_a}{P_v}\right)^{\frac{1}{\gamma}} \tag{4.148}$$

where a denotes atmospheric conditions.

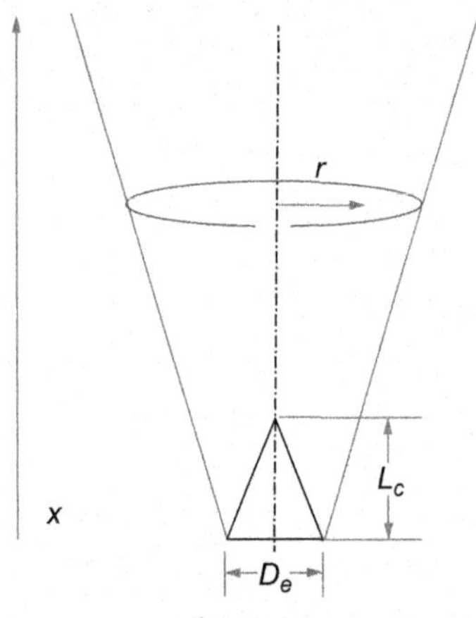

Fig. 4.31
Subsonic flow.

Immediately downstream of the outlet, a specific distance is needed for the flow to become fully laminar. This distance is called *potential core*. The extension of the potential core in the laminar regime, L_C, is related to Reynolds number, R_E, according to the following equation of Boussinesq, suggested by Akaike and Nemoto (1988) (Fig. 4.31):

$$L_C = 0.03 \cdot R_E \cdot D_e \tag{4.149}$$

Benintendi (2011), on the basis of Schlichting's approach (1968), carried out a gas laminar jet analysis classifying the substances into three different categories, in terms of their transport properties with respect to air's properties:

- Equivalent gases, such as ethane, ethylene, acetylene
- Light gases such as hydrogen and ammonia
- Heavy gas such as propane

4.6.8.1 Equivalent Gases

For equivalent gases the following remarkable results have been obtained:

$$\frac{u_x}{u_e}(x,\eta) = \frac{3 \cdot Q_e \cdot}{8 \cdot \pi \cdot \nu \cdot x} \cdot \left(\frac{1}{1+\eta^2}\right)^2 \tag{4.150}$$

$$y(x,\eta) = \frac{3 \cdot Q_e \cdot}{8 \cdot \pi \cdot D_d \cdot x} \cdot \left(\frac{1}{1+\eta^2}\right)^2 \tag{4.151}$$

$$x_{CL} = \frac{3 \cdot u_e \cdot R_e^2}{8 \cdot D_d \cdot y_{CL}} \tag{4.152}$$

$$\eta = \sqrt{\frac{3 \cdot \rho_e \cdot M_e}{64 \cdot \pi}} \cdot \frac{1}{\mu} \cdot \frac{r}{x} \tag{4.153}$$

here:

- y is the molar fraction.
- M_e is the initial momentum:

$$M_e = \rho_e \cdot u_e^2 \cdot \pi \cdot R_e^2 \tag{4.154}$$

- ν is the kinematic viscosity μ/ρ.
- D_d is the substance diffusivity.
- R_e is the outlet radius.
- x is the jet axis.

The contour of a specific molar fraction y_s marking a specific volume is given by the solutions of the equation:

$$y_s = \frac{3 \cdot \nu_e \cdot R_e^2}{8 \cdot D_d \cdot x} \cdot \left(\frac{1}{1 + \eta_s^2}\right)^2 \tag{4.155}$$

If the substance in the jet is flammable, the contour in the plane of the volume defined by the points where the concentration coincides with the LEL is given by the following equation:

$$Y_{LEL} = \frac{3 \cdot \nu_e \cdot R_e^2}{8 \cdot D_d \cdot x} \cdot \left(\frac{1}{1 + \eta_{LEL}{}^2}\right)^2 \tag{4.156}$$

and the corresponding radial flammable width $R_{LEL}(x)$:

$$R_{LEL}(x) = x \cdot \sqrt{\sqrt{\frac{3 \cdot \nu_e \cdot R_e^2}{8 \cdot y_{LEL} \cdot D_d \cdot x}} - 1} \cdot \sqrt{\frac{64 \cdot \pi}{3 \cdot \rho_e \cdot M_e}} \cdot \mu \tag{4.157}$$

And for the UEL:

$$Y_{UEL} = \frac{3 \cdot \nu_e \cdot R_e^2}{8 \cdot D \cdot x} \cdot \left(\frac{1}{1 + \eta_{UEL}^2}\right)^2 \tag{4.158}$$

$$R_{UEL}(x) = x \cdot \sqrt{\sqrt{\frac{3 \cdot \nu_e \cdot R_e^2}{8 \cdot y_{UEL} \cdot D \cdot x}} - 1} \cdot \sqrt{\frac{64 \cdot \pi}{3 \cdot \rho_e \cdot M_e}} \cdot \mu \tag{4.159}$$

For the stoichiometric molar fraction:

$$Y_{ST} = \frac{3 \cdot \nu_e \cdot R_e^2}{8 \cdot D_d \cdot x} \cdot \left(\frac{1}{1 + \eta_{ST}^2}\right)^2 \tag{4.160}$$

$$R_{ST}(x) = x \cdot \sqrt{\sqrt{\frac{3 \cdot v_e \cdot R_e^2}{8 \cdot y_{ST} \cdot D_d \cdot x}} - 1 \cdot \sqrt{\frac{64 \cdot \pi}{3 \cdot \rho_e \cdot M_e}} \cdot \mu} \tag{4.161}$$

4.6.8.2 Light Gases

For light gases, Benintendi (2011) has provided the following values:

Centreline mass fraction decay:

$$y(0, x) = \frac{R_E \cdot R_e}{8 \cdot x} \tag{4.162}$$

where R_E is the Reynolds number at the outlet.

Maximum distance from the outlet where a specific molar concentration is attained, for example the Lower Explosion Limit:

$$x_{\max} = \frac{\rho_e \cdot v_e \cdot R_e^2}{8 \cdot \mu \cdot y_{LEL}} \tag{4.163}$$

Radial Flammable Width

$$R_{LEL}(x) = R_e \cdot \sqrt{\left[\sqrt{\left(\frac{\frac{\varepsilon}{y_{LEL}} + 1}{1 + 8 \cdot \varepsilon \cdot \frac{x}{R_E \cdot R_e}} \right)^{\frac{1}{\varepsilon \cdot S_{ch}}}} - 1 \right] \cdot \frac{8}{3} \cdot \frac{x}{R_E \cdot R_e}} \tag{4.164}$$

where:

- ε is the light gas to air density ratio.
- S_{ch} is the Schmidt number:

$$S_{ch} = \frac{\mu_e}{\rho_e \cdot D_d} \tag{4.165}$$

4.6.8.3 Heavy Gases

For heavy gases, Benintendi (2011) has provided the maximum distance to a specified molar fraction y_s:

$$L_s = \frac{3 \cdot C_s \cdot R_E \cdot S_{ch} \cdot}{32 \cdot y_s} \tag{4.166}$$

where C_f is an empirical coefficient ≈ 1, and the other terms are all known, and the maximum radius for a given molar fraction y_s:

$$R_{S_{MAX}} = \frac{9 \cdot R_e}{16 \cdot y_s} \tag{4.167}$$

4.6.9 Turbulent to Laminar Jet Comparison

Entrainment for laminar jets is much less effective than for turbulent jets with comparable release characteristics. Table 4.8 shows, for a specific case, the different length of laminar and turbulent jets for different hydrocarbons (Benintendi, 2011).

4.7 Buoyancy

Buoyancy is the behaviour of a gas/vapour, depending on its density (true or apparent) to air density ratio. *Apparent* density refers to gases or vapours whose molecular weight is smaller than air molecular weight, but whose temperature is much lower, behaving as a heavier than air fluid, for example, LNG and refrigerated ammonia.

Gases and vapours are classified as:

- *Positively buoyant gas*: true or apparent density is smaller than air density.
- *Neutrally buoyant gas*: true or apparent density is comparable to air density.
- *Negatively buoyant gas*: Gas true or apparent density is greater than air density.

The densimetric Froude number is the parameter used to establish whether buoyancy is important and how much it affects the jet/plume behaviour:

$$Fr = \frac{u}{\sqrt{g \cdot D_e \cdot \frac{|\rho_e - \rho_a|}{\rho_a}}} \tag{4.168}$$

where all the notations are known.

Table 4.8 Laminar vs turbulent jet maximum length

	Methane	Ethylene	Propane
Laminar jet maximum length (m)	1.67	2.48	2.8
Turbulent jet maximum length (m)	1.18	1.28	1.06

4.8 Flashing Liquids

Liquid or flashing jets can result in spray and droplet formation, or in falling jets to the ground that are not immediately dispersed and form a pool.

4.8.1 Jet Shattering by Flashing

According to Van den Bosch and Duijm (2005), liquid jet shattering by flashing occurs if the exit temperature exceeds a shatter temperature defined as:

$$C_\varepsilon = \frac{T_{sh} - T_b}{T_{sh}} \tag{4.169}$$

where T_b is the normal boiling temperature, and for C_ε an average value of 0.085 can be assumed.

Example 4.13

Iso-pentane is stored at 20°C. Evaluate whether or not a flow will result in jet shattering.

Solution

The normal boiling point of iso-pentane is 301.1 K

$$T_{sh} = \frac{T_b}{1 - C_\varepsilon} = \frac{301.1}{0.915} = 329\,\text{K} \tag{4.170}$$

Iso-pentane liquid jet will remain well formed.

Example 4.14

Calculate the shattering temperature of LNG, assuming normal boiling temperature of methane, −161.5°C.

Solution

$$T_{sh} = \frac{T_b}{1 - C_\varepsilon} = \frac{111.66}{0.915} = 122\,\text{K} \tag{4.171}$$

Exit plane

Transition from stagnation to the outlet is accompanied by a first flash of the pressurised liquid. This must not be confused with the fully developed flashing occurring outside the vessel.

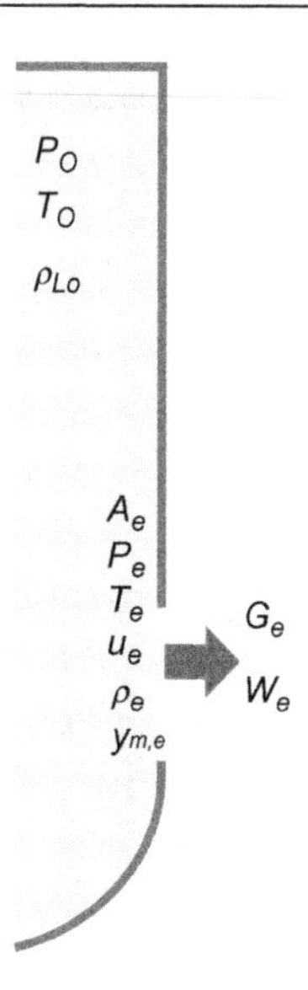

Flashing jets from saturated stagnation at the exit plane are properly modelled according to the ω (omega) method (Leung, 1994). This approach is useful to calculate the mass flow rate of the release.

The omega parameter is defined as:

$$\omega = \frac{\frac{\rho_{Lo}}{\rho_e} - 1}{\frac{P_o}{P_e} - 1} \tag{4.172}$$

where subscript o refers to stagnation condition, e refers to the flash result expected around the outlet, and L refers to the liquid phase. For pressurised saturated liquids, a flash pressure at 90% of the stagnation value generally provides a reasonable correlation parameter (API 520, 2014). The saturated omega parameter is calculated:

$$\omega_s = 9 \cdot \left(\frac{\rho_{Lo}}{\rho_{L9}} - 1\right) \tag{4.173}$$

where:

- ρ_{Lo} is the density at saturation (stagnation) condition.
- ρ_{L9} is the specific volume evaluated at 90% of the saturation pressure, P_o.

Example 4.15

Omega method is illustrated considering release of propane, stored at 15°C, through a 50 mm hole.

Solution

The assumption of flashing corresponding to the 90% of the stagnation pressure would result in a pressure equal to 658,350 Pa.

Saturation Pressure (Pa)	Saturation Temperature (K)	Liquid Density (kg/m^3)	Vapour Density (kg/m^3)	Latent Heat of Vaporisation (J/kg)	Specific Heat (J/kg K)
731,500	288.16	507.5	15.80	352,466	2637
658,350	284.34	513.1	14.24	358,553	2600

Adopting the average specific heat, the vapour quality at the exit is calculated according to Eq. (2.160):

$$y_{m,e} = \frac{2618 \cdot (288.16 - 284.34)}{358{,}553} = 0.0278 \tag{4.174}$$

The flashing density at the outlet is:

$$\rho_e = \frac{1}{\left(\frac{1 - y_{m,e}}{\rho_{L,e}}\right) + \frac{y_{m,e}}{\rho_{v,e}}} = \frac{1}{\left(\frac{1 - 0.0278}{513.1}\right) + \frac{0.0278}{14.24}} = 260\,\text{kg/m}^3 \tag{4.175}$$

and:

$$\omega_s = 9 \cdot \left(\frac{\rho_{Lo}}{\rho_{L9}} - 1\right) = 9 \cdot \left(\frac{507.5}{260} - 1\right) = 8.55 \tag{4.176}$$

It must be determined if the flow is critical (choked) or subcritical. The critical pressure is defined as:

$$P_c = \eta_c \cdot P_o \tag{4.177}$$

where η_c is the critical pressure ratio, which can be calculated through Fig. 4.32 (Leung, 1994):

Entering with $w = 8.55 \rightarrow \eta_c \cong 0.83$.

$P_c \geq P_a$ (downstream pressure) $\rightarrow$ critical flow

$P_c < P_a \rightarrow$ subcritical flow

$$P_c = \eta_c \cdot P_o = 0.83 \cdot 731{,}500 = 607{,}145 > P_a \tag{4.178}$$

Flow is critical.

A flashing liquid can present a high degree of subcooling (high subcooling region) or a low subcooling region (close to saturation curve). The subcooling zone must be determined. Defining:

$$\eta_{st} = \frac{2 \cdot \omega_s}{1 + 2 \cdot \omega_s} \tag{4.179}$$

It is:

$P_o \geq \eta_{st} \cdot P_o \rightarrow 1 \geq \eta_{st} \rightarrow$ low subcooling flow (saturated)
$P_o < \eta_{st} \cdot P_o \rightarrow 1 < \eta_{st} \rightarrow$ high subcooling flow

$$\eta_{st} = \frac{2 \cdot \omega_s}{1 + 2 \cdot \omega_s} = \frac{2 \cdot 8.55}{1 + 2 \cdot 8.55} = 0.9447 \tag{4.180}$$

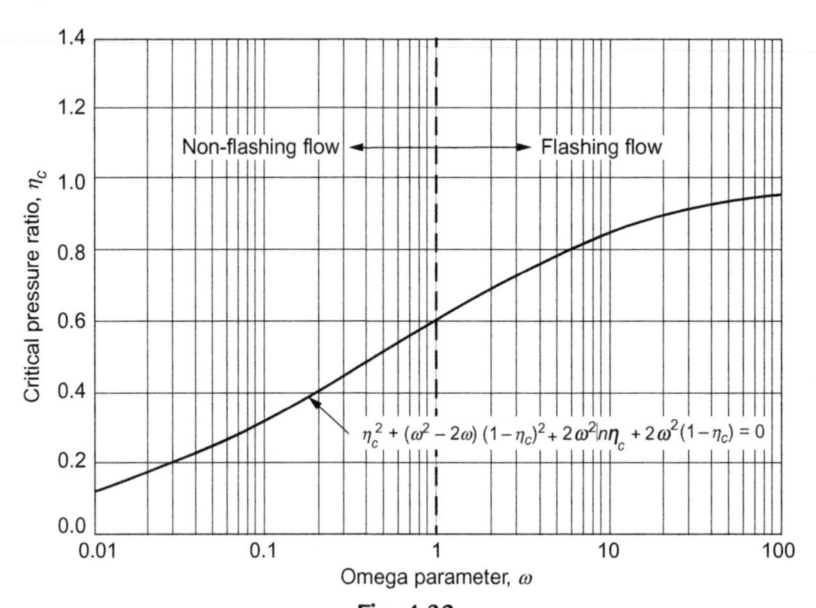

Fig. 4.32
Critical pressure ratio vs omega parameter.

For low subcooling region, the mass flux can be calculated by means of the equation (SI units):

$$G_e = \frac{\sqrt{\left\{2\cdot(1-\eta_s)+2\cdot\left[\omega_s\cdot\eta_s\cdot\ \ln\left(\frac{\eta_s}{\eta_c}\right)-(\omega_s-1)\cdot(\eta_s-\eta_c)\right]\right\}\cdot P_o\cdot\rho_{Lo}}}{\omega_s\cdot\left(\frac{\eta_s}{\eta_c}-1\right)+1} \tag{4.181}$$

$$G_e = \frac{\sqrt{\left\{2\cdot(1-1)+2\cdot\left[8.55\cdot1\cdot\ \ln\left(\frac{1}{0.83}\right)-(8.55-1)\cdot(1-0.83)\right]\right\}\cdot 731{,}500\cdot 507.5}}{8.55\cdot\left(\frac{1}{0.83}-1\right)+1}$$
$$= 5511\,\text{kg/s}\times\text{m}^2 \tag{4.182}$$

Including the calculated values:

$$W_e = G_e\cdot A_e = 5511\cdot\pi\cdot\frac{D_e^{\ 2}}{4} = 10.8\,\text{kg/s} \tag{4.183}$$

$$u_e = \frac{G_e}{\rho_{Le}} = \frac{5511}{260} = 21.2\,\text{m/s} \tag{4.184}$$

A highly subcooled liquid's mass flow rate can be calculated according to the Leung formula, provided liquid has not the time to reach two phase equilibrium. According to Crowl and Louvar (2011), this condition is met if the wall thickness is shorter than 10 cm:

$$W_e = \sqrt{2\cdot\rho_L\cdot(P_o-P_e)} \tag{4.185}$$

Plane of fully developed flash

The jet has the same temperature as the ambient one so flashing liquid will accommodate to the ambient pressure and temperature by vaporising a liquid fraction at the expenses of the remaining part.

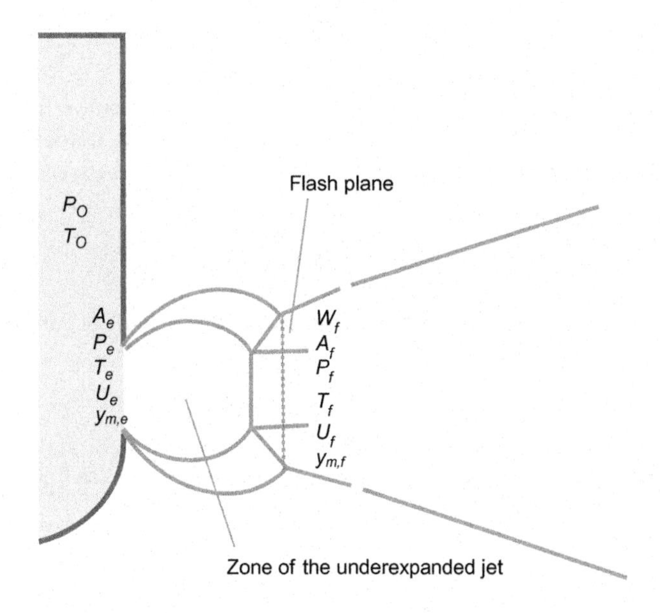

This is expressed by the thermal balance:

$$y_{m,f} = -\frac{c_{P_L} \cdot (T_f - T_e)}{\lambda(T_f)} \tag{4.186}$$

where:

- T_f is the flash liquid vapour equilibrium temperature corresponding to the liquid normal boiling temperature.
- $Y_{m,f}$ is the vapour mass fraction.
- $\lambda(T_f)$ is the latent heat of vaporisation at T_f.

Example 4.16

Find the flash temperature, the vapour fraction, and the mixture density for the propane flashing jet discussed in the previous example, neglecting the small vapour fraction at the exit plane.

Solution

T_f (K)	T_e (K)	Latent Heat of Vaporisation $\lambda(T_f)$ (J/kg)	Specific Heat at T_f (J/kg K)	Specific Heat at T_e (J/kg K)	Vapour Density at T_f (kg/m^3)	Liquid Density at T_f (kg/m^3)
231.16	284.34	425,590	2258	2600	2.5	581

$$y_{m,f} = -\frac{\left(\frac{2258+2600}{2}\right)\cdot(231.16-284.34)}{425,590} = 0.3 \tag{4.187}$$

$$\rho_f = \frac{1}{\left(\frac{0.7}{581}\right)+\frac{0.3}{2.7}} = \frac{1}{0.0012+0.1111} = 8.9\,\text{kg/m}^3 \tag{4.188}$$

Flashing jet at the flash plane has not entrained air yet, and vaporisation has occurred adiabatically at the expenses of liquid fraction, so its velocity can be immediately calculated applying the energy conservation equation, assuming no slip effect between the phases:

$$h_e + \frac{u_e^2}{2} = h_f + \frac{u_f^2}{2} \tag{4.189}$$

where enthalpy terms and jet velocity at the outlet plane are known. Accordingly:

$$u_f = \sqrt{2\cdot\left(h_e + \frac{u_e^2}{2} - h_f\right)} \tag{4.190}$$

Then, the area of the flash section can be calculated knowing the mass flow rate through the vessel hole, applying the mass conservation equation:

$$A_e = \frac{W_e}{\rho_f \cdot u_f} \tag{4.191}$$

Example 4.17

For the propane flashing jet, find the jet velocity at the flash plane and at the section area (Fig. 4.33).

Solution

T_e (K)	Vapour Quality at T_e (kg/kg)	T_f (K)	Vapour Quality at T_f (kg/kg)	Enthalpy of Vapour at T_e (J/kg)	Enthalpy of Liquid at T_e (J/kg)
284.34	0.0278	231.16	0.3	286,000	−75,000

Enthalpy of Vapour at T_f (J/kg)	Enthalpy of Liquid at T_f (J/kg)	Velocity at Exit (m/s)	Mass Flow Rate (kg/s)	Vapour Density at T_f (kg/m³)	Liquid Density at T_f (kg/m³)
226,000	−196,000	21.2	−10.8	2.5	581

$$h_e = 0.0278\cdot 286,000 - 0.9722\cdot 75,000 = -64,964\,\text{J/kg} \tag{4.192}$$

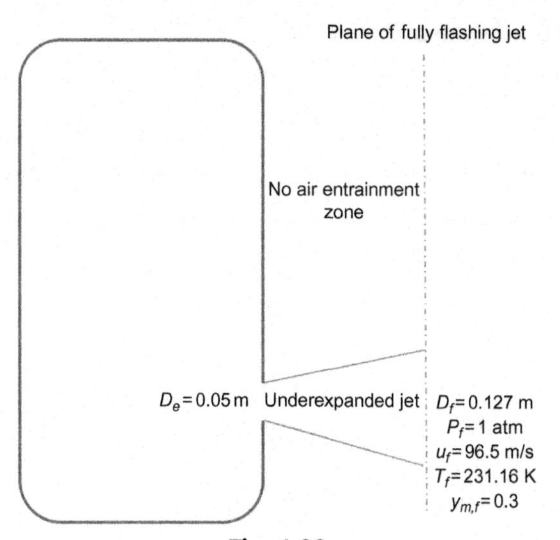

Fig. 4.33
Flashing jet of Example 4.17.

$$h_f = 0.3 \cdot 226{,}000 - 0.7 \cdot 196{,}000 = -69{,}400\,\mathrm{J/kg} \tag{4.193}$$

$$u_f = \sqrt{2 \cdot \left(-64{,}964 + \frac{21.2^2}{2} + 69{,}400\right)} = 96.5\,\mathrm{m/s} \tag{4.194}$$

$$A_f = \frac{10.8}{8.9 \cdot 96.5} = 0.0126\,\mathrm{m}^2 \tag{4.195}$$

$$D_f = 0.127\,\mathrm{m} \tag{4.196}$$

4.9 Spray Release and Droplet Dynamics

Aerosols are formed according to one of three mechanisms:

- capillary breakup
- aerodynamic breakup
- flashing breakup.

4.9.1 Capillary Breakup

Capillary breakup is related to liquids discharged through holes with diameters smaller than 2 mm (Woodward, 1998), and the mean droplet diameter is proportional to the orifice diameter (Lienhard and Day, 1970):

$$d_p = 1.9 \cdot D_e \tag{4.197}$$

4.9.2 Flashing and Aerodynamic Break Up

Flashing break up is related to the shattering action of flashing. A temperature threshold has been provided in Eq. (4.169). The droplet size after flashing is calculated using the critical Weber number for aerodynamic break up. The Weber number is defined as:

$$We_f = \frac{D_f \cdot u_f^2 \cdot \rho_{f,L}}{\sigma_{f,L}} \tag{4.198}$$

With the droplet Reynolds number:

$$Re_f = \frac{\rho_f \cdot D_f \cdot u_f}{\mu_{f,L}} \tag{4.199}$$

where all terms are known, $\rho_{f,L}$ is the density, $\mu_{f,L}$ is the dynamic viscosity of the liquid after flashing, and σ_f is the surface tension between the liquid and the vapour. According to van den Bosch, the droplet size can be estimated according to the following correlations (Van den Bosch and Duijm, 2005):

$$d_d = 1.89 \cdot D_f \cdot \sqrt{1 + \frac{\sqrt{We_f}}{Re_f}} \tag{4.200}$$

if $We_f < Re_f^{-0.45} < 10^6$ and $T_e < 1.11\ T_b$, where T_e is the exit temperature, and T_b is the normal boiling temperature. If not:

$$d_d = C_d \cdot \frac{\sigma_f}{u_f^2 \cdot \rho_a} \tag{4.201}$$

where C_d is a unitless constant that can be assumed as equal to 15, and ρ_a is air density. The initial drops formed in the jet may break further due to aerodynamic forces. A Weber critical number can be assumed to establish the breaking threshold. According to Van den Bosch and Duijm (2005), critical values between 10 and 20 are adopted.

Example 4.18

For the propane flashing jet, calculate the droplet diameter if air temperature is 20°C, and the jet is shattering due to the flash action.

Solution

T_a (K)	Air Density (kg/m^3)	u_f (m/s)	σ_f (m/s)
293.16	1.2	96.5	0.015418

In this case $T_e > 1.11\ T_b$.

$$d_d = C_d \cdot \frac{\sigma_f}{u_f^2 \cdot \rho_a} = 15 \cdot \frac{0.015418}{96.5^2 \cdot 1.2} = 0.02\,\text{mm} \tag{4.202}$$

Droplet rain-out velocity

The droplet rain-out velocity can be calculated applying Stoke's formula, which is strictly valid for $Re_d < 1$, even if results are not sensitive to this assumption

$$u_d = \frac{\rho_L \cdot g}{18 \cdot \nu_a \cdot \rho_a} \cdot d_d^{\,2} \tag{4.203}$$

where:

- ρ_L is the liquid density at boiling temperature after the flashing.
- ρ_a is the air density.
- g is the acceleration of gravity.
- ν_a is the air kinematic viscosity.

Example 4.19

For the propane flashing jet, calculate the free fall velocity if air temperature is 20°C.

Solution

T_f (K)	T_a (K)	Air Density (kg/m^3)	Air Kinematic Viscosity	Droplet Diameter (m)	Liquid Density at T_f (kg/m^3)
231.16	293.16	1.2	0.000015 m^2/s	0.00002	581

$$u_d = \frac{\rho_L \cdot g}{18 \cdot \nu_a \cdot \rho_a} \cdot d_d^2 = 7\,\text{mm/s} \tag{4.204}$$

4.9.3 Prediction of the Rain-Out Fraction

For screening purposes relating to continuous and discontinuous releases, the following rule of thumb equation authored by Kletz (1977), and referenced by Woodward (1998), can be used to estimate the rain out fraction η_r:

If $y_f < 0.5$

$$\eta_r = 1 - 2 \cdot y_f \tag{4.205}$$

If $y_f > 0.5$:

$$\eta_r = 0 \tag{4.206}$$

Example 4.20

For the propane flashing jet, determine if rainout takes place, determine the rain out fraction and calculate time taken to form a 1 m^3 pool, assuming a constant mass flow rate W_e.

Solution

T_f (K)	Liquid Density at T_f (kg/m^3)	y_f	W_e (kg/s)
231.16	581	0.3	10.8

$$\eta_r = 1 - 2 \cdot 0.3 = 0.4 \tag{4.207}$$

$$W_{f,L} = \eta_r \cdot W_e = 0.4 \cdot 10.8 = 4.32\,\text{kg/s} \tag{4.208}$$

$$t_{pool} = \frac{\rho_{f,l}}{W_{f,L}} = \frac{581}{4.32} = 135\,\text{s} \tag{4.209}$$

4.10 Pool Evaporation

Some pool evaporation and specific gas stripping formulas are provided in this section. Additional more general formulas have been included in chapter 7.

4.10.1 Simplified Formula for a Single Component Pool Evaporation

A simplified formula for non-cryogenic liquid has been provided by Cox et al. (1990):

$$G_{ev} = \beta \cdot u_w \cdot \rho_v \cdot \frac{P_v^o}{P_a} \cdot \left(\frac{2}{u_w^2 \cdot D_{pool}}\right)^{n_c} \tag{4.210}$$

where:

- G_{ev} is the evaporation mass flux (kg/m s^2).
- u_w is the wind velocity (m/s) to be assumed as 0.5 as minimum.
- P_v^o is the liquid vapour pressure.
- P_a is the ambient pressure.
- ρ_v is the vapour density (kg/m^3).
- D_{pool} is the pool diameter (m).
- β and n_c are weather stability class coefficients included in Table 4.9 (Woodward, 1998) with:

$$n_c = \frac{n}{2+n} \tag{4.211}$$

Table 4.9 Stability class coefficients

Stability Class	Sutton Index n	Exponent n_c	β
A	0.17	0.078	0.0010
B	0.20	0.091	0.0012
C	0.25	0.111	0.0015
D	0.30	0.130	0.0017
E	0.35	0.149	0.0018
F	0.44	0.180	0.0018

Example 4.21

Find the evaporation mass flux from a 10 m^2 pool for n-pentane at 20°C and stability class F. Assume no wind, i.e. according to the ATEX standard EN 60079-10-1, when no wind is perceivable, at least 0.5 m/s has to be assumed.

Solution

T_a (K)	Vapour Pressure (Pa)	Vapour Density (kg/m³)	Wind Velocity (m/s)	β	n_c	D_{pool} (m)
298.16	57,900	3	0.5	0.00180	0.18	3.5

$$G_{ev} = 0.0018 \cdot 0.5 \cdot 3 \cdot \frac{57,900}{101,325} \cdot \left(\frac{2}{5^2 \cdot 3.5}\right)^{0.18} = 0.00078\,\text{kg/m}^2 \times \text{s} \tag{4.212}$$

4.10.2 Evaporation Flux From a Cryogenic Pool

According to Cox et al. (1990), cryogenic fluid evaporation flux depends on the time square root:

$$G_{ev} = \frac{2 \cdot X \cdot k}{\lambda} \cdot \sqrt{\left(\frac{1}{\pi \cdot \alpha}\right)} \cdot (T_G - T_L) \cdot \sqrt{t} \tag{4.213}$$

- G_{ev} is the evaporation mass flux (kg/m s²).
- k is the thermal conductivity of the substrate (W/m K).
- T_L is the normal boiling point.
- λ *is* the heat of vaporisation (J/kg).
- α is the thermal diffusivity of the substrate (m²/s).
- T_G is the ground temperature (K).
- X is a surface correction factor:

$X = 1$ for nonporous ground

$X = 3$ for porous (gravel, loose sand, grassy) ground

4.11 Hydrogen Sulphide Release From Free Surfaces

Pomeroy's equation can be used to calculate the release of H_2S from free surface flows (U.S. EPA, 1985) (Fig. 4.34):

$$R_{sf} = \frac{\varphi_{sf}}{d_m} \tag{4.214}$$

and:

$$\varphi_{sf} = 0.69 \cdot C_A \cdot T \cdot (s \cdot u)^{3/8} \cdot (1 - q) \cdot [DS] \tag{4.215}$$

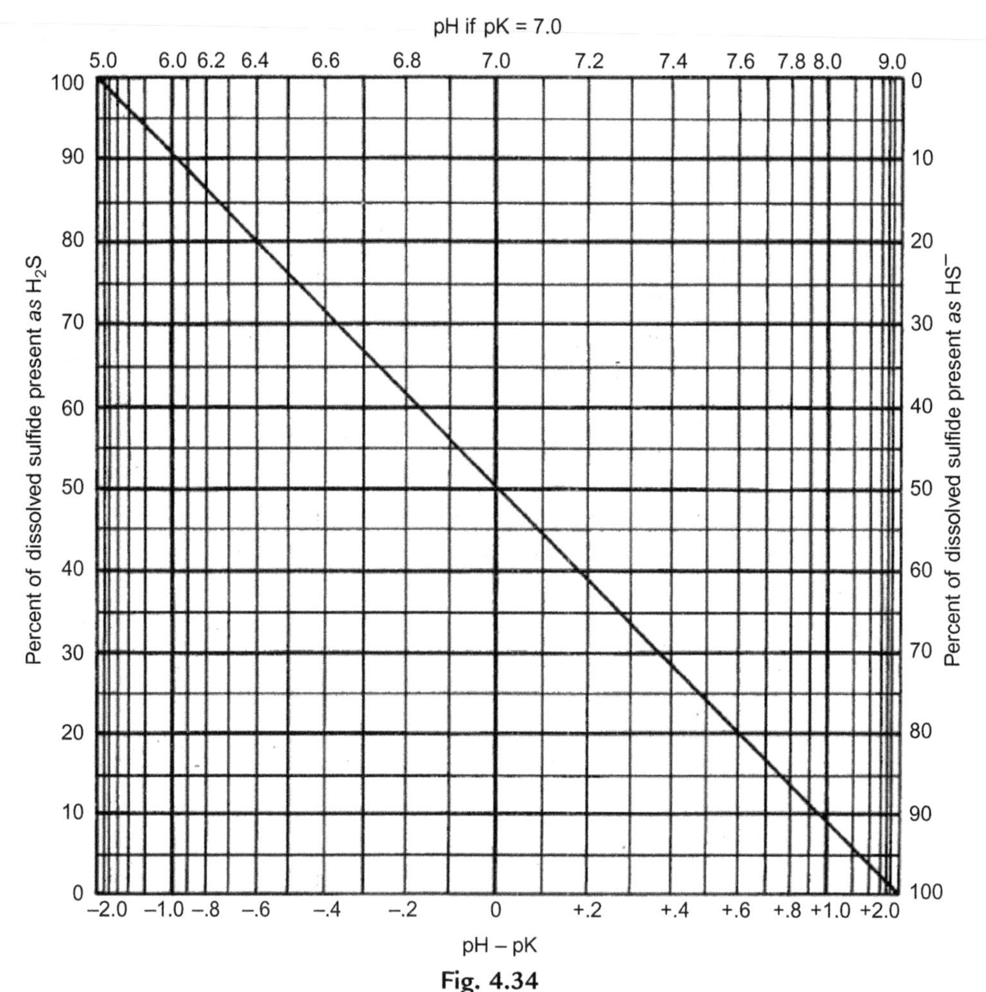

Fig. 4.34
Dissolved fractions of hydrogen sulphide in water (U.S. EPA, 1985).

where:

- R_{sf} is the sulphide depletion in the stream due to the escape of H_2S, mg/L-h.
- Φ_{sf} is the flux of H_2S from the stream surface, grams of H_2S/m^2h.
- d_m is the mean hydraulic depth (defined as the cross-sectional area of the stream divided by its surface width), m.
- u is the stream velocity, m/s.
- C_A is a factor representing the effect of turbulence in comparison to a slow stream:

$$C_A = 1 + \left[\frac{0.17 \cdot u^2}{g \cdot d_m}\right] \tag{4.216}$$

- T is a temperature coefficient, equal to unity at 20°C.
- s is the slope of energy grade line of the stream m/m.
- q is the relative H_2S saturation in the air compared to equilibrium concentration (typically 2%–20%).
- [DS] is the dissolved sulphide concentration in wastewater, mg/L.
- j is the proportion of dissolved sulphide present as H_2S, to be collected from the following diagram (U.S. EPA, 1985).

References

Akaike, S., Nemoto, M., 1988. Potential core of a submerged laminar jet. J. Fluids Eng. 110, 392–398. American Society of Mechanical Engineers.

API Standard 520, 2014. Sizing, Selection, and Installation of Pressure-relieving Devices Part I—Sizing and Selection, ninth ed. American Petroleum Institute.

API Standard 521, 2014. Pressure-Relieving and Depressuring Systems, sixth ed.

Ashkenas, H., Sherman, F.S., 1966. Rarefied gas dynamics. In: De Leeuw, J.H. (Ed.), Proc. 4th Int. Symp. Rarefied Gas Dynamics, vol. II. Academic Press, New York, pp. 84–105.

Avallone, E.A., Baumeister, T., Sadegh, A.M., 2007. Mark's Standard Handbook for Mechanical Engineers, 11th ed. McGrawHill.

Bair, S., 2014. The pressure and temperature dependence of volume and viscosity of four Diesel fuels. Fuel 135, 112–119. Elsevier.

Benintendi, R., 2010. Turbulent jet modelling for hazardous area classification. J. Loss Prev. Process Ind. 23, 373–378.

Benintendi, R., 2011. 2011, Laminar jet modelling for hazardous area classification. J. Loss Prev. Process Ind. 24, 123–130.

Bird, R.B., Stewart, W.E., Lightfoot, E.N., 2002. Transport Phenomena, second ed. Wiley & Sons, Inc.

Bogey, C., Bailly, C., 2006. Computation of the Self-Similarity Region of a Turbulent Round Jet Using Large-Eddy Simulation. Direct and Large-Eddy Simulation VI. Springer, Netherlands.

Boguslawski, L., Popiel, C.O., 1979. Flow structure of the free round turbulent jet in the initial region. J. Fluid Mech. 90 (3), 531–539. Cambridge University Press.

CCPS, 1999. Guidelines for Chemical Process Quantitative Risk Analysis, second ed. AIChE, New York.

Chen, C., Rodi, W., 1980. Vertical Turbulent Buoyant Jets: A Review of Experimental Data. Pergamon Press, Oxford.

Cox, A., Lees, F., Ang, M., 1990. Classification of hazardous locations. Report of the Inter-Institutional Group on the Classification of Hazardous Locations (IIGCHL).

Crowl, D.A., Louvar, J.F., 2011. Chemical Process Safety: Fundamentals with Applications, third ed. Prentice Hall International Series in Physical and Chemical Engineering.

Din, F., 1962. Thermodynamic Functions of Gases, Volume 1: Ammonia, Carbon Dioxide, and Carbon Monoxide; Volume 2: Air, Acetylene, Ethylene, Propane and Argon, Volume 3. Ethane, Methane, and Nitrogen. Butterworths.

Forsythe, W.E., 2003. Smithsonian Tables, ninth ed. Smithsonian Institution Press.

Goodfellow, H.D., Tähti, E., 2001. Industrial Ventilation Design Guidebook. Academic Press.

Hill, B., 1972. Measurement of local entrainment rate in the initial region of axisymmetric turbulent air jets. J. Fluid Mech. 51, 773–779.

Hess, K., Leuckel, W., Stoeckel, A., 1973. Ausbildung von Explosiblen Gaswolken bei Uberdachentspannung und MaBnahmen zu deren Vermeidung. Chem. Ing. Tech. 45 (73), 5323–5329.

Joukowsky, N., 1898. Über den Hydraulischen Stoss in Wasserleitungsröhren. (On the hydraulic hammer in water supply pipes). Mém. Acad. Imp. Sci. St.-Pétersbourg (1900) Ser. 9 (5), 1–71 (in German).

Kletz, T.A., 1977. Unconfined Vapor Cloud Explosions. AIChE Loss Prevention, 11, p. 50, Appendix 1.

Leung, J.C., 1994. The omega method for discharge rate evaluation. In: Fauske & Associates, AIChE Meeting, Denver, Colorado, August 16.
Li, Z.H., Zhang, J.S., Zhivov, A.M., Christianson, L.L., 1993. Characteristics of diffuser air jets and airflow in the occupied regions of mechanically ventilated rooms—a literature review. ASHRAE Trans. 99, 1119–1127.
Lienhard, J.H., Day, J.B., 1970. The breakup of superheated liquid jets. J. Basic Eng. 92 (3), 515–521.
Oosthuizen, P.H., 1983. An experimental study of low reynolds number turbulent circular jet flow. In: American Society of Mechanical Engineers, Applied Mechanics, Bioengineering, and Fluids Engineering Conference, Houston, TX, June 20–22.
Patrick, M.A., 1967. Experimental investigation of the mixing and penetration of a round turbulent jet injected perpendicularly into a transverse stream. Trans. Inst. Chem. Eng. 45, T16–T31.
Plank, R., Kuprianoff, J., 1929. Die Thermischen Eigenschaften der Kohlensaure im Gasformigen, Flussigen und Festen Zustand, Beihefte zur Zeitschrift fur die gesamte Kalte-Industrie, Reihe 1, Heft 1.
Pope, S.B., 2000. Turbulent Flows. Cambridge University Press.
Ramskill, P.K., 1985. A Study of Axisymmetric Underexpanded Gas Jets, SRD R-302. U.K. Atomic Energy Authority.
Rankin, G.W., Sridhar, K., Arulraja, M., Kumar, K.R., 1983. An experimental investigation of laminar submerged jets. J. Fluid Mech. 133, 217–231.
Ricou, F.P., Spalding, D.B., 1961. Measurements of entrainment by axisymmetrical turbulent jets. J. Fluid Mech. 8, 21–32.
Schlichting, H., 1968. Boundary Layer Theory, sixth ed. McGraw Hill, New York.
Shashi Menon, E., 2004. Liquid Pipeline Hydraulics. Marcel Dekker Inc.
Shepelev, I., 1961. Air Supply Ventilation Jets and Air Fountains. 4 Proceedings of the Academy of Construction and Architecture of the USSR.
Spouge, J., 1999. A Guide To Quantitative Risk Assessment for Offshore Installations. DNV Technica, CMPT.
Thring, M.W., Newby, M.P., 1953. Combustion length of enclosed turbulent jet flame. In: Proceedings of 4th International Symposium on Combustion, Pittsburgh, PA, pp. 789–796.
Totten, G.E., De Negri, V.J., 2011. Handbook of Hydraulic Fluid Technology, second ed. CRC Press.
U.S. EPA, 1985. Odor and Corrosion Control in Sanitary Sewerage Systems and Treatment Plants. Design Manual. EPA/62571-85-018.
Van den Bosch, C.J.H., Duijm, N.J., 2005. Methods for the calculation of physical effects. In: van den Bosch, C.J.H. (Ed.), Yellow Book. R.A.P.M. Weterings.
White, F.M., 2011. Fluid Mechanics, seventh ed. McGrawHill.
Woodward, J.L., 1998. Estimating the Flammable Mass of a Vapor Cloud. CCPS, AIChE.
Younglove, B.A., Ely, J.F., 1987. Thermophysical properties of fluids, II, Methane, Ethane, Propane, Isobutane and Normalbutane. J. Phys. Chem. Ref. Data 16 (4). Thermophysics Division, National Engineering Laboratory, National Bureau of Standards.
Yue, Z., 1999. Installationsteknik Bullettin, n. 48, Air Jet Velocity Decay in Ventilation Applications. ISSN 0248-141X.

Further Reading

De Vaull, G.E., King, J.A., 1992. Similarity scaling of droplet evaporation and liquid rain-out following the release of a superheated flashing liquid to the environment. In: 85th Annual Meeting, Air & Waste Management Assoc., Kansas City, MO, June 21–26.
Hussein, J.H., Capp, S.P., George, W.K., 1994. Velocity measurements in a high-reynolds number, momentum-conserving, axisymmetric, turbulent jet. J. Fluid Mech. 258, 31–75. Cambridge University Press.
Mannan, S. (Ed.), 2012. Lees' Loss Prevention in the Process Industries, fourth ed. Butterworth-Heinemann.
Panchapakesan, N.R., Lumley, J.L., 1993. Turbulence measurements in axisymmetric jets of air and helium. I: air jet. J. Fluid Mech. 246, 197–223. Cambridge University Press.

CHAPTER 5

Loads and Stress Analysis for Process Safety

Ut tensio, sic vis (Robert Hooke, 1653–1703)

5.1 Structural Failure Scenarios in Process Safety

On June 1st 1974 a catastrophic explosion occurred at the Nypro plant near Flixborough, North Lincolnshire. A significant root cause of the explosion was the full lack of awareness of the structural consequences of an improper plant configuration change implemented in an emergency situation. This specific case has been analysed in detail in this chapter along with the fundamental concepts of equipment structural analysis, with the aim to show the importance of their implications in process safety (Fig. 5.1 and Table 5.1).

5.2 Key Concepts

5.2.1 Burst

Violent and rapid loss of containment of an enclosure system due to a significant internal pressure or caused by a chemical reaction (CCPS, 2005).

5.2.2 Explosion

A release of energy that causes a pressure discontinuity or blast wave (CCPS, 2005).

5.2.3 Static Pressure

The pressure that would exist at a point in a medium if no sound waves were present (Bjerketvedt et al., 1997).

Process Safety Calculations. https://doi.org/10.1016/B978-0-08-101228-4.00005-8

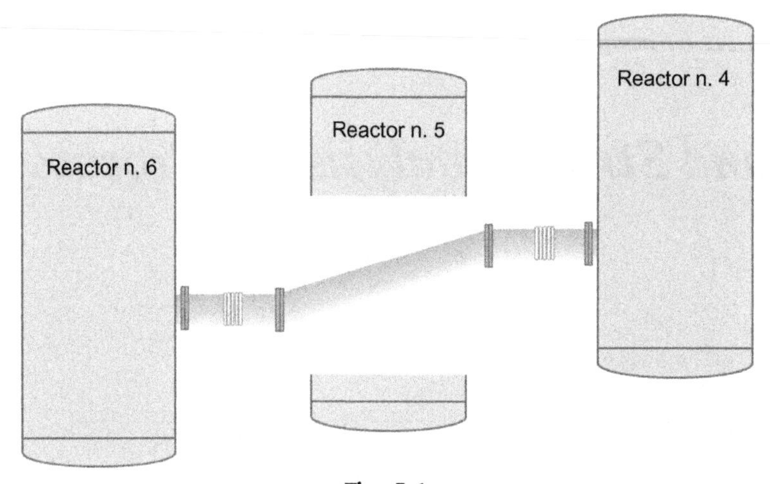

Fig. 5.1
Configuration of Flixborough reactors after removal of reactor n. 5.

Table 5.1 Structural failure scenarios

Affecting Factor	Incident Scenario	Structural Target			
		Vessels	Piping	Buildings	Rack/Stick Built
Thermal radiation	Fire/explosion	✓	✓	✓	✓
Pressure wave	Explosion	✓	–	✓	–
Shock wave	Explosion	✓	–	✓	–
Buckling	Misoperation	✓	✓	–	–
	Design failure	✓	✓	–	–
Drag load	Explosion	–	✓	–	✓
Corrosion	Design failure	✓	✓	✓	✓
Material failure		✓	✓	✓	✓

5.2.4 Dynamic Pressure

The pressure increase that a moving fluid would have if it were brought to rest by isentropic flow against a pressure gradient (Bjerketvedt et al., 1997). The dynamic pressure can also be expressed by the flow velocity, u, and the density, ρ:

$$P_D = \frac{1}{2} \cdot \rho \cdot u^2 \tag{5.1}$$

5.2.5 Shock and Pressure Wave

A transient change occurs in the gas density, pressure, and velocity of the air surrounding an explosion point. The initial change can be either discontinuous or gradual. A discontinuous change is referred to as a shock wave, and a gradual change is known as a pressure wave. A shock wave moves faster than sound speed (CCPS, 2005).

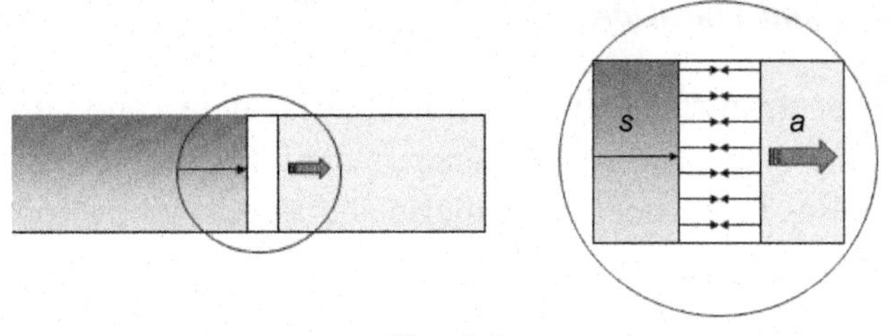

Fig. 5.2
Flame front velocity and speed velocity in deflagration.

5.2.6 Deflagration and Detonation

Flame motion produces a shock wave in the unburnt mixture, which consists of an adiabatic compression that moves at sound velocity a in the unreacted mixture (S is the flame front velocity). In the ordinary flame, velocity S is much lower than a, and the adiabatic perturbation moves through the unburnt mixture at a velocity much lower than it (Fig. 5.2). This is a deflagration. If a exceeds S, the phenomenon is a detonation.

Propagation velocities in the range 1000–3500 m/s may be observed depending on the gas mixture, initial temperature, pressure, and type of detonation (CCPS, 2005).

5.2.7 Deflagration to Detonation Transition (DDT)

The transition phenomenon resulting from the acceleration of a deflagration flame to detonation via flame-generated turbulent flow and compressive heating effects. At the instant of transition a volume of pre-compressed, turbulent gas ahead of the flame front detonates at unusually high velocity and overpressure (CCPS, 2005).

5.2.8 Physical Explosion

The catastrophic rupture of a pressurised gas/vapour-filled vessel by means other than reaction, or the sudden phase-change from liquid to vapour of a superheated liquid (CCPS, 2005).

5.2.9 Confined Explosion

An explosion of fuel-oxidant mixture inside a closed system (e.g., vessel or building) (CCPS, 2005).

5.2.10 Unconfined Vapour Cloud Explosion (UCVE)

An explosion occurring when a flammable vapour is released, and its mixture with air will form a flammable vapour cloud. If ignited, the flame speed may accelerate to high velocities and produce significant blast overpressure.

5.2.11 Overpressure and Duration

Release of mechanical or chemical energy of a fluid may result in a pressure or shock wave. The wave travels through the space, and two parameters are identified for its definition: overpressure and impulse. The concept of impulse and associated overpressure is based on Newton's second law of dynamics. In scalar notations:

$$F = \frac{d(m \times u)}{dt} = m \times \frac{du}{dt} \tag{5.2}$$

The impulse is defined as:

$$I = \int_{\Theta} F(t) \cdot dt \tag{5.3}$$

where Θ is the duration of the action performed by the force F.

For a pressure or for a shock wave, replacing $F(t)$ with the associated transient overpressure $P(t)$, it will be:

$$I = \int_{\Theta} P(t) \cdot dt \tag{5.4}$$

The time history of $P(t)$ can be depicted as shown in Fig. 5.3, where a maximum overpressure ΔP_{max} can be detected at any location, after which pressure drops down to ambient pressure. For practical scopes, the transient pressure behaviour is acceptably approximated with triangular diagrams, which allows relating the impulse to duration (Fig. 5.4).

So:

$$I \cong \frac{\Delta P_{max} \times \Theta}{2} \tag{5.5}$$

Pressure and impulse can be watched as the static and dynamic components of a wave, with both playing a significant role in terms of damage of targets such as buildings or

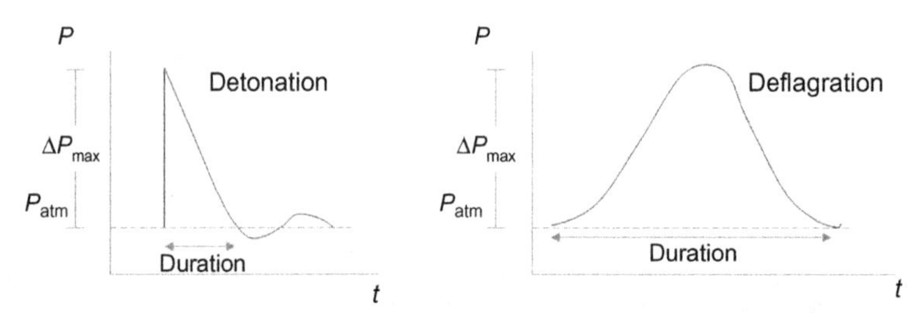

Fig. 5.3
Time history of pressure in detonation and deflagration.

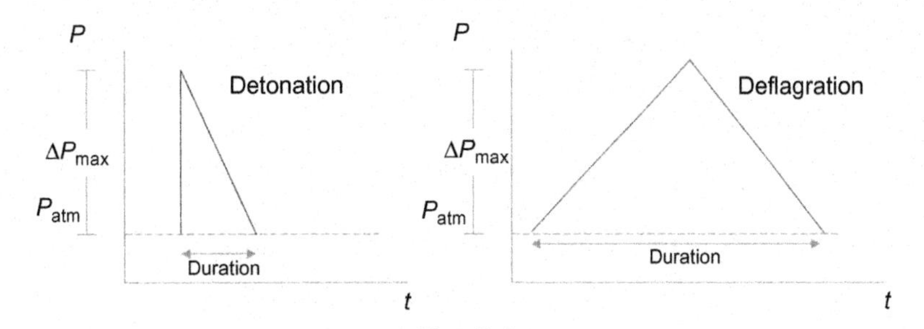

Fig. 5.4
Simplified approximation of shock and pressure wave.

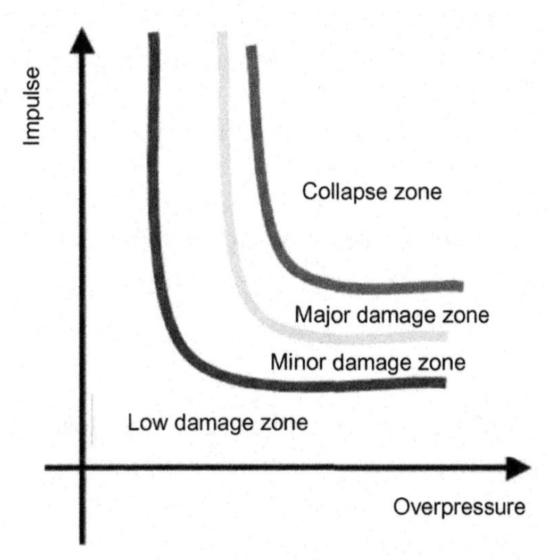

Fig. 5.5
Baker's diagram.

equipment. This is typically represented with Baker's diagram, where probability zoning of specified damages is related both to overpressure and duration (Fig. 5.5).

5.2.12 Stagnation Pressure

Stagnation pressure is the pressure that a moving fluid would have if it were brought isentropically to rest against a pressure gradient (Bjerketvedt et al., 1997). The stagnation pressure is the sum of the static and the dynamic pressures.

5.2.13 Side-on Pressure, Reflected Pressure, Diffracted Pressure

- The *side-on pressure* is the static pressure behind a wave.

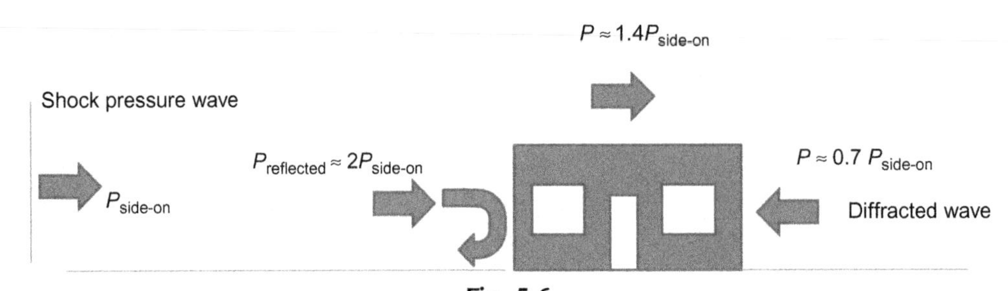

Fig. 5.6
Side-on pressure, reflected pressure, reflected pressure.

- The *reflected pressure* is measured when the wave hits an object like a wall head-on. The reflection is not an isentropic process, so stagnation and reflected pressure don't coincide.
- The *diffracted pressure* is the pressure measured on the wall opposite to the reflected pressure wall (Fig. 5.6).

For practical calculations, the approximate ratios shown in Fig. 5.6 can be adopted (Bjerketvedt et al., 1997).

5.2.14 Wind or Drag Load

The wind force generated by the explosion will act as a drag load on smaller equipment, such as piping. It is the dynamic pressure associated with the wind velocity, which is the prevailing load for small objects. The drag force is a combination of tangential and normal stresses according to Baker's formula:

$$F_D = A \cdot C_D \cdot \frac{1}{2} \cdot \rho \cdot u^2 \tag{5.6}$$

where:

- C_D is the drag coefficient, depending on the shape of the obstacle, on the Reynolds number and on the local Mach number.
- ρ is the fluid density.
- u is the large scale fluid velocity, whose distribution close to the obstacle may significantly vary.
- A is the maximum cross sectional area of the object in a plane normal to u.

The drag force Baker formula is valid for obstacles with reference size or diameter lower than 0.3 m and, for these items, other loads are negligible.

Table 5.2 (Baker et al., 1983) includes the drag coefficients for some configurations.

Table 5.2 Drag coefficients (Baker et al., 1983)

Item	Shape	CD
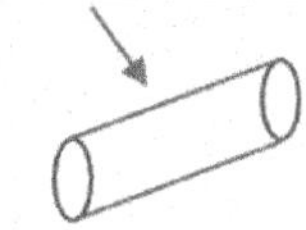	Long circular cylinder—Side on	1.20
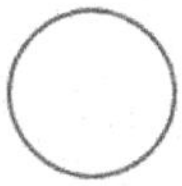	Sphere	0.47
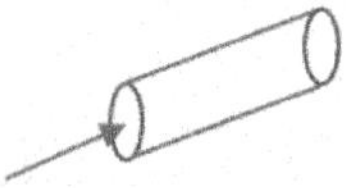	Circular cylinder—End on	0.82
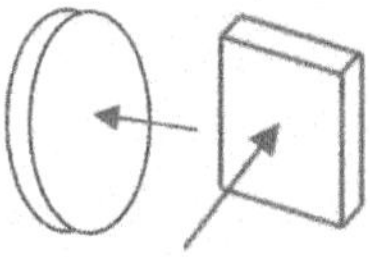	Circular or square disc	1.17
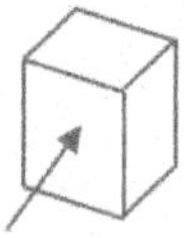	Cube—Face on	1.05
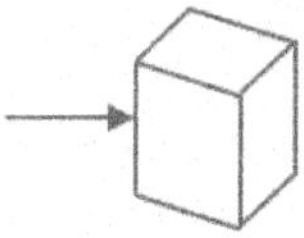	Cube—Face on	0.80
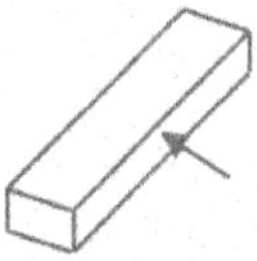	Long rectangular member—Face on	2.05
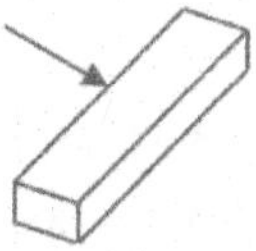	Long rectangular member—Edge on	1.55
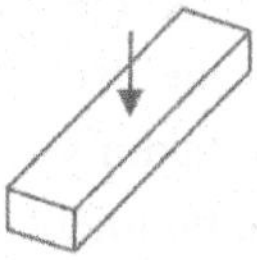	Narrow strip—Face on	1.98

Example 5.1

An aircraft is provided with a series of air brakes installed on the wings. Each air brake is approximately square, with 1.75 m sides, and makes an angle of 60° with the horizontal plane. On landing, the aircraft decelerates with the speed of 175 km/h. Find the drag force acting on each air brake if ground temperature is 20°C (Fig. 5.7).

Solution

C_D	Speed (km/h)	Wind velocity (m/s)	α	T_{amb} (K)	ρ_a (kg/m^3)
1.17	175	0.5	60°	293.16	1.2

$$F_D = \sin 60° \cdot 1.75^2 \cdot 1.17 \cdot \frac{1}{2} \cdot 1.2 \cdot \left(\frac{175{,}000}{3600}\right)^2 = 4400\,\text{N} \tag{5.7}$$

Fig. 5.7
Air brake of Example 5.1.

5.2.15 Buckling

Buckling is a structural failure due to the instability generally related to the aspect ratio of equipment and structural elements. For the purpose of process safety, buckling is defined as a structural failure caused by overstress or instability of the wall under compressive loading, typically under uniform external pressure, of pressure vessels and pipes. For this cause, thin-wall vessels under external pressure fail at stresses much lower than yield stress. Proper vacuum design, correct operation, and management of this potential issue are essential to prevent an occurrence of catastrophic failure. Two kinds of buckling failures are theoretically possible for membrane structures, the first being the common stability failure of beams and columns subject to axial loads, and related to the slenderness ratio analysed by the traditional Euler method. This instability is not dealt with in this handbook. The second one, that is specifically related to the containment function of vessels and piping used in process plants, is dependent on the thickness to diameter ratio. Ordinary vessels and atmospheric tanks are subject just to external atmospheric pressure, and internal pressure may drop down by design or, mistakenly, to vacuum. Cavitation at a pump delivery or surge can result in buckling and loss of containment.

Buckling affecting factors are:

- Material features
- Operating and design temperature
- Design configuration

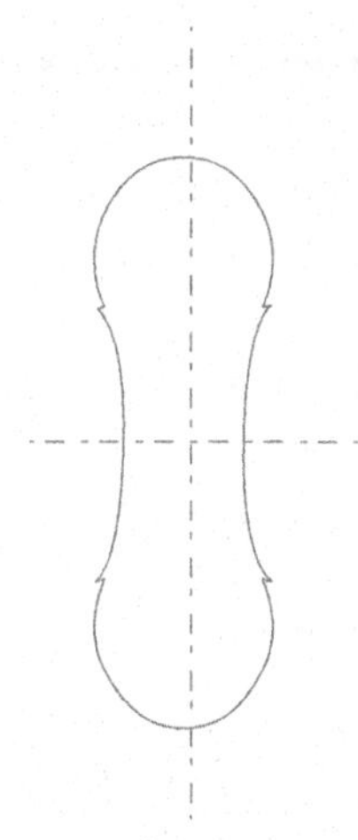

Fig. 5.8
Buckling of very long cylinders.

However, instability and critical (collapsing) pressure P_c of vessels and piping are related to geometrical and constructional factors, which are:

- The unsupported shell/pipe length L
- Thickness t
- Outside diameter D_o

Length categories differentiate the identification of the affecting parameters.

Very long cylinders

Pipes and some deaerators belong to this category and collapse into two lobes (Fig. 5.8).

This outcome derives from a lack of stiffness of the structure, which results in a catastrophic loss of integrity. The collapsing pressure is given by the following formula (Timoshenko and Gere, 1961):

$$P_c = \left(\frac{2 \cdot E}{1-\nu^2}\right) \cdot \left(\frac{t}{D_o}\right)^3 \tag{5.8}$$

where:

- E is the modulus of elasticity, 200×10^9 N/m^2 for structural steel
- ν is the coefficient of Poisson $= 0.3$ for steel.

The minimum unsupported length beyond which P_c is independent of L is the critical length:

$$L_c = 1.14 \cdot \sqrt[4]{1-\nu^2} D_o \cdot \sqrt{\frac{D_o}{t}} \tag{5.9}$$

Example 5.2

For a 500 mm diameter steel pipeline, find the maximum thickness for which buckling is independent of the length and calculate the corresponding critical length L_c.

Solution

The problem consists of calculating the thickness t_c corresponding to a critical pressure equal to 1 atm (Fig. 5.9).

D_o (mm)	t_c (mm)	P_c (N/mm^2)	E (N/mm^2)	ν
500	Unknown	0.101325	210,000	0.3

$$P_c = \left(\frac{2 \cdot E}{1-\nu^2}\right) \cdot \left(\frac{t}{D_o}\right)^3 \tag{5.10}$$

$$t_c = D_o \cdot \sqrt[3]{P_c \cdot \left(\frac{1-\nu^2}{2 \cdot E}\right)} = 500 \cdot \sqrt[3]{0.101325 \cdot \left(\frac{1-0.3^2}{2 \cdot 210{,}000}\right)} = 3\,\text{mm} \tag{5.11}$$

$$L_c = 1.14 \cdot \sqrt[4]{1-0.3^2} \cdot 500 \cdot \sqrt{\frac{500}{3}} = 7187\,\text{mm} \tag{5.12}$$

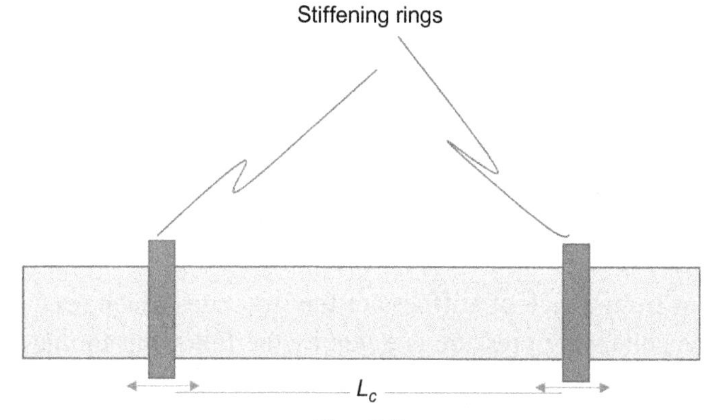

Fig. 5.9
Critical length of pipeline of Example 5.2.

Short cylinders

Below L_c, both the critical pressure P_c and the number of lobes n increase, and they depend also on the length. Notably, the following correlation has been proposed by Sturm (1941):

$$P_c = K \cdot E \cdot \left(\frac{t}{D_o}\right)^3 = 0.27\,\text{N/mm}^2 \tag{5.13}$$

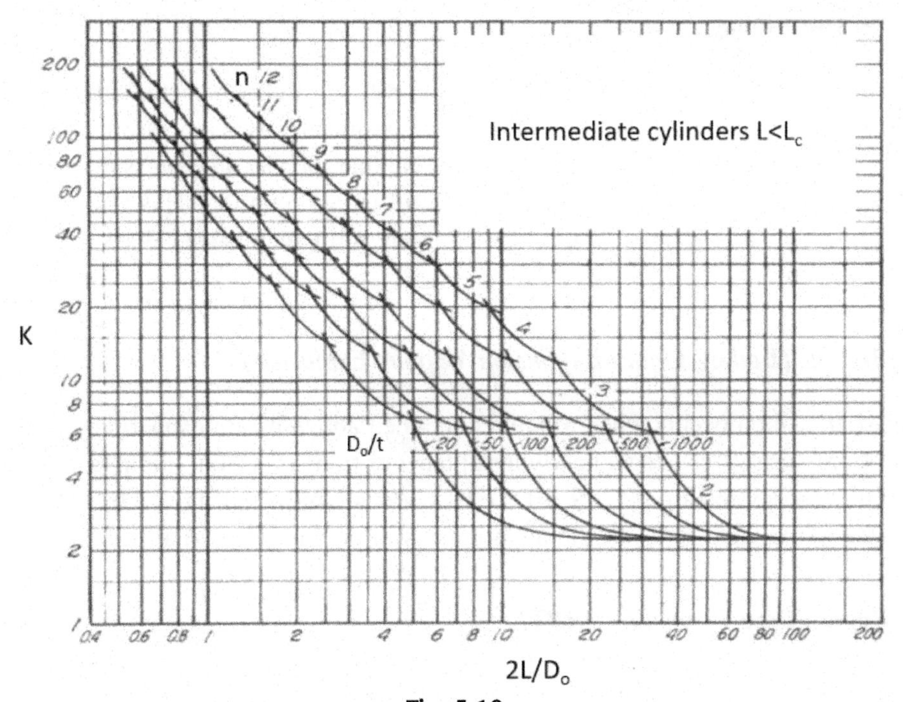

Fig. 5.10
Coefficients for the Sturm correlation.

where K and n can be read on the following diagram (Sturm, 1941) as a function of ratios of the thickness, the diameter, and the length (Fig. 5.10).

Example 5.3

For the 500 mm diameter steel pipeline, STD schedule, find the critical pressure and the number of lobes for a length of 3000 mm.

Solution

Entering the diagram with:

$$\frac{2 \cdot L}{D_o} = \frac{2 \cdot 3000}{500} = 12 \tag{5.14}$$

and:

$$\frac{D_o}{t} = \frac{500}{3} = 166.7 \tag{5.15}$$

$K = 6$, $n = 2$

$$P_c = 6 \cdot 210{,}000 \cdot \left(\frac{3}{500}\right)^3 = 0.27 \mathrm{N/mm^2} \tag{5.16}$$

5.2.16 Pressure Piling

In a compartmented system in which there are separate but interconnected volumes, the pressure developed by the deflagration in one compartment causes a pressure rise in the unburned gas in the interconnected compartment, so that the elevated pressure in the latter compartment becomes the starting pressure for a further deflagration. This effect is known as pressure piling, or cascading. (CCPS, 2005). Pressure piling can trigger deflagration to detonation transition.

5.2.17 BLEVE (Boiling Liquid Expansion Vapour Explosion)

Rapid phase transition occurs when a liquid contained above its atmospheric boiling point is rapidly depressurised, causing a nearly instantaneous transition from liquid to vapour with a corresponding energy release (CCPS, 2005). BLEVE is typically triggered by the structural collapse, due to the yield stress of the membrane wall due to an external fire.

5.2.18 Rapid Phase Transition (RPT)

RPT is a particular BLEVE affecting LNG due to its expansion coefficient (1–600) causing potential overpressure up to 30 bars and significant overpressure detectable up to 1 mile from the event.

5.3 Stresses

Stresses are forces acting on the per unit area of a body. Their units are the same as pressure units.

5.3.1 Tensile and Compression Stresses

Tensile and compression stresses act axially to a beam element, normally to the cross section (Fig. 5.11).

Normal compression or tensile stress is produced by normal loads or bending moments (Fig. 5.12).

Fig. 5.11
Compression and tensile stresses.

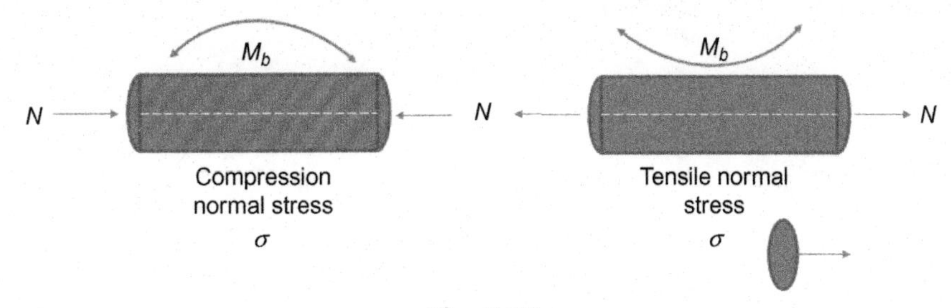

Fig. 5.12
Compression and tensile stresses due to bending moment.

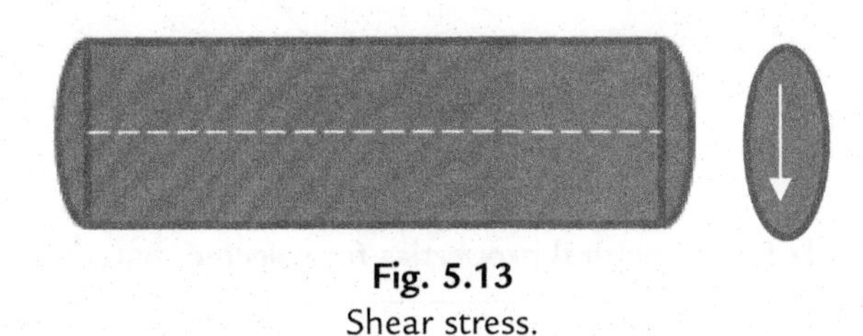

Fig. 5.13
Shear stress.

Liquids are only subject to compression stresses, which are coincident with pressure.

5.3.2 Shear Stresses

Shear stresses are the load component parallel to the cross section (Fig. 5.13).

According to Budynas and Nisbett (2014), a rule of thumb for shear stress yield stress is that it can be assumed as 60% of the normal yield stress.

5.3.3 Elastic and Plastic Stresses

Elasticity is the property of many materials, including metals, to deform under applied loads and to revert to their initial state when loads are removed, with no permanent deformation. Plastic behaviour is the stress-strain field in which loads are associated with permanent deformation. Elastic behaviour is described by Hooke's law[1]:

$$\sigma = E \times \varepsilon \tag{5.17}$$

where σ is the tensile stress, ε is extension, and E is the coefficient of the elasticity (Young's modulus).

[1] Robert Hooke (1635–3 March 1703). English scientist and philosopher. He announced the law of elasticity using the Latin sentence: Ut tension, sic vis (as the extension, so the force), anagrammed as CEIIINOSSSTTUV.

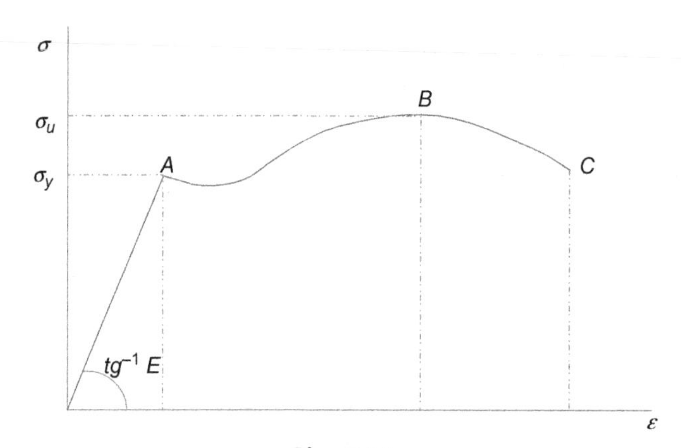

Fig. 5.14
Stress-strain diagram.

Table 5.3 Structural properties for selected materials

Material	σ_y (MPa)	σ_u (MPa)	Young's Modulus (GPa)
Cast irons	217–790	350–1000	165–180
Carbon steel	250–1155	345–1640	200–215
Stainless steel	170–1000	480–2240	189–210
Aluminium alloys	30–500	58–550	68–82
Copper alloys	30–500	100–550	112–148
Nickel alloys	70–1100	345–1200	190–220
Magnesium alloys	70–400	185–475	42–47
Lead alloys	8–14	12–20	12.5–15
Zinc alloys	80–450	135–520	68–95

The stress-strain diagram for steels and other metals has the behaviour shown in Fig. 5.14.

- s_y is the yield stress that is the nominal stress at the limit of elasticity in a tensile test (point A).
- s_u is the Ultimate Tensile Stress (UTS) that is the nominal stress at maximum load in a tensile test (point B).
- C is the fracture point.

Metallic materials data have been included in Table 5.3 (Materials Data Book, 2003):

Example 5.4

A batch reactor is closed by a 2 m diameter lid supported by four carbon steel flanges, each taking four M10 bolts. The reactor undergoes thermal runaway, and its pressure and temperature increase. Find the pressure at which the carbon steel yield stress (assumed as 300 MPa) is attained in the flange bolts (Fig. 5.15).

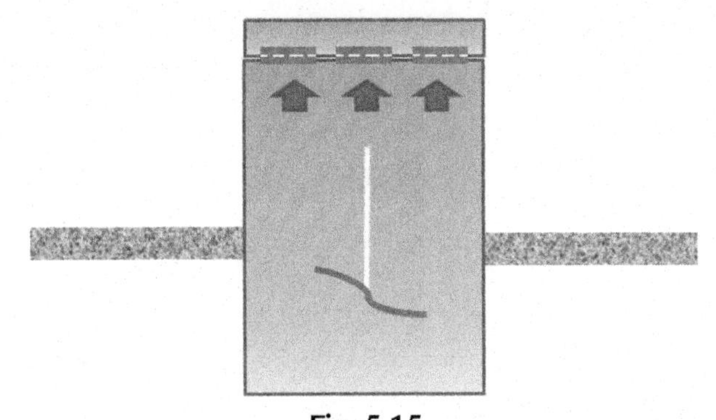

Fig. 5.15
Overheating and overpressurisation of the batch of Example 5.4.

Solution

σ_y (MPa)	Reactor D (m)	Reactor A (m^2)	Bolt d (m)	Bolt A (m^2)
300	2	3,14	0.01	0.0000785

The force acting on the single bolt subject to the yield stress is:

$$f_b = 300,000,000 \cdot 0.0000785 = 23,550\,\text{N} \tag{5.18}$$

The force acting on all the bolts is:

$$F_b = 23,550\,\text{N} \times 16 = 376,800\,\text{N} \tag{5.19}$$

And pressure is:

$$P = \frac{376,800\,\text{N}}{\frac{\pi}{4} \cdot 2^2} = 120,000\,\text{Pag} \tag{5.20}$$

5.3.4 Viscous Creep

Viscous creep is the tendency of a solid material to move (flow) slowly or deform permanently under the influence of mechanical stresses at given temperatures. Steel displays a significant creep phenomenon at temperatures over 450°C (Fig. 5.16).

5.4 Membrane Stresses in Thin-Shell Structures

Structures such as vessels, reactors, tanks, and pipes are the main elements met in process safety. The rigorous analysis of the stress acting on these items is very complex and, due to the severe loads scenario, specific skills are needed. However, in most practical cases, for

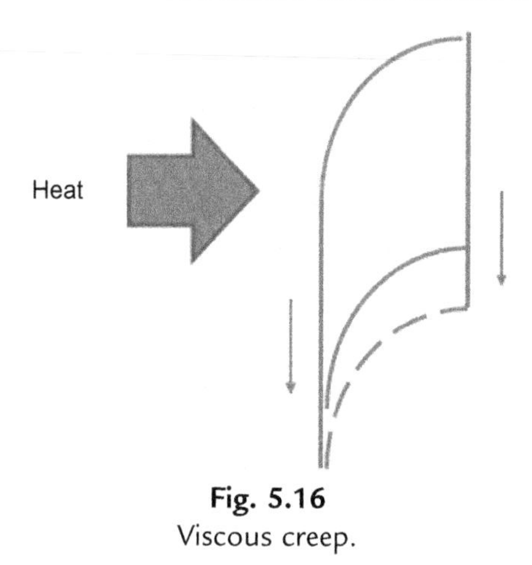

Fig. 5.16
Viscous creep.

screening purposes, the membrane stress theory allows to neglect the bending, and transverse shear forces can be neglected, so the resulting stress scenario is considerably simplified. This section includes some definitions and frequent geometrical configurations with the identification of the simplified stress scenario (Bednar, 1986).

5.4.1 Cylindrical Shell

Cylindrical shells under pressure can simplistically be subject to longitudinal stress and circumferential stress (Figs. 5.17 and 5.18).

Longitudinal stress

For the equilibrium it is:

$$P \times \pi \times \frac{D^2}{4} = \sigma_L \times \pi \times D \times t \tag{5.21}$$

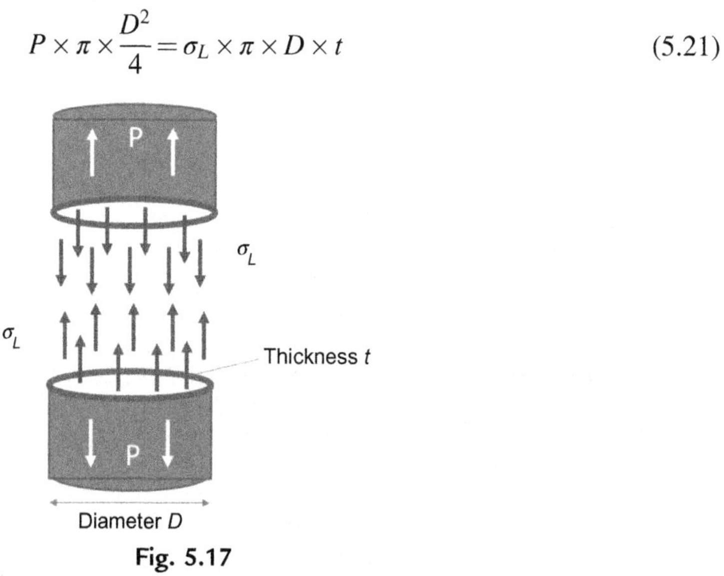

Fig. 5.17
Longitudinal stress.

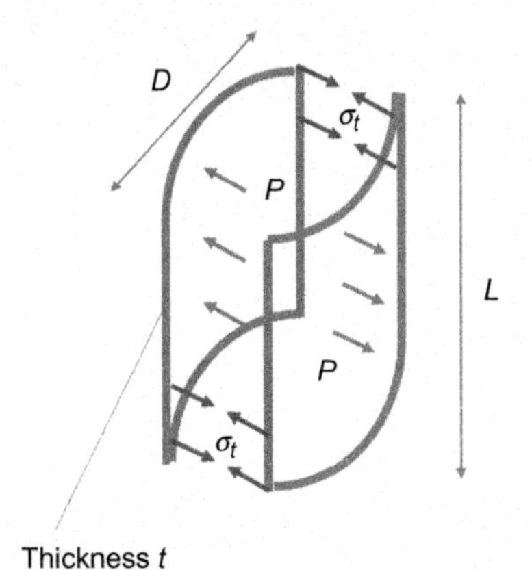

Fig. 5.18
Circumferential stress.

So:

$$\sigma_L = \frac{P \times D}{4 \times t} \tag{5.22}$$

Circumferential stress

$$P \times D \times L = 2 \times \sigma_t \times \pi \times L \times t \tag{5.23}$$

$$\sigma_t = \frac{P \times D}{2 \times t} \tag{5.24}$$

5.4.2 Spherical Shell

For symmetry (Fig. 5.19):

$$\sigma_t = \sigma_L = \frac{P \times D}{2 \times t} \tag{5.25}$$

5.5 Forces in Piping Bends

Forces in piping bends can be calculated writing the equation of momentum along the axis x and y (Fig. 5.20):

$$F_x + P_1 \times A_1 - P_2 \times A_2 \times \cos\theta = \rho \times Q \times (u_2 \times \cos\theta - u_1) \tag{5.26}$$

$$-m \times g + F_y - P_2 \times A_2 \times \sin\theta = \rho \times Q \times (u_2 \times \sin\theta) \tag{5.27}$$

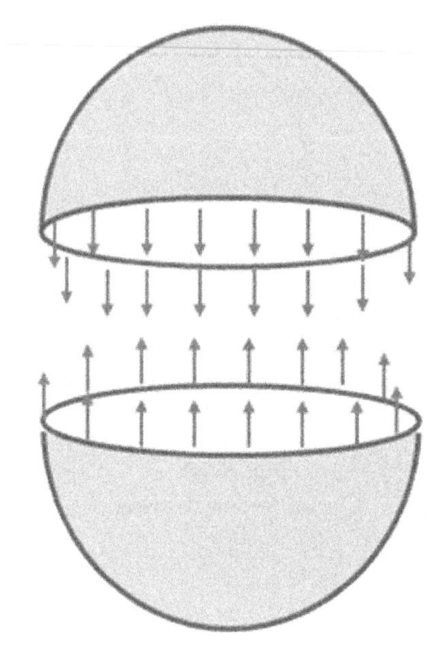

Fig. 5.19
Stress distribution in spherical shell.

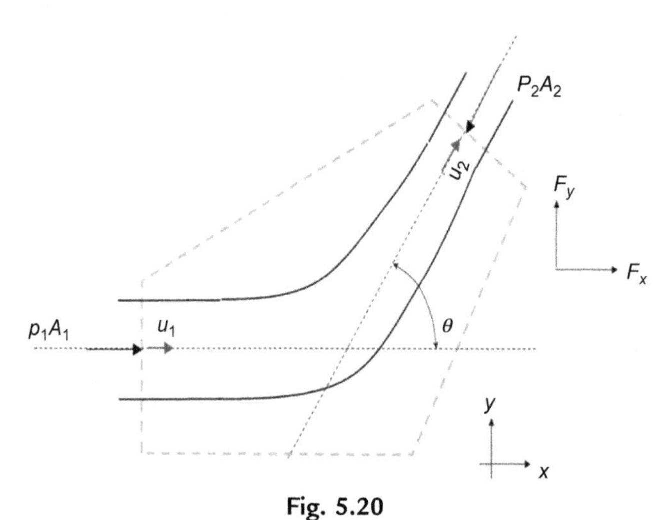

Fig. 5.20
Forces acting on piping bends.

where:

- F_x and F_y are the horizontal and vertical components of the action exerted by the bend on the fluid (opposite to the one exerted by the fluid on the bend).
- Q is the fluid flow rate.
- ρ is the fluid density.
- u is the fluid velocity.
- m is the mass of fluid contained in any vertical pipe segment.

Example 5.5

Water at 20°C is flowing at a rate of 0.0117 m^3/s through a 60° bend, in which there is a contraction from 100 to 50 mm internal diameter. Compute the force exerted by the bend if the pressure at the downstream end is 1.1 atm (111,457.5 Pa), the water mass in the control volume is 2 kg, and the pressure drop is negligible. Assume the pipe is routed on an horizontal plane (Fig. 5.21).

Solution

Q (m^3/s)	u_1 (m/s)	θ (deg)	A_1 (m^2)	A_2 (m^2)	D_1 (m)	D_2 (m)	ρ (m^3/kg)	P_2 (Pa)	2 (kg)
0.0117	1.5	60	0.00785	0.0019625	0.1	0.05	1000	111,457.5	2

Equation of continuity states:

$$u_2 = u_1 \cdot \frac{A_1}{A_2} = 6\,\text{m/s} \tag{5.28}$$

Energy balance equation:

$$\frac{1}{2} \cdot u_1^2 + \frac{P_1}{\rho} = \frac{1}{2} \cdot u_2^2 + \frac{P_2}{\rho} \tag{5.29}$$

$$P_1 = P_2 + \frac{\rho}{2} \cdot u_2^2 - \frac{\rho}{2} \cdot u_1 = 111{,}457.5 + 500 \cdot 36 - 500 \cdot 2.25 = 128{,}332.5\,\text{Pa} \tag{5.30}$$

Equation of momentum requires:

$$F_x = -P_1 \times A_1 + P_2 \times A_2 \times \cos\theta + \rho \times Q \times (u_2 \times \cos\theta - u_1) \tag{5.31}$$

$$F_x = -128{,}332.5 \cdot 0.00785 + 111{,}457.5 \cdot 0.0019625 \times 0.5 + 1000 \times 0.0117 \times (6 \times 0.5 - 1.5) = -880.5\,\text{N} \tag{5.32}$$

$$-m \times g + F_y - P_2 \times A_2 \times \sin\theta = \rho \times Q \times (u_2 \times \sin\theta) \tag{5.33}$$

$$F_y = P_2 \times A_2 \times \sin\theta + \rho \times Q \times (u_2 \times \sin\theta) + m \cdot g \tag{5.34}$$

$$F_y = 111{,}457.5 \cdot 0.0019625 \cdot 0.866 + 1000 \cdot 0.0117 \times (6 \times 0.866) + 2 \cdot 9.81 = 269.8\,\text{N} \tag{5.35}$$

The total force is:

$$F = \sqrt{880^2 + 269.8^2} = 920.43\,\text{N} \tag{5.36}$$

At an angle with the horizontal axis equal to:

$$\alpha = \arctan\frac{F_y}{F_x} = \arctan\left(-\frac{269.8}{880.5}\right) = 163° \tag{5.37}$$

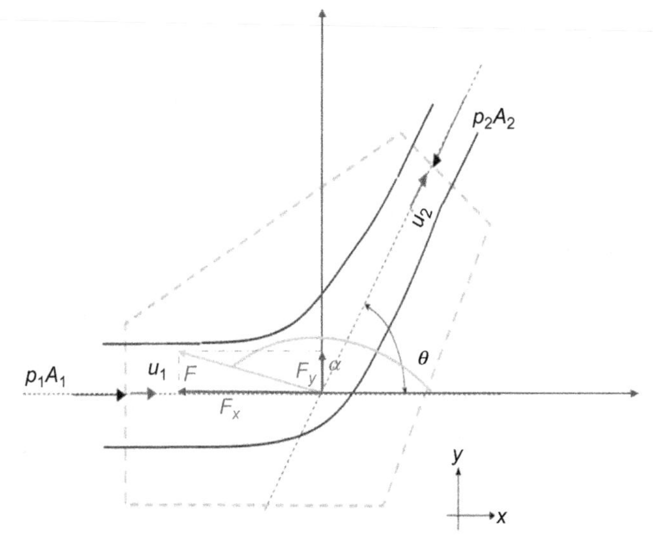

Fig. 5.21
Balance of forces for Example 5.5.

5.6 Thermal Loads

A discussed in Chapter 4, the tendencies of liquids to expand with significant pressure rise can pose significant risks in piping and equipment. Loss of containment can occur also when trapped liquids expand and pressure can reach rupture stress. Prediction of thermal stress can be made according to the guidelines provided by Karcher (1976), recommended by API 521 (2014). Final gauge pressure following a temperature increase in a blocked liquid is given by the formula:

$$P_2 = P_1 + \frac{(T_2 - T_1)\cdot(\alpha_v - 3\cdot\alpha_l)}{\chi + \left(\frac{d}{2\cdot E\cdot\delta_W}\right)\cdot(2.5 - 2\cdot\nu)} \tag{5.38}$$

where:

- P_1 is the initial gauge pressure (kPa).
- T_1 is the initial temperature (°C).
- T_2 is the final temperature (°C).
- α_v is the cubic expansion coefficient of liquid (1/°C).
- α_l is the linear expansion coefficient of the metal wall (1/°C).
- χ is isothermal compressibility of the liquid (1/kPa).
- D is the internal pipe diameter (m).

Table 5.4 Recommended data for Karcher's equation (API 521, 2014)

Material	α_l (1/°C)	E (kPa)
Carbon steel	1.21×10^{-5}	207×10^{6}
304 Stainless steel	1.73×10^{-5}	193×10^{6}
316 Stainless steel	1.60×10^{-5}	193×10^{6}
Nickel-copper alloy	1.01×10^{-5}	169×10^{6}

- E is Young's modulus (kPa).
- δ_W is the metal wall thickness (m).
- ν is Poisson's ratio, 0.3.

and α_v can be calculated as (API 521, 2014):

$$\alpha_v = \frac{\rho_1^2 - \rho_2^2}{2 \cdot (T_2 - T_1) \cdot \rho_1 \cdot \rho_2} \chi \tag{5.39}$$

where ρ_s are liquid densities at the respective temperatures, and:

$$\chi = \rho_1 \cdot \frac{\frac{1}{\rho_1} - \frac{1}{\rho_2}}{P_2 - P_1} \tag{5.40}$$

where ρ_s are liquid densities at their respective absolute pressure (Table 5.4).

Example 5.6

Crude oil is trapped in a 100 mm diameter, 5 mm thick carbon steel pipe at 25°C and 1 atm gauge (101,325 Pa g). Find the value of the solar radiation at which the pipe is subject to the yield tensile stress if air temperature rises to 40°C.

Solution

α_v (1/°C)	α_l (1/°C)	χ (1/kPa)	d (m)	δ_W (m)	E (kPa)	h (W/m^2)	ε	P_1 (Pa)	T_1 (°C)	T_a (°C)	σ_y (MPa)
0.001	1.21×10^{-5}	1.45×10^{-6}	0.1	0.005	207×10^{6}	11.3 (Coker, 2015)	0.9	101,325	25	40	300

Final pressure can be calculated assuming the maximum σ:

$$\sigma_t = \frac{P \times D}{2 \times t} \tag{5.41}$$

Rearranging:

$$P_2 = \frac{2 \cdot 0.005 \cdot 300}{0.1} = 30 \, \text{MPag} \tag{5.42}$$

Solving Eq. (5.38) for T_2:

$$T_2 = 25 + \frac{(30{,}000 - 101.325)\cdot \left[1.45\times10^{-6} + \left(\frac{0.1}{2\cdot 207\times10^{6}\times\cdot 0.005}\right)\cdot(2.5-2\cdot 0.3)\right]}{(0.001 - 3\cdot 1.21\times10^{-5})} = 73°\text{C} \tag{5.43}$$

Thermal balance on the pipe:

$$\begin{aligned} W_{sun} &= h\cdot(T_2 - T_a) + \varepsilon\cdot\sigma\cdot\left(T_2^4 - T_a^4\right) = 11.3\cdot(73-40) \\ &+0.8\cdot 5.67\cdot 10^{-8}\left(346^4 - 313^4\right) = 587.6\text{W/m}^2 \end{aligned} \tag{5.44}$$

5.7 Flixborough UVCE: Analysis of the Structural Causes

Fig. 5.22 shows the sketch of the Nypro-Plant reactors 4 and 6, after the removal of reactor 5, which had been discovered to have a crack. The by-pass had to accommodate the difference level between the outlets of the connected reactors and a 20 in. pipe was used instead of the 28 in. branches pipe. The presence of two stainless steel expansion bellows,

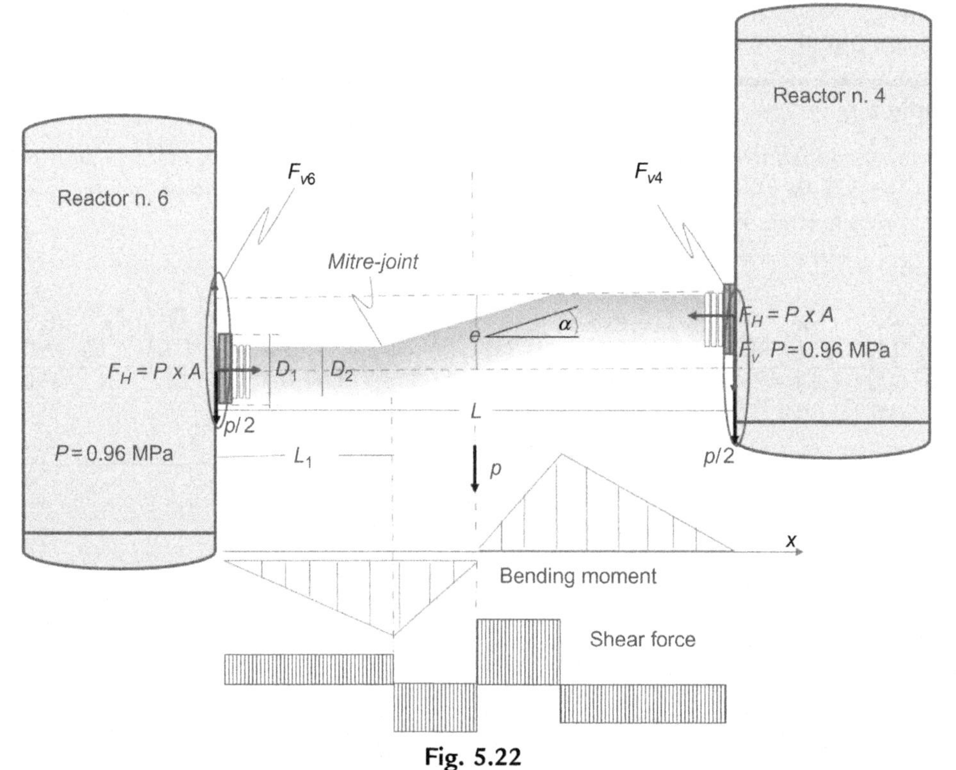

Fig. 5.22
Forces and bending moments acting on reactors 4 and 6.

Table 5.5 Process, structural and geometric data of Flixborough incident

P (Pa)	Temperature (°C)	D_1 (m)	D_2 (m)	L (m)	L_1 (m)	Pipe Bridge Thickness, t_2 (mm)	Pipe Bridge Weight (Pa)
960,000	155	0.75	0.5	6.17	2.5	4.8	7850

e (m)	Bellow Thickness, t_1 (mm)	α (deg)	Yield Strength (MPa)	Ultimate Strength (MPa)	Shear to Tensile Strength Ratio
0.356	20	1	648	766	0.6

and the existence of an eccentricity e between the axis of the reactor outlet, were ignored by people on-site.

Immediately before pipe rupture, the equipment and structural scenario was the one depicted in Fig. 5.22. In the same picture the bending moment and the shear force diagrams are shown (Table 5.5).

The two disaligned pressure thrusts, and the related turning moment, was compensated by two opposite shear forces and by the resulting moment. It has to be considered that the dog-leg pipe weight has to be added to the reactor 4 shear load and subtracted from the reactor 6 shear load.

Balance of forces

$$F_H = 960{,}000 \cdot \frac{\pi}{4} \cdot 0.75^2 = 423{,}900\text{N} \tag{5.45}$$

$$F_V = \frac{423{,}900 \cdot 0.356}{6.17} = 24{,}458\text{N} \tag{5.46}$$

$$F_{V4} = 24{,}458 + 3925 = 28{,}383\text{N} \tag{5.47}$$

$$F_{V6} = 24{,}458 - 3925 = 20{,}506\text{N} \tag{5.48}$$

For $0 > x > L_1$

$$S(x) = F_V \tag{5.49}$$

$$M_b(x) = F_V \cdot x \tag{5.50}$$

For the most loaded side, Reactor n. 4:

$$S_4(x) = F_{V4} = 28{,}384\text{N} \tag{5.51}$$

$$M_b(x) = 28{,}384 \cdot x \tag{5.52}$$

For $x > L_1$

$$S(x) = F_V \cdot \cos\alpha - F_H \cdot \sin\alpha \tag{5.53}$$

$$M_b(x) = F_V \cdot x - F_H \cdot tg\alpha \cdot (x - L_1) \tag{5.54}$$

For the most overloaded side, Reactor n. 4:

$$S_4(x) = F_{V4} \cdot \cos\alpha - F_{H4} \cdot \sin\alpha \tag{5.55}$$

$$S_4(x) = 28{,}384 \cdot 0.94 - 424{,}900 \cdot 0.34 = -117{,}785\,\text{N} \tag{5.56}$$

$$M_{b4}(x) = F_{V4} \cdot x - F_H \cdot tg\alpha \cdot (x - L_1) \tag{5.57}$$

$$M_b(x) = 28{,}384 \cdot x - 423{,}900 \cdot 0.36 \cdot (x - 2.5) \tag{5.58}$$

5.7.1 Stress Analysis

Bellow shear stress

$$\tau_4(x) = \frac{F_{V4}}{\pi \cdot D_1 \cdot t} = \frac{2 \cdot 28{,}384}{\pi \cdot 0.75 \cdot 0.001} = 21.1\,\text{MPa} \tag{5.59}$$

Shear stress at the mitre-joint point

$$\tau_4(x) = \frac{F_{V4}}{\pi \cdot D_1 \cdot t} = \frac{28{,}384}{\pi \cdot 0.5 \cdot 0.0048} = 3.77\,\text{MPa} \tag{5.60}$$

Tensile stress at the mitre-joint point

$$\tau_4(x) = \frac{F_{V4} \cdot \left(\frac{D_2}{2}\right)}{\pi \cdot \left(\frac{D_2}{2}\right)^3 \cdot t} = \frac{28{,}384 \cdot 2.5}{\pi \cdot \left(\frac{0.5}{2}\right)^2 \cdot 0.0048} = 75.33\,\text{MPa} \tag{5.61}$$

5.7.2 Analysis and Conclusions

Shear stresses in two potential sensitive sections, and bending tensile stress in the most loaded section, the mitre-joint point, have been calculated. These values coincide with those reported by the Court of Inquiry (Report of the Court of Inquiry, 1975). All these values are lower than yield stresses for steel, nor has any significant plastic effect been identified in the calculations. This has been confirmed by Marshall (1987), who has concluded that a much bigger pressure is required for the bellow to burst. The Report of the Court of Inquiry included some pictures taken on-site.

Fig. 5.23
Bellow on reactor 4 after the explosion (Court of inquiry, 1975).

Fig. 5.23 shows that that no rupture occurred on the bellow; however, the shear force has been considered capable of causing squirm (large distortion) of the bellow on the reactor 4 side, which had 12 convolutions (Venart), and then the bending stresses around the mitre-joint point, resulting in buckling and jack-knifing of the by-pass pipe. It is worth noting that people on site, in setting up the connection of reactor 4 to reactor 6, accounted just for the longitudinal stress (Fig. 5.24):

Fig. 5.24
Jack-knifed by-pass pipe (Court of inquiry, 1975).

$$\sigma_L = \frac{P \cdot D}{4 \cdot t} \tag{5.62}$$

ignoring the importance of other loads, along with the implications of the eccentricity of the bridge-pipe and the buckling hazard.

References

American Petroleum Institute, 2014. API Standard 521: Pressure-relieving and Depressuring Systems, sixth ed. January.

Baker, W.E., Cox, P.A., Westine, P.S., Kulesz, J.J., Strehlow, R.A., 1983. Explosion Hazards and Evaluation. Elsevier Scientific Publishing Company.

Bednar, H., 1986. Pressure Vessel Design Handbook, second ed. Krieger Publishing Company.

Bjerketvedt, D., Bakke, J.R., van Wingerden, K., 1997. Gas explosion handbook. J. Hazard. Mater. 52, 1–150. Elsevier.

Budynas, R.G., Nisbett, K.J., 2014. In: Shigley's Mechanical Engineering Design.McGraw-Hill Series in Mechanical Engineering.

CCPS, 2005. Glossary of Terms. American Institute of Chemical Engineers.

Coker, A., 2015. Ludwig's Applied Process Design for Chemical and Petrochemical Plants, fourth ed. Gulf Professional Publishing.

Karcher, G.C., 1976. Pressure Changes in Liquid Filled Vessels or Piping due to Temperature Changes, ExxonMobil Research and Engineering Mechanical Newsletter EE.84E.76, August.

Marshall, V.C., 1987. Major Chemical Hazards, John Wiley and Sons.

Materials Data Book, 2003. Cambridge University Engineering Department, http://www-mdp.eng.cam.ac.uk/web/library/enginfo/cueddatabooks/materials.pdf.

Report of the Court of Inquiry, 1975. The Flixborough Disaster, Her Majesty's Stationery Office, Department of Employment.

Sturm, R.G., 1941. A Study of the Collapsing Pressure of Thin-Wall Cylinders, University of Illinois Bulletin, Engineering Experiment Station, Bulletin Series No 329, November 11.

Timoshenko, S.P., Gere, J.M., 1961. Theory of Elastic Stability. McGraw-Hill Book Co., New York.

CHAPTER 6

Statistics and Reliability for Process Safety

God doesn't play dice with the world.

(Albert Einstein, 1879–1955)

6.1 Background

6.1.1 Gaussian Function

Gaussian function is described by the mathematical formula:

$$f(x) = A \cdot e^{-\frac{(x-b)^2}{2 \cdot c^2}} \tag{6.1}$$

where A, b, and c are real constants. It is worth noting:

$$e^{-\frac{(x-b)^2}{2 \cdot c^2}} > 0 \forall x \in R \tag{6.2}$$

$$\lim_{x \to \pm\infty} e^{-\frac{(x-b)^2}{2 \cdot c^2}} = 0 \tag{6.3}$$

Being:

$$\int_{-\infty}^{+\infty} e^{-x^2} \cdot dx = \sqrt{\pi} \tag{6.4}$$

Assuming:

$$\frac{x-b}{\sqrt{2} \cdot c} = y \tag{6.5}$$

It is:

$$dx = \sqrt{2} \cdot c \cdot dy \tag{6.6}$$

and:

$$\int_{-\infty}^{+\infty} A \cdot e^{-\frac{(x-b)^2}{2 \cdot c^2}} \cdot dx = A \cdot \sqrt{2} \cdot c \cdot \int_{-\infty}^{+\infty} e^{-y^2} \cdot dy = A \cdot \sqrt{2 \cdot \pi} \cdot c \tag{6.7}$$

Process Safety Calculations. https://doi.org/10.1016/B978-0-08-101228-4.00006-X

Defining A, b, and c as:

$$A = \frac{1}{\sqrt{2 \cdot \pi \cdot \sigma}} \tag{6.8}$$

$$b = \sigma \tag{6.9}$$

$$c = \sigma \tag{6.10}$$

a normalised Gaussian function is defined as:

$$g(x) = \frac{1}{\sqrt{2 \cdot \pi \cdot \sigma}} \cdot e^{-\frac{(x-\mu)^2}{2 \cdot \sigma^2}} \tag{6.11}$$

for which:

$$\int_{-\infty}^{+\infty} g(x) \cdot dx = 1 \tag{6.12}$$

6.1.2 Gaussian Probability Distribution

A random variable x is said to be normally distributed with mean μ and variance σ^2 (standard deviation σ) if its probability density function is described by the Gaussian function (Fig. 6.1):

$$f_n(x) = \frac{1}{\sqrt{2 \cdot \pi \cdot \sigma}} \cdot e^{-\frac{(x-\mu)^2}{2 \cdot \sigma^2}} \tag{6.13}$$

It's useful to introduce a unitless variable X:

$$X = \frac{x - \mu}{\sigma} \tag{6.14}$$

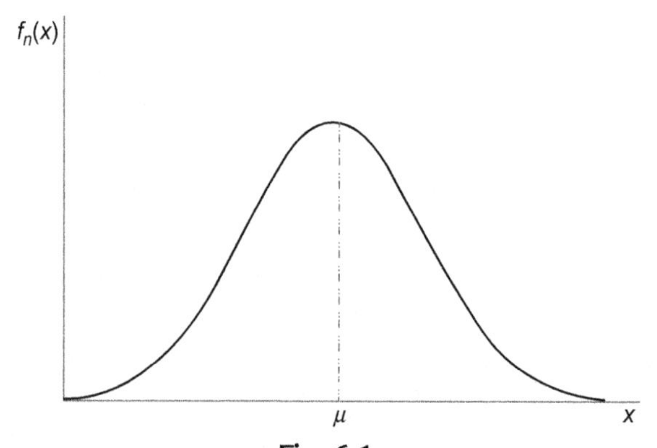

Fig. 6.1
Gaussian function.

This variable has distribution density equal to:

$$f_n(x) = \frac{1}{\sqrt{2 \cdot \pi}} \cdot e^{-\frac{X^2}{2}} \tag{6.15}$$

so it corresponds to a Gaussian variable of mean = 0 and variance = 1. This is the so-called standard normal distribution. The normal distribution:

$$F_n(Y) = \frac{1}{\sqrt{2 \cdot \pi}} \cdot \int_{-\infty}^{Y} e^{-\frac{X^2}{2}} \cdot dX \tag{6.16}$$

gives the probability that a standard normal variable assumes a value in the interval [0, Y] (Fig. 6.2).

Of course:

$$F_n(+\infty) = \frac{1}{\sqrt{2 \cdot \pi}} \cdot \int_{-\infty}^{+\infty} e^{-\frac{X^2}{2}} \cdot dX = 1 \tag{6.17}$$

The rationale for adopting the variable X is because it is:

$$F_n(X) = \frac{1}{\sqrt{2 \cdot \pi}} \cdot \int_{-\infty}^{X} e^{--\frac{X^2}{2}} \cdot dX = \frac{1}{2} \cdot \left(erf \frac{X}{\sqrt{2}} \right) \tag{6.18}$$

where $\left(erf \frac{X}{\sqrt{2}} \right)$ is the function error. Probabilities for the normal variable to assume specific values to the left of X are tabulated for calculation use (Tables 6.1 and 6.2).

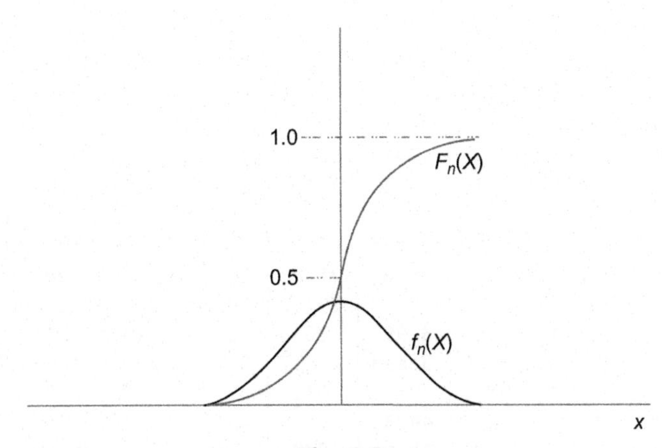

Fig. 6.2
Gaussian distribution.

Table 6.1 Standard normal distribution for $-3.9 \leq X \leq 0$

X	.00	.01	.02	.03	.04	.05	.06	.07	.08	.09	Area to the LEFT of the X Score
−3.9	.00005	.00005	.00004	.00004	.00004	.00004	.00004	.00004	.00003	.00003	
−3.8	.00007	.00007	.00007	.00006	.00006	.00006	.00006	.00005	.00005	.00005	
−3.7	.00011	.00010	.00010	.00010	.00009	.00009	.00008	.00008	.00008	.00008	
−3.6	.00016	.00015	.00015	.00014	.00014	.00013	.00013	.00012	.00012	.00011	
−3.5	.00023	.00022	.00022	.00021	.00020	.00019	.00019	.00018	.00017	.00017	
−3.4	.00034	.00032	.00031	.00030	.00029	.00028	.00027	.00026	.00025	.00024	
−3.3	.00048	.00047	.00045	.00043	.00042	.00040	.00039	.00038	.00036	.00035	
−3.2	.00069	.00066	.00064	.00062	.00060	.00058	.00056	.00054	.00052	.00050	
−3.1	.00097	.00094	.00090	.00087	.00084	.00082	.00079	.00076	.00074	.00071	
−3.0	.00135	.00131	.00126	.00122	.00118	.00114	.00111	.00107	.00104	.00100	
−2.9	.00187	.00181	.00175	.00169	.00164	.00159	.00154	.00149	.00144	.00139	
−2.8	.00256	.00248	.00240	.00233	.00226	.00219	.00212	.00205	.00199	.00193	
−2.7	.00347	.00336	.00326	.00317	.00307	.00298	.00289	.00280	.00272	.00264	
−2.6	.00466	.00453	.00440	.00427	.00415	.00402	.00391	.00379	.00368	.00357	
−2.5	.00621	.00604	.00587	.00570	.00554	.00539	.00523	.00508	.00494	.00480	
−2.4	.00820	.00798	.00776	.00755	.00734	.00714	.00695	.00676	.00657	.00639	
−2.3	.01072	.01044	.01017	.00990	.00964	.00939	.00914	.00889	.00866	.00842	
−2.2	.01390	.01355	.01321	.01287	.01255	.01222	.01191	.01160	.01130	.01101	
−2.1	.01786	.01743	.01700	.01659	.01618	.01578	.01539	.01500	.01463	.01426	
−2.0	.02275	.02222	.02169	.02118	.02068	.02018	.01970	.01923	.01876	.01831	
−1.9	.02872	.02807	.02743	.02680	.02619	.02559	.02500	.02442	.02385	.02330	
−1.8	.03593	.03515	.03438	.03362	.03288	.03216	.03144	.03074	.03005	.02938	
−1.7	.04457	.04363	.04272	.04182	.04093	.04006	.03920	.03836	.03754	.03673	
−1.6	.05480	.05370	.05262	.05155	.05050	.04947	.04846	.04746	.04648	.04551	
−1.5	.06681	.06552	.06426	.06301	.06178	.06057	.05938	.05821	.05705	.05592	
−1.4	.08076	.07927	.07780	.07636	.07493	.07353	.07215	.07078	.06944	.06811	
−1.3	.09680	.09510	.09342	.09176	.09012	.08851	.08691	.08534	.08379	.08226	
−1.2	.11507	.11314	.11123	.10935	.10749	.10565	.10383	.10204	.10027	.09853	
−1.1	.13567	.13350	.13136	.12924	.12714	.12507	.12302	.12100	.11900	.11702	
−1.0	.15866	.15625	.15386	.15151	.14917	.14686	.14457	.14231	.14007	.13786	
−0.9	.18406	.18141	.17879	.17619	.17361	.17106	.16853	.16602	.16354	.16109	
−0.8	.21186	.20897	.20611	.20327	.20045	.19766	.19489	.19215	.18943	.18673	
−0.7	.24196	.23885	.23576	.23270	.22965	.22663	.22363	.22065	.21770	.21476	
−0.6	.27425	.27093	.26763	.26435	.26109	.25785	.25463	.25143	.24825	.24510	
−0.5	.30854	.30503	.30153	.29806	.29460	.29116	.28774	.28434	.28096	.27760	
−0.4	.34458	.34090	.33724	.33360	.32997	.32636	.32276	.31918	.31561	.31207	
−0.3	.38209	.37828	.37448	.37070	.36693	.36317	.35942	.35569	.35197	.34827	
−0.2	.42074	.41683	.41294	.40905	.40517	.40129	.39743	.39358	.38974	.38591	
−0.1	.46017	.45620	.45224	.44828	.44433	.44038	.43644	.43251	.42858	.42465	
−0.0	.50000	.49601	.49202	.48803	.48405	.48006	.47608	.47210	.46812	.46414	

Table 6.2 Standard normal distribution for $0 \leq X \leq 3.9$

X	.00	.01	.02	.03	.04	.05	.06	.07	.08	.09	Area to the LEFT of the X Score
0.0	.50000	.50399	.50798	.51197	.51595	.51994	.52392	.52790	.53188	.53586	
0.1	.53983	.54380	.54776	.55172	.55567	.55962	.56356	.56749	.57142	.57535	
0.2	.57926	.58317	.58706	.59095	.59483	.59871	.60257	.60642	.61026	.61409	
0.3	.61791	.62172	.62552	.62930	.63307	.63683	.64058	.64431	.64803	.65173	
0.4	.65542	.65910	.66276	.66640	.67003	.67364	.67724	.68082	.68439	.68793	
0.5	.69146	.69497	.69847	.70194	.70540	.70884	.71226	.71566	.71904	.72240	
0.6	.72575	.72907	.73237	.73565	.73891	.74215	.74537	.74857	.75175	.75490	
0.7	.75804	.76115	.76424	.76730	.77035	.77337	.77637	.77935	.78230	.78524	
0.8	.78814	.79103	.79389	.79673	.79955	.80234	.80511	.80785	.81057	.81327	
0.9	.81594	.81859	.82121	.82381	.82639	.82894	.83147	.83398	.83646	.83891	
1.0	.84134	.84375	.84614	.84849	.85083	.85314	.85543	.85769	.85993	.86214	
1.1	.86433	.86650	.86864	.87076	.87286	.87493	.87698	.87900	.88100	.88298	
1.2	.88493	.88686	.88877	.89065	.89251	.89435	.89617	.89796	.89973	.90147	
1.3	.90320	.90490	.90658	.90824	.90988	.91149	.91309	.91466	.91621	.91774	
1.4	.91924	.92073	.92220	.92364	.92507	.92647	.92785	.92922	.93056	.93189	
1.5	.93319	.93448	.93574	.93699	.93822	.93943	.94062	.94179	.94295	.94408	
1.6	.94520	.94630	.94738	.94845	.94950	.95053	.95154	.95254	.95352	.95449	
1.7	.95543	.95637	.95728	.95818	.95907	.95994	.96080	.96164	.96246	.96327	
1.8	.96407	.96485	.96562	.96638	.96712	.96784	.96856	.96926	.96995	.97062	
1.9	.97128	.97193	.97257	.97320	.97381	.97441	.97500	.97558	.97615	.97670	
2.0	.97725	.97778	.97831	.97882	.97932	.97982	.98030	.98077	.98124	.98169	
2.1	.98214	.98257	.98300	.98341	.98382	.98422	.98461	.98500	.98537	.98574	
2.2	.98610	.98645	.98679	.98713	.98745	.98778	.98809	.98840	.98870	.98899	
2.3	.98928	.98956	.98983	.99010	.99036	.99061	.99086	.99111	.99134	.99158	
2.4	.99180	.99202	.99224	.99245	.99266	.99286	.99305	.99324	.99343	.99361	
2.5	.99379	.99396	.99413	.99430	.99446	.99461	.99477	.99492	.99506	.99520	
2.6	.99534	.99547	.99560	.99573	.99585	.99598	.99609	.99621	.99632	.99643	
2.7	.99653	.99664	.99674	.99683	.99693	.99702	.99711	.99720	.99728	.99736	
2.8	.99744	.99752	.99760	.99767	.99774	.99781	.99788	.99795	.99801	.99807	
2.9	.99813	.99819	.99825	.99831	.99836	.99841	.99846	.99851	.99856	.998610	
3.0	.99865	.99869	.99874	.99878	.99882	.99886	.99889	.99893	.99896	.99900	

Continued

Table 6.2 Standard normal distribution for $0 \leq X \leq 3.9$—cont'd

X	.00	.01	.02	.03	.04	.05	.06	.07	.08	.09	Area to the LEFT of the *X* Score
3.1	.99903	.99906	.99910	.99913	.99916	.99918	.99921	.99924	.99926	.99929	
3.2	.99931	.99934	.99936	.99938	.99940	.99942	.99944	.99946	.99948	.99950	
3.3	.99952	.99953	.99955	.99957	.99958	.99960	.99961	.99962	.99964	.99965	
3.4	.99966	.99968	.99969	.99970	.99971	.99972	.99973	.99974	.99975	.99976	
3.5	.99977	.99978	.99978	.99979	.99980	.99981	.99981	.99982	.99983	.99983	
3.6	.99984	.99985	.99985	.99986	.99986	.99987	.99987	.99988	.99988	.99989	
3.7	.99989	.99990	.99990	.99990	.99991	.99991	.99992	.99992	.99992	.99992	
3.8	.99993	.99993	.99993	.99994	.99994	.99994	.99994	.99995	.99995	.99995	
3.9	.99995	.99995	.99996	.99996	.99996	.99996	.99996	.99996	.99997	.99997	

Example 6.1

H_2S concentration leaks from compressor seals are statistically described by a Gaussian density function, with mean and standard deviation equal to 5 ppm. Find:

1. The probability that the concentration is lower than 5 ppm.
2. The probability that the concentration falls within the range 5–7 ppm.

Solution

1. The standard variable is:

$$X = \frac{5-5}{5} = 0 \tag{6.19}$$

Enter this value in the table:

$$F_n(0) = 0.5 \tag{6.20}$$

So the probability will be 50%. This result is intuitive because 5 is the mean.

2. This is a differential probability that can be calculated as:

$$F_n\left(\frac{7-5}{5}\right) - F_n\left(\frac{5-5}{5}\right) = 0.6554 - 0.5 = 0.1554 \tag{6.21}$$

The probability will be 15.54%.

6.1.3 Probit Function

Bliss (1934) discovered that the dosage–response curve of a toxic agent upon the mortality of an organism could be expressed as an asymmetrical S-shaped curve, ranging between 0% and 100% of killed organisms. Conceptually, this was very similar to the function $F_n(X)$. Brown (1978) confirmed that toxicological dose–response tests frequently showed that the observed proportion of responders monotonically increases with dose, according to a sigmoid relation with some function of the exposure level. This evidence led to the development of probit dose–tolerance distribution by Finney (1952). The model assumes that the population distribution of tolerance is given by the Gaussian function:

$$f[x(D)] = \frac{1}{\sqrt{2 \cdot \pi \cdot \sigma}} \cdot e^{-\frac{[x(D)-\mu]}{2 \cdot \sigma^2}} \tag{6.22}$$

where D represents di dose level. The unitless variable $X(D)$:

$$X(D) = \frac{x(D) - \mu}{\sigma} \tag{6.23}$$

can be written as:

$$X(D) = -\frac{\mu}{\sigma} + \frac{x(D)}{\sigma} \tag{6.24}$$

Assuming:

$$-\frac{\mu}{\sigma}=a \tag{6.25}$$

and:

$$+\frac{1}{\sigma}=b \tag{6.26}$$

it is worth noting that probit is a random variable with mean 5 and standard deviation 1. Introducing a and b:

$$X(D)=a+b\cdot x(D) \tag{6.27}$$

Brown (1978) reports the following function commonly used for $x(D)$:

$$x(D)=\log_{10}(D) \tag{6.28}$$

Finally, the unitless variable becomes:

$$X(D)=a+b\cdot \log_{10}(D) \tag{6.29}$$

Moving from the decimal to the natural logarithms, the conventional equation adopted in process safety is:

$$X(D)=a+b\cdot \ln(D) \tag{6.30}$$

In this equation $X(D)$, it is the probit corresponding to the probability of death P, which can be read on Table 6.3 of probit.

6.2 Probit Functions for Process Safety

Evidence and experimental research have shown that a wide range of CAUSE-IMPACT phenomena, not only belonging to toxicological scenarios, occurs in a way that can be satisfactorily described by the general probit function:

$$Pr=a+b\cdot \ln(A) \tag{6.31}$$

Table 6.3 Table of probit

P	0	0.01	0.02	0.03	0.04	0.05	0.06	0.07	0.08	0.09
0	–	2.67	2.95	3.12	3.25	3.36	3.45	3.52	3.59	3.66
0.1	3.72	3.77	3.82	3.87	3.92	3.96	4.01	4.05	4.08	4.12
0.2	4.16	4.19	4.23	4.26	4.29	4.33	4.36	4.39	4.42	4.45
0.3	4.48	4.50	4.53	4.56	4.59	4.61	4.64	4.67	4.69	4.72
0.4	4.75	4.77	4.80	4.82	4.85	4.87	4.90	4.92	4.95	4.97
0.5	5.00	5.03	5.05	5.08	5.10	5.13	5.15	5.18	5.20	5.23
0.6	5.25	5.28	5.31	5.33	5.36	5.39	5.41	5.44	5.47	5.50
0.7	5.52	5.55	5.58	5.61	5.64	5.67	5.71	5.74	5.77	5.81
0.8	5.84	5.88	5.92	5.95	5.99	6.04	6.08	6.13	6.18	6.23
0.9	6.28	6.34	6.41	6.48	6.55	6.64	6.75	6.88	7.05	7.33

where Pr is the *probit function* related to the impact, and A is a physical or chemical agent eventually associated with a time of action or exposure. Table 6.4 summarises the most frequent probit functions adopted in process safety with the definition of the impact scenario, the parameters of influence, and the reference. Constants have been referenced and included in Tables 6.5–6.7.

Table 6.4 Probit equations for process safety

No	Scenario	Damage	Probit Equation	Parameters	Institutions	Reference
1	Toxicity	Contamination	$Pr=a+b\cdot\ln(C^n\cdot t)$	C=concentration (mg/m^3) t=exposure time (min) a, b, n=constants of the substance	TNO	Constant a and b by Uijt de Haag and Ale (2005)
2	Toxicity	Contamination	$C^n\cdot t$=constant (SLOT AND SLOD in probit function)	C=concentration (mg/m^3) t=exposure time (min) n=exponent of the substance	HSE	http://www.hse.gov.uk/chemicals/haztox.htm (HSE)
3	Fire	Burns	$Pr=a+b\cdot\ln(W^n\cdot t/10^4)$	W=radiation intensity (W/m^2) t=exposure time (min)	US Dept. of Transportation	Eisenberg et al. (1975)
4	Explosion	Lung haemorrhage	$Pr=a+b\cdot\ln(P^o)$	P^o=overpressure (N/m^2)	US Dept. of Transportation	Eisenberg et al. (1975)
5	Explosion	Ear drum rupture	$Pr=a+b\cdot\ln(P^o)$			
6	Explosion	Glass breakage	$Pr=a+b\cdot\ln(P^o)$			
7	Explosion	Death from impact	$Pr=a+b\cdot\ln(I^o)$	I^o=Impulse (Ns/m^2)	US Dept. of Transportation	Eisenberg et al. (1975)
8	Explosion	Injuries from impact	$Pr=a+b\cdot\ln(I^o)$			
9	Explosion	Injuries from flying fragments	$Pr=a+b\cdot\ln(I^o)$			
10	Explosion	Minor damage to buildings	$Pr=a+b\cdot\ln(V)$			
11	Explosion	Major damage to buildings	$V=(C/P^o)^c+(D/I^o)^d$	P^o=overpressure (N/m^2) I^o=Impulse (Ns/m^2)	TNO	CPR 16E–Green Book (1992)
12	Explosion	Buildings collapse				

SLOT, specified level of toxicity; SLOD, significant likelihood of death.

Table 6.5 Constants of probit equations for scenario 1

Scenario No	Substance	*a*	*b*	*n*
1	Acrolein	−4.1	1.00	1.00
	Acrylonitrile	−8.6	1.00	1.30
	Ammonia	−11.7	1.00	2.00
	Benzene (HSE)	−109.78	5.30	2.00
	Bromine	−12.4	1.00	2.00
	Carbon monoxide	−7.4	1.00	1.00
	Chlorine	−6.35	0.50	2.75
	Ethylene oxide	−6.8	1.00	1.00
	Hydrogen chloride	−37.3	3.69	1.00
	Hydrogen cyanide	−9.8	1.00	2.40
	Hydrogen fluoride	−8.4	1.00	1.50
	Hydrogen sulphide	−11.5	1.00	1.90
	Methyl bromide	−7.3	1.00	1.10
	Methyl isocyanate	−1.2	1.00	0.70
	Nitrogen dioxide	−18.6	1.00	3.70
	Phosgene	−10.6	2.00	1.00
	Phosphine	−6.8	1.00	2.00
	Sulphur dioxide	−19.2	1.00	2.40
	Toluene (HSE)	−6.794	0.41	2.50

Table 6.6 Constants for SLOT and SLOD probit equations

Scenario No	Substance	SLOT ($ppm^n \cdot min$)	SLOD ($ppm^n \cdot min$)	*n*
2	Acetic acid	7.50×10^4	3.00×10^5	1.00
	Acetonitrile	1.08×10^4	1.60×10^5	1.00
	Acrylamide	1.30×10^5	5.20×10^5	1.00
	Ammonia	3.78×10^8	1.03×10^8	2.00
	Carbon dioxide	1.50×10^{40}	1.50×10^{41}	8.00
	Carbon disulphide	9.60×10^4	3.84×10^5	1.00
	Carbon monoxide	4.01×10^4	5.70×10^4	1.00
	Carbon tetrachloride	7.20×10^5	2.60×10^6	1.00
	Chloride	1.08×10^5	4.84×10^5	2.00
	1,2-Dicholoroethane	9.00×10^4	3.60×10^5	1.00
	Ethyl mercaptan	1.66×10^5	6.62×10^5	1.00
	Ethylene oxide	4.68×10^4	1.87×10^5	1.00
	Fluorine	3.80×10^5	1.50×10^6	2.00
	Formaldehyde	5700	8100	1.00
	Glutaraldehyde	1410	5640	1.00
	Hydrazine	1.51×10^4	6.05×10^4	1.00
	Hydrogen bromine	1.22×10^4	4.88×10^4	1.00
	Hydrogen chloride	2.37×10^4	7.65×10^4	1.00
	Hydrogen cyanide	1.92×10^5	4.32×10^5	2.00
	Hydrogen fluoride	1.20×10^4	2.10×10^4	1.00
	Hydrogen peroxide (solution)	8.60×10^4	3.44×10^5	1.00

Table 6.6 Constants for SLOT and SLOD probit equations—cont'd

Scenario No	Substance	SLOT ($ppm^n \cdot min$)	SLOD ($ppm^n \cdot min$)	*n*
	Hydrogen sulphide	2.00×10^{12}	2.10×10^{13}	4.00
	Iodine	2055	8220	1.00
	Methanol	8.02×10^{5}	2.67×10^{6}	1.00
	Methyl chloride	2.02×10^{5}	8.10×10^{5}	1.00
	Methyl isocyanate	750	1680	1.00
	Nitric oxide	2.09×10^{4}	2.43×10^{4}	1.00
	Nitrobenzene	8.54×10^{4}	3.41×10^{5}	1.00
	m-Nitrotoluene	5.06×10^{4}	2.02×10^{5}	1.00
	o-Nitrotoluene	1.38×10^{5}	5.52×10^{5}	1.00
	m-Nitrotoluene	5.06×10^{4}	2.02×10^{5}	1.00
	p-Nitrotoluene	1.89×10^{5}	7.65×10^{5}	1.00
	Ozone	1980	3600	1.00
	Phenol	1.50×10^{4}	6.00×10^{4}	1.00
	Phosgene	108	348	1.00
	Pyridine	1.35×10^{5}	5.40×10^{5}	1.00
	Sulphur dioxide	4.65×10^{6}	7.45×10^{7}	2.00
	Toluene diisocyanate	176	480	1.00
	Vinyl chloride	3.39×10^{6}	1.36×10^{7}	2.00

Table 6.7 Constants of probit equations for scenarios 3–12

Scenario No	*a*	*b*	*N*	*C*	*c*	*D*	*d*
3	−14.9	2.56	4/3	–	–	–	–
4	−77.1	6.91	–	–	–	–	–
5	−15.6	1.93	–	–	–	–	–
6	−18.1	2.79	–	–	–	–	–
7	−46.1	4.82	–	–	–	–	–
8	−39.1	4.45	–	–	–	–	–
9	−27.1	4.26	–	–	–	–	–
10	5	−0.26	–	4600	3.9	110	5.0
11	5	−0.26	–	17,500	8.4	290	9.3
12	5	−0.22	–	40,000	7.4	460	11.3

6.3 Failure Frequency and Probability

Events occurring random at constant rate in a given time interval can be described by Poisson's distribution. If λ is the constant frequency at which the event occurs, the probability of event occurring n times at time t is given by the formula:

$$P_n(t) = \frac{(\lambda \cdot t)^n \cdot e^{-\lambda \cdot t}}{n!} \tag{6.32}$$

So, the probability of occurring zero times is:

$$P_0(t) = e^{-\lambda \cdot t} \tag{6.33}$$

and this relationship defines the reliability $R(t)$ of a system affected by the constant failure rate λ, i.e., the probability that no failure occurs from $t=0$ to time$=t$:

$$R(t) = e^{-\lambda \cdot t} \tag{6.34}$$

The complement of the reliability is the failure probability P_f, that is given by the formula:

$$P_f(t) = 1 - e^{-\lambda \cdot t} \tag{6.35}$$

By definition of probability density function, it is:

$$f_f(t) = \frac{dP_f(t)}{dt} = \lambda \cdot e^{-\lambda \cdot t} \tag{6.36}$$

So, the probability of failure in the time period t_o to T:

$$P_f(t_o \rightarrow T) = \lambda \cdot \int_{t_o}^{T} \lambda \cdot e^{-\lambda \cdot t} = e^{-\lambda \cdot t_o} - e^{-\lambda \cdot T} \tag{6.37}$$

which coincides with Eq. (6.35) if t_o is equal to zero. If the probability $P(0 \rightarrow T)$ is known, the failure rate can be calculated as:

$$P_f(t_o \rightarrow T) = \lambda \cdot \int_{t_o}^{T} \lambda \cdot e^{-\lambda \cdot t} = e^{-\lambda \cdot t_o} - e^{-\lambda \cdot T} \tag{6.38}$$

$$P_f(0 \rightarrow T) - 1 = -e^{-\lambda \cdot T} \tag{6.39}$$

$$-P_f(0 \rightarrow T) + 1 = e^{-\lambda \cdot T} \tag{6.40}$$

$$\ln\left[1 - P_f(0 \rightarrow T)\right] = -\lambda \cdot T \tag{6.41}$$

$$\lambda = -\frac{\ln\left[1 - P_f(0 \rightarrow T)\right]}{T} \tag{6.42}$$

Poisson distribution allows one to convert frequency into probability and vice versa.

Example 6.2

The failure rate of a centrifugal pump is 2.6^{-1}/year. Find the probability of failure in 6 months' time.

Solution

$$P_f(0 \rightarrow 0.5) = 1 - e^{-0.26 \cdot 0.5} = 0.12 \tag{6.43}$$

Example 6.3

The probability of failure of a compressor in 3 months' time is 0.28. Find the failure rate.

Solution

$$P_f(0 \rightarrow 0.25) = 0.28 = 1 - e^{-\lambda \cdot 0.25} \tag{6.44}$$

$$\lambda = -\frac{\ln\left[1 - P_f(0 \rightarrow 0.25)\right]}{0.25} = -\frac{\ln[1 - 0.28]}{0.25} = 1.314/y \tag{6.45}$$

The ratio:

$$\frac{1}{\lambda} \tag{6.46}$$

is the time requested for a failure to occur. It is called Mean Time To Failure (MTTF).

6.4 Failures and Faults

6.4.1 Definitions

- Failure: termination of the ability of a functional unit to perform a required function (IEC 61511-3)
- Fault: abnormal condition that may cause a reduction in, or loss of, the capability of a functional unit to perform a required function (IEC 61511-3)
- All failures are faults, not all faults are failures.
- Failure: termination of the ability of a functional unit to perform a required function (IEC 61511-3)
- Failure rate: measurement of the rate of loss of functionality by a given component expressed in terms of the fraction of the test sample. From the mathematical point of view:

$$\frac{dN}{N \times dt} = \lambda \tag{6.47}$$

where N is the component number. Integrating, with No components at time $t = 0$ (assumed as all working):

$$\frac{N(t)}{N_o} = e^{-\lambda \times t} \tag{6.48}$$

that is consistent with the Poisson's distribution.

- Failure rates of items that are in continuous service: rate of loss of functionality of components whose service is continuously requested within the selected time interval
- Failure rates of items that work on demand: rate of loss of functionality of components whose service is not continuously requested, but only when necessary or specifically required within the selected time interval

- Composite failure rate: failure rate of items that are operated only a fraction of time are operated, and/or during operation, at different load levels.

6.4.2 Failure Rates

Failure rates for instrumentation and equipment (Tables 6.8–6.11):

6.4.3 Composite Failure Rate

According to the definition, a composite failure rate is given by the following general formula:

$$\lambda_C = \lambda_1 \cdot x_1 + \lambda_2 \cdot \prod x_i \cdot s_i \tag{6.49}$$

where:

- λ_c is the composite failure rate.
- λ_1 is the failure rate during nonoperating time.
- λ_2 is the failure rate during operating time.
- x_1 is the fraction of time during which the system is not operating.
- x_i are the fractions of time during which the system is operating, corresponding to the load levels s_i.

Table 6.8 Fault rates for instrumentation

Instrumentation (Anyakora et al., 1971)	
Item	**Faults/Year**
Controller	0.29
Control valve	0.60
Solenoid valve	0.42
I/P converter	0.49
Valve positioner	0.44
Magnetic flowmeter	2.18
Differential pressure flowmeter	1.73
Load cell solid flowmeter	3.75
Belt speed measurement	15.3
Liquid level measurement	1.70
Solid level measurement	6.86
Temperature measurement (excluding pyrometers)	0.35
Thermocouple	0.52
Resistance thermometer	0.41
Pressure switch	0.34
Flow switch	1.12
Optical pyrometer	9.70
PH-meter	5.88
Oxygen analyser	5.65
Carbon dioxide analyser	10.5
Hydrogen analyser	0.99
Controller settings	0.14

Table 6.9 Fault rates for equipment

Equipment (OREDA, 2002)	
Item	**Faults/Year**
Centrifugal pump	
Breakdown	0.0100
Internal leakage	0.0014
External leakage (process side)	0.0200
External leakage (utility side)	0.0140
Fail to start on demand	0.0200
Fail to stop on demand	0.0011
Compressor	
Breakdown	0.0110
Internal leakage	0.0120
External leakage (process side)	0.0899
External leakage (utility side)	0.1033
Fail to start on demand	0.1966
Fail to stop on demand	0.0126
Gas turbine	
Breakdown	0.0800
Internal leakage	0.0175
External leakage (process side)	0.1640
External leakage (utility side)	0.0676
Fail to start on demand	0.9700
Fail to stop on demand	0.0260
Combustion engine	
Diesel	
External leakage	0.3570
Fail to start on demand	0.3570
Low output	0.3570
Electric generator	
Breakdown	0.0094
External leakage	0.0019
Fail to start on demand	0.1253
Fail to stop on demand	0.0120
Low output	0.0068
Faulty output voltage	0.0097
Electric motor (pump)	
Breakdown	0.0178
Fail to start on demand	0.0300
Low output	0.0036
Heat exchanger	
External leakage (process side)	0.0376
External leakage (utility side)	0.0116
Plugged/choked	0.0050
Valve	
External leakage	0.0021
Fail to close on demand	0.0202
Fail to open on demand	0.0246
PSV	
Leakage in open position	0.1339
Fail to close on demand	0.0050
Fail to open on demand	0.0215
ESDV	
Leakage in open position	0.1850
Fail to close on demand	0.1394
Fail to open on demand	0.0500

Tanks and vessels (Health and Safety Executive, 2012)		
Item	**Faults/Year**	
Large atmospheric tanks (>450 m^3)		
Catastrophic	5.0×10^{-6}	
Major	1.0×10^{-4}	
Minor	2.5×10^{-3}	
Roof	2.0×10^{-3}	
Small and medium atmospheric tanks (<450 m^3)	**Flammable contents**	**Non flammable contents**
Catastrophic	1.6×10^{-5}	8.0×10^{-6}
Major	1.0×10^{-4}	5.0×10^{-5}
Minor	1.0×10^{-3}	5.0×10^{-4}

Equipment (OREDA, 2002)		
Item	**Faults/Year**	
Refrigerated ambient pressure tanks	**Single walled**	**Double walled**
Catastrophic	4.0×10^{-5}	5.0×10^{-7}
Major	1.0×10^{-5}	1.0×10^{-5}
Minor	8.0×10^{-5}	3.0×10^{-5}
Failure with release of vapour only	2.0×10^{-4}	4.0×10^{-4}
LNG refrigerated tanks		**Double walled**
Catastrophic	–	5.0×10^{-8}
Major	–	1.0×10^{-6}
Minor	–	3.0×10^{-6}
Failure with release of vapour only	–	4.0×10^{-5}
Pressure vessels		
Catastrophic (upper failure)	6.0×10^{-6}	
Catastrophic (median failure)	4.0×10^{-6}	
Catastrophic (lower failure)	2.0×10^{-6}	
50 mm hole	5.0×10^{-6}	
25 mm hole	5.0×10^{-6}	
13 mm hole	5.0×10^{-6}	
6.0 mm hole	1.0×10^{-5}	
	4.0×10^{-5}	
LPG pressure vessels		
Catastrophic	2.0×10^{-6}	
BLEVE	1.0×10^{-5}	
50 mm hole	5.0×10^{-6}	
25 mm hole	5.0×10^{-6}	
13 mm hole	1.0×10^{-5}	
Spherical vessels		
Catastrophic (upper failure)	6.0×10^{-6}	
Catastrophic (median failure)	4.0×10^{-6}	
Catastrophic (lower failure)	2.0×10^{-6}	
50 mm hole	5.0×10^{-6}	
25 mm hole	5.0×10^{-6}	
13 mm hole	1.0×10^{-5}	
6.0 mm hole	4.0×10^{-5}	
Chemical reactors		
Catastrophic	1.0×10^{-5}	
Catastrophic (potential for thermal runaway)	5.0×10^{-5}	
Catastrophic (no potential for thermal runaway)	3.0×10^{-6}	
50 mm hole	5.0×10^{-6}	
25 mm hole	5.0×10^{-6}	
13 mm hole	1.0×10^{-5}	
6.0 mm hole	4.0×10^{-5}	

Table 6.10 Failure rates for pipework

Failure Rates (per m per year) For Pipework Diameter (mm) (HSE, 2012)					
Hole Diameter Size (mm)	**0–49**	**50–149**	**150–299**	**300–499**	**500–1000**
3.00	1×10^{-5}	2×10^{-6}			
4.00			1×10^{-6}	8×10^{-7}	7×10^{-7}
25.0	5×10^{-6}	1×10^{-6}	7×10^{-7}	5×10^{-7}	4×10^{-7}
1/3 Pipework diameter			4×10^{-7}	2×10^{-7}	1×10^{-7}
Guillotine	1×10^{-6}	5×10^{-7}	2×10^{-7}	7×10^{-8}	4×10^{-8}

Table 6.11 Failure rates for above-ground pipelines

Failure Rates (per m per year) For Above Ground Pipelines (HSE, 2012)	
Failure Category	**Faults/Year**
Rupture	6.5×10^{-9}
Large hole (110 mm $\geq d >$ 75 mm)	3.3×10^{-8}
Small hole (75 mm $\geq d >$ 25 mm)	6.7×10^{-8}
Pin hole ($d \leq$ 25 mm)	1.6×10^{-7}

Example 6.4

A unit works according to the following daily load levels:

Load Percentage	Hours per Day	Load Levels	Fraction of Time
0	4	0	0.17
25	8	1	0.33
75	6	3	0.25
100	6	4	0.25

The failure rate during nonoperating time = 150 failure/million hours. The failure rate during operating time is 300 failure/million hours. Find the composite failure rate (Fig. 6.3).

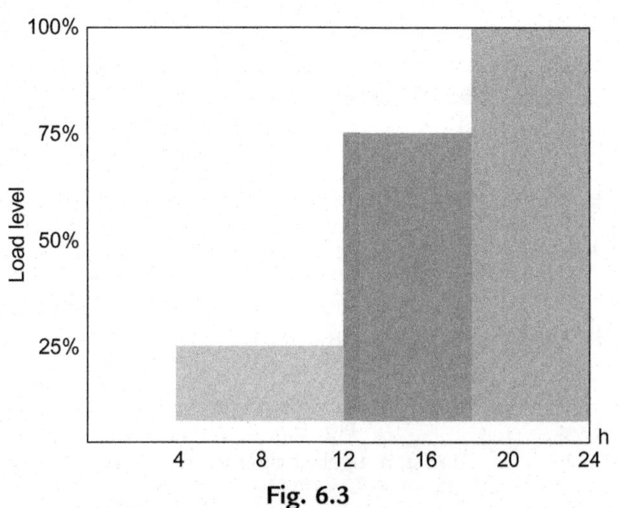

Fig. 6.3
Load levels with time for Example 6.4.

Solution

$$\lambda_c = 0.17 \cdot 150 + 300 \cdot (0.33 \cdot 1 + 0.25 \cdot 3 + 0.25 \cdot 4) = 649.5/\text{million\_hours} \tag{6.50}$$

6.5 Boolean Algebra

Application of Boolean algebra in process safety consists of the binary identification of the frequency or probability of occurrence of a given event, on the basis of the combination of probabilities/frequencies of two or more causing events (Fig. 6.4).

An *AND GATE* states that for the consequence to occur, it is necessary that all of the causing events occur at the same time.

An *OR GATE* states that for the consequence to occur, it is enough that just one of the causing events occurs (Fig. 6.5).

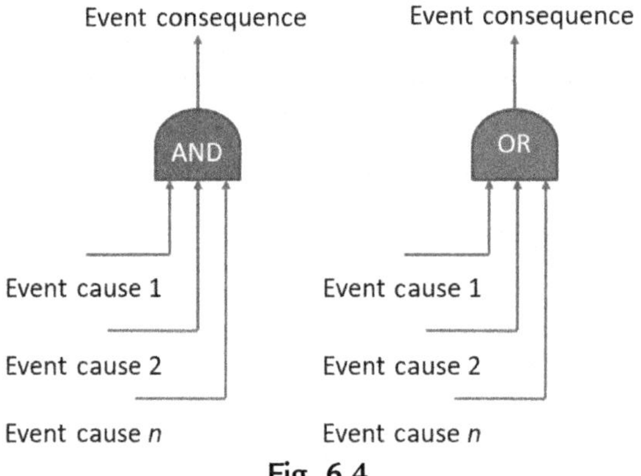

Fig. 6.4
Gates AND and OR.

GATE	Permitted Operations	Non-Permitted Operations
 AND	$f = f_k \times \Pi P_i$ $P = \Pi P_i$	$f \neq \Pi f_i$
 OR	$f = \Sigma f_i$ $P = 1 - \Pi(1 - P_i)$	$\Sigma P_i + \Sigma f_i$

Fig. 6.5
Boolean algebra operations.

Example 6.5

Two failure events relevant to the components A and B occur at a frequency equal to 0.1 y^{-1} and 0.5 y^{-1}, respectively. The functionality of at least one of them is essential for the integrity of the system. Analyse the case study.

Solution

System loss of integrity is caused by the simultaneous failure of two components, so the applicable gate is AND. On the other hand, we have two frequencies to combine, but no operations are permitted. Consequently, we have to convert one frequency into a probability:

$$P_A = 1 - e^{-\lambda_A \cdot t} = 1 - e^{-0.1} = 0.095 \tag{6.51}$$

or:

$$P_B = 1 - e^{-\lambda_B \cdot t} = 1 - e^{-0.5} = 0.39 \tag{6.52}$$

Option	Probability	Frequency (y^{-1})	$f_{failure}$ (y^{-1})
1	$P_A = 0.095$	$f_B = 0.5$	$f = 0.047$
2	$f_A = 0.1$	$P_B = 0.4$	$f = 0.040$

Example 6.6

A loop consisting of one sensor, one logical solver and one actuator presents respective failure rates of 1.41, 0.29, and 0.6 y^{-1}. Find the loop failure probability and the overall failure rate.

Solution

Probability

$$P_{ACT} = 1 - e^{-0.6\times 1} = 0.45 \tag{6.53}$$

$$P_{LS} = 1 - e^{-0.29\times 1} = 0.25 \tag{6.54}$$

$$P_{SENSOR} = 1 - e^{-1.41\times 1} = 0.76 \tag{6.55}$$

$$P = 1 - \Pi(1 - P_i) \tag{6.56}$$

$$P = 1 - \Pi(1 - P_i) = 1 - (1 - 0.45) \times (1 - 0.25) \times (1 - 0.76) = 0.9 \tag{6.57}$$

Frequency

$$f = 1.41 + 0.6 + 0.29 = 2.3y^{-1} \tag{6.58}$$

6.6 Boolean Algebra in Functional Safety

6.6.1 Probability and Frequency of Failure on Demand

The definition of probability of failure on demand is given by the IEC 61508-4 (2010) as *the safety unavailability of a system to perform the specified safety function when a demand occurs*. If PFD is the Probability of Failure on Demand, a Probability of Availability on Demand (PAD) is:

$$PAD = (1 - PFD) \tag{6.59}$$

According to Eq. (6.35), If one assumes that the probability of failure is very low and tends to zero, it can be written:

$$\lim_{\lambda \cdot t \to 0} 1 - e^{-\lambda \cdot t} = \lambda \cdot t \tag{6.60}$$

In functional safety, this approximation is generally adopted.

6.6.2 Common Cause Failure

Common cause failure is defined as *the failure of more than one component, item, or system due to the same cause* (CCPS, 2005). According to IEC 61508-61511 and to ISA-TR84.00.02, if not better specified or known, a beta (probability)-factor (β) equal to 0.1 is assigned for CCF. This has to be considered as an inherent system failure affecting all of the involved items, in parallel to the general failure rate relevant to a single item (Figs. 6.6 and 6.7).

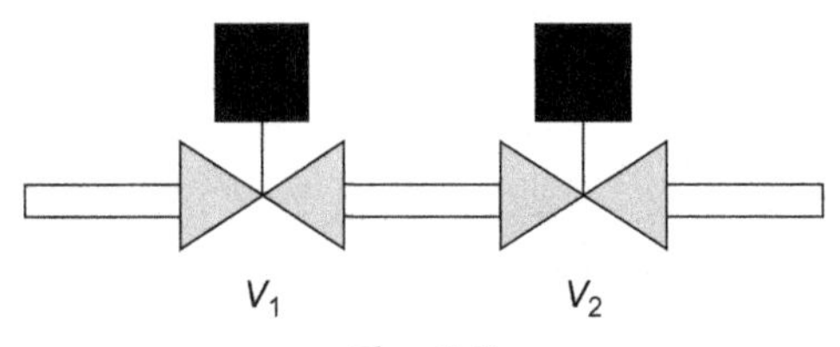

Fig. 6.6
Common cause failure.

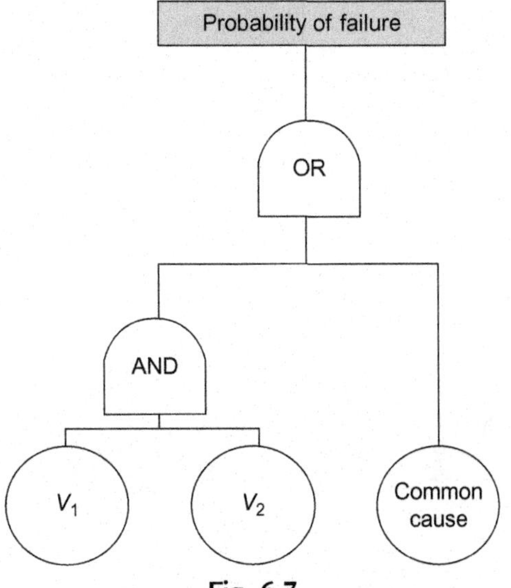

Fig. 6.7
Common cause failure—fault tree.

Example 6.7

Find the probability failure for a couple of identical electric generators (Fig. 6.8).

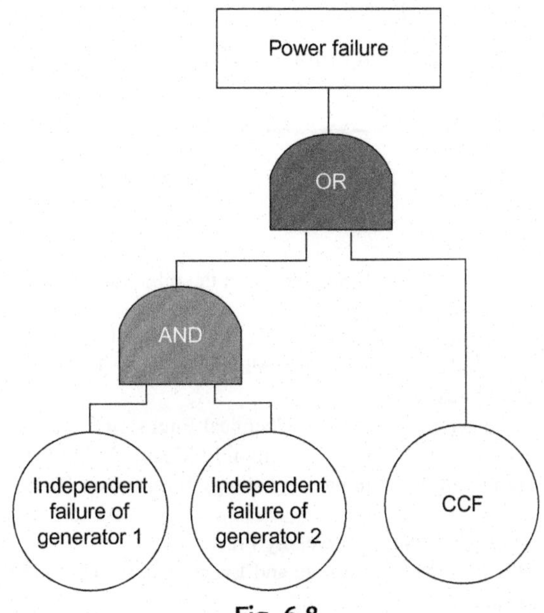

Fig. 6.8
Fault tree for Example 6.7.

Solution

Item	Number	Failure Rate, λ (y^{-1})	CCF
Electric generator	2	0.01	$\beta=0.1$

The probability of failure on demand can be calculated as shown in the following figure.

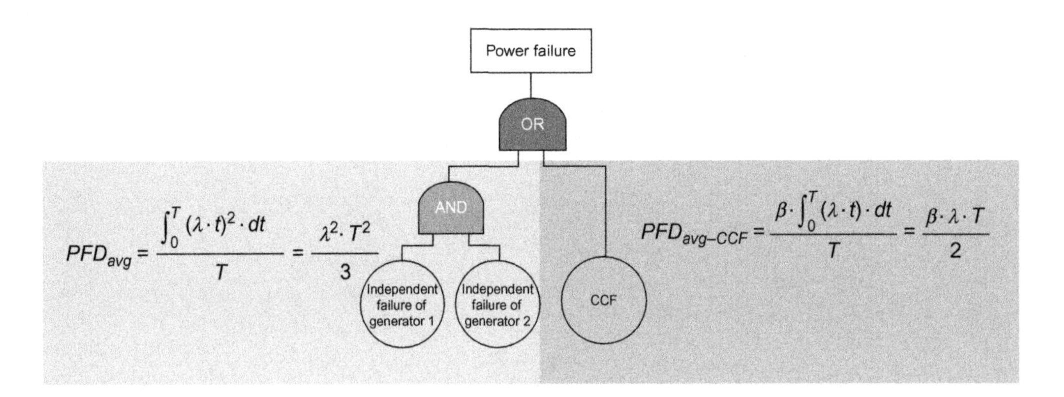

$$PFD_{avg} = \frac{\int_0^T (\lambda \cdot t)^2 \cdot dt}{T} = \frac{\lambda^2 \cdot T^2}{3} = \frac{0.01^2 \cdot 1}{3} = 3.3 \times 10^{-5} \tag{6.61}$$

$$PFD_{avg-CCF} = \frac{\beta \cdot \int_0^T \lambda \cdot t \cdot dt}{T} = \frac{\beta \cdot \lambda \cdot T}{2} = \frac{0.1 \cdot 0.01 \cdot 1}{2} = 5 \times 10^{-4} \tag{6.62}$$

$$P = 1 - \Pi(1 - P_i) = 1 - (1 - 0.000033) \cdot (1 - 0.0005) = 0.00053 \tag{6.63}$$

References

Anyakora, S.N., Engel, G.F.M., Lees, F.P., 1971. Some data on the reliability of instruments in the chemical plant environment. Chem. Engr. 255, 396.

Bliss, C.I., 1934. The method of probits. Science 79 (2053), 409–410.

Brown, C.C., 1978. The statistical analysis of dose-effect relationships. In: Butler, G.C. (Ed.), Principles of Ecotoxicology. John Wiley & Sons (Chapter 6).

CCPS, 2005. Glossary of Terms. American Institute of Chemical Engineers.

Eisenberg, N.A., Lynch, C.J., Breeding, R.J., 1975. Vulnerability Model. A Simulation System for Assessing Damage Resulting from Marine Spills. National Technology Information Service Report, AD-A015-245, Springfield, MA.

Finney, D.J., 1952. Probit Analysis. Cambridge University Press.

Health and Safety Executive (HSE), 2012. Failure Rate and Event Data for Use within Risk Assessments, http://www.hse.gov.uk/landuseplanning/failure-rates.pdf.

Health and Safety Executive (HSE), http://www.hse.gov.uk/chemicals/haztox.htm.

International Electrotechnical Commission, 2010. Functional safety of electrical/electronic/programmable electronic safety-related systems – Part 4: Definitions and abbreviations.

International Electrotechnical Commission, 2016. IEC 61511-3. Functional safety - Safety instrumented systems for the process industry sector - Part 3: Guidance for the determination of the required safety integrity levels.

ISA-TR84.00.02, 2002. Safety Instrumented Functions (SIF)-Safety Integrity Level (SIL) Evaluation Techniques Part 2: Determining the SIL of a SIF via Simplified Equations.

OREDA, 2002. Offshore Reliability Data Handbook, fourth ed., SINTEF, Det Norske Veritas.

TNO, 1992. Methods for the Determination of Possible Damage, CPR 16E, Green Book.

Uijt de Haag, P.A.M., Ale, B.J.M., 2005. Guideline for Quantitative Risk Assessment, CPR 18E, Purple Book, RVIM.

PART 2

Consequence Assessment

CHAPTER 7

Source Models

Πάντα ῥεῖ (Heraclitus, 535–475 BC.)

7.1 Summary of Scenarios

A general summary of the source models has been included in Table 7.1.

7.2 Subcooled Liquids

7.2.1 Unpressurised Liquid Discharge From Tanks

Unpressurised liquids are systems stored and processed at pressure comparable to the ambient pressure. This is typically atmospheric pressure. A so called tank model can be adopted when the containment aspect ratio (diameter to liquid ratio) order of magnitude is close to 1.

Mass flow rate can be calculated according to Bernoulli equation:

$$W(t) = C_D \cdot \rho_L \cdot \pi \cdot \frac{d^2}{4} \sqrt{2 \cdot g \cdot h(t)} \tag{7.1}$$

where:

- ρ_L is liquid density.
- d is the hole diameter.
- g is acceleration of gravity.
- $h(t)$ is the static head related to the relative pressure (gauge).
- C_D is the coefficient of discharge.

C_D can be assumed as 0.61 for sharp-edged orifices and for Reynolds numbers greater than 30,000 (Crowl and Louvar, 2011). In case of uncertainty, a value of 1 shall be adopted.

The transient level during the discharge will be calculated integrating Eq. (7.1) as follows.

Process Safety Calculations. https://doi.org/10.1016/B978-0-08-101228-4.00007-1

Table 7.1 Summary of source model scenarios

Phase	Process Condition	Storage Scenario	Source		Section
Liquid	Subcooled	Atmosph. pressure	Vertical Tank		7.2.1
			Horizontal tank/Pipelines		7.2.2
			Pool formation (transient radius)		7.7.1
			Pool evaporation		7.7.2
			Sources from EPA OCA models		7.8
		Pressurised	Vessel	Elastic-to-Torricellian transition (bulk modulus effect)	7.2.3
				Driving force: process pressure (highly volatile liquid)	
				Driving force: headspace gas pressure	
			Pipelines	Bulk modulus	7.2.3
	Boiling	Atmosph. pressure			7.3.1
		Pressurised			7.3.2 7.3.3
	Transition and transient behaviour				7.3.4 7.3.5 7.3.6
	Flash				7.4.1
	Rainout				7.4.2 7.4.3
	Droplet formation				7.4.4
	Carbon dioxide	Liquid to solid transition including snow-out			7.5
Gases Vapours			Choked/unchoked flow		7.6.1
			Releases from pipelines/blowdown		7.6.2
All	Simplified models				7.9

$$\frac{dM(t)}{dt} = \rho_L \cdot \pi \cdot \frac{D^2}{4} \cdot \frac{dh(t)}{dt} = -C_D \cdot \rho_L \cdot \pi \cdot \frac{d^2}{4} \cdot \sqrt{2 \cdot g \cdot h(t)} \tag{7.2}$$

Rearranging:

$$D^2 \cdot \frac{dh(t)}{dt} = -C_D \cdot d^2 \cdot \sqrt{2 \cdot g \cdot h(t)} \tag{7.3}$$

Integrating:

$$\sqrt{h(t)}-\sqrt{ho}=\frac{\sqrt{2\cdot g}}{2}\cdot C_D\cdot\left(\frac{d}{D}\right)^2\cdot t^2 \quad (7.4)$$

and:

$$h(t)=\left[\sqrt{ho}-\sqrt{\frac{g}{2}}\cdot C_D\cdot\left(\frac{d}{D}\right)^2\cdot t\right]^2 \quad (7.5)$$

Emptying time is:

$$t_e=\left[\frac{\sqrt{ho}}{\sqrt{\frac{g}{2}}\cdot C_D\cdot\left(\frac{d}{D}\right)^2}\right] \quad (7.6)$$

Example 7.1

Water from an atmospheric tank 15 m high is released through a 50 mm diameter hole located 3 m above the bottom. Tank diameter is 5 m. Find:

- The maximum discharge velocity
- The law of emptying
- The emptying time

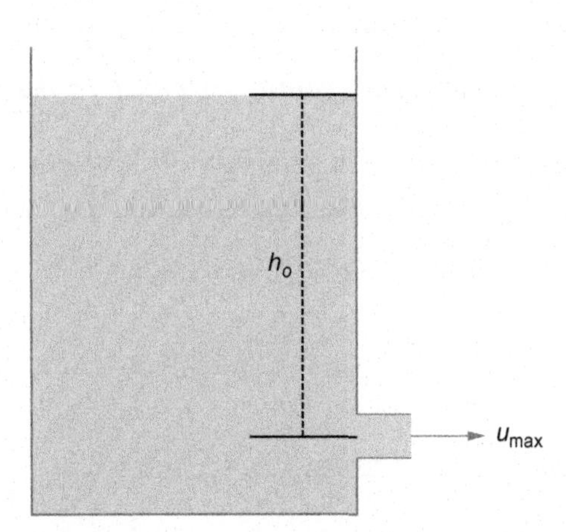

Solution

The maximum discharge velocity is found writing:

$$u_{max} = C_D \cdot \sqrt{2 \cdot g \cdot h_o} = 0.61 \cdot \sqrt{2 \cdot 9.81 \cdot 12} = 9.36 \text{ m/s} \tag{7.7}$$

The emptying law is included in the following figure:

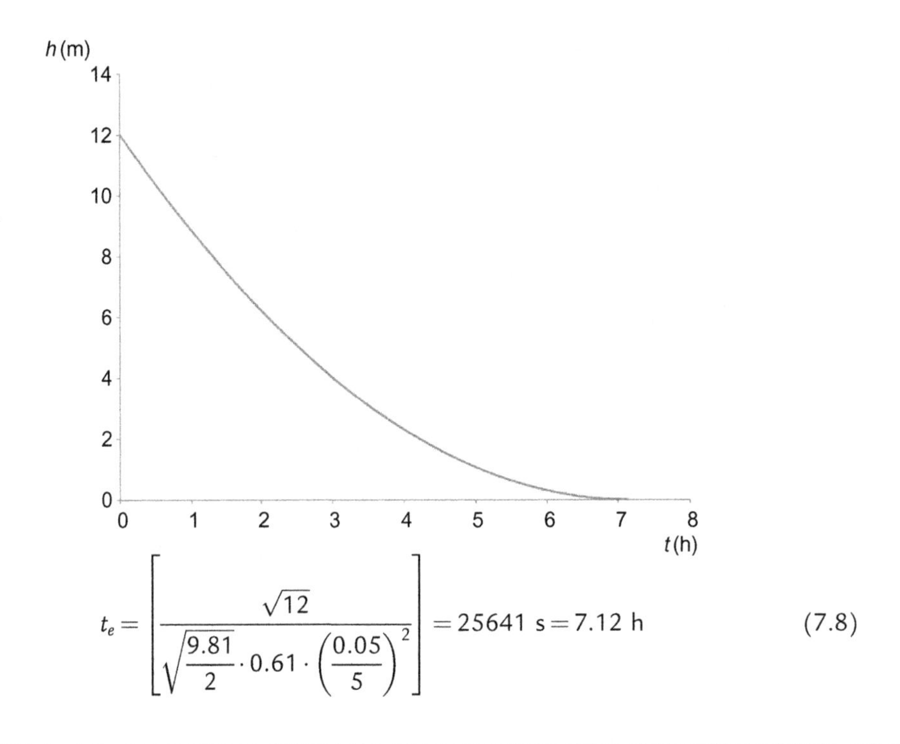

$$t_e = \left[\frac{\sqrt{12}}{\sqrt{\frac{9.81}{2}} \cdot 0.61 \cdot \left(\frac{0.05}{5} \right)^2} \right] = 25641 \text{ s} = 7.12 \text{ h} \tag{7.8}$$

7.2.2 Unpressurised Liquid Discharge From Horizontal Tanks and Pipelines

Horizontal tanks or pipelines have a different geometry. With reference to Fig. 7.1, the transient mass balance can be written after defining the cross section area as a function of level $h(t)$:

$$A = 0.5 \cdot R^2 \cdot (2 \cdot \alpha - \sin 2\alpha) \tag{7.9}$$

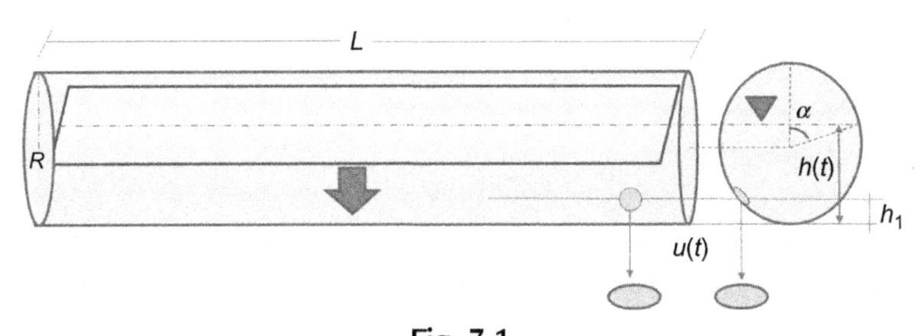

Fig. 7.1
Transient flow from unpressurised horizontal tank.

$$h(t) = R + R \cdot \cos\alpha \tag{7.10}$$

$$\alpha = ar\cos\left(\frac{h(t)}{R} - 1\right) \tag{7.11}$$

$$A = R^2 \cdot \left\{ ar\cos\left(\frac{h(t)}{R} - 1\right) - 0.5 \cdot \sin\left[2 \cdot ar\cos\left(\frac{h(t)}{R} - 1\right)\right]\right\} \tag{7.12}$$

$$dV = L \cdot R^2 \cdot d\left[ar\cos\left(\frac{h(t)}{R} - 1\right)\right] - 0.5 \cdot L \cdot R^2 \cdot d\sin\left[2 \cdot ar\cos\left(\frac{h(t)}{R} - 1\right)\right] \tag{7.13}$$

$$dV = L \cdot R^2 \cdot d\alpha - 0.5 \cdot L \cdot R^2 \cdot d\sin\alpha \tag{7.14}$$

Solving the mass balance:

$$\frac{dM(t)}{dt} = \rho_L \cdot \frac{dV[h(t)]}{dt} = -C_D \cdot \rho_L \cdot \pi \cdot \frac{d^2}{4} \cdot \sqrt{2 \cdot g \cdot h(t)} \tag{7.15}$$

The integration of the differential equation will include two terms, in addition to other constants:

$$\int_1^{\frac{h_1(t)}{R} - 1} \frac{d\alpha}{\sqrt{R \cdot (\cos\alpha + 1)}} \tag{7.16}$$

and:

$$\int_1^{\frac{h_1(t)}{R} - 1} \frac{2 \cdot \cos\alpha \cdot d\alpha}{\sqrt{R \cdot (\cos\alpha + 1)}} \tag{7.17}$$

7.2.3 Pressurised Liquids

Elastic-to-Torricellian transition (non-volatile liquids)

Liquid bulk modulus B_n has been defined in Section 4.4 as:

$$B = -v \cdot \left(\frac{\partial P}{\partial v}\right)_n \tag{7.18}$$

where:

- v is the specific volume.
- P is liquid pressure.

Benintendi (2016) has analysed outflow phenomena from pressurised vessels where liquid elasticity is the driving force. In fact, should a pressurised non-boiling liquid be affected by a sudden pressure drop due to the discharge through a hole in the containment wall, liquid elasticity is expected to play a significant role until the system reaches pure Torricelli's regime at atmospheric pressure. Temperature will substantially remain constant, whereas pressure will undergo a decay driven by liquid elasticity.

Vessel-type systems

For containment whose aspect ratio is not too far from 1, the mass flow rate through a hole of diameter d is given by following the formula, which is a Torricellian equation (Fig. 7.2):

$$M(t) = \int_0^t c_D \cdot \frac{\pi \cdot d^2}{4} \sqrt{2 \cdot \rho(t) \cdot (P(t) - P_{atm})} dt \tag{7.19}$$

Being:

$$B \cdot \ln \frac{\rho_o}{\rho(t)} = (P_o - P(t)) \tag{7.20}$$

It will be:

$$M_o - V \cdot \rho(t) = \int_0^t c_D \cdot \frac{\pi \cdot d^2}{4} \cdot \sqrt{2 \cdot \rho(t) \cdot B \cdot \ln \frac{\rho(t)}{\rho_{atm}}} dt \tag{7.21}$$

where M_o is the initial mass. In the previous equation, P is the absolute pressure.

Differentiating:

$$-V \cdot d\rho(t) = c_D \cdot \frac{\pi \cdot d^2}{4} \cdot \sqrt{2 \cdot \rho(t) \cdot B \cdot \ln \frac{\rho(t)}{\rho_{atm}}} dt \tag{7.22}$$

and being:

$$\frac{\rho(t)}{\rho_{atm}} = e^{\frac{(P(t) - P_{atm})}{B}} \tag{7.23}$$

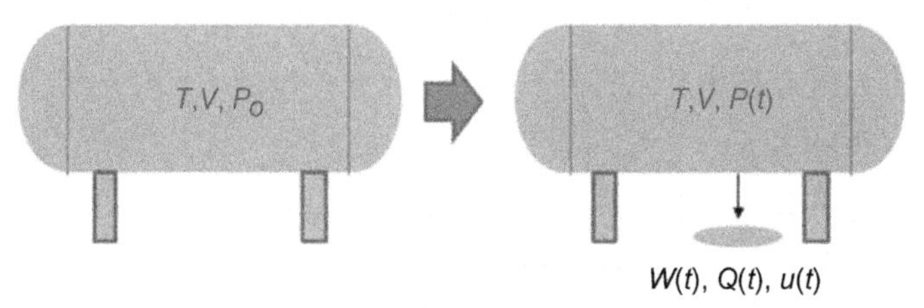

Fig. 7.2
Vessel-type system.

$$\frac{d\left(e^{\frac{P(t)-P_{atm}}{B}}\right)}{\sqrt{e^{\frac{P(t)-P_{atm}}{B}} \cdot \ln e^{\frac{P(t)-P_{atm}}{B}}}} = -c_D \cdot \frac{\pi \cdot d^2}{V \cdot 4} \cdot \sqrt{\frac{2 \cdot B}{\rho_{atm}}} \cdot dt \tag{7.24}$$

The integral of the first term of Eq. (7.24) between:

- time $= 0 \rightarrow P(0) = P_o$ (storage pressure)
- time $= t_e \rightarrow P(t_e) = P_{atm}$ (t_e = elastic to Torricellian transition time)

is:

$$\int_{e^{\frac{P_o-P_{atm}}{B}}}^{0} \frac{d\left(e^{\frac{P(t)-P_{atm}}{B}}\right)}{\sqrt{e^{\frac{P(t)-P_{atm}}{B}} \cdot \ln\left(e^{\frac{P(t)-P_{atm}}{B}}\right)}} = -\sqrt{2 \cdot \pi} \cdot \left[erfi\left(\sqrt{\frac{P_o - P_{atm}}{2 \cdot B}}\right)\right] \tag{7.25}$$

where *erfi* is the imaginary error function.

$$\sqrt{2 \cdot \pi} \cdot \left[erfi\left(\sqrt{\frac{P_o - P_{atm}}{2 \cdot B}}\right)\right] = C_D \cdot \frac{\pi \cdot d^2}{4 \cdot V} \cdot \sqrt{\frac{2 \cdot B}{\rho_{atm}}} \cdot t_e \tag{7.26}$$

Rearranging the elastic-to-Torricellian-phase transition time (end of the elastic phase) t_e of a vessel with initial pressure P, volume V, and liquid bulk modulus B is given by the definite integral:

$$\left[erfi\left(\sqrt{\frac{P_o - P_{atm}}{2 \cdot B}}\right)\right] = C_D \cdot \frac{d^2}{2 \cdot V} \cdot \sqrt{\frac{B \cdot \pi}{\rho_{atm}}} \cdot t_e \tag{7.27}$$

and:

$$t_e = \frac{\left[erfi\left(\sqrt{\frac{P_o - P_{atm}}{2 \cdot B}}\right)\right]}{C_D \cdot \frac{d^2}{2 \cdot V} \cdot \sqrt{\frac{B \cdot \pi}{\rho_{atm}}}} \tag{7.28}$$

Fig. 7.3 can be used for the calculation in the specified pressure range.

Example 7.2

Liquid kerosene, contained in a 25 m^3 vessel at an initial pressure of 200 bar, is released through a 50 mm diameter hole. Find:

- The initial mass flow rate
- The elastic-to-Torricellian-transition time, t_e
- The mass flow rate at half t_e

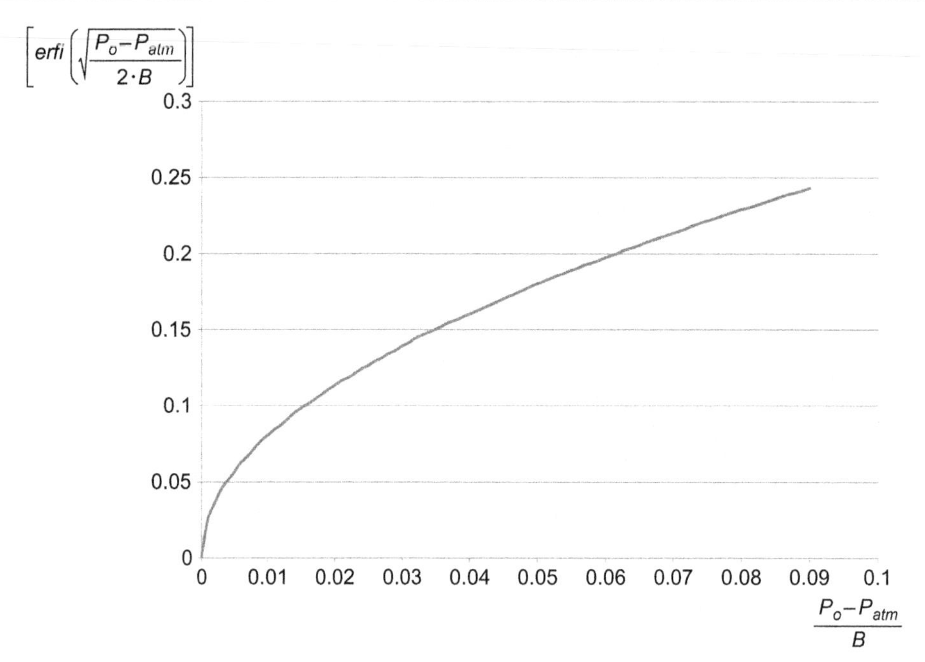

Fig. 7.3
Imaginary error function.

Solution

Average Bulk Modulus (bar)	Initial pressure (bar)	Volume (m^3)	Average density (kg/m^3)
13,000	200	25	790

Initial mass flow rate

The initial pressure has to be converted into a static head:

$$h_o = \frac{20{,}000{,}000\ \text{Pa}}{790 \cdot 9.81} = 2581\ \text{m} \tag{7.29}$$

The Torricellian velocity corresponding to this head is:

$$u_{\max} = C_D \cdot \sqrt{2 \cdot g \cdot h_o} = 0.61 \cdot \sqrt{2 \cdot 9.81 \cdot 2581} = 137\ \text{m/s} \tag{7.30}$$

This is less than kerosene sound speed, 1324 m/s. Flow is not choked.

$$W(0) = 790 \cdot \pi \cdot \frac{0.05^2}{4} \cdot 137 = 212.4\ \text{kg/s} \tag{7.31}$$

Elastic to Torricellian phase transition time, t_e

From Fig. 7.2 with:

$$t_e = \frac{\left[erfi\left(\sqrt{\frac{200-1}{2\cdot 13,000}}\right)\right]}{0.61\cdot\frac{0.05^2}{2\cdot 25}\cdot\sqrt{\frac{13,000\cdot\pi}{790}}} = \frac{0.099}{0.00022} = 451\ \text{s} \tag{7.32}$$

mass flow at half t_e:

$$\left[erfi\left(\sqrt{\frac{P_{1/2}-P_{atm}}{2\cdot B}}\right)\right] = C_D\cdot\frac{d^2}{2\cdot V}\cdot\sqrt{\frac{B\cdot\pi}{\rho_{atm}}}\cdot\frac{t_e}{2} = 0.00022\cdot\frac{451}{2} = 0.0496 \tag{7.33}$$

$$\frac{P_{1/2}-P_{atm}}{B} = 0.004 \tag{7.34}$$

$$P_{1/2} = 53\ \text{bar} \tag{7.35}$$

$$h_{1/2} = \frac{5,300,000\ \text{Pa}}{790\cdot 9.81} = 684\ \text{m} \tag{7.36}$$

$$u_{1/2} = C_D\cdot\sqrt{2\cdot g\cdot h_{1/2}} = 0.61\cdot\sqrt{2\cdot 9.81\cdot 684} = 70.6\ \text{m/s} \tag{7.37}$$

$$W(t_{1/2}) = 790\cdot\pi\cdot\frac{0.05^2}{4}\cdot 70.6 = 109.25\ \text{kg/s} \tag{7.38}$$

Pipeline-type systems

A distributed parameter (pipeline-type) system is typically a long pipeline of cross sectional diameter D and length L, initially at pressure P_o, which is assumed to be subject to a leak through a hole at one end. As liquid begins to flow out, a pressure drop is detected by just a small part around the hole. Unlike vessels, the rest of the pipeline will remain at the initial pressure and will not be impacted by the equilibrium change occurring at one end. This scenario can be easily approached, considering that the physical phenomenon belongs to the fluid-dynamics of pressure surge, just assuming that the system undergoes a pressure drop instead of a pressure rise along the pipeline. As the liquid flows out, a decompression wave will travel upstream from the hole at the speed of sound (wave speed), c, generally in the range of 600 and 1500 m/s (Shashi Menon, 2004), while liquid will move with a cross sectional velocity w_1 (see Fig. 7.4).

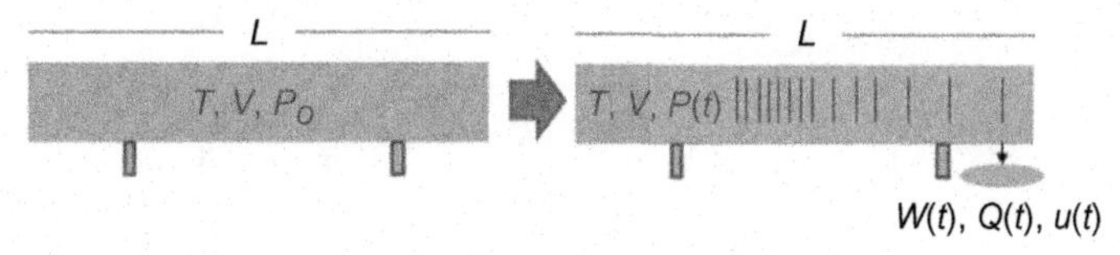

Fig. 7.4
Pipeline-type system.

Soon after the hole formation, a decompression wave J_1 will travel, and a corresponding liquid discharge will occur at velocity u_1, while liquid pressure and density will be reduced from the stagnant P_o and ρ_o to P_1 and ρ_1. Applying Allievi equation (Citrini and Noseda, 1987) for the decompression wave:

$$P_o - P_1 = \rho_1 \cdot c \cdot w_1 \tag{7.39}$$

and:

$$u_1 = C_D \cdot \sqrt{2 \cdot \frac{P_1 - P_{atm}}{\rho_1}} \tag{7.40}$$

Equation of continuity requires:

$$\frac{u_1}{w_1} = \left(\frac{D}{d}\right)^2 \tag{7.41}$$

The decompression wave will last L/C seconds, L being the pipeline length, so after this time, the initial mass M_o will have become:

$$M_1 = M_o - C_D \cdot \sqrt{2 \cdot (P_1 - P_{atm}) \cdot \rho_1} \cdot \pi \cdot \frac{d^2}{4} \cdot \frac{L}{c} \tag{7.42}$$

and the density:

$$\rho_1 = \frac{M_o - C_D \cdot \sqrt{2 \cdot (P_1 - P_{atm}) \cdot \rho_1} \cdot \pi \cdot \frac{d^2}{4} \cdot \frac{L}{c}}{V} \tag{7.43}$$

The system includes five unknown variables, P_1, u_1, w_1, ρ_1, M_o, and five independent equations. In general, N decompression waves travel until Torricelli regime is attained. The generic compression waves and relevant unknowns can be generalised:

$$P_{j+1} - P_j = \rho_j \cdot c \cdot w_j \tag{7.44}$$

$$u_j = C_D \cdot \sqrt{2 \cdot \frac{(P_j - P_{atm})}{\rho_j}} \tag{7.45}$$

$$\frac{u_j}{w_j} = \left(\frac{D}{d}\right)^2 \tag{7.46}$$

$$M_j = M_{j+1} - C_D \cdot \sqrt{2 \cdot g \cdot (P_j - P_{atm}) \rho_j} \cdot \pi \cdot \frac{d^2}{4} \cdot \frac{L}{c} \tag{7.47}$$

$$\rho_j = \frac{M_{j+1} - C_D \cdot \sqrt{2 \cdot g \cdot (P_j - P_{atm}) \rho_j} \cdot \pi \cdot \frac{d^2}{4} \cdot \frac{L}{c}}{V} \tag{7.48}$$

The release scenario has been fully resolved, and the same calculation presented for the vessel-type system can be carried out.

Example 7.3

A water pipeline, 2500 m long, with a 500 mm diameter, is operated at 50 bar (absolute pressure) and 20°C. Find the initial release rate and the elastic to Torricellian phase transition time t_e through a 50 mm hole.

Solution

Initial pressure (bar)	Temperature (°C)	Initial density (kg/m^3)	Volume (m^3)	Initial mass (kg)	Sound speed (m/s)
50	20	1000.4	490.6	490,821	1500

The calculation can significantly be simplified solving, by iteration, the two equations:

$$P_{j+1} - P_j = \rho_j \cdot 1500 \cdot w_j \tag{7.49}$$

$$u_j = 0.61 \cdot \sqrt{2 \cdot \frac{(P_j - P_{atm})}{\rho_j}} \tag{7.50}$$

and considering that density variation is very small, and that:

$$\frac{u_j}{w_j} = \left(\frac{D}{d}\right)^2 = \left(\frac{0.5}{0.05}\right)^2 = 100 \tag{7.51}$$

Wave trip No	Time	P_j (bar abs)	u_j (m/s)	w_j (m/s)	ρ_j (kg/m^3)
0	0	50	0	0	1000.4
1	$1 \times \frac{L}{c} = 1 \times \frac{2500}{1500} = 1.67\,\text{s}$	41.7	55	0.55	1000.06
2	$2 \times \frac{L}{c} = 2 \times \frac{2500}{1500} = 3.33\,\text{s}$	34.3	49	0.49	999.7
3	$3 \times \frac{L}{c} = 3 \times \frac{2500}{1500} = 5.0\,\text{s}$	27.7	44	0.44	999.4
4	$4 \times \frac{L}{c} = 4 \times \frac{2500}{1500} = 6.7\,\text{s}$	21.8	39	0.39	999.1
5	$5 \times \frac{L}{c} = 5 \times \frac{2500}{1500} = 8.3\,\text{s}$	16.7	34	0.34	998.9
6	$6 \times \frac{L}{c} = 6 \times \frac{2500}{1500} = 10.0\,\text{s}$	12.3	29	0.29	998.7
7	$7 \times \frac{L}{c} = 7 \times \frac{2500}{1500} = 11.6\,\text{s}$	8.7	24	0.24	998.5
8	$8 \times \frac{L}{c} = 8 \times \frac{2500}{1500} = 13.3\,\text{s}$	5.8	19	0.19	998.4
9	$9 \times \frac{L}{c} = 9 \times \frac{2500}{1500} = 15.0\,\text{s}$	3.7	14	0.14	998.3

10	$10\times\frac{L}{c}=10\times\frac{2500}{1500}=16.7\,\text{s}$	2.25	9.6	0.096	998.2
11	$11\times\frac{L}{c}=11\times\frac{2500}{1500}=18.3\,\text{s}$	1.42	5.6	0.056	998.2
12	$12\times\frac{L}{c}=12\times\frac{2500}{1500}=20\,\text{s}$	1.07	2.3	0.023	998.2

The initial mass flow rate is:

$$W_o=1000\cdot 55\cdot\frac{\pi}{4}\cdot 0.05^2=108\,\text{kg/s} \tag{7.52}$$

During the elastic-to-Torricelli-phase transition the system has released:

$$M=(1000.4-998.2)\cdot 490.6=1079.32\,\text{kg} \tag{7.53}$$

t_e is equal to 20 s.

Driving force: Process pressure (highly volatile subcooled liquids) – API 520 method

As discussed in Chapter 4, subcooled and saturated liquids can be modelled according to the ω (omega) method (Leung, 1994). For saturated liquids, for example LPG, the application of this method has already been shown in Example 4.15. The following procedure has been collected from Appendix C of the API Standard 520 (2014).

Step 1: Omega and saturation pressure ratio parameters

Omega parameter is defined as:

$$\omega=\frac{\frac{\rho_{Lo}}{\rho_{L9}}-1}{\frac{P_o}{P_{L9}}-1} \tag{7.54}$$

Subscript "o" refers to stagnation condition and:

- ρ_{Lo} is the density at stagnation condition.
- ρ_{L9} is the specific volume attained by the vapour–liquid mixture assuming an adiabatic flash occurs from the saturation point to the 90% of the saturation pressure P_s corresponding to the stagnation temperature, T_o..

The saturated omega parameter is defined as:

$$\omega_s=9\cdot\left(\frac{\rho_{Lo}}{\rho_{L9}}-1\right) \tag{7.55}$$

Subscript "o" refers to stagnation condition and:

- ρ_{Lo} is the density at stagnation condition.

- ρ_{L9} is the specific volume attained by the vapour–liquid mixture assuming an adiabatic flash occurs from the stagnation condition to the 90% of the stagnation pressure, P_o.

The saturation pressure ratio η_s is defined as:

$$\eta_s = \frac{P_s}{P_o} \tag{7.56}$$

where P_o is the stagnation pressure and P_s is the saturation temperature at stagnation temperature.

Step 2: Subcooling region

Defining η_{st} as:

$$\eta_{st} = \frac{2 \cdot \omega_s}{1 + 2 \cdot \omega_s} \tag{7.57}$$

two cooling regions are determined.

Low subcooling region

$$P_s \geq \eta st \cdot P_o \tag{7.58}$$

In this case the flash is expected to occur upstream of the outlet.

High subcooling region

$$P_s \leq \eta st \cdot P_o \tag{7.59}$$

In this case the flash is expected to occur at the outlet.

Step 3: Critical flow

The critical pressure is defined as:

$$P_c = \eta_c \cdot P_o \tag{7.60}$$

Where the critical pressure ratio η_c is:

$$\eta_c = \eta_s \tag{7.61}$$

if:

$$\eta_s \leq \eta_{st} \tag{7.62}$$

and can approximately calculated according to the following equation:

$$\eta_c = \eta_s \cdot \frac{2\cdot\omega}{2\cdot\omega-1} \cdot \left[1-\sqrt{1-\frac{1}{\eta_s}\cdot\left(\frac{2\cdot\omega-1}{2\cdot\omega}\right)}\right] \tag{7.63}$$

if:

$$\eta_s > \eta_{st} \tag{7.64}$$

Low subcooling region Flow has to be determined as critical (choked):

$$P_c \geq P_a \tag{7.65}$$

or subcritical:

$$P_c < P_a \tag{7.66}$$

where P_a is the downstream backpressure and P_c is the critical (choked) pressure.

High subcooling region Flow is critical (choked):

$$P_s \geq P_a \tag{7.67}$$

or subcritical:

$$P_s < P_a \tag{7.68}$$

where P_s is the saturation pressure corresponding to the stagnation temperature T_o.

Step 4 Mass flux calculation

Low subcooling region

$$G = \frac{\sqrt{\left\{2\cdot(1-\eta_s)+2\cdot\left[\omega_s\cdot\eta_s\cdot\ \ln\left(\frac{\eta_s}{\eta}\right)-(\omega_s-1)\cdot(\eta_s-\eta)\right]\right\}}}{\omega_s\cdot\left(\frac{\eta_s}{\eta}-1\right)+1}\cdot\sqrt{P_o\cdot\rho_{Lo}} \tag{7.69}$$

where η_c has to be used for η if the flow is critical, and η_a if the flow is subcritical.

High subcooling region

$$G = \sqrt{2\cdot\rho_{Lo}\cdot(P_o-P)} \tag{7.70}$$

If the flow is critical, P_s is to be used for P, and if the flow is subcritical (all liquid flow), P_a will be used for P.

According to the standard API 520, Annex C – Section C.2.3.1, a discharge coefficient of 0.65 for subcooled liquids and 0.85 for saturated liquid can be used.

Example 7.4

Liquid carbon dioxide is stored at 100 bar abs and 0°C. Find the mass flux through a submerged hole. Data have been collected from Plank and Kuprianov (1929), and Anwar and Carroll (2011). The discharge coefficient has been assumed equal to 1. The overpressure related to the liquid level is not considered in the calculation.

Solution

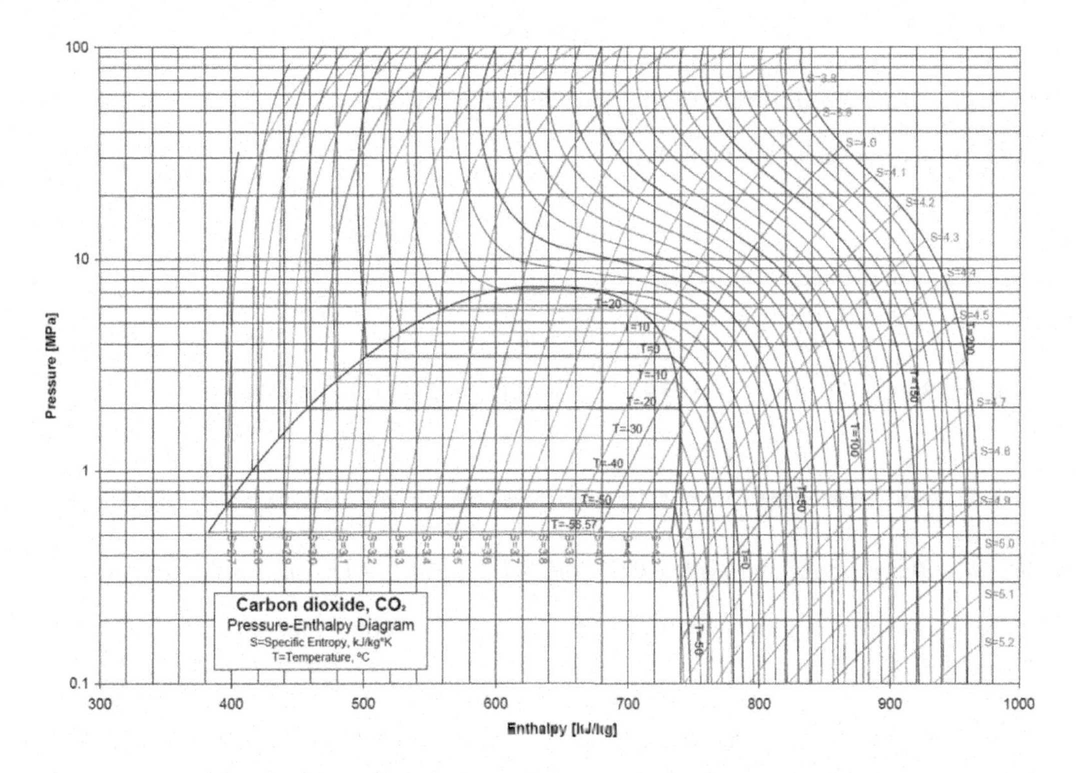

Calculation of ω_s

Saturation pressure P_s at 0°C = 34.85 bar abs

Saturation pressure at $0.9 \times P_s = 0.9 \times 34.85 = 31.365$ bar abs

Saturation temperature at 31.365 bar abs = −3.92°C

Latent heat at −3.92°C = 242,344 J/kg

Specific heat at 0.5·(273.16 + 269.24) = 271.2 K = 2485 J/kg K

$$y_{CO_2} = \frac{c_{p_L} \cdot (T_s - T_{0.9 \cdot T_S})}{\lambda_{CO_2}} = \frac{2485 \cdot 3.92}{242{,}344} = 0.04 \qquad (7.71)$$

$\rho_{CO2L}(0°C) = 974.1\ kg/m^3$

$\rho_{CO2L}(-3.92°C) = 950.21\ kg/m^3$

$\rho_{CO2V}(-3.92°C) = 86.23\ kg/m^3$

$$\rho_{L9} = \frac{1}{\frac{0.96}{950.21} + \frac{0.04}{86.23}} = 678.34\,\text{kg/m}^3 \tag{7.72}$$

$$\omega_s = 9 \cdot \left(\frac{\rho_{Lo}}{\rho_{L9}} - 1\right) = 9 \cdot \left(\frac{974.1}{678.34} - 1\right) = 3.92 \tag{7.73}$$

Calculation of subcooling region

$$\eta_{st} = \frac{2 \cdot \omega_s}{1 + 2 \cdot \omega_s} = \frac{2 \cdot 3.92}{1 + 2 \cdot 3.92} = 0.88 \tag{7.74}$$

$$P_s = 34.85\,\text{bar} < \eta_{st} \cdot P_o = 0.88 \cdot 100 = 88\,\text{bar}$$

The liquid is determined to fall into the high subcooling region.

Determination of critical or subcritical flow

$$\eta_s = 34.85/100 = 0.3485 \le \eta_{st} = 0.88 \tag{7.75}$$

The flow is critical since:

$$\text{P}_\text{s} > \text{P}_\text{a} \tag{7.76}$$

$$G = \sqrt{2 \cdot \rho_{Lo} \cdot (P_o - P_s)} = \sqrt{2 \cdot 974.1 \cdot (10{,}000{,}000 - 3{,}485{,}000)} = 112{,}661\,\text{kg/s} \cdot \text{m}^2 \tag{7.77}$$

Driving force: Process pressure (highly volatile subcooled liquids) – Transient behaviour

The release rate calculated in the previous section is the maximum mass flow rate occurring at $t=0$. As liquid is released, pressure decreases according to the bulk modulus (ref. Driving force: bulk modulus), until the saturation at atmospheric pressure is reached. This final configuration coincides with the release of a saturated (boiling) liquid at the initial stagnation temperature, which falls within the calculation methods, omega-API 520 and Fauske-Epstein, as described at Sections 7.2.5 and 7.3.3.

Driving force: Head-space gas pressure

This case consists of the following configuration (see Fig. 7.5):

The liquid is over-pressurised by a gas cushion like in a surge tank. Should gas pressure be kept constant during liquid release, the study case would fall within the pure Torricellian scenario with an additional static head corresponding to the relative gas pressure increment. Should the pressure decrease according to liquid volume variation, the case has to be treated as a transient process.

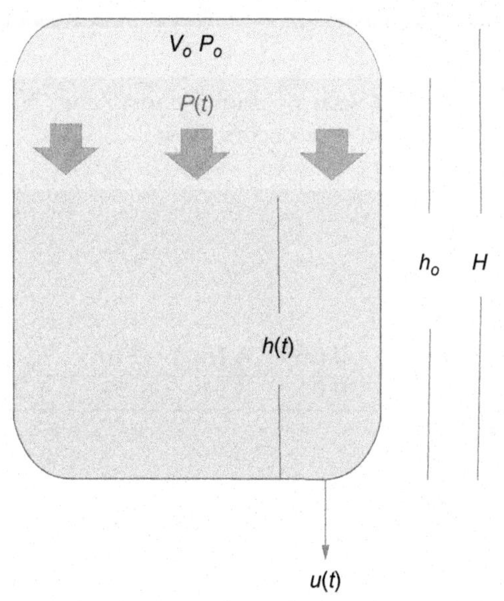

Fig. 7.5
Head-space gas pressurised vessel.

$$V(t) = V_o + [h_o - h(t)] \cdot \frac{\pi}{4} \cdot D^2 \tag{7.78}$$

$$\frac{P_G(t)}{\rho_L \cdot g} + h(t) = \frac{P_{atm}}{\rho_L \cdot g} - \frac{u(t)^2}{2 \cdot g} \tag{7.79}$$

$$u(t) = C_D \cdot \sqrt{2 \cdot \left(\frac{P_G(t)}{\rho_L} - \frac{P_{atm}}{\rho_L} + g \cdot h(t) \right)} \tag{7.80}$$

$$\frac{dV(t)}{dt} = C_D \cdot \sqrt{2 \cdot \left(\frac{P_G(t)}{\rho_L} - \frac{P_{atm}}{\rho_L} + g \cdot h(t) \right)} \cdot \frac{\pi}{4} \cdot d^2 \tag{7.81}$$

$$P_G(t) = \frac{n \cdot R \cdot T}{V(t)} = \frac{n \cdot R \cdot T}{\frac{\pi}{4} \cdot D^2 [h_o - h(t)]} \tag{7.82}$$

$$\frac{\pi}{4} \cdot D^2 \cdot \frac{d[h_o - h(t)]}{dt} = C_D \cdot \sqrt{2 \cdot \left(\frac{n \cdot R \cdot T}{\left\{ V_o + [h_o - h(t)] \cdot \frac{\pi}{4} \cdot D^2 \right\} \cdot \rho_L} - \frac{P_{atm}}{\rho_L} - g \cdot [h_o - h(t)] + g \cdot h_o \right)} \cdot \frac{\pi}{4} \cdot d^2 \tag{7.83}$$

Example 7.5

A 5 m^3 water surge drum pressurised with air, with 2 m diameter, has the functional data included in the next table. If a 10 mm hole occurs, find:

- The initial release velocity
- The transient liquid level

P_o (bar)	T (°C)	Initial Liquid Volume (m^3)	D (m)	d (m)	A (m^2)	H (m)	h_o (m)	V_o (m^3)	n_{air} (kmole)
5	20	4	2	0.01	3.14	1.59	1.274	1	0.2

Solution

$$u(0) = 0.61 \cdot \sqrt{2 \cdot \left(\frac{500{,}000}{1000} - \frac{100{,}000}{1000} + 9.81 \cdot 1.274\right)} = 17.5\,\text{m/s} \tag{7.84}$$

The integration of the differential equation:

$$\frac{d[h_o - h(t)]}{C_D \cdot \sqrt{2 \cdot \left(\frac{n \cdot R \cdot T}{\left\{V_o + [h_o - h(t)] \cdot \frac{\pi}{4} \cdot D^2\right\} \cdot \rho_L} - \frac{P_{atm}}{\rho_L} - g \cdot [h_o - h(t)] + g \cdot h_o\right)} \cdot \left(\frac{d}{D}\right)^2} = dt \tag{7.85}$$

can be done considering progressive gaseous volume increments corresponding to a very small time interval, order of magnitude of 1 second (see Fig. 7.6).

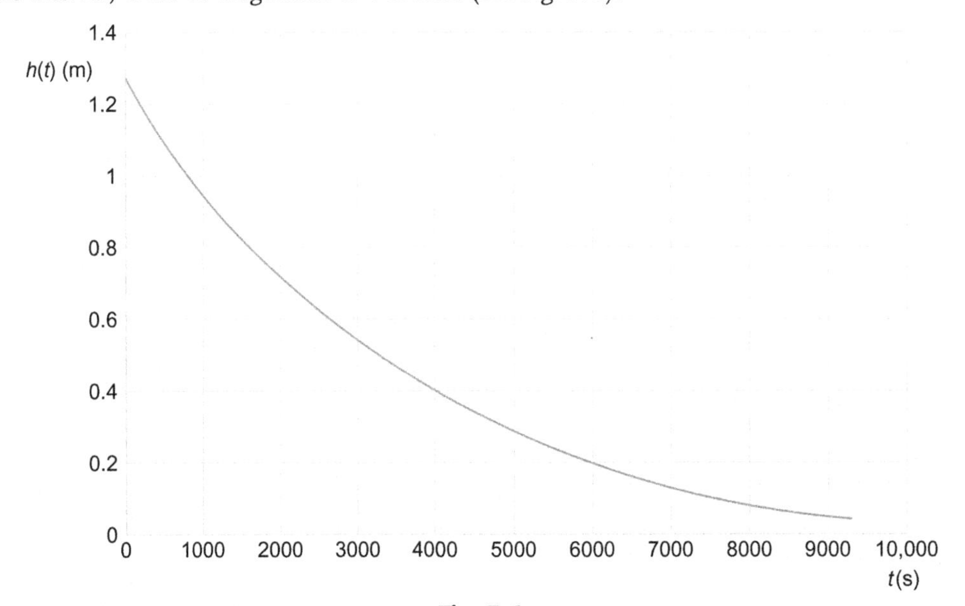

Fig. 7.6
Transient liquid level.

Δt (s)	u (m/s)	Discharged volume (m^3)	Air volume (m^3)	Liquid level (m)
0	17.50000	0.000000	1.000000	1.274000
1	17.24262	0.001354	1.001354	1.273454
2	17.22825	0.001352	1.002706	1.273024
3	17.21398	0.001351	1.004057	1.272593
4	17.19976	0.001350	1.005407	1.272163
5	17.18557	0.001349	1.006757	1.271734
6	17.17142	0.001348	1.008104	1.271304
7	17.1573	0.001347	1.009451	1.270875
8	17.14323	0.001346	1.010797	1.270447
9	17.12919	0.001345	1.012142	1.270019
10	17.11519	0.001344	1.013485	1.269591
9008	0.884465	6.94E-05	4.842937	0.05002

7.3 Boiling Liquids

Saturated liquids are in equilibrium with their vapour at storage temperature. They can be stored at atmospheric pressure temperature, or they can be pressurised. The two different scenarios are discussed separately.

7.3.1 Boiling Liquid Stored at Atmospheric Pressure

A relatively simple procedure can be followed (U.S. EPA, 1996). This procedure is based on the following assumptions:

Step 1: Fluid will be released as a pure liquid.

Step 2: Any jet fragmentation and significant jet vaporisation will be assessed depending on the hole height with respect to grade and fluid-dynamic jet properties.

Step 3: A pool will be formed.

Step 4: Liquid will be vaporised according to the liquid and ambient heat transfer.

The first two steps are dealt with in the present section.

Mass flow rate

The mass flow rate is given by the Torricellian formula:

$$W(t) = C_D \cdot \rho_L \cdot \pi \cdot \frac{d^2}{4}\sqrt{2 \cdot g \cdot h(t)} \tag{7.86}$$

Jet fragmentation and droplet formation

Due to the isobaric nature of the release, the liquid will be assumed as non-flashing (US.EPA). However, the jet can be fragmented, and, depending on droplet size, a significant liquid fraction can be vaporised prior to impinging on grade.

Liquids under pressure form droplets through an orifice according to two consecutive steps:

- primary breakup, during which liquid is fragmented into large droplets.
- secondary breakup, which is causes by aerodynamic forces and mutual collisions.

According to Taylor (1940), the following formula may be used to determine whether primary spray breakup is completed prior to the liquid impingement on a nearby surface:

$$L_c = d \cdot C_C \cdot \left(\frac{\rho_L}{\rho_a}\right)^{1/2} \tag{7.87}$$

where:

- L_c is the length of the unbroken liquid core.
- d is the orifice diameter.
- ρ_L is the liquid density.
- ρ_a is air density.
- C_C is the model constant, which, for round jets, ranges between 7 and 16, according to Chehroudi et al. (1985).

This formula is valid for non-turbulent jets.

Should the spray break-up take place, the Sauter mean diameter d_{SMD}, for Weber number >40 can be calculated as (Ruff et al., 1990):

$$d_{SMD} = \frac{2 \cdot \pi \cdot C_B \cdot \sigma}{\rho_a \cdot u^2} \tag{7.88}$$

where u is the relative speed of the droplets in the air, σ is the surface tension, and C_B is a constant which has a value of approximately 0.4.

Formulas (7.87) and (7.88) allow one to predict whether mechanical break-up takes place and, if so, the average droplet diameter size.

Atomisation within a primary break up regime can be considered developed, provided the Weber number:

$$We = \frac{\rho_L \cdot u \cdot d^2}{\sigma} \tag{7.89}$$

is greater than 40 (Health and Safety Laboratory, 2013).

Droplet evaporation

Droplet evaporation can be assessed by means of the following simple thermal balance:

$$M_v = \frac{1}{6} \cdot \pi \cdot \rho_L \cdot \frac{d(d_{SMD})^3}{dt} = \frac{A_d \cdot h_m \cdot (T_\infty - T_{LNG})}{\lambda} \tag{7.90}$$

where:

- M_v is the mass vaporisation rate.
- ρ_L is liquid density.
- A_d is the droplet area, $\pi \cdot d_{SMD}{}^2/4$.
- T_∞ is air temperature far from the drop.
- T_{LNG} is the droplet saturation temperature.
- h_m is the inter-phase heat transfer coefficient.
- λ is latent heat of evaporation at saturation condition.

With the aim at carrying out an order-of-magnitude evaluation of the liquid to air heat exchange, the following relation can be used (Bird et al., 2002), which is valid for a forced convection around a sphere:

$$h_m = \frac{k_{t_f}}{d_{SMD}} \cdot \left[2 + 0.6 \cdot \left(\frac{d_{SMD} \cdot u_\infty \cdot \rho_f}{\mu_f} \right)^{1/2} \cdot \left(\frac{\mu_f}{c_{Pf} \cdot k_{t_f}} \right)^{1/3} \right] \tag{7.91}$$

where:

- k_t is gaseous phase thermal conductivity.
- μ is air viscosity.
- u_∞ is the droplet velocity relative to air.
- c_P is the specific heat at constant pressure.
- subscript f refers to the arithmetic mean temperature T_f between liquid temperature and air temperature at which the related properties have to be estimated.

The final droplet velocity can be assumed to coincide with Stokes's velocity (Van den Bosch and Duijm, 2005):

$$u_\infty = \frac{g \cdot (\rho_L - \rho_a)}{18 \cdot \mu} \cdot d_{SMD}{}^2 \tag{7.92}$$

Combining the equations:

$$h_m = \frac{k_{t_f}}{d_{SMD}} \cdot \left[2 + 0.6 \cdot \left(\frac{d_{SMD}{}^3 \cdot \frac{g \cdot (\rho_L - \rho_a)}{18 \cdot \mu} \cdot \rho_f}{\mu_f} \right)^{1/2} \cdot \left(\frac{\mu_f}{c_{Pf} \cdot k_{t_f}} \right)^{1/3} \right] \tag{7.93}$$

it gives:

$$\frac{dd_{SMD}}{dt} = \frac{k_{tf}}{\rho_L \cdot d_{SMD}} \cdot \left[2 + 0.6 \cdot \left(\frac{d_{SMD}{}^3 \cdot \frac{g \cdot (\rho_L - \rho_a)}{18 \cdot \mu} \cdot \rho_f}{\mu_f} \right)^{1/2} \cdot \left(\frac{\mu_f}{c_{pf} \cdot k_{tf}} \right)^{1/3} \right] \cdot \frac{(T_\infty - T_{LNG})}{\lambda} \tag{7.94}$$

This equation can be integrated, and the likelihood of pool formation from liquid released at different heights can be known.

Rain out distance from the outlet

A liquid jet prior to breaking up will be subject to gravity, and its pathway will be described by a parabolic curve (see Fig. 7.7):

$$y = \frac{g}{2 \cdot u_o{}^2} \cdot x^2 \tag{7.95}$$

where:

- y is the vertical distance from the outlet at which primary breakup is complete.
- u_o is the horizontal outlet (Torricellian) velocity.
- x is the horizontal distance from the outlet.

This equation is valid until the gravitational field is taken over by the Stoke' regime, that is, until the liquid is fragmented. Therefore, in Eq. (7.95) x can be replaced by the result of Eq. (7.87) and u_o by the Torricellian velocity related to the liquid level:

$$y = \frac{1}{4 \cdot h} \cdot d^2 \cdot C_C{}^2 \cdot \left(\frac{\rho_L}{\rho_a} \right) \tag{7.96}$$

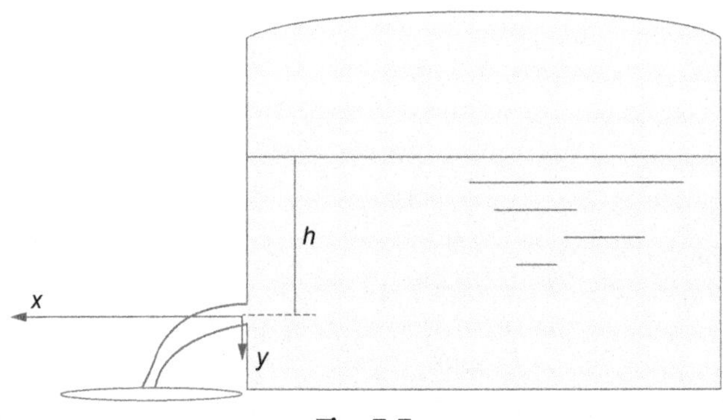

Fig. 7.7
Liquid jet pathway.

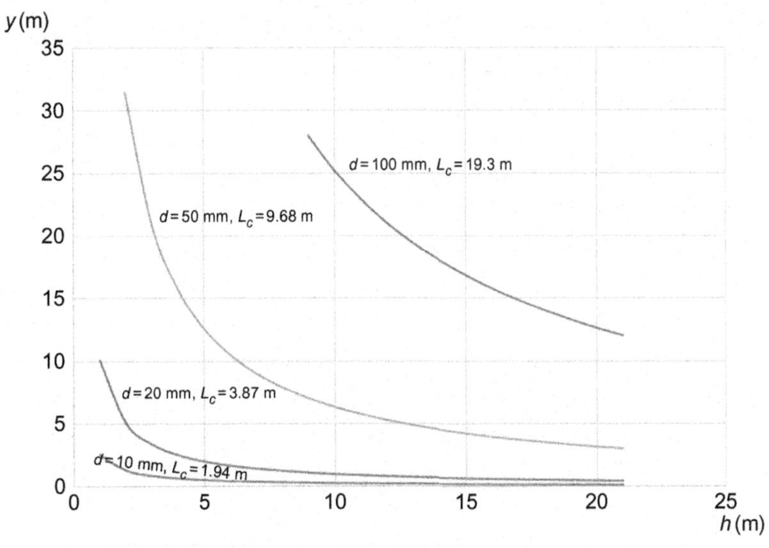

Fig. 7.8
Distance from the outlet at which primary breakup is complete as a function of liquid level and hole size.

Fig. 7.8 presents the behaviour of y with h for atmospheric LNG at air temperature of 20°C and various values of d, in correspondence of an average C_c equal to 10.

Under Stoke's regime, if the jet has not impinged on a surface, its further pathway will be travelled at Stoke's velocity. It is:

$$u_\infty = \frac{g \cdot (\rho_L - \rho_a)}{18 \cdot \mu_a} \cdot {d_{SMD}}^2 \tag{7.97}$$

or:

$$\frac{dy}{dt} = \frac{g \cdot (\rho_L - \rho_a)}{18 \cdot \mu_a} \cdot {d_{SMD}}^2 \tag{7.98}$$

$$dy = \frac{g \cdot (\rho_L - \rho_a)}{18 \cdot \mu_a} \cdot {d_{SMD}}^2 \cdot dt \tag{7.99}$$

Being:

$$dt = \frac{d\left(d_{SMD}\right)}{\frac{k_{tf}}{\rho_L \cdot d_{SMD}} \cdot \left[2 + 0.6 \cdot \left(\frac{{d_{SMD}}^3 \cdot \frac{g \cdot (\rho_L - \rho_a)}{18 \cdot a = \frac{g \cdot (\rho_L - \rho_a)}{18 \cdot \mu} \cdot \frac{\rho_L}{k_{tf}} \cdot \frac{\lambda}{(T_\infty - T_{LNG})}} \cdot \rho_f}{\mu_f}\right)^{1/2} \cdot \left(\frac{\mu_f}{c_{pf} \cdot k_{tf}}\right)^{1/3}\right] \cdot \frac{(T_\infty - T_{LNG})}{\lambda}} \tag{7.100}$$

$$dy = \frac{g\cdot(\rho_L-\rho_a)}{18\cdot\mu}\cdot\frac{{d_{SMD}}^3\cdot d\left(d_{SMD}\right)}{\frac{k_{t_f}}{\rho_L}\cdot\left[2+0.6\cdot\left(\frac{{d_{SMD}}^3\cdot\frac{g\cdot(\rho_L-\rho_a)}{18\cdot\mu}\cdot\rho_f}{\mu_f}\right)^{1/2}\cdot\left(\frac{\mu_f}{c_{p_f}\cdot k_{t_f}}\right)^{1/3}\right]\cdot\frac{(T_\infty-T_{LNG})}{\lambda}} \tag{7.101}$$

This differential equation allows one to know the vertical distance covered by the droplet until vaporisation is completed. It can be simplified as:

$$dy = a\cdot\frac{{d_{SMD}}^3\cdot d(d_{SMD})}{\left[2+0.6\cdot b\cdot {d_{SMD}}^{1.5}\right]} \tag{7.102}$$

with:

$$a = \frac{g\cdot(\rho_L-\rho_a)}{18\cdot\mu}\cdot\frac{\rho_L}{k_{t_f}}\cdot\frac{\lambda}{(T_\infty-T_{LNG})} \tag{7.103}$$

$$b = \left(\frac{\frac{g\cdot(\rho_L-\rho_a)}{18\cdot\mu_a}\cdot\rho_f}{\mu_f}\right)^{1/2}\cdot\left(\frac{\mu_f}{c_{p_f}\cdot k_{t_f}}\right)^{1/3} \tag{7.104}$$

Integrating:

$$\int_0^{y_{ev}} dy = \int_{d_{SMD_o}}^{0} a\cdot\frac{{d_{SMD}}^3\cdot d(d_{SMD})}{\left[2+0.6\cdot b\cdot {d_{SMD}}^{1.5}\right]} \tag{7.105}$$

$$\int_{d_{SMD_o}}^{0} a\cdot\frac{{d_{SMD}}^3\cdot d(d_{SMD})}{\left[2+0.6\cdot b\cdot {d_{SMD}}^{1.5}\right]} = a\cdot b^{-\frac{8}{3}}\cdot\left[\begin{array}{l}-14.31\cdot\tan^{-1}\left(0.577-0.77\cdot\sqrt[3]{b}\cdot\sqrt{d_{SMD}}\right)+\\ 0.66\cdot b^{\frac{5}{3}}\cdot d_{SMD}^{\frac{5}{2}}-5.55\cdot b^{\frac{5}{3}}+4.13\cdot\ln\left(4.48\cdot b^{\frac{2}{3}}\cdot d_{SMD}-6.69\cdot\sqrt[3]{b}\cdot\sqrt{d_{SMD}}+10\right)\\ -8.26\cdot\ln\left(6.69\cdot\sqrt[3]{b}\cdot\sqrt{d_{SMD}}+10\right)\end{array}\right]_{d_{SMD_o}}^{0} \tag{7.106}$$

Example 7.6

Release from a 100 mm diameter hole in an LNG atmospheric tank at 30 m from grade. The hole static head is 10 m. Transport properties except density (450 kg/m^3) assumed as for pure methane. Air temperature is 20°C. Find the maximum mass flow rate and the outcome of the jet.

T_{LNG} (°C)	ρ_L (kg/m³)	T_{air} (°C)	ρ_{air} (kg/m³)	C_C	h (m)	D (m)	T_f (K)	Air viscosity (Pa s)
−161.4	450	20	1.2	10	10	0.1	$(T_{LNG}+T_{air})/2=202.16$ K	1.10E-05

σ (N/m)	k_{tf}(W/mK)	λ (J/kg)	c_{pf} (J/kg)	Hole elevation (m)	Air viscosity at T_f (Pa s)
0.014	0.0188	0.510	1398	30	1.82E-05

Solution

Maximum mass flow rate

$$W = C_D \cdot \rho_L \cdot \pi \cdot \frac{d^2}{4}\sqrt{2 \cdot g \cdot h} = 0.61 \cdot 450 \cdot \pi \cdot \frac{0.01}{4} \cdot \sqrt{2 \cdot 9.81 \cdot 10} = 30.18 \text{ kg/s} \tag{7.107}$$

Horizontal maximum velocity

$$u_o = C_D \cdot \sqrt{2 \cdot g \cdot h} = 0.61 \cdot \sqrt{2 \cdot 9.81 \cdot 10} = 8.5 \text{ m/s} \tag{7.108}$$

Length of the unbroken liquid core

$$L_c = d \cdot C_C \cdot \left(\frac{\rho_L}{\rho_a}\right)^{1/2} = 0.1 \cdot 10 \cdot \left(\frac{450}{1.2}\right)^{0.5} = 19.3 \text{ m} \tag{7.109}$$

Vertical elevation at L_C

$$y_c = 30 - \frac{1}{4 \cdot C_D{}^2 \cdot h} \cdot d^2 \cdot C_C{}^2 \cdot \left(\frac{\rho_L}{\rho_a}\right) = 30 - \frac{1}{4 \cdot 0.61^2 \cdot 10} \cdot 0.01 \cdot 100 \cdot 375 = 30 - 25 = 5 \text{ m} \tag{7.110}$$

Droplets will be formed.

Liquid velocity at L_c

$$u = u_o +^o \sqrt{2 \cdot g \cdot y_c} = 8.5 + \sqrt{2 \cdot 9.81 \cdot 19.3} = 28 \text{ m/s} \tag{7.111}$$

Droplet size

$$d_{SMD} = \frac{2 \cdot \pi \cdot C_B \cdot \sigma}{\rho_a \cdot u^2} = \frac{2 \cdot \pi \cdot 0.4 \cdot 0.014}{1.2 \cdot 28^2} = 0.037 \text{ mm} \tag{7.112}$$

and:

$$We = \frac{\rho_L \cdot u \cdot d^2}{\sigma} = \frac{450 \cdot 28 \cdot 0.01}{0.014} > 40 \tag{7.113}$$

Vaporisation distance

$$a = \frac{g \cdot (\rho_L - \rho_a)}{18 \cdot \mu_a} \cdot \frac{\rho_L}{k_{tf}} \cdot \frac{\lambda}{(T_\infty - T_{LNG})} = \frac{9.81 \cdot (450 - 1.2)}{18 \cdot 1.1\text{E}-05} \cdot \frac{450}{0.0188} \cdot \frac{0.510}{(293.16 + 161.4)} \tag{7.114}$$

$$= 597{,}159{,}422$$

$$b=\left(\frac{\frac{g\cdot(\rho_L-\rho_a)}{18\cdot\mu_a}\cdot\rho_f}{\mu_f}\right)^{1/2}\cdot\left(\frac{\mu_f}{c_{p_f}\cdot k_{t_f}}\right)^{1/3}=\left(\frac{\frac{9.81\cdot(450-1.2)}{18\cdot 1.1E-05}\cdot 450}{1.82E-05}\right)^{1/2}\cdot \tag{7.115}$$

$$\left(\frac{1.82E-05}{1398\cdot 0.0188}\right)^{1/3}=207{,}444$$

$$We=\frac{\rho_L\cdot u\cdot d^2}{\sigma}=\frac{450\cdot 28\cdot 0.01}{0.014}>40 \tag{7.116}$$

$$a\cdot b^{-\frac{8}{3}}=597{,}159{,}422\cdot 207{,}444^{-\frac{8}{3}}=48.63 \tag{7.117}$$

$$\int_{d_{SMD_o}}^{0} a\cdot\frac{d_{SMD}{}^3\cdot d(d_{SMD})}{\left[2+0.6\cdot b\cdot d_{SMD}{}^{1.5}\right]}=3.96E-06\cdot \left\{\begin{array}{l}\left[\begin{array}{l}-14.31\cdot\tan^{-1}(0.577)-\\-19{,}448.83+9.5\end{array}\right]-\\ \left[\begin{array}{l}-14.31\cdot\tan^{-1}\left(0.577-0.77\cdot 59.2\cdot\sqrt{0.000037}\right)+\\0.66\cdot 726{,}944{,}848.7\cdot 0.000037^{\frac{5}{2}}-19{,}4448.83+\\4.13\cdot\ln\left(4.48\cdot 3{,}504.29\cdot 0.000037-6.69\cdot 59.197\cdot\sqrt{0.000037}+10\right)-\\8.26\cdot\ln\left(6.69\cdot 59.197\cdot\sqrt{0.000037}+10\right)\end{array}\right]\end{array}\right\} \tag{7.118}$$

$$\int_{d_{SMD_o}}^{0} a\cdot\frac{d_{SMD}{}^3\cdot d(d_{SMD})}{\left[2+0.6\cdot b\cdot d_{SMD}{}^{1.5}\right]}=a\cdot b^{-\frac{8}{3}}\cdot \left[\begin{array}{l}-14.31\cdot\tan^{-1}\left(0.577-0.77\cdot\sqrt[3]{b}\cdot\sqrt{d_{SMD}}\right)+\\0.66\cdot b^{\frac{5}{3}}\cdot d_{SMD}{}^{\frac{5}{2}}-5.55\cdot b^{\frac{2}{3}}+4.13\cdot\ln\left(4.48\cdot b^{\frac{2}{3}}\cdot d_{SMD}-6.69\cdot\sqrt[3]{b}\cdot\sqrt{d_{SMD}}+10\right)\\-8.26\cdot\ln\left(6.69\cdot\sqrt[3]{b}\cdot\sqrt{d_{SMD}}+10\right)\end{array}\right]_{d_{SMD_o}}^{0} \tag{7.119}$$

$$\begin{aligned}y_{ev}&=\int_{d_{SMD_o}}^{0} a\cdot\frac{d_{SMD}{}^3\cdot d(d_{SMD})}{\left[2+0.6\cdot b\cdot d_{SMD}{}^{1.5}\right]}=3.96E-06\cdot\\&\quad[-19{,}456.32-(-4.17+0.004-19{,}448.83+8.68-20.8)]\\&=0.14\,\text{mm}\end{aligned} \tag{7.120}$$

Since $y_c > y_{ev}$ no rainout is expected.

7.3.2 Boiling Liquids Stored Under Pressure – API 520 Method

Release of boiling liquids stored under pressure is properly described through the omega method (API Standard 520, 2014), assuming a low subcooling region. The API 520 procedure has been included in Section 7.2.3.

Example 7.7

Liquid propane is stored at 10 bar abs in a spherical tank 85% full. The discharge coefficient has been assumed equal to 1, and the overpressure related to the liquid level is not considered in the calculation.

Solution

The liquid is saturated at 10 bar. The corresponding saturation temperature is 27°C.

Calculation of ω_s

Saturation pressure P_s at 27°C = 10 bar abs

Saturation temperature at 9 bar abs = 22.8°C

Latent heat at 22.8°C = 339,552 J/kg

Specific heat at 0.5·(296 + 300) = 298 K = 2717 J/kg K

$$y_{CO_2} = \frac{c_{p_L} \cdot (T_s - T_{0.9 \cdot T_S})}{\lambda_{CO_2}} = \frac{2717 \cdot 4.2}{339,552} = 0.034 \quad (7.121)$$

$\rho_{PROP\text{-}L}(27°C) = 489.5\ kg/m^3$

$\rho_{PROP\text{-}L}(22.8°C) = 495.8\ kg/m^3$

$\rho_{PROP\text{-}V}(22.8°C) = 19.5\ kg/m^3$

$$\rho_{L9} = \frac{1}{\dfrac{0.966}{495.8} + \dfrac{0.034}{19.5}} = 270.86\ kg/m^3 \quad (7.122)$$

$$\omega_s = 9 \cdot \left(\frac{\rho_{Lo}}{\rho_{L9}} - 1\right) = 9 \cdot \left(\frac{489.5}{270.86} - 1\right) = 7.26 \quad (7.123)$$

Calculation of subcooling region

$$\eta_{st} = \frac{2 \cdot \omega_s}{1 + 2 \cdot \omega_s} = \frac{2 \cdot 7.26}{1 + 2 \cdot 7.26} = 0.93 \quad (7.124)$$

$$P_s = 10\ bar > \eta_{st} \cdot 10 = 0.93 \cdot 10 = 9.3\ bar \quad (7.125)$$

Flash is expected to occur upstream of the outlet, and the region is low subcooling.

Of course, this result could inherently be assumed from the nature of the propane saturation state. However, it is worth noting that, depending on the liquid static head, a certain subcooling degree exists around the hole.

Determination of critical or subcritical flow

$$\eta_s = 1 > \eta_{st} = 0.93 \quad (7.126)$$

In this case $\omega = \omega_s$.

$$\eta_c = \eta_s \cdot \frac{2 \cdot \omega}{2 \cdot \omega - 1} \cdot \left[1 - \sqrt{1 - \frac{1}{\eta_s} \cdot \left(\frac{2 \cdot \omega - 1}{2 \cdot \omega}\right)}\right] = 1 \cdot 1.073 \cdot \left[1 - \sqrt{1 - 0.93}\right] = 0.79 \quad (7.127)$$

$$P_c = \eta_c \cdot P_o = 0.79 \cdot 10 = 7.9\ bar > P_a \quad (7.128)$$

$$G = \frac{\sqrt{\left\{2 \cdot (1-\eta_s) + 2 \cdot \left[\omega_s \cdot \eta_s \cdot \ln\left(\frac{\eta_s}{\eta_c}\right) - (\omega_s - 1) \cdot (\eta_s - \eta_c)\right]\right\}}}{\omega_s \cdot \left(\frac{\eta_s}{\eta_c} - 1\right) + 1} \cdot \sqrt{P_o \cdot \rho_{Lo}} \quad (7.129)$$

$$G = \frac{\sqrt{\left\{2 \cdot \left[7.26 \cdot \ln\left(\frac{1}{0.79}\right) - (7.26 - 1) \cdot (1 - 0.79)\right]\right\}}}{7.26 \cdot \left(\frac{1}{0.79} - 1\right) + 1} \cdot \sqrt{1000{,}000 \cdot 489.5} \quad (7.130)$$

$$G = \frac{\sqrt{\{2 \cdot [1.71 - 1.31]\}}}{2.93} \cdot 22124.65 = 6{,}754 \, \text{kg/sm}^2 \quad (7.131)$$

7.3.3 Boiling Liquids Stored Under Pressure – Fauske and Epstein Method

Fauske and Epstein (1988) have proposed the following simple formula for the choked mass flux:

$$G = \frac{\lambda}{v_{GL}} \cdot \sqrt{\frac{1}{T_o \cdot c_{PL}}} \quad (7.132)$$

where:

- λ is the latent heat of vaporisation at ambient temperature.
- v_{GL} is the difference between specific volumes of gas and liquid at ambient temperature.
- T_o is the stagnation temperature.
- c_{PL} is liquid the specific heat at stagnation temperature.

Example 7.8

Liquid propane is stored at 10 bar abs in a spherical tank 85% full. The discharge coefficient has been assumed equal to 1, and the overpressure related to the liquid level is not considered in the calculation.

The liquid is saturated at 10 bar. The corresponding saturation temperature is 27°C.

$$G = \frac{328{,}600}{0.044} \cdot \sqrt{\frac{1}{300 \cdot 2250}} = 9090 \, \text{kg/sm}^2 \quad (7.133)$$

It is remarked that the mass flux calculated through Fauske equation overestimates by 30% the one calculated through the API 520 method.

7.3.4 Intermediate Situation Between Subcooled and Saturated Stagnation

In the intermediate regime between subcooled and saturated stagnation condition, Fauske and Epstein (1988) suggest the following combined equation:

$$G=\sqrt{2\cdot\rho_{Lo}\cdot(P_o-P_s)+\left(\frac{\lambda}{v_{GL}}\right)^2\cdot\frac{1}{T_o\cdot c_{PL}}} \tag{7.134}$$

However, for any subcooled and saturated conditions, omega method provides a much more accurate and tuned calculation procedure.

7.3.5 Transition From Subcooled to Saturated Stagnation

Release of a subcooled liquid to the atmosphere through a short pipe connected to the vessel undergoes a transition from the pure subcooled zone to the saturation regime, depending on the length of pipe. Notably, for length greater than 10 cm, the equilibrium may be assumed as attained, whereas at the vessel outlet subcooled regime calculation applies. In between, Fauske and Epstein (1988) introduced a non equilibrium parameter as:

$$N=\frac{\lambda^2}{2\cdot\Delta P\cdot\rho_L\cdot C_D{}^2\cdot v_{GL}{}^2\cdot T\cdot C_{PL}}+10\cdot L \tag{7.135}$$

where ΔP is the pressure drop, L is the pipe length, and G is generalised as:

$$G=\sqrt{2\cdot\rho_{Lo}\cdot(P_o-P_s)+\frac{\left(\frac{\lambda}{v_{GL}}\right)^2\cdot\frac{1}{T_o\cdot c_{PL}}}{N}} \tag{7.136}$$

7.3.6 Boiling Liquids Stored Under Pressure – Transient Behaviour

The evolution of the liquid release after $t=0$ will be driven by the temperature, which will remain constant, forcing new liquid to vaporise as part of it is released. Consequently, just liquid level will change, whereas saturation pressure will not change, and, apart from the level reduction, the initial mass flow rate will remain constant until the last liquid layer has disappeared.

7.4 Rainout and Flashing – Liquid Droplets Formation

7.4.1 Flash Vapour Fraction

Rainout from pressurised or atmospheric liquids is a complex phenomenon. In this section, liquid fragmentation, droplet formation, and vaporisation modelling have been presented.

Simple correlations with low accuracy can be used for screening purposes, considering the following simplified formula for vapour fraction:

$$y \cong \frac{c_{pL} \cdot (T_o - T_s)}{\lambda_s} \tag{7.137}$$

where:

- y is the vapour fraction.
- T_o is liquid stagnation temperature.
- T_s is normal boiling temperature.
- c_{pL} is the specific heat at constant pressure calculated at $0.5 \cdot (T_o + T_s)$.
- λ_s is the latent heat of vaporisation at T_s.

Example 7.9

Saturated liquid chlorine, stored at 6.84 bar abs in a bottle at 20°C, is released to the atmosphere, at an ambient temperature of 20°C. Find the flashed vapour fraction.

Solution

Storage Temperature (°C)	Storage Pressure (bar)	Normal Boiling Temperature (°C)	Liquid Specific Heat (J/kg K)	Latent Heat of Evaporation (J/kg)
20	6.84	−34.04	926	287,840

$$y \cong \frac{926 \cdot 54.04}{287,840} = 0.174 \tag{7.138}$$

7.4.2 Rainout for Continuous Releases of Superheated Liquids

Kletz (1997) correlation

The rainout fraction η_R is:

$$\eta_R = 1 - 2 \cdot y \tag{7.139}$$

if $y < 0.5$, and:

$$\eta_R = 0 \tag{7.140}$$

If $x > 0.5$.

Kletz correlation is recommended only for rainout prediction of superheated water. As a matter of fact, Johnson and Woodward (1999) have stated that experimental data are largely lower than those predicted through Kletz formula, except for superheated steam.

DeVaull and King refitted correlation

Lautkaski (2008) refitted the more accurate DeVaull and King correlation (DeVaull and King, 1992) for volatile liquids, which fits CCPS experimental data better than Kltez formula and for $y < 0.224$, is defined as:

$$\eta_R = \eta_{R*} \cdot \left[1 - \left(\frac{y}{0.224}\right)^{1.69}\right] \tag{7.141}$$

where η_{R*} is a scaling factor defined as:

$$\eta_{R*} = 1 - 2.33 \cdot \frac{T_a - T_{as}}{T_a} \tag{7.142}$$

where T_a is the ambient temperature, and T_{as}, is a thermodynamic property, is the adiabatic saturation temperature. The requirement of this datum is a limit for a screening application of this correlation. Low volatile liquids, according to DeVaull, are those for which volatility:

$$\frac{T_a - T_{as}}{T_a} \geq 0.114 \tag{7.143}$$

Example 7.10

Find the rainout fraction for chlorine described in the previous example. Volatility for this example is 0.317.

Solution

$$\eta_{R*} = 1 - 2.33 \cdot 0.317 = 0.26 \tag{7.144}$$

The vapour fraction is 0.174, so:

$$\eta_R = 0.26 \cdot \left[1 - \left(\frac{0.174}{0.224}\right)^{1.69}\right] = 0.09 \tag{7.145}$$

This value fits well the CCPS experimental data for chlorine.

Lautkaski correlation

Lautkaski (2008) correlation does not require calculating the adiabatic saturation temperature and is given by the following formula:

$$\eta_R = 0.6 \cdot \left[1 - \left(\frac{Ja}{93}\right)^{1.36}\right] \tag{7.146}$$

where Ja is the Jacob number:

$$Ja = \frac{c_{pL} \cdot (T_o - T_s)}{\lambda_s} \cdot \frac{\rho_L}{\rho_V} \tag{7.147}$$

and the ρ are the densities of liquid and vapour.

Example 7.11

Find the rainout fraction for chlorine described in the previous example.

Solution

$$Ja = \frac{926 \cdot 54.4}{287,840} \cdot 395 = 69.13 \quad (7.148)$$

$$\eta_R = 0.6 \cdot \left[1 - \left(\frac{69.13}{93}\right)^{1.36}\right] = 0.19 \quad (7.149)$$

The presented correlations are valid for superheated liquid only. For low volatility subcooled liquids, according to Tickle (2015), the results obtained applying the DeVaull and King correlation are reasonable.

7.4.3 Rainout for Instant Releases of Superheated Liquids

Prugh (1987) correlation

$$\eta_R = 1 - \frac{\min\left(y, \frac{1-y}{2}\right)}{1-y} \quad (7.150)$$

Mudan and Croce (1988)

$$\eta_R = \begin{cases} 1 \ if \ y \leq 01.5 \\ 0 \ if \ y \geq 0.30 \end{cases} \quad (7.151)$$

η_R is linearly interpolated if $0.15 < y < 0.30$.

7.4.4 Mean Drop Size

Capillary breakup occurs when subcooled liquids discharged through a hole with a diameter smaller than 2 mm. In this case, simple Lienhard and Day (1970) can be used:

$$d_p = 1.9 \cdot d_o \quad (7.152)$$

where d_o is the orifice diameter.

Aerodynamic breakup occurs with subcooled and slightly superheated liquids at larger diameters and at critical Weber numbers between 12 and 22. Within this range (Woodward, 1998):

$$d_p = \frac{2 \cdot \sigma \cdot We_c}{\rho_g \cdot {u_\infty}^2} \quad (7.153)$$

The following correlation relates the drop diameter against the excess temperature above the saturation temperature (Johnson and Woodward, 1999):

$$d_p = \frac{3}{1 + 0.4 \cdot \Delta T_{SH}} \tag{7.154}$$

where:

- d_p is the drop diameter (mm).
- ΔT_{SH} is the temperature increment with respect to the saturation temperature at the discharge pressure (K).

Example 7.12

Saturated liquid chlorine, stored at 6.84 bar abs in a bottle at 20°C, is released to the atmosphere, at an ambient temperature of 20°C. Find the Weber number.

Solution

$$d_p = \frac{3}{1 + 0.4 \cdot (20 + 34.04)} = 0.132 \text{ mm} \tag{7.155}$$

The Weber number is defined as:

$$We = \frac{d_p \cdot \rho_g \cdot u_\infty^2}{2 \cdot \sigma} \tag{7.156}$$

Drop velocity has to be calculated according to Stoke's equation:

$$u_\infty = \frac{g \cdot (\rho_L - \rho_a)}{18 \cdot \mu} \cdot d_p^2 = \frac{9.81 \cdot (1554 - 1.2)}{18 \cdot 0.000011} \cdot 0.000132^2 = 1.34 \text{ m/s} \tag{7.157}$$

$$We = \frac{0.000132 \cdot 1554 \cdot 1.34^2}{2 \cdot 0.02655} = 6.9 \tag{7.158}$$

The droplet size distribution in a spray is:

$$P(d) = 1 - \exp\left[-0.422 \cdot \left(\frac{d}{d_{SMD}}\right)^{5.32}\right] \tag{7.159}$$

where d_{SMD} is the Sauter mean diameter calculated with formulas provided above.

7.5 Carbon Dioxide: Liquid to Solid Transition

Carbon dioxide may not exist as a liquid below its triple point, that is a three-phases domain, across which liquid–vapour phase turns rapidly to solid–vapour phase. In this section, further expansion of a subcooled carbon dioxide example analysed in the Example 7.4 studied (see Figs 7.9 and 7.10).

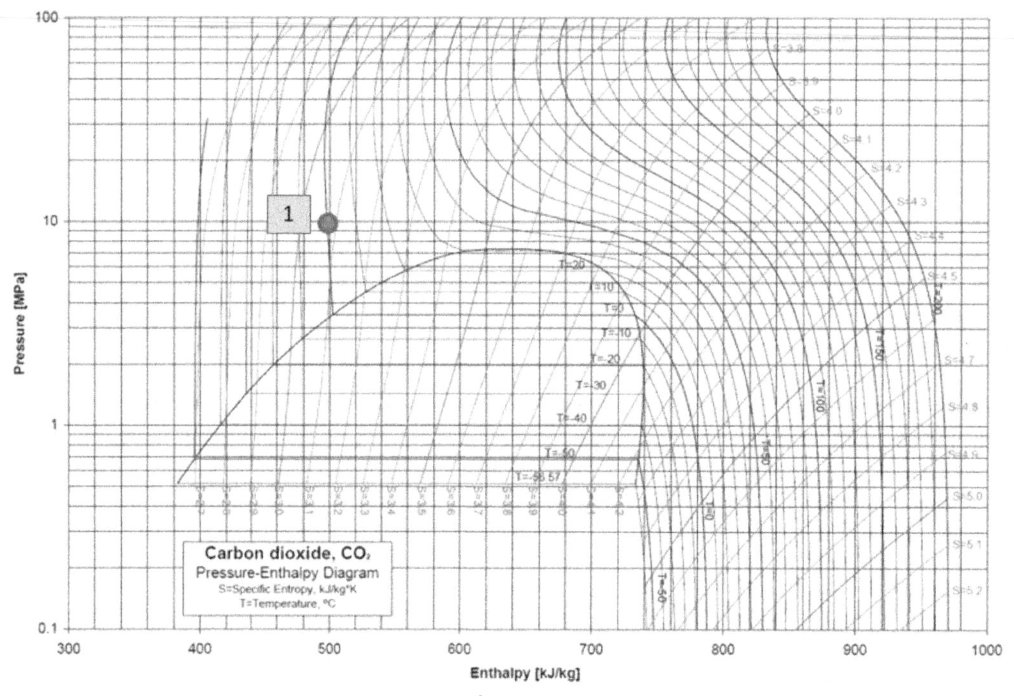

Fig. 7.9
Pressure enthalpy diagram for carbon dioxide of Example 7.5 – Stagnation point.

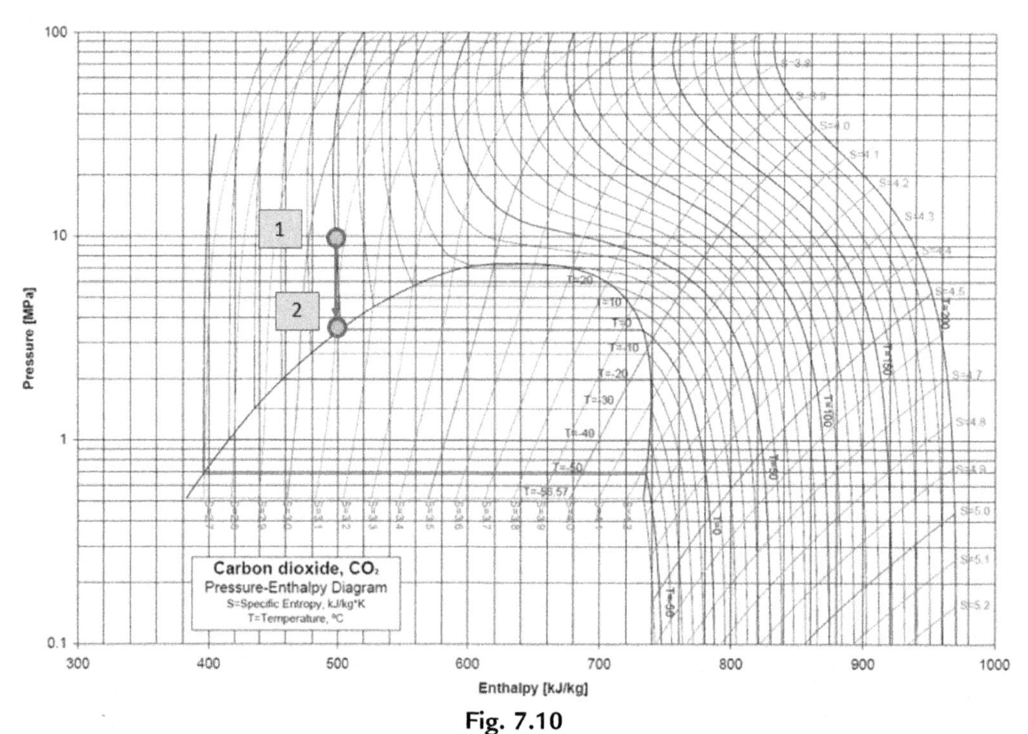

Fig. 7.10
Pressure enthalpy diagram for carbon dioxide of Example 7.5 – Hole outlet.

7.5.1 State 1: Stagnation

Storage Temperature (°C)	Storage Pressure (bar)	Enthalpy (J/kg K)	Entropy (J/kg K)
0	100	503,000	926

7.5.2 State 2: Hole Outlet (Ref. Example 7.4)

Temperature (°C)	Saturation Pressure (bar)	Enthalpy (J/kg)	Entropy (J/kg K)	Mass Flux (kg/h s^2)	Liquid Density
0	34.85	503,000	926	112,661	974.1

Fluid is subcooled at stagnation condition and at the outlet, so vapour is not produced. Outlet velocity is:

$$u_1 = \frac{112,661}{974.1} = 115.6\,\text{m/s} \tag{7.160}$$

From the equation of conservation of energy:

$$H_1 = H_o - \frac{u_1{}^2}{2} = 503,000 - \frac{115.6^2}{2} = 496318\,\text{J/kg} \tag{7.161}$$

It results that transition from state 0 to state 1 is practically isentropic.

7.5.3 State 3: Triple Point/Liquid Vapour Equilibrium

From the saturation temperature, vaporisation begins and continues through the liquid–vapour zone until the triple point is reached, 216.55 K at 5.18 bar. The inter-phase differential energy balance between saturation and triple points for liquid vapour transition, assuming no external heat transfer occurs, can be written as:

$$x(T) \cdot c_{pL}(T) \cdot dT = \lambda_v(T) \cdot dx \tag{7.162}$$

where x is the liquid fraction, $c_{PLl}(T)$, (J/kmol K), and $\lambda_v(T)$, (J/kmol), are given by the following expressions (Green and Perry, 2007):

$$c_{P_L}(T) = -8,304,300 + 104,370 \cdot T - 433.33 \cdot T^2 + 0.60052 \cdot T^3 \tag{7.163}$$

$$\lambda_v(T) = 21,730,000 \cdot (1 - T_r)^{(0.382 - 0.4339\,T_r + 0.42213\,T_r{}^2)} \tag{7.164}$$

where T_r is carbon dioxide reduced temperature, and T_c is carbon dioxide critical temperature:

$$T_r = \frac{T}{T_c} = \frac{T}{301.4} \tag{7.165}$$

$$x(T) \cdot c_{pL}(T) \cdot dT = \lambda_v(T) \cdot dx \tag{7.166}$$

The definite integral from saturation temperature T_s to triple point temperature T_{TP}, 216.55 K gives the liquid mass fraction at the triple point X_{LTP}:

$$\int_{T_s}^{T_{TP}} \frac{c_{P_L}(T)}{\lambda_v(T)} \cdot dT = \int_1^{x_{TP}} \frac{dx}{x} \tag{7.167}$$

Solving from the saturation condition, $x=1$ to $x=x_{TP}$ (see Fig. 7.11)

- TTP = 216.55 K
- PTP = 5.18 bar
- Liquid mass fraction, at triple point x_{TP}

$$e^{\left(\int_{273.16}^{216.55} \frac{-8304300+104300T-433.33T^2+0.60052T^3}{21730000\left(1-\frac{T}{304.1}\right)^{\left(0.382-0.4339\frac{T}{304.1}+0.42213\left(\frac{T}{304.1}\right)^2\right)}} dT\right)} = 0.71 \tag{7.168}$$

- Vapour mass fraction, $y_{TP}=0.29$
- Enthalpy = 483,000 J/kg
- Entropy = 3100 J/kg

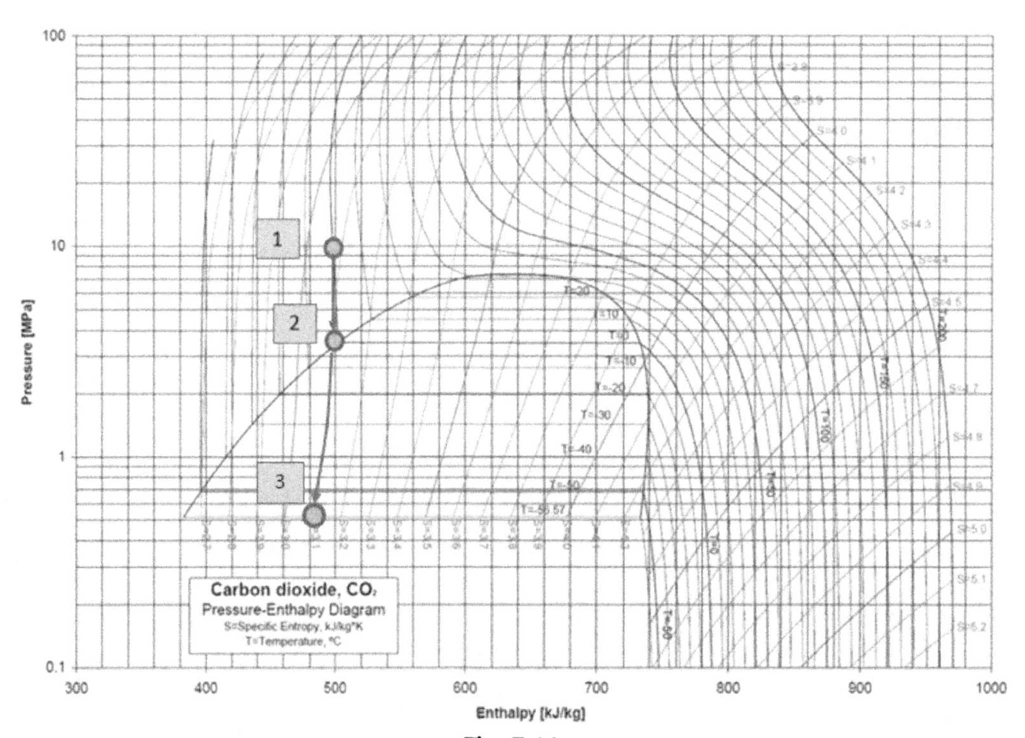

Fig. 7.11
Pressure enthalpy diagram for carbon dioxide of Example 7.5 – Triple point: liquid vapour equilibrium.

7.5.4 State 4: Triple Point: Solid Vapour Non-Equilibrium

At the triple point, the liquid vapour system should undergo a transition to solid-vapour equilibrium. The triple point should also represent the vapour–liquid to vapour–solid transition equilibrium point. This scenario does not correspond to what really happens, due both to thermodynamic and thermo-kinetic reasons. As for the liquid–vapour transition, nucleation requires overcoming a free energy barrier, and so the expected equilibrium cannot be achieved. A quantitative assessment of the liquid-to-solid transition around the triple point has been carried out by Martynov et al. (2012) with reference to carbon dioxide liquid expansion from a stagnation condition of 140 bar and 20°C. Around the triple point there is a sudden increase in vapour mass fraction from 0.38, consistent with the vapour–liquid equilibrium at the triple point, to 0.73, approximately at the same temperature and pressure. Afterwards, vapour mass fraction values move back to the expected vapour–solid equilibrium value, which is around 0.63. The first value, 0.73, corresponds to higher entropy and enthalpy that are not compatible with the vapour–solid equilibrium.

The only way to justify and explain this evidence results is to assume:

- Point 3 on the triple point axis does not represent an equilibrium state, but it is representative of a metastable condition of carbon dioxide; enthalpy increase corresponding to vapour mass fraction equal to 73 is to be considered apparent, otherwise the principle of conservation of energy would not be met; in fact, the values on the state diagrams are representative of equilibrium states. That is not the case of triple point transition.
- entropy change corresponding to vapour mass fraction equal to 73 is to be considered definitive, since the nuclei formation is accompanied by a significant free energy barrier; this new entropic state is necessary, due to the specific behaviour of the isentropic curves in the vapour-solid zone below the triple point, to meet the energy conservation requirement; furthermore, over-evaporation is the way that the system can increase the overall entropy.
- entropy increase can be assessed using Richard's rule as reported by Naterer (2002), applied to the vapour phase, which states that the nucleation entropy change can be calculated through the formula:

$$\Delta S_{nucl} = \frac{\lambda_{vTP}}{T_{TP}} \tag{7.169}$$

where λ_{vTP} is the heat of evaporation at the triple point temperature T_{TP}, 83.12 kcal/kg, according to Plank and Kuprianoff. Richard's equation can be applied considering the non equilibrium steeped incremental vapour maximum fraction from Martynov that is about 0.1. Multiplying this fraction by the ratio in the previous equation, it will result that 0.16 kJ/kg K is the entropic increment. Comparing this result to both the non equilibrium ΔS obtained by Martynov relative to the vapour phase (i.e. stagnation point 140 bar and 273.16 K and 140 bar and 253.16 K), we find the same value. We can argue that is the entropy increase corresponding to the liquid to solid transition phase and to the nucleation free energy barrier (see Fig. 7.12).

- TTP = 216.55 K
- PTP = 5.18 bar
- Vapour mass fraction, $y_{VP} = 0.62$
- Solid mass fraction, $z_{STP} = 0.38$
- Entropy = 3260 J/kg K

7.5.5 State 5: Solid Vapour Final Expansion

We would consider cooling of forming solid carbon dioxide along the previous isentropic curve causing vapour deposition, that is:

$$\int_{Tf}^{T_{TP}} \frac{C_s(T)}{\lambda_s(T)} \cdot dT = \int_{X_f}^{X_{TP}} \frac{dX_s}{X_s} \tag{7.170}$$

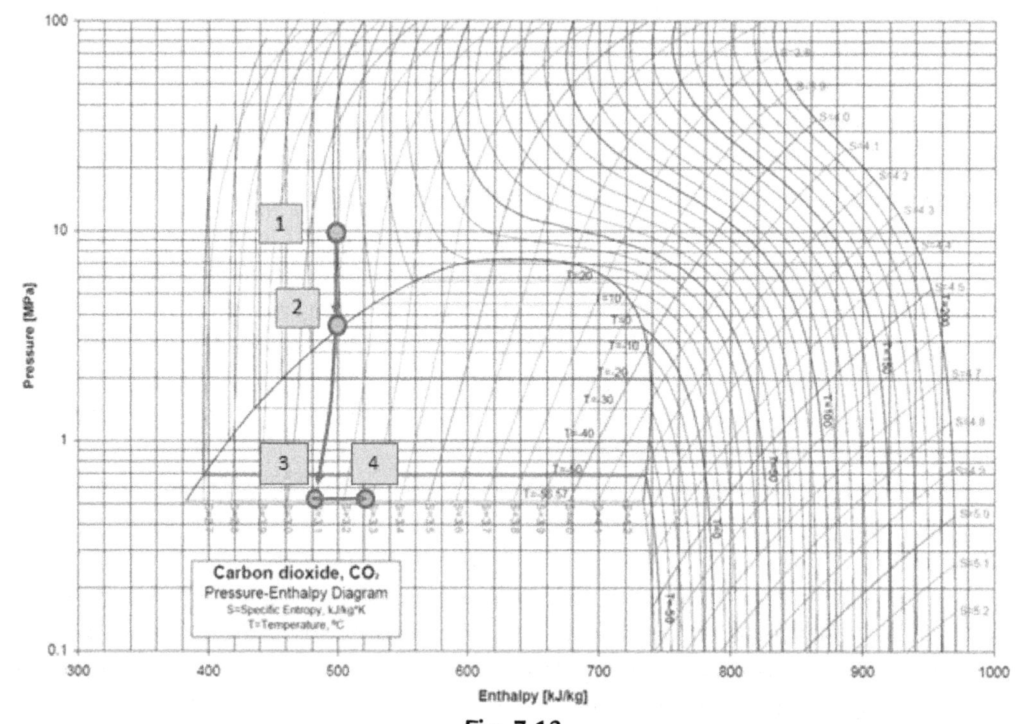

Fig. 7.12
Pressure enthalpy diagram for carbon dioxide of Example 7.5 – Triple point: solid liquid non-equilibrium.

where T_f is the final expansion temperature generally assumed as the equilibrium temperature at 1 atm. However, this is definitely arguable. Woolley et al. (2011) have presented experimental temperatures definitely lower than the atmospheric saturation pressure, up to 20÷50°C according to Erbland et al. (2000). Mazzoldi et al. (2011), using a constant Joule Thomson coefficient = 13 kMPa, with reference to an initial pressure equal to 10 MPa, assume a temperature drop up to 130 K. Through the equations of conservation, energy, and entropy, taking into account the kinetic effect and the isentropic increase, the following equation system has been resolved (subscript M is related to Mach disk, where it is assumed the jet has attained the sound speed u_M):

$$S_M = 4.35\,\text{kJ/kg K} \tag{7.171}$$

$$H_o = H_{Mv} \cdot (1 - x_{Ms}) + H_{Ms} \cdot x_{Ms} + {u_m}^2/2 \tag{7.172}$$

$$S_M = S_{Mv} \cdot (1 - x_{Ms}) + S_{Ms} \cdot x_{Ms} \tag{7.173}$$

$$u_M = (k \cdot R \cdot T_M)^{0.5} \tag{7.174}$$

$$H_v = 169.34 + (0.1965 + 0.000115 \cdot t) \cdot t - 8.3724 \cdot \frac{p}{\left(\frac{T}{100}\right)^{\frac{2}{3}}} \cdot (1 + 0.007424 \cdot p) \tag{7.175}$$

$$H_{Mv} = f(P_M, T_M) \tag{7.176}$$

$$S_{Mv} = g(P_M, T_M) \tag{7.177}$$

$$\lambda_s = 158.96 - 0.2409 \cdot T + 0.0014957 \cdot T^2 - 0.00000431 \cdot T^3 \tag{7.178}$$

$$S_{Ms} = S_{Mv} - \lambda_{Ms}/T_M \tag{7.179}$$

$$H_{Ms} = H_{Mv} - \lambda_{Ms} \tag{7.180}$$

$$x_{sTP} \times e^{-\int_{216.55}^{T_f} 216.55 \frac{0.44 - 0.00283 \times T + 0.0000125 \times T^2}{158.96 - 0.2409 \times T + 0.0014957 \times T^2 - 4.31 \times 10^{-6} \cdot T^3} dT} = x_{Ms} \tag{7.181}$$

where:

- S_M is the entropy of state 4.
- H_o is the liquid enthalpy at stagnation.
- Subscripts v and s refer to vapour and solid.
- λ_s is the latent heat of sublimation kcal/kg (Plank and Kuprianoff, 1929) with T in kelvin.
- H_v is vapour enthalpy in kcal/kg (Plank and Kuprianoff, 1929), p is pressure in kg/sq cm^2, t is the temperature in °C, and T is the temperature in Kelvin degrees.
- u_M is the sound velocity at the Mach disk.

- k is the carbon dioxide specific heats ratio.
- R is the ideal gas constant.
- x is the solid mass fraction.

The solution of the systems shows that final temperature is 179 K, much below the saturation point at 1 atm, as foreseen by Erbland et al. (2000). The corresponding saturation pressure is 0.42 bar absolute, the solid mass fraction is 0.42, and the final velocity is 212 m/s.

A full coverage of the non-equilibrium features of carbon dioxide expansion can be found in Benintendi (2014).

Particle size and snow-out

Theoretical and experimental studies such as Hulsbosch-Dam et al. (2012), Liu et al. (2010) have shown that particle size order of magnitude falls in the range 1–5 μ. In this situation, no snow-out occurs after horizontal release, and the sublimation rate is very fast.

7.6 Gases and Vapours

7.6.1 Choked and Unchoked Jet

Choked and unchoked flows have been defined at Section 4.6.2. The flow rate for an adiabatic expansion through an orifice can be calculated by the formula:

$$W = P_o \cdot A \cdot C_D \cdot \sqrt{2 \cdot \frac{[g \cdot MW \cdot \gamma]}{R \cdot T_o \cdot (\gamma - 1)} \cdot \left[\left(\frac{P}{P_o}\right)^{2/\gamma} - \left(\frac{P}{P_o}\right)\right]^{(\gamma-1)/\gamma}} \tag{7.182}$$

where:

- P_o: upstream pressure
- P: downstream pressure
- MW: molecular weight
- T_o: downstream temperature, K
- R: ideal gas constant, 8.31446 J/K mole
- γ: specific heats ratio = 1.4
- g: acceleration of gravity
- C_D: coefficient of discharge

and, for choked flow:

$$W_c = P_o \cdot A \cdot C_D \cdot \sqrt{\frac{MW \cdot \gamma}{R \cdot T_o} \cdot \left[\frac{2}{\gamma + 1}\right]^{(\gamma+1)/(\gamma-1)}} \tag{7.183}$$

In order to decide which formula is to be used, P_{choked} will be preliminarily calculated and compared to the downstream pressure. If the latter is equal or lower, then the choked flow rate will be adopted. According to CCPS (1999), a threshold pressure for real gas can be assumed at 1.4 barg (20 psig). A range of 0.61–0.85 is suggested for C_D (CCPS), and a value of 1 is suggested for screening purposes.

Example 7.13

Methane is stored at 10 bar absolute and 25°C in a 10 m^3 pressure vessel. Find the pressure decay rate and the related mass flux if the vessel is depressurised through a 100 mm rupture disk. Find also the time for the system to leave the choked condition. Assume that temperature in the vessel remains constant and evaluate the correctness of this assumption.

Solution

Temperature (T_o °C)	Stagnation pressure (P_o bar abs)	C_D	MW	R (J/K kmole)	γ	A (m^2)	V (m^3)
25	10	0.85	16	8314.46	1.32	0.0785	10

$$P_{\text{choked}} = P_o \cdot \left(\frac{2}{\gamma+1}\right)^{\frac{\gamma}{\gamma-1}} = 1000{,}000 \cdot \left(\frac{2}{2.32}\right)^{\frac{1.32}{0.32}} = 542{,}139.2 > P_{atm} \tag{7.184}$$

Flow is choked.

$$W(t) = P(t) \cdot A \cdot C_D \cdot \sqrt{\frac{g \cdot MW \cdot \gamma}{R \cdot T} \cdot \left[\frac{2}{\gamma+1}\right]^{(\gamma+1)/(\gamma-1)}}$$
$$= P(t) \cdot 0.00785 \cdot 0.85 \cdot \sqrt{0.000084 \cdot 0.34} = 0.000036 \cdot P(t) \tag{7.185}$$

A differential mass balance on the vessel gives:

$$dn = -\frac{W(t)}{MW} \cdot dt \tag{7.186}$$

where n is the number of kmoles in the vessel and $W(t)$ the mass flow. Applying the ideal gas law:

$$dn = \frac{V}{RT} \cdot dP \tag{7.187}$$

$$\frac{V}{RT} \cdot dP = -\frac{W(t)}{MW} \cdot dt \tag{7.188}$$

$$\frac{V}{RT} \cdot dP = \frac{10}{8314.46 \cdot 298.16} \cdot dP = 0.000004 \cdot dP \tag{7.189}$$

$$\frac{W(t)}{MW} \cdot dt = 0.00000225 \cdot P(t) \cdot dt \tag{7.190}$$

Equating:

$$\frac{dP}{P(t)} = -0.5625 \cdot dt \tag{7.191}$$

Integrating:

$$P(t) = P(0) \cdot \exp[-0.5625 \cdot t] \tag{7.192}$$

Fig. 7.13 shows the decay rate of pressure and flow mass within the choked range.

The calculation is based on the assumption that temperature is constant during the release. In reality, methane inside the vessel undergoes an expansion which implies a significant cooling effect. A simple evaluation of the temperature drop can be approached, assuming that the depressurisation is so fast that no significant heat transfer occurs between the ambient and the vessel, so that the polytrophic expansion can be considered, and the final temperature is:

$$T_f = \frac{T_o}{\left(\frac{P_o}{P_{atm}}\right)^{\frac{\gamma-1}{\gamma}}} = \frac{298.16}{\left(\frac{10}{1.013}\right)^{\frac{0.32}{1.32}}} = 171\,K \tag{7.193}$$

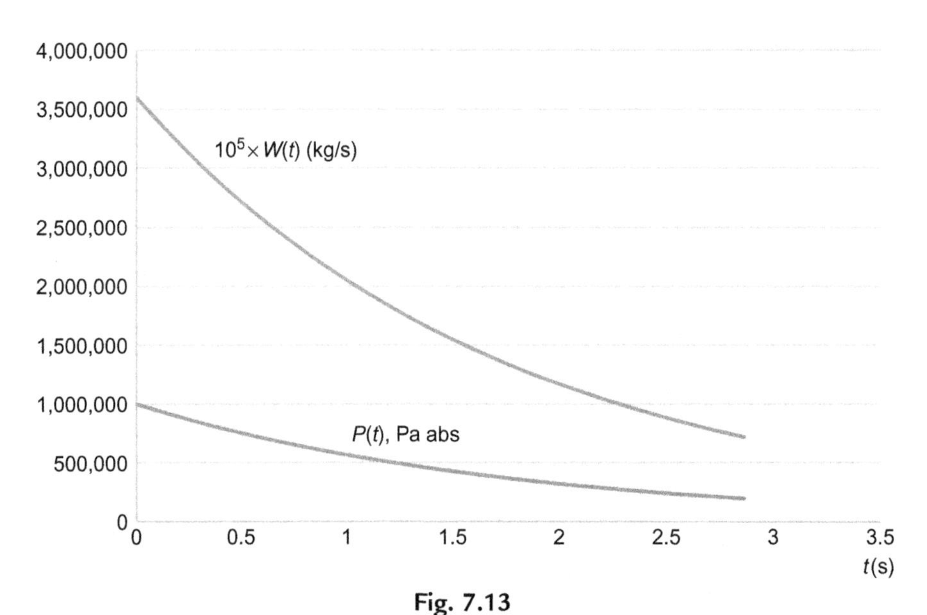

Fig. 7.13
Methane mass flow and pressure decay rate for Example 7.12.

Example 7.14

Liquid pentane is stored at ambient temperature in a pressure vessel. The ambient temperature is 45°C. Determine the release conditions from a hole.

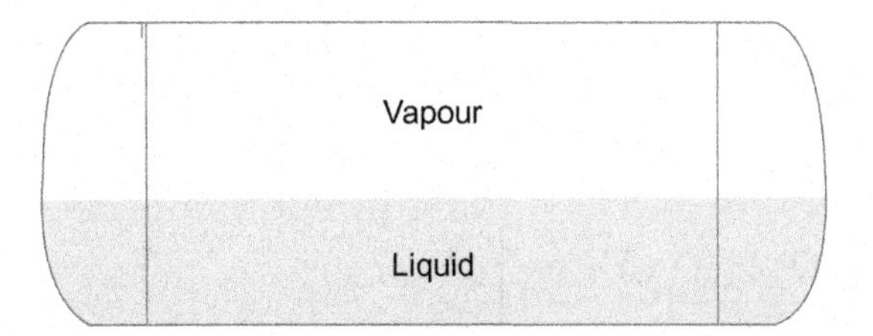

Solution

Ambient Temperature (T_a °C)	Vapour Pressure at (T_a bar)	γ
45	1.33	1.07

Internal pressure is significantly greater than atmospheric pressure.

$$P_{\text{choked}} = 133{,}000 \cdot \left(\frac{2}{2.07}\right)^{\frac{1.07}{0.07}} = 78{,}610 < P_{atm} \tag{7.194}$$

Flow is unchoked. Mass flow release will be calculated by means of the unchoked flow equation:

$$W = P_o \cdot A \cdot C_D \cdot \sqrt{2 \cdot \frac{[g \cdot MW \cdot \gamma]}{R \cdot T_o \cdot (\gamma - 1)} \cdot \left[\left(\frac{P}{P_o}\right)^{2/\gamma} - \left(\frac{P}{P_o}\right)\right]^{(\gamma-1)/\gamma}} \tag{7.195}$$

7.6.2 Effect of Pressure and Temperature: Gas Release From Pipelines

Calculation of gas release from pipelines is a particularly complex task, due to several factors. including temperature, pressure, friction effects, pipeline length, and geometrical features. In the literature, three main cases are approached, consisting of:

A. Pipelines for which the product of the hole cross-sectional area and the discharge coefficient is less than 3% of the line cross-sectional area.
B. Pipelines for which the product of the hole cross-sectional area and the discharge coefficient is greater than 3% and less than 30% of the line cross-sectional area.
C. Full bore releases

A. Pipelines for which the product of hole cross-sectional area and the discharge coefficient is less than 3% of the line cross-sectional area.

This approach, which assumes as negligible both the friction effects down the pipeline and the heat exchange, gives the transient flow rate W(t) from a pipeline blocked between two valves, as in:

$$W(t)=W_o\cdot\left[1+0.5\cdot(\gamma-1)\cdot C_D\cdot A\cdot\frac{t}{W_o}\cdot\sqrt{P_o\cdot\rho_g\cdot\gamma\cdot\left(\frac{2}{\gamma+1}\right)^{\frac{\gamma+1}{\gamma-1}}}\right]^{\frac{\gamma+1}{\gamma-1}} \tag{7.196}$$

Time taken for the system for transition from choked to unchoked is given by the formula:

$$t_c=\frac{2\cdot\left[\left(\frac{P_{\text{choked}}}{P_o}\right)^{-\frac{\gamma-1}{2\cdot\gamma}}-1\right]}{\gamma-1}\cdot\frac{W_o}{C_D\cdot A\cdot\sqrt{P_o\cdot\rho_g\cdot\gamma\cdot\left(\frac{2}{\gamma+1}\right)^{\frac{\gamma+1}{\gamma-1}}}} \tag{7.197}$$

In these formulas:

- W_o is the initial mass flow rate, kg/s.
- A is the hole area, m^2.

Example 7.15

Recalculate the example of methane stored at 10 bar absolute and 25°C in a 10 m^3 pressure vessel, assuming that the product of the hole cross-sectional area and the discharge coefficient is less than 3% of the line cross-sectional area (see Fig. 7.14).

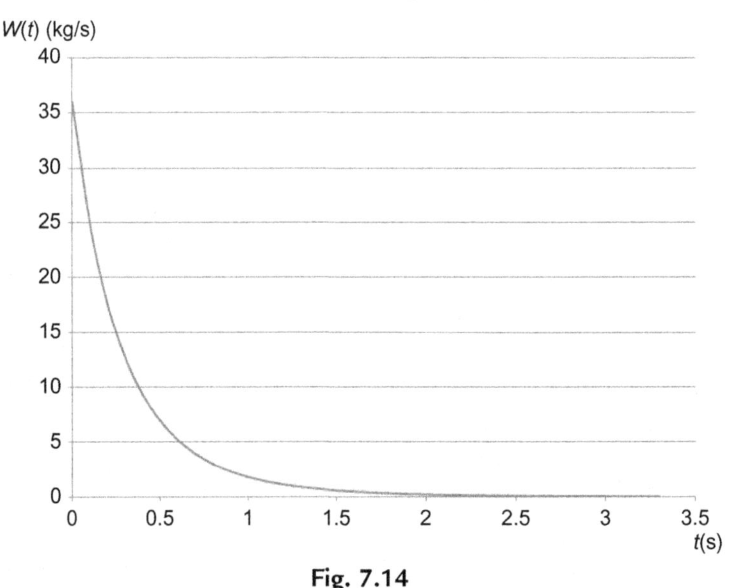

Fig. 7.14
Methane mass flow decay rate for Example 7.13.

Solution

Temperature (T_o °C)	Stagnation Pressure (P_o bar abs)	C_D	MW	R (J/K kmole)	γ	A (m^2)	ρ_g (Kg/m^3)
25	10	0.85	16	8314.46	1.32	0.0785	6.54

$$W_o = 0.000036 \cdot 1{,}000{,}000 = 36\ \text{kg/s} \tag{7.198}$$

$$W(t) = 36 \cdot \left[1 + 0.5 \cdot 0.32 \cdot 0.85 \cdot 0.0785 \cdot \frac{t}{36} \cdot \sqrt{1{,}000{,}000 \cdot 6.54 \cdot 1.32 \cdot 0.341}\right]^{\frac{2.32}{0.32}} \tag{7.199}$$

Comparing the isothermal case to the non-isothermal and much more realistic case, it is worth noticing how the cooling effect, due to the internal expansion, results in a drastic mass flow rate reduction caused by the exponential pressure reduction.

B. Pipelines for which the product of hole cross-sectional area and the discharge coefficient is greater than 3% and less than 30% of the line cross-sectional area

The method for predicting blow-down times has been presented by Weiss et al. (1988). It consists of the following steps:

1. Determine the sound speed and the isentropic exponent, γ.

$$c = \sqrt{z \cdot \gamma \cdot R \cdot T} \tag{7.200}$$

Where c is the sound speed and z is the compressibility factor.

2. Calculate the following dimensionless sonic blowdown time, t_c:

$$t_c = \ln\left(\frac{P_o}{P_a}\right) - \left(\frac{\gamma}{\gamma - 1}\right) \cdot \ln\left(\frac{\gamma + 1}{2}\right) \tag{7.201}$$

3. Determine the dimensionless subsonic blowdown time, t_s, entering P_c into Table 7.2:

Table 7.2 Dimensionless subsonic blowdown time

γ	Critical Pressure Ratio (P_c)	t_s
1.20	0.5645	0.7371
1.25	0.5549	0.7605
1.30	0.5457	0.7833
1.35	0.5368	0.8058
1.40	0.5283	0.8278
1.45	0.5200	0.8495
1.50	0.5120	0.8707
1.55	0.5041	0.8916
1.60	0.4968	0.9122

$$P_c = \left(\frac{2}{\gamma + 1}\right)^{\frac{\gamma}{\gamma - 1}} \tag{7.202}$$

4. Estimate the correction factor, C_f:

$$C_f = a_1 + a_2 \cdot \log_{10}\left(\frac{f \cdot L}{D}\right) + a_3 \cdot \left[\log_{10}\left(\frac{f \cdot L}{D}\right)\right]^2 + a_4 \cdot \left[\log_{10}\left(\frac{f \cdot L}{D}\right)\right]^4 \tag{7.203}$$

$$a_1 = 1.0319 - 5.2735 \cdot \left(\frac{1}{A_r}\right) + 25.680 \cdot \left(\frac{1}{A_r}\right)^2 - 38.409 \cdot \left(\frac{1}{A_r}\right)^3 \tag{7.204}$$

$$a_2 = -0.26994 + 17.304 \cdot \left(\frac{1}{A_r}\right) - 86.415 \cdot \left(\frac{1}{A_r}\right)^2 + 144.77 \cdot \left(\frac{1}{A_r}\right)^3 \tag{7.205}$$

$$a_3 = 0.24175 - 12.637 \cdot \left(\frac{1}{A_r}\right) + 56.772 \cdot \left(\frac{1}{A_r}\right)^2 - 88.351 \cdot \left(\frac{1}{A_r}\right)^3 \tag{7.206}$$

$$a_4 = -0.054856 + 2.6258 \cdot \left(\frac{1}{A_r}\right) - 8.9593 \cdot \left(\frac{1}{A_r}\right)^2 + 12.139 \cdot \left(\frac{1}{A_r}\right)^3 \tag{7.207}$$

$$Ar = \frac{A_p}{A_{th} \cdot C_D} \tag{7.208}$$

with:

- f, Darcy friction factor
- D, Inside pipeline diameter, m
- L, pipeline length, m
- A_p = pipeline cross section, m^2
- A_{th}, blowdown discharge area, m^2
- C_D, coefficient of discharge

5. Calculate the time constant, τ_v:

$$\tau_v = \frac{V \cdot \left(\frac{\gamma + 1}{2}\right)^{\frac{\gamma+1}{2 \cdot (\gamma - 1)}}}{c \cdot A_{th} \cdot C_D} \tag{7.209}$$

6. Find the blowdown time, t_{BD}:

$$t_{BD} = (t_c + t_s) \cdot \tau_v \cdot C_f \tag{7.210}$$

Example 7.16

Estimate the blowdown time for a methane pipeline 10 km long, internal diameter 1 m, friction factor 0.01, temperature 20°C, pressure 50 bar abs, with a discharge area of 0.1 m^2.

Solution

Pressure (P_o, bar)	Ambient Temperature (T_a °C)	Pipeline Length (m)	D (m)	A_p (m^2)	A_{th} (m^2)	γ	C_D	f	z	R (J/K kmole)
50	20	10,000	0.1	0.785	0.1	1.32	0.85	0.01	0.88	8314.86

1. Determine the sound speed and the isentropic exponent:

$$c = \sqrt{\frac{0.88 \cdot 1.32 \cdot 8314.86 \cdot 293.16}{16}} = 421\,\text{m/s} \tag{7.211}$$

2. Calculate the following dimensionless sonic blowdown time, t_c:

$$t_c = \ln\left(\frac{P_o}{P_a}\right) - \left(\frac{\gamma}{\gamma - 1}\right) \cdot \ln\left(\frac{\gamma + 1}{2}\right) = \ln(50) - \frac{1.32}{0.32} \cdot \ln\left(\frac{2.32}{2}\right) = 3.3 \tag{7.212}$$

3. Determine the dimensionless subsonic blowdown time, t_s:

$$P_c = \left(\frac{2}{2.32}\right)^{\frac{1.32}{0.32}} = 0.54 \rightarrow t_s = 0.784\,s \tag{7.213}$$

4. Estimate the correction factor, C_f

$$Ar = \frac{A_p}{A_{th} \cdot C_D} = \frac{0.785}{0.1 \cdot 0.85} = 9.24 \tag{7.214}$$

$$\frac{f \cdot L}{D} = \frac{0.01 \cdot 10{,}000}{0.1} = 1000 \tag{7.215}$$

$$C_f = 1.5 \tag{7.216}$$

5. Calculate the time constant, τ_v:

$$V = L \cdot A = 10{,}000 \cdot 0.785 = 7{,}850\,\text{m}^3 \tag{7.217}$$

$$\tau_v = \frac{V \cdot \left(\frac{\gamma+1}{2}\right)^{\frac{\gamma+1}{2\cdot(\gamma-1)}}}{c \cdot A_{th} \cdot C_D} = \frac{7850 \cdot \left(\frac{2.32}{2}\right)^{\frac{2.32}{0.64}}}{421 \cdot 0.1 \cdot 0.85} = 376\ \mathrm{s} \tag{7.218}$$

6. Find blowdown time, t_{BD}:

$$t_{BD} = \left(t_c + t_s\right) \cdot \tau_v \cdot C_f = (3.3 + 0.784) \cdot 376 \cdot 1.5 = 2303\ \mathrm{s} \tag{7.219}$$

C. Pipeline full bore releases

When the discharge area is greater than 30%, a full bure rupture is assumed. A simple double *exponential model* has been developed by Bell (1978), which has proposed the following formula for the calculation of the release rate with time:

$$W(t) = \frac{W_o}{1 + \frac{M_o}{W_o \cdot B}} \cdot \left[\exp\left(-\frac{B \cdot W_o^{\,2}}{M_o^2} \cdot t\right) + \frac{M_o}{W_o \cdot B} \cdot \exp\left(-\frac{t}{B}\right)\right] \tag{7.220}$$

where:

$$B = 0.67 \cdot \frac{L}{c} \cdot \sqrt{\frac{\gamma \cdot f \cdot L}{D}} \tag{7.221}$$

with the same definitions of the previous case, and:

$$W_o = A_{th} \cdot \sqrt{P_o \cdot \rho_o} \cdot \left(\frac{2}{\gamma+1}\right)^{\left(\frac{\gamma+1}{\gamma-1}\right)} \tag{7.222}$$

7.7 Pool Formation and Liquid Evaporation

7.7.1 Transient Pool Radius – Wu and Schroy Formula

Wu and Schroy (1979) proposed the following equation to find the pool radius growth with the liquid flow, on the basis of the assumption that the area is not bunded and is round-shaped:

$$r(t) = \sqrt[5]{\frac{t^3 \cdot \rho_L \cdot Q \cdot \sin 2\alpha}{2 \cdot C^3 \cdot \mu_L \cdot \pi^2/6 \cdot g}} \tag{7.223}$$

where:

- r(is) is the pool radius (m).
- t is the spill time (s).
- g is the acceleration of gravity (m/s^2).
- Q is the spill flow rate (m^3/s).
- ρ_L is the liquid density (m^3/kg).
- μ_L is the liquid viscosity (Pa·s).
- C is a constant which is 2 for a pool Reynolds number greater than 25 and 5 for different pool Reynolds numbers:

$$Re_P = \frac{2 \cdot \rho_L \cdot Q}{\pi \cdot r \cdot \mu_L} \tag{7.224}$$

- α is the angle between the pool surface and the axis perpendicular to the ground:

$$\alpha = \tan^{-1} \sqrt{\sqrt{0.25 + r^4 \cdot \frac{22.489 \cdot \rho_L}{Q \cdot \mu_L}} - 0.5} \tag{7.225}$$

The calculation is iterative.

According to EPA, the worst case assumption for the height of an unconfined pool is 1 cm.

7.7.2 Pool Evaporation Formulas

Equations of Cox et al. for non-cryogenic and cryogenic pools

These equations have been included in Sections 4.10.1 and 4.10.2.

Recknagel equation

This very simple equation (Recknagel-Sprenger, 1961) can be used for screening purposes.

$$W = 6.3 \cdot 10^{-7} \cdot (2.92 + u) \cdot A \cdot P_v{}^o \cdot WM \tag{7.226}$$

with:

- W: lb$_m$/min of evaporating liquid
- u: wind speed (miles/h)
- A: spill area (ft^2)
- P_v^o: vapour pressure of liquid at the surface at average ambient temperature (mm Hg)
- WM: molecular weight

Spill area can be calculated with the formulas included in the previous and following sections.

EPA equation for mixture pools

This equation has been included at Section 7.8.5.

Sutton equation

This equation is valid for liquid with a vapour pressure below 0.4 bar (Sutton, 1953):

$$W = 0.002 \cdot u^{0.78} \cdot r^{-0.11} \cdot \frac{MW \cdot 10^5}{R \cdot T} \cdot \ln\left(\frac{1}{1 - 10^{-5} P_v{}^o}\right) \quad (7.227)$$

with:

- W, evaporation mass flux, kg/m^2 s
- u, wind speed at 10 m height, m/s
- MW, molecular weight, kg/kmole
- R, ideal gas constant, 8314 J/kmole K
- T, liquid temperature, K
- P^v_o, vapour pressure, Pa
- r, pool radius, m

Matthiesen (1986) equation

Matthiesen (1986) proposed the following evaporation rate equation:

$$W = \frac{MW \cdot A \cdot P_o{}^v}{R \cdot T} \cdot 0.0083 \cdot \left(\frac{18}{WM}\right)^{\frac{1}{3}} \quad (7.228)$$

with:

- MW, molecular weight, kg/kmole
- A, pool area , m^2
- P^v_o, vapour pressure, Pa
- R, ideal gas constant, 8314 J/kmole K
- T, liquid temperature, K

This formula is based on the assumption that the vapour concentration in the surrounding bulk is significantly lower than the one corresponding to the vapour pressure, that is, wind influence is very small. This is not always true, and should be considered.

Equations of Kawamura and MacKay

Kawamura and MacKay (1987) have modelled one of the most accurate pool evaporation formulas including solar radiation, evaporative cooling, and heat transfer from ground:

$$Q_v = q_v \cdot A \tag{7.229}$$

with:

$$q_v = k_m \cdot P_v{}^{\circ}(T_{ps}) \cdot \frac{WM}{R \cdot T_{ps}} \tag{7.230}$$

and:

$$k_m = 0.004886 \cdot u_w{}^{0.78} \cdot (2 \cdot r)^{-0.11} \cdot 0.8^{-0.67} \tag{7.231}$$

where:

- k_m is the mass transfer coefficient, m/s.
- $P_v{}^{\circ}(T)$ is the vapour pressure at temperature T, Pa.
- r is the liquid pool, m.
- R is the gas constant, J/mol K.
- T_{ps} is the liquid temperature at the liquid–vapour interface.
- u_w is the wind velocity, as usually taken at 10 m above grade, m/s.
- WM is the molecular weight.
- Q_v is in kg/s units.
- q_v is in kg/m^{2}s units.

7.8 U.S. EPA Offsite Consequence Analysis (OCA) Models

7.8.1 Release Rate From Pool at Ambient Temperature

The following equation applies for release rate from pools spreading out to a depth of 1 cm at 25°C (U.S. EPA, 2009):

$$W = M \cdot 1.4 \cdot LFA \cdot DF \tag{7.232}$$

- W is the release rate (pounds per minute).
- M is the released quantity (pounds).
- 1.4 is the wind speed factor $= 1.5^{0.78}$, where 1.5 m/s is the worst case.
- LFA is the Liquid Factor Ambient (Table 7.3).
- DF is the Density Factor (Table 7.3)

Example 7.18

20,000 pounds of sulphur trioxide are spilled at ambient temperature, and a pool is formed 1 cm tick. Find the release rate.

Solution

$$W = 20{,}000 \cdot 1.4 \cdot 0.057 \cdot 0.26 = 415 \text{ pounds/min} \tag{7.233}$$

7.8.2 Release Rate From Pool at Elevated Temperature (50°C > T > 25°C)

The following equation applies for release rate from pools spreading out to a depth of 1 cm between 25°C and 50°C (U.S. EPA, 2009):

$$W_C = W \cdot TCF \tag{7.234}$$

where W is the release rate as calculated at Section 7.8.1, and TCF is the closest temperature correction factor with respect the applicable temperature collected from Table 7.4.

7.8.3 Release Rate From Pool at Elevated Temperature (>50°C)

The following equation applies for the release rate from pools spreading out to a depth of 1 cm at temperatures > 50°C (U.S. EPA, 2009):

$$W = M \cdot 1.4 \cdot LFB \cdot DF \tag{7.235}$$

7.8.4 Pool Spreading Area (Other Than Diked)

Pool areas other, or smaller, than diked can be predicted according to the following equation (U.S. EPA, 2009):

$$A = QS \cdot DF \tag{7.236}$$

where:

- A is the pool area (square feet) assumed to be 1 cm thick.
- QS is the quantity released (pounds).
- DF is a density factor included in Tablex 7.3.

Table 7.3 Liquid and density factors (U.S. EPA, 2009)

CAS Number	Chemical Name	Molecular Weight	Vapour Pressure at 25°C (mm Hg)	Toxic Endpoint[a]			Liquid Factors		Density Factor (DF)	Liquid Leak Factor (LLF)[b]	Reference Table[c]	
				mg/L	ppm	Basis	Ambient (LFA)	Boiling (LFB)			Worst Case	Alternative Case
107-02-8	Acrolein	56.06	274	0.0011	0.5	ERPG-2	0.047	0.12	0.58	40	Dense	Dense
107-13-1	Acrylonitrile	53.06	108	0.076	35	ERPG-2	0.018	0.11	0.61	39	Dense	Dense
814-68-6	Acryloyl chloride	90.51	110	0.00090	0.2	EHS-LOC (Tox[d])	0.026	0.15	0.44	54	Dense	Dense
107-18-6	Allyl alcohol	58.08	26.1	0.036	15	EHS-LOC (IDLH)	0.0046	0.11	0.58	41	Dense	Buoyant[e]
107-11-9	Allylamine	57.10	242	0.0032	1	EHS-LOC (Tox[d])	0.042	0.12	0.64	36	Dense	Dense
7784-34-1	Arsenous trichloride	181.28	10	0.01	1	EHS-LOC (Tox[d])	0.0037	0.21	0.23	100	Dense	Buoyant[e]
353-42-4	Boron trifluoride compound with methyl ether (1:1)	113.89	11	0.023	5	EHS-LOC (Tox[d])	0.0030	0.16	0.49	48	Dense	Buoyant[e]
7726-95-6	Bromine	159.81	212	0.0065	1	ERPG-2	0.073	0.23	0.16	150	Dense	Dense
75-15-0	Carbon disulfide	76.14	359	0.16	50	ERPG-2	0.075	0.15	0.39	60	Dense	Dense
67-66-3	Chloroform	119.38	196	0.49	100	EHS-LOC (IDLH)	0.055	0.19	0.33	71	Dense	Dense
542-88-1	Chloromethyl ether	114.96	29.4	0.00025	0.05	EHS-LOC (Tox[d])	0.0080	0.17	0.37	63	Dense	Dense
107-30-2	Chloromethyl methyl ether	80.51	199	0.0018	0.6	EHS-LOC (Tos[d])	0.043	0.15	0.46	51	Dense	Dense
4170-30-3	Crotonaldehyde	70.09	33.1	0.029	10	ERPG-2	0.0066	0.12	0.58	41	Dense	Buoyant[e]
123-73-9	Crotonaldehyde, (E)-	70.09	33.1	0.029	10	ERPG-2	0.0066	0.12	0.58	41	Dense	Buoyant[e]
108-91-8	Cyclohexylamine	99.18	10.1	0.16	39	EHS-LOC (Tox[d])	0.0025	0.14	0.56	41	Dense	Buoyant[e]
75-78-5	Dimethyldichlorosilane	129.06	141	0.026	5	ERPG-2	0.042	0.20	0.46	51	Dense	Dense
57-14-7	1,1-Dimethylhydrazine	60.10	157	0.012	5	EHS-LOC (IDLH)	0.028	012	0.62	38	Dense	Dense
106-89-8	Epichlorohydrin	92.53	17.0	0.076	20	ERPG-2	0.0040	0.14	0.42	57	Dense	Buoyant[e]
107-15-3	Ethylenediamine	60.10	12.2	0.49	200	EHS-LOC (IDLH)	0.0022	0.13	0.54	43	Dense	Buoyant[e]
151-56-4	Ethyleneimine	43.07	211	0.018	10	EHS-LOC (IDLH)	0.030	0.10	0.58	40	Dense	Dense
110-00-9	Furan	68.08	600	0.0012	0.4	EHS-LOC (Tox[d])	0.12	0.14	0.52	45	Dense	Dense
302-01-2	Hydrazine	32.05	14.4	0.011	8	EHS-LOC (IDLH)	0.0017	0.069	0.48	48	Buoyant[d]	Buoyant[e]
13463-40-6	Iron, pentacarbonyl-	195.90	40	0.00044	0.05	EHS-LOC (Tox[d])	0.016	0.24	0.33	70	Dense	Dense
78-82-0	Isobutyronitrile	69.11	32.7	0.14	50	ERPG-2	0.0064	0.12	0.63	37	Dense	Buoyant[e]

Continued

Table 7.3 Liquid and density factors (U.S. EPA, 2009)—cont'd

CAS Number	Chemical Name	Molecular Weight	Vapour Pressure at 25°C (mm Hg)	Toxic Endpoint[a]			Liquid Factors		Density Factor (DF)	Liquid Leak Factor (LLF)[b]	Reference Table[c]	
				mg/L	ppm	Basis	Ambient (LFA)	Boiling (LFB)			Worst Case	Alternative Case
108-23-6	Isopropyl chloroformate	122.55	28	0.10	20	EHS-LOC (Tox[d])	0.0080	0.17	0.45	52	Dense	Dense
126-98-7	Methacrylonitrile	67.09	71.2	0.0027	1	EHS-LOC (TLV[f])	0.014	0.12	0.61	38	Dense	Dense
79-22-1	Methyl chloroformate	94.50	108	0.0019	0.5	EHS-LOC (Tox[d])	0.026	0.16	0.40	58	Dense	Dense
60-34-4	Methyl hydrazine	46 07	49.6	0.0094	5	EHS-LOC (IDLH)	0.0074	0 094	0.56	42	Dense	Buoyant[e]
624-83-9	Methyl isocyanate	57.05	457	0.0012	0.5	ERPG-2	0.079	0.13	0.52	45	Dense	Dense
556-64-9	Methyl thiocyanate	73.12	10	0.085	29	EHS-LOC (Tox[d])	0.0020	0.11	0.45	51	Dense	Buoyant[e]
75-79-6	Methyltrichlorosilane	149.48	173	0.018	3	ERPG-2	0.057	0.22	0.38	61	Dense	Dense
13463-39-3	Nickel carbonyl	170.73	400	0.00067	0.1	EHS-LOC (Tox[d])	0.14	0.26	0.37	63	Dense	Dense
7697-37-2	Nitric acid (100%)[g]	63.01	63.0	0.026	10	EHS-LOC (Tox[d])	0.012	0.12	0.32	73	Dense	Dense
79-21-0	Peracetic acid	76.05	13.9	0.0045	1.5	EHS-LOC (Tox[d])	0.0029	0.12	0.40	58	Dense	Buoyant[e]
594-42-3	Perchloromethyl mercaptan	185.87	6	0.0076	1	EHS-LOC (IDLH)	0.0023	0.20	0.29	81	Dense	Buoyant[e]
10025-87-3	Phosphorus oxychloride	153.33	35.8	0.0030	0.5	EHS-LOC (Tox[d])	0.012	0.20	0.29	80	Dense	Dense
7719-12-2	Phosphorus trichloride	137.33	120	0.028	5	EHS-LOC (IDLH)	0.037	0.20	0.31	75	Dense	Dense
110-89-4	Piperidine	85.15	32.1	0.022	6	EHS-LOC (Tox[d])	0.0072	0.13	0.57	41	Dense	Buoyant[e]
107-12-0	Propionitrile	55.08	47.3	0.0037	1.6	EHS-LOC (Tox[d])	0.0080	0.10	0.63	37	Dense	Buoyant[e]
109-61-5	Propyl chloroformate	122.56	20.0	0.010	2	EHS-LOC (Tox[d])	0.0058	0.17	0.45	52	Dense	Buoyant[e]
75-55-8	Propyleneimine	57.10	187	0.12	50	EHS-LOC (IDLH)	0.032	0.12	0.61	39	Dense	Dense
75-56-9	Propylene oxide	58.08	533	0.59	250	ERPG-2	0.093	0.13	0.59	40	Dense	Dense
7446-11-9	Sulphur trioxide	80.06	263	0.010	3	ERPG-2	0.057	0.15	0.26	91	Dense	Dense

Table 7.3 Liquid and density factors (U.S. EPA, 2009)—cont'd

CAS Number	Chemical Name	Molecular Weight	Vapour Pressure at 25°C (mm Hg)	Toxic Endpoint[a]			Liquid Factors		Density Factor (DF)	Liquid Leak Factor (LLF)[b]	Reference Table[c]	
				mg/L	ppm	Basis	Ambient (LFA)	Boiling (LFB)			Worst Case	Alternative Case
75-74-1	Tetramethyllead	267.33	22.5	0.0040	0.4	EHS-LOC (IDLH)	0.011	0.29	0.24	96	Dense	Dense
509-14-8	Tetranitromethane	196.04	11.4	0.0040	0.5	EHS-LOC (IDLH)	0.0045	0.22	0.30	78	Dense	Buoyant[e]
7550-45-0	Titanium tetrachloride	189.69	12.4	0.020	2.6	ERPG-2	0.0048	0.21	0.28	82	Dense	Buoyant[e]
584-84-9	Toluene 2,4-diisocyanate	174.16	0.017	0.0070	1	EHS-LOC (IDLH)	0.000006	0.16	0.40	59	Buoyant[d]	Buoyant[e]
91-08-7	Toluene 2,6-diisocyanate	174.16	0.05	0.0070	1	EHS-LOC (IDLH[h])	0.000018	0.16	0.40	59	Buoyant[d]	Buoyant[e]
26471-62-5	Toluene diisocyanate (unspecified isomer)	174.16	0.017	0.0070	1	EHS-LOC equivalent (IDLH[i])	0.000006	0.16	0.40	59	Buoyant[d]	Buoyant[e]
75-77-4	Trimethylchlorosilane	108.64	231	0.050	11	EHS-LOC (Tox[d])	0.061	0.18	0.57	41	Dense	Dense
108-05-4	Vinyl acetate monomer	86.09	113	0.26	75	ERPG-2	0.026	0.15	0.53	45	Dense	Dense

LOC is based on the IDLH-equivalent level estimated from toxicity data. GF to be used for gas leaks under choked (maximum) flow conditions.

[a]Toxic endpoints are specified in Appendix A to 40 CFR part 68 in units of mg/L.

[b]Hydrogen fluoride is lighter than air, but may behave as a dense gas upon release under some circumstances (e.g., release under pressure, high concentration in the released cloud) because of hydrogen bonding; consider the conditions of release when choosing the appropriate table.

[c]"Buoyant" in the Reference Table column refers to the tables for neutrally buoyant gases and vapors; "Dense" refers to the tables for dense gases and vapors. Ref. Appendix D, Section D.4.4, of the cited EPA standard for more information on the choice of reference tables.

[d]Ref. Exhibit B-3 of the mentioned EPA standard for data on water solutions.

[e]Gases that are lighter than air may behave as dense gases upon release if liquefied under pressure or cold; consider the conditions of release when choosing the appropriate table.

[f]LOC based on Threshold Limit Value (TLV)—Time-weighted average (TWA) developed by the American Conference of Governmental Industrial Hygienists (ACGIH).

[g]Cannot be liquefied at 25°C.

[h]Not an EHS; LOC-equivalent value estimated from one-tenth of the IDLH.

[i]Not an EHS; LOC-equivalent value estimated from one-tenth of the IDLH-equivalent level estimated from toxicity data.

Table 7.4 Temperature correction factors for release from pools between 25°C and 50°C (U.S. EPA, 2009)

CAS Number	Chemical Name	Boiling Point (°C)	Temperature Correction Factor (TCF)				
			30°C (86°F)	35°C (95°F)	40°C (104°F)	45°C (113°F)	50°C (122°F)
107-02-8	Acrolein	52.69	1.2	1.4	1.7	2.0	2.3
107-13-1	Acrylonitrile	77.35	1.2	1.5	1.8	2.1	2.5
814-68-6	Acryloyl chloride	75.00	ND	ND	ND	ND	ND
107-18-6	Allyl alcohol	97.08	1.3	1.7	2.2	2.9	3.6
107-11-9	Allylamine	53.30	1.2	1.5	1.8	2.1	2.5
7784-34-1	Arsenous trichloride	130.06	ND	ND	ND	ND	ND
353-42-4	Boron trifluoride compound with methyl ether (1:1)	126.85	ND	ND	ND	ND	ND
7726-95-6	Bromine	58.75	1.2	1.5	1.7	2.1	2.5
75-15-0	Carbon disulfide	46.22	1.2	1.4	1.6	1.9	LFB
67-66-3	Chloroform	61.18	1.2	1.5	1.8	2.1	2.5
542-88-1	Chloromethyl ether	104.85	1.3	1.6	2.0	2.5	3.1
107-30-2	Chloromethyl methyl ether	59.50	12	1.5	1.8	2.1	2.5
4170-30-3	Crotonaldehyde	104.10	1.3	1.6	2.0	2.5	3.1
123-73-9	Crotonaldehyde, (E)-	102.22	1.3	1.6	2.0	2.5	3.1
108-91-8	Cyclohexylamine	134.50	1.3	1.7	2.1	2.7	3.4
75-78-5	Dimethyldichlorosilane	70.20	1.2	1.5	1.8	2.1	2.5
57-14-7	1,1-Dimethylhydrazine	63.90	ND	ND	ND	ND	ND
106-89-8	Epichlorohydrin	118.50	1.3	1.7	2.1	2.7	3.4
107-15-3	Ethylenediamine	36.26	1.3	1.8	LFB	LFB	LFB
151-56-4	Ethyleneimine	55.85	1.2	1.5	1.8	2.2	2.7
110-00-9	Furan	31.35	1.2	LFB	LFB	LFB	LFB
302-01-2	Hydrazine	113.50	1.3	1.7	2.2	2.9	3.6
13463-40-6	Iron, pentacarbonyl-	102.65	ND	ND	ND	ND	ND
78-82-0	Isobutyronitrile	103.61	1.3	1.6	2.0	2.5	3.1
108-23-6	Isopropyl chloroformate	104.60	ND	ND	ND	ND	ND
126-98-7	Methacrylonitrile	90.30	1.2	1.5	1.8	2.2	2.6
79-22-1	Methyl chloroformate	70.85	1.3	1.6	1.9	2.4	2.9
60-34-4	Methyl hydrazine	87.50	ND	ND	ND	ND	ND
624-83-9	Methyl isocyanate	38.85	1.2	1.4	LFB	LFB	LFB
556-64-9	Methyl thiocyanate	130.00	ND	ND	ND	ND	ND
75-79-6	Methyltrichlorosilane	66.40	1.2	1.4	1.7	2.0	2.4
13463-39-3	Nickel carbonyl	42.85	ND	ND	ND	ND	ND
7697-37-2	Nitric acid	83.00	1.3	1.6	2.0	2.5	3.1
79-21-0	Peracetic acid	109.85	1.3	1.8	2.3	3.0	3.8
594-42-3	Perchloromethylmercaptan	147.00	ND	ND	ND	ND	ND
10025-87-3	Phosphorus oxychloride	105.50	1.3	1.6	1.9	2.4	2.9
7719-12-2	Phosphorus trichloride	76.10	1.2	1.5	1.8	2.1	2.5
110-89-4	Piperidine	106.40	1.3	1.6	2.0	2.4	3.0
107-12-0	Propionitrile	97.35	1.3	1.6	1.9	2.3	2.8
109-61-5	Propyl chloroformate	112.40	ND	ND	ND	ND	ND
75-55-8	Propyleneimine	60.85	1.2	1.5	1.8	2.1	2.5
75-56-9	Propylene oxide	33.90	1.2	LFB	LFB	LFB	LFB

Table 7.4 Temperature correction factors for release from pools between 25°C and 50°C (U.S. EPA, 2009)—cont'd

CAS Number	Chemical Name	Boiling Point (°C)	Temperature Correction Factor (TCF)				
			30°C (86°F)	35°C (95°F)	40°C (104°F)	45°C (113°F)	50°C (122°F)
7446-11-9	Sulphur trioxide	44.75	1.3	1.7	LFB	LFB	LFB
75-74-1	Tetramethyllead	110.00	ND	ND	ND	ND	ND
509-14-8	Tetranitromethane	125.70	1.3	1.7	2.2	2.8	3.5
7550-45-0	Titanium tetrachloride	135.85	1.3	1.6	2.0	2.6	3.2
584-84-9	Toluene 2,4-diisocyanate	251.00	1.6	2.4	3.6	5.3	7.7
91-08-7	Toluene 2,6-diisocyanate	244.85	ND	ND	ND	ND	ND
26471-62-5	Toluene diisocyanate (unspecified isomer)	250.00	1.6	2.4	3.6	5.3	7.7
75-77-4	Trimethylchlorosilane	57.60	1.2	1.4	1.7	2.0	2.3
108-05-4	Vinyl acetate monomer	72.50	1.2	1.5	1.9	2.3	2.7

Example 7.18

20,000 pounds of sulphur trioxide are spilled into an unbunded area. Find the area.

Solution

$$A = 20{,}000 \cdot 0.26 = 5200\,\mathrm{ft}_2 \tag{7.237}$$

7.8.5 Release Rate From Pool for Mixture Components

Release rate of mixture components from pools can be calculated applying Raoult's law to the general evaporation rate equation (U.S. EPA, 2009):

$$Wi = \frac{0.035 \cdot u^{0.78} \cdot MW_i^{\frac{2}{3}} \cdot A \cdot P_i}{T} \tag{7.238}$$

where:

- W_i is the release rate of component i (pounds per minute).
- u is the wind speed (m/s).
- MW_i is the molecular weight of component -i.
- A is the pool area calculated using the density factor of component -i (ft^2).
- P_i is the partial pressure of component -i calculated multiplying its vapour pressure by molar fraction (mmHg).

7.9 Simplified Formulas

The following simple formulas are very useful when a quick estimation is sufficient, and should be used for screening purposes.

7.9.1 Liquid Releases

Subcooled liquids

Release rate (lb_m/min) of a liquid of density ρ_L from a hole of diameter d(inches), under constant pressure P (gauge, lb_f/in^2) and head, (Winegarden, 1987).

$$W = 20 \cdot d^2 \cdot \sqrt{(P \cdot \rho_L)} \tag{7.239}$$

Subcooled liquids in vertical cylindrical tanks

Release rate (lb_m/min) of a liquid of density ρ_L from a hole of diameter d(inches) in a vertical tank with a liquid height h_L (ft) and pressure P (absolute, lb_f/in^2), (Winegarden, 1987).

$$W = 1.628 \cdot d^2 \cdot \rho_L \cdot \sqrt{h_L + \frac{144 \cdot P}{\rho_L}} \tag{7.240}$$

The emptying time (min) for this tank of diameter D (ft), if the initial level is h_{Lo} (ft), and pressure is constant is:

$$W = \frac{0.965}{d^2} \cdot D^2 \cdot \left(\sqrt{h_{Lo} + \frac{144 \cdot P}{\rho_L}} - \sqrt{\frac{144 \cdot P}{\rho_L}} \right) \tag{7.241}$$

Boiling liquids

The following simplified formulas for mass flow (kg/s) are based on Leung (1994) omega method as proposed by Hanna and Drivas (1987):

For $w \leq 4$

$$W = A \cdot \sqrt{(P_o \cdot \rho_o)} \cdot \frac{0.6055 + 0.1356 \cdot \ln\omega - 0.0131 \cdot (\ln\omega)^2}{\sqrt{\omega}} \tag{7.242}$$

where:

P_o and ρ_o are, respectively, stagnation pressure and vapour density (units are according to the international system).

For $w > 4$

$$W = A \cdot \sqrt{(P_o \cdot \rho_o)} \cdot 0.66 \cdot \omega^{-0.39} \tag{7.243}$$

7.9.2 Vapour Flow

For subsonic vapour flow (with approximately upstream pressure $P_o < 2 \times$ downstream pressure) at constant upstream pressure P_1 (lb_f/in^2), and average density ρ_v(lb/ft^3), the total vapour amount (lb) flowing in 15 min through a hole with diameter d (inches) is (Brasie, 1977):

$$M_{15} = 300 \cdot d^2 \cdot \sqrt{\rho_v \cdot (P_o - P_1)} \tag{7.244}$$

For sonic vapour flow (with approximately upstream pressure $P_o > 2 \times$ downstream pressure):

$$M_{15} = 212 \cdot d^2 \cdot \sqrt{\rho_v \cdot P_o} \tag{7.245}$$

References

Anwar, S., Carroll, J.J., 2011. Carbon Dioxide Thermodynamic Properties Handbook: Covering Temperatures from −20 Degrees to 250 Degrees Celsius and Pressures up to 1000 bar. Wiley & Sons, Hoboken, NJ.

API Standard 520, 2014. Sizing, Selection, and Installation of Pressure-relieving Devices Part I—Sizing and Selection, nineth ed. American Petroleum Institute.

Bell, R.P., 1978. Isopleth calculations for ruptures in sour gas pipelines. Energy Process. Canada, 36–39.

Benintendi, R., 2014. Non-equilibrium phenomena in carbon dioxide expansion. Process Saf. Environ. Prot. 92 (1), 47–59.

Benintendi, R., 2016. Influence of the elastic behaviour of liquids on transient flow from pressurised containments. J. Loss Prev. Process Ind. 40 (2016), 537–545.

Bird, R.B., Stewart, W.E., Lightfoot, E.N., 2002. Transport Phenomena, second ed. John Wiley & Sons, Inc.

Brasie, W.C., 1977. Guidelines for Estimating the Hazard Potential of Chemicals, Loss Prevention, Vol. 10, A CEP Manual, AIChE, New York.

CCPS, 1999. Guidelines for Chemical Process Quantitative Risk Analysis, second ed.

Chehroudi, B., Onuma, Y., Chen, S.H., Bracco, F V., 1985. On the Intact Core of Full-cone Sprays. SAE technical paper series, 850126.

Citrini, D., Noseda, G., 1987. Idraulica, Casa Editrice Ambrosiana, 2nd ed.

Crowl, D.A., Louvar, J.F., 2011. In: Chemical Process Safety: Fundamentals with Applications. third ed. International Series in Physical and Chemical Engineering, Prentice Hall, Boston, MA.

DeVaull, G.F., King, I.A., 1992. Similarity scaling of droplet evaporation and liquid rain-out following a release of superheated flashing liquid to the environment. In: 85th Annual Meeting, Air and Waste Management Association, Kansas City, MO, June 21-26, 1992. Paper 92-155.08.

Erbland, P.J., Rizzetta, D.P., Miles, R.B., 2000. Numerical and experimental investigation of CO_2 condensate behavior in hypersonic flow, AIAA-200-2379. In: 21st AIAA Aerodynamic Measurement Technology & Ground Testing Conference, Denver, CO.

Fauske, H.K., Epstein, M., 1988. Source Term Considerations in Connection with Chemical Accidents and Vapor Cloud Modeling.

Green, D., Perry, R., 2007. Perry's Chemical Engineers' Handbook, eighth ed. McGraw-Hill Professional Publishing.

Hanna, S.R., Drivas, P.J., 1987. Guidelines for the Use of Vapor Cloud Dispersion Models, CCPS. AIChE, New York.

Health and Safety Laboratory, 2013. Generation of Flammable Mists from High Flashpoint Fluids: Literature Review.

Hulsbosch-Dam, C.E.C., Spruijt, M.P.N., Necci, A., Cozzani, V., 2012. Assessment of particle size distribution in CO_2 accidental releases. J. Loss Prev. Process Ind. 25, 254–262.

Johnson, D.W., Woodward, J.L., 1999. RELEASE: A model with data to predict aerosol rainout in accidental releases. CCPS AIChE.

Kawamura, P.I., MacKay, D., 1987. The evaporation of volatile liquids. J. Hazard. Mater. 15, 365–376.

Kletz, T.A., 1977. Unconfined vapor cloud explosions. AIChE Loss Prev. 11, 50.

Lautkaski, R., 2008. Experimental correlations for the estimation of the rainout of flashing liquid releases – revisited. J. Loss Prev. Process Ind. 21, 506–511.

Leung, J.C., 1994. The omega method for discharge rate evaluation, Fauske & Associates. In: AIChE Meeting, Denver, Colorado, August 16.

Lienhard, J.H., Day, J.B., 1970. The breakup of superheated liquid jets. J. Basic Eng. 92 (3), 515–521.

Liu, Y.-H., Maruyama, H., Matsusaka, S., 2010. Agglomeration process of dry ice particles produced by expanding liquid carbon dioxide. Adv. Powder Technol. 21, 652–657.

Martynov, S., Brown, S., Mahgerefteh, H., 2012. Modelling dry ice formation following rapid decompression of CO_2 pipelines. In: IChemE Safety & Loss Prevention Special Interest Group Workshop, Birmingham, UK, pp. 22–23.

Matthiesen, R.C., 1986. Estimating chemical exposure levels in the workplace. Chem. Eng. Prog. 30.

Mazzoldi, A., Hill, T., Colls, J.J., 2011. Assessing the risk for CO_2 transportation within CCS projects, CFD modelling. Int. J. Greenh. Gas Control 5, 816–825.

Mudan, K., Croce, P., 1988. Fire hazard calculations for large open hydrocarbon fires. In: DiNenno, P., et al., (Eds.), SFPE Handbook of Fire Protection Engineering. National Fire Protection Association, Quincy MA, pp. 45–87, Section 2 (Chapter 4).

Naterer, G.F., 2002. Heat Transfer in Single and Multiphase Systems. CRC Press, Boca Raton.

Plank, R., Kuprianoff, J., 1929. Die thermischen Eigenschaften der Kohlensäure im gasförmigen, flüssigen und festen Zustand, beihefte zur zeitschrift fur die gesamte kalte-industrie, Reihe 1, Heft 1.

Prugh, R.W., 1987. Evaluation of unconfined vapor cloud explosion hazards. In: International Conference on Vapor Cloud Modeling, Cambridge, MA. AIChE, New York, pp. 713–755.

Recknagel-Sprenger, 1961. Taschenbuch für Heizung, Lüfting und Klimatecknik. Oldenburg-München-Wein, p. 115.

Ruff, G.A., Wu, P.K., Bernal, L.P., Faeth, G.M., 1990. Continuous and dispersed-phase structure of dense non evaporating pressure-atomized sprays. In: AIAA Aerospace Sciences Meeting, AIAA Paper 90-464, Reno.

Shashi Menon, E., 2004. Liquid Pipeline Hydraulics. Marcel Dekker Inc., New York.

Sutton, O.G., 1953. Micrometeorology. McGrawHill, New York.

Taylor, G.I., 1940. Generation of Ripples by the Wind Blowing Over a Viscous Liquid, Collected Works of G. I. Taylor 3. Cambridge University Press.

Tickle, G., 2015. Rainout Correlations for Continuous Releases, GT Science & Software Limited under sub-contract from the Health and Safety Laboratory.

U.S.EPA, 1996. Guidance on the Application of Refined Dispersion Models to Hazardous/Toxic Air Pollutant Releases, EPA 454/R-93-002.

U.S. EPA, Risk Management for Offsite Consequence Analysis, 2009.

van den Bosch, C.J.H., Duijm, N.J., 2005. Methods for the calculation of physical effects. In: van den Bosch, C.J.H., Weterings, R.A.P.M. (Eds.), Yellow Book.http://content.publicatiereeksgevaarlijkestoffen.nl/documents/PGS2/PGS2-1997-v0.1-physical-effects.pdf.

Weiss, M.H., Botros, K.K., Jungowski, W.M., 1988. Simple method predicts gas-line blowdown times, technology. Oil Gas J. 12, 55–58 (December).

winegarden, D., 1987. Michigan Division, Process Engineering, The Dow Chemical Co., Nidland, MI, Personal Communication, Nov. 1.

Woodward, J.L., 1998. Estimating the Flammable Mass of a Vapor Cloud: A CCPS Concept Book. American Institute of Chemical Engineers, New York.

Woolley R.M., Falle, S.A.E.G., Fairweather M., 2011. Development of Near-field Dispersion Model, CO2 PipeHaz News, Autumn.

Wu, J. M., Schroy, J.M., (1979). Emissions from Spills, 55 APCA and WPCF Joint Conference on Control of Specific (Toxic) Pollutants. Feb. 13-16, Gainesville, Air Pollution Control Association, Florida Section.

Further Reading

Midland, Michigan, Presented to 81st National Loss Prevension Symposium, AIChE, Kansas City, Missouri, April 1976.

U.S. Environmental Protection Agency, 1996. Risk Management Program Rule.

CHAPTER 8

Dispersion Models

Atque in ventos vita recessit (Virgil, Aeneid)

8.1 Summary of Scenarios

See Table 8.1.

Fig. 8.1 provides a summary of the dispersion scenarios analysed in this chapter.

8.2 Dispersion Key Drivers

Dispersion key drivers have been depicted as separate tiles in the following drawing, and specifically examined in this chapter. The understanding of their nature, and their influence over a dispersion outcome, is essential to properly frame the dispersion phenomenon from the release to the expansion outcome. The following paragraphs define the essence of these key drivers and then integrate them in the applicable models (Figs 8.2 and 8.3).

8.3 Meteorology

Meteorology has a strong influence on the outcome of a release, as well as on the logistic design of a hazardous process plant. Consequently, the knowledge and full understanding of some fundamental meteorological parameters are important for the most accurate dispersion calculation. This paragraph provides some guidance on the definition of some meteorological entities, their selection, and utilisation in consequence assessment calculations.

8.3.1 Weather Stability Classes

Atmospheric stability is referred to the vertical temperature gradient and the atmospheric turbulence. Pasquill-Gifford weather stability classes were developed in the sixties and in the seventies within the nuclear industry safety frame and are still the most used weather categories in environmental and process safety engineering. They are defined as follows (Gifford, 1976) (Table 8.2):

Process Safety Calculations. https://doi.org/10.1016/B978-0-08-101228-4.00008-3

Table 8.1 Summary of dispersion scenarios

Dispersion Scenario	Section
Momentum dominated jet	8.5
- Turbulent jet	8.5.1
- Transitional laminar jet	8.5.2
- Momentum to buoyancy transition	8.5.3
Density and thermal positively buoyant plumes	8.6
- Momentum-dominated positive buoyant plume	8.6.1
- Combination of momentum and positive buoyant plume	8.6.2
Thermal plume rise	8.7
Negatively buoyant plume (dense gas)	8.8
- Onset of dense gas dominating regime	8.8.1
- Decision of dense gas model application	8.8.2
- Britter McQuaid model	8.8.3
- Hoehne and Luce model	8.8.4
Transition to passive dispersion	8.9
Gaussian dispersion	8.10

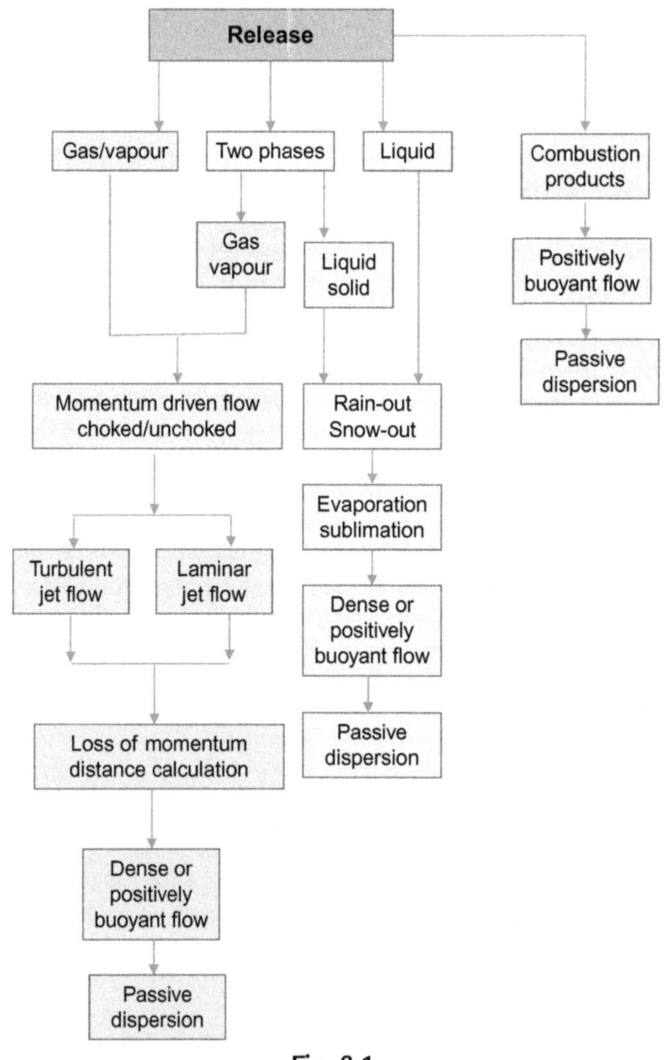

Fig. 8.1
Dispersion tree.

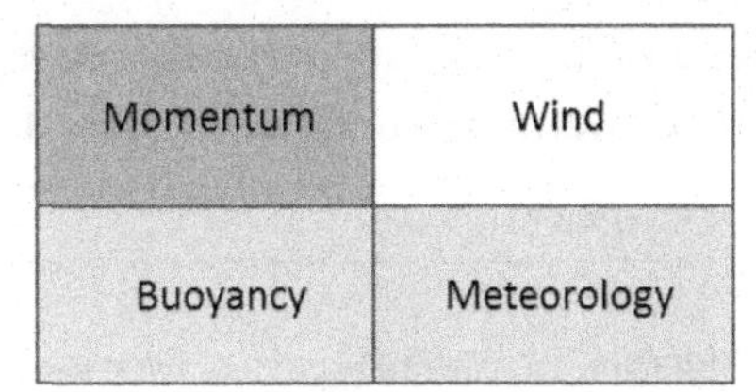

Fig. 8.2
Key-drivers tiles.

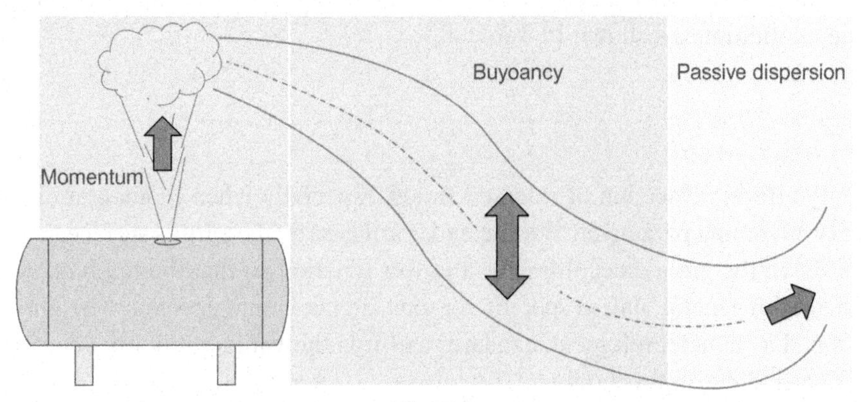

Fig. 8.3
Dispersion key-drivers in the physical scenario.

Table 8.2 Weather stability classes (Gifford, 1976)

P-G Class	Definition	Definition
A B C	Very unstable Moderately unstable Slightly unstable	Unstable conditions are related to atmospheric buoyancy forces which enhance turbulence. In these conditions, dispersion is promoted
D	Neutral	Neutral conditions occur when there is no significant effect in terms of atmospheric turbulence enhancement or reduction
E F G	Slightly stable Moderately stable Extremely stable	Stable conditions are related to atmospheric buoyancy forces which destroy turbulence. In these condition, dispersion is not promoted

Table 8.3 Relationships of weather stability classes with wind and daytime/overnight conditions (Gifford, 1976)

Surface Wind Speed (m/s)	Daytime Insolation				Night Time	
	Strong	Moderate	Slight	Overcast	>4/8 Low Clouds	<3/8 Cloudiness
<2	A	A–B	B	C	E	F or G
2–3	A–B	B	C	C	E	F
2–5	B	B–C	C	C	D	E
5–6	C	C–D	D	D	D	D
>6	C	D	D	D	D	D

Category G is not used in dispersion modelling. Stability classes are related to wind, daytime, and overnight conditions as shown in Table 8.3.

8.3.2 Wind

Wind strongly affects dispersion of released fluids, especially when momentum is lost, or is not the only governing parameter. Wind speed is affected by the earth's surface, and increases with height above the ground according to a power function so that the reference height should be stated: in general, unless specific reasons are taken into account, a height of 10 m is usually assumed as a meteorological standard, and it is the value implicitly adopted in the calculations carried out in this book.

A general equation is provided by API (1996) for near-neutral and stable wind profile:

$$\frac{u_w}{u^*} = \frac{1}{k} \cdot \left(\ln \frac{z}{z_o} + 4.5 \cdot \frac{z}{L} \right) \tag{8.1}$$

where:

- u_w is the wind speed, m/s.
- u^* is friction velocity constant (m/s), which is assumed as equivalent to about 10% of the wind speed at 10 m.
- k is von Karman's constant, 0.41.
- z is the height, m.
- z_o is the surface roughness length parameter, m.
- L is the Monin-Obukhov length, m.

The Monin-Obukhov length is positive during stable conditions (night time) and negative during unstable conditions (daytime). It is defined as:

$$L = \frac{u^{*3}}{0.41 \cdot g \cdot (H/T)} \tag{8.2}$$

Table 8.4 Obuchov length relationships (CCPS, 1996)

P-G Class	Wind Speed (m/s)	M-O Length (m)
F	<3	10
E	2–4	50
D	Any	$>\lvert 100 \rvert$
B or C	2–6	−50
A	<3	−10

where g is the acceleration of gravity, T is the absolute temperature, and H is the surface heat flux (J/m^3).

Monin-Obukhov length is related to the Pasquill-Gifford and to the wind speed as shown in Table 8.4 (CCPS, 1996) and the typical surface roughness as shown in Table 8.5 (CCPS, 1996)

API (1996) states that Eq. (8.3) can be simplified for neutral stability as:

$$\frac{u_W}{u*} = \frac{1}{k} \cdot \left(\ln \frac{z}{z_o} \right) \tag{8.3}$$

Theoretical calculation methods of heat surface flux and of friction velocity are given by Oke (1978) and Panofsky and Dutton (1984), respectively, and have been validated by the US EPA.

If the unknown wind speed can be compared to the speed at a fixed height, a power law applies (Hanna et al., 1982):

$$u(y) = u_{10} \cdot \left(\frac{y}{10} \right)^p \tag{8.4}$$

Table 8.5 Surface roughness identification (CCPS, 1996)

Terrain Classification	Terrain Description	Surface Roughness z_o (m)
Highly urban	Centres of cities with tall buildings, very hilly or mountainous area	3–10
Urban area	Centres of towns, villages, fairly level wooded country	1–3
Residential area	Area with dense but low buildings, wooded area, industrial site without large obstacles	1
Large refineries	Distillation columns and other tall equipment pieces	1
Small refineries	Smaller equipment over a small area	0.5
Cultivated land	Open area with great overgrowth and scattered houses	0.3
Flat land	Few trees, long grass, fairly level grass plains	0.1
Open water	Large expanses ofwater, desert flats	0.001
Sea	Calm open sea, snow covered flat, rolling land	0.0001

Table 8.6 Power law exponent (CCPS, 1999a)

P-G Class	Power Law Exponent p	
	Urban	Rural
A	0.15	0.07
B	0.15	0.07
C	0.20	010
D	0.25	0.15
E	0.40	0.35
F	0.60	0.55

where u_{10} is the speed at 10 m height, and p is an exponent which depends on the weather stability and on the terrain. Table 8.6 includes these exponents with reference to urban and rural areas and to the Pasquill and Guifford stability classes (CCPS, 1999a).

8.3.3 Wind Rose

A wind rose is a simple graphical tool, generally divided into 16 bands, which relates frequency (probability), magnitude, and orientation of local wind fields in order to allow one to collect the most significant data to utilise in QRA (Quantitative Risk Assessment) studies. Knowledge of prevailing wind direction is an essential datum also for design purposes, such as the proper localization of potentially hazardous process plants, such as reactors, compressors, and vessels. A typical rose wind is depicted in Fig. 8.4. Each colour represents a specific wind speed.

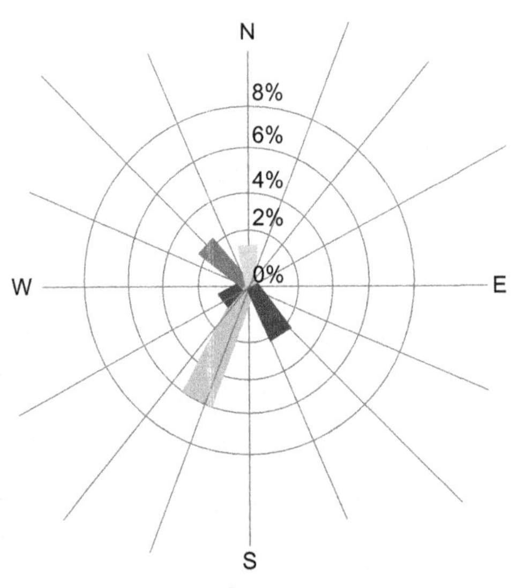

Fig. 8.4
Typical wind rose.

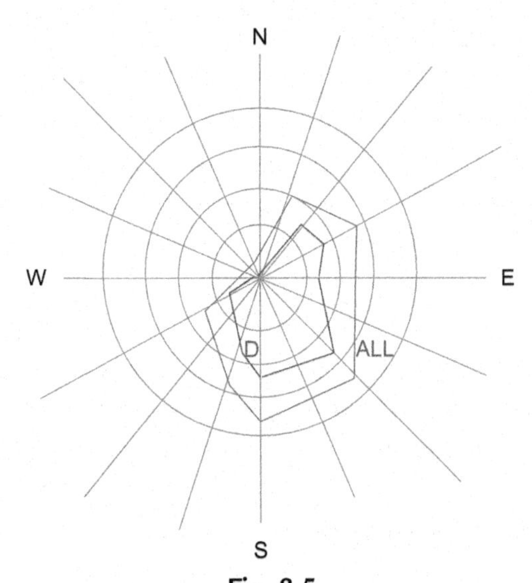

Fig. 8.5
Wind rose with prevailing weather classes.

A useful version of the wind rose includes the probability wind profiles related to the P-G category class. This is important because it shows the weight of each weather class (see Fig. 8.5).

8.4 Buoyancy

Buoyancy is the behaviour of a gas/vapour depending on its density (true or apparent) to air density ratio. *Apparent* density refers to gases or vapours whose molecular weight is smaller than air molecular weight, but whose temperature is much lower, forcing them to behave as a heavier than air fluid, for example, LNG and refrigerated ammonia.

Gases and vapours are classified as:

- *Positively buoyant gas*: true or apparent density is smaller than air density.
- *Neutrally buoyant gas*: true or apparent density is comparable to air density.
- *Negatively buoyant gas*: Gas true or apparent density is greater than air density.

Buoyancy is the behaviour of a gas/vapour depending on the relationship of its density (true or apparent) with respect to air density.

$$Fr = \frac{u}{\sqrt{g \cdot D_e \cdot \left(\frac{\rho_e}{\rho_a} - 1\right)}} \tag{8.5}$$

where:

- u is the flow speed.
- g is the acceleration due to gravity.

- D_e is the outlet diameter.
- ρ_a is air density.
- ρ_e is fluid density at the outlet.

Ammonia released from refrigerated plants behaves as a heavier gas and must be modelled as a heavy gas (EPA). The ratio:

$$B_A = \left(\frac{MW}{T}\right) \tag{8.6}$$

has been defined in this book as the buoyancy factor. Its comparison to air B_A tells us about the buoyancy behaviour of the substance.

Example 8.1

Categorise the buoyancy of ammonia and propane with respect to air at the same temperature and pressure. Find the state of ammonia and propane in order for these gases to become neutrally buoyant with respect to air at 20°C.

Solution

Buoyancy is defined in terms of density. Gas density at atmospheric pressure and temperature can be written as:

$$\rho = \frac{n \cdot MW}{V} = \frac{P \cdot MW}{R \cdot T} \tag{8.7}$$

At the same pressure and temperature, just the molecular weight affects buoyancy.

Substance	Molecular Weight	Buoyancy
Ammonia	17.031	Positive
Air	28.966	–
Propane	44.096	Negative

At different temperatures, the governing factor is the buoyancy factor B_A: in order for a substance to show the same (temporary) behaviour as air, it has to possess the same buoyancy factor:

$$T_{NH_3} = \frac{MW_{NH_3}}{MW_{AIR}} \cdot T_{AIR} = \frac{17.031}{28.966} \cdot 293.16 = 172.37\,\text{K} \tag{8.8}$$

$$T_{C_3H_8} = \frac{MW_{C_3H_8}}{MW_{AIR}} \cdot T_{AIR} = \frac{44.096}{28.966} \cdot 293.16 = 446.28\,\text{K} \tag{8.9}$$

Example 8.2

Liquid hydrogen sulphide stored in a vessel at ambient temperature of 294 K is released to the atmosphere. Find the buoyancy of the vapour phase.

Solution

Substance	Molecular Weight	Storage Temperature (K)	Storage Pressure (Bar)
Hydrogen sulphide	34	294	17.4

At the same temperature, hydrogen sulphide is slightly heavier than air, so in general its behaviour would be almost neutral. After release, this substance will flash, and liquid will be cooled until the normal boiling temperature is attained, which is around 213 K. Accordingly:

$$B_{A-H_2S} = \left[\frac{MW}{T}\right]_{ammonia} = \frac{34}{213} = 0.16 \tag{8.10}$$

$$B_{A-Air} = \left[\frac{MW}{T}\right]_{air} = \frac{28.966}{294} = 0.098 \tag{8.11}$$

Hydrogen sulphide vapour will be apparently 1.6 times heavier than air, so it is expected to slump down to the ground.

After loss of low or high jet momentum, a competition between buoyancy and wind will be engaged. When both momentum and buoyancy effects have expired, the cloud will be dispersed according to a neutral or passive dispersion, where density and momentum will no longer affect the cloud behaviour. Passive dispersion is also called "Gaussian dispersion" because velocity, concentration, and temperature transversal profiles can be described according to a Gauss Function.

8.5 Momentum Dominated Jet Behaviour

Turbulent and laminar jet theories have been described in Chapter 4, along with thermal and dilution effects. The following practical criteria can be adopted to assess the importance of momentum on a release.

8.5.1 Turbulent Jet

- If Reynolds number (Re) at the outlet condition is:

$$Re \geq 15{,}400 \cdot \frac{\rho_g}{\rho_{air}} \tag{8.12}$$

where ρ_g and ρ_{air} are gas and air density respectively, the jet will be momentum dominated (API Standard 521, 2014), and its outcome has to be assessed prior to considering the influence of other mechanisms. If Re is significantly smaller than 10,000, gas velocity is lower than 12 m/s, or jet to wind velocity ratio is less than 50, influence of wind and/or buoyancy properties can dominate (API Standard 521, 2014), and the dilution promoted by turbulence-driven is low or negligible.

- Length of flow establishment zone can be assumed as a distance equal to 4 to 6 hole diameters from the outlet. Lee and Chu (2003) recommend 5.2 diameters.
- Equations of the effects of air entrainment of turbulent jet on average jet velocity, temperature, and concentration have been provided in Chapter 4.
- The axis of the wind-deflected jet, that is, the locus where concentrations and velocities are maximum, can be identified according to Patrick's equation included in Chapter 4.
- According to Lee and Chu (2003), the distance at which the jet stops being momentum-driven, and where buoyancy and wind become the prevailing mechanisms is

$$x_{M-B} = \frac{\left(\frac{M_o}{\rho_o}\right)^{3/4}}{\left(\frac{F_o}{\rho_o}\right)^{1/2}} \tag{8.13}$$

where:

- M_o is the momentum calculated as:

$$M_o = A_e \cdot \rho_e \cdot {u_e}^2 \tag{8.14}$$

- F_o is the buoyancy flux defined as:

$$F_o = A_e \cdot \frac{\rho_e - \rho_a}{\rho_a} \cdot u_e \cdot g \cdot \rho_e \tag{8.15}$$

Example 8.3

Carbon dioxide is released at 220 m/s, through a 50 mm diameter hole, at a temperature of 255 K, in a direction forming an angle of 30° with wind direction, whose speed is 10 m/s. Find the distance from the outlet where momentum flow is taken over by buoyant flow, assuming air density of 1.2 kg/m^3.

Solution

Substance	Molecular Weight	Temperature (K)	Density (kg/m^3)	Diameter (m)	Velocity (m/s)
Carbon dioxide	44	255	2.1	0.05	220

Carbon dioxide at 255 K and 1 atm is superheated vapour, so no solid formation is expected.

The axis of the deflected jet, the locus where concentrations and velocities are maximum, can be identified according to Patrick (1967):

$$y = D_e \cdot \alpha^{0.85} \cdot \left(\frac{x}{D_e}\right)^{0.36} \tag{8.16}$$

where α is the jet to wind velocity ratio, falling in the range from 6 to 50 (see Figs 8.6 and 8.7).

Transversal wind effect can be assessed as follows:

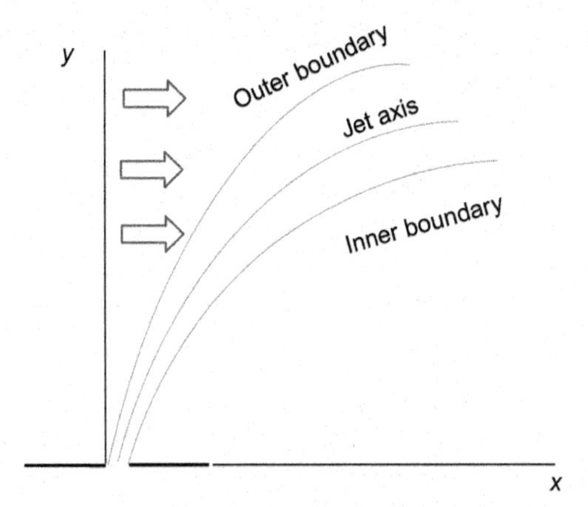

Fig. 8.6
Axis and boundaries of deflected jet.

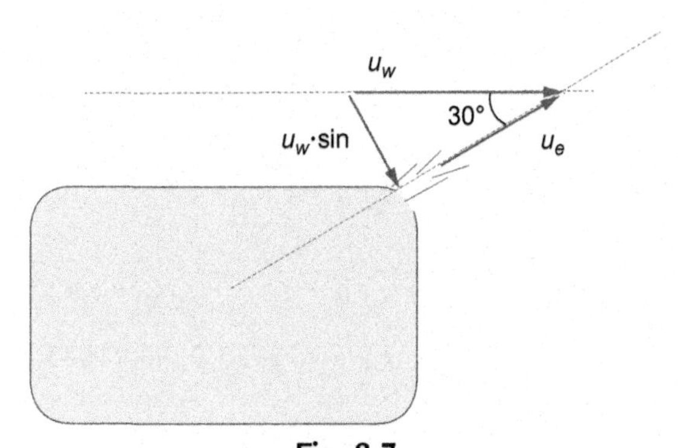

Fig. 8.7
Jet and wind orientation for Example 8.3.

- where u_W is wind velocity, and u_e is jet velocity at the outlet. The active wind velocity component is $u_W \cdot \sin\theta$. By definition, the distance from the outlet where momentum flow is taken over by buoyant flow is calculated by equating the curve length to the distance at which the jet stops to be momentum-driven, and buoyancy and the wind become the prevailing mechanisms, as given by equation:

$$x_{M-B} = \frac{\left(\frac{M_o}{\rho_o}\right)^{3/4}}{\left(\frac{F_o}{\rho_o}\right)^{1/2}} \tag{8.17}$$

$$M_o = A_e \cdot \rho_e \cdot u_e^2 = \frac{\pi}{4} \cdot 0.05^2 \cdot 2.1 \cdot 220^2 = 199.47\,\text{kg} \cdot \text{m/s}^2 \tag{8.18}$$

$$F_o = A_e \cdot \frac{\rho_e - \rho_a}{\rho_a} \cdot u_e \cdot g \cdot \rho_e = \frac{\pi}{4} \cdot 0.05^2 \cdot \frac{2.1 - 1.2}{1.2} \cdot 220 \cdot 9.81 \cdot 2.1 = 6.68\,\mathrm{kg \cdot m/s^3} \tag{8.19}$$

$$x_{M-B} = \frac{\left(\frac{M_o}{\rho_o}\right)^{3/4}}{\left(\frac{F_o}{\rho_o}\right)^{1/2}} = \frac{\left(\frac{199.47}{2.1}\right)^{3/4}}{\left(\frac{6.68}{2.1}\right)^{1/2}} = 17\ \mathrm{m} \tag{8.20}$$

The length of the curve $y = f(x)$, between two x values x_1 and x_2, is calculated according to the following formula:

$$L = \int_{x_1}^{x_2} \sqrt{f'(x)^2 + 1} \cdot dx \tag{8.21}$$

The deflected jet curve is represented by the formula:

$$y = D_e \cdot \alpha^{0.85} \cdot \left(\frac{x}{D_e}\right)^{0.36} = 0.05 \cdot \left(\frac{220}{10 \cdot 0.5}\right)^{0.85} \cdot \left(\frac{x}{0.05}\right)^{0.36} = 3.66 \cdot x^{0.36} \tag{8.22}$$

$$\frac{dy}{dx} = 1.32 \cdot x^{-0.64} \tag{8.23}$$

So

$$17 = \int_0^{x_2} \sqrt{1.736 \cdot x^{-1.28} + 1} \cdot dx \tag{8.24}$$

Solving:

$$2.77778 \cdot x \cdot \sqrt{1 + \frac{1.736}{x^{1.28}}} - 1.77778 \cdot x \cdot F\left(-0.78125, 0.5; 0.21875; -\frac{1.736}{x^{1.28}}\right) \tag{8.25}$$

where F is the hypergeometric function.

$$x_2 = 12.75\,\mathrm{m} \tag{8.26}$$

$$y_2 = 3.66 \cdot 12.75^{0.36} = 9.15\ \mathrm{m} \tag{8.27}$$

8.5.2 Transitional and Laminar Jet

According to Oosthuizen's findings (1983), a gas jet showing significant instabilities is expected in the Reynolds range between 5000 and 10,000. For practical applications, Reynolds numbers below 5000 can be assumed as representative of a well-established laminar regime. Jets belonging to this regime have been presented in Chapter 4. Both cases can be assessed by calculating the momentum-driven to buoyancy/wind-driven transition distance, so that loss of momentum distance can be quantitatively estimated.

8.5.3 Momentum to Buoyancy Transition

Papanicolaou and List (1988) have proposed a momentum-to-buoyancy transition range based on the densimetric Froude number defined by Eq. (8.5):

$$Fr = \frac{u}{\sqrt{g \cdot D_e \cdot \frac{|\rho_e - \rho_a|}{\rho_a}}} \tag{8.28}$$

Notably, the transition range between the two regimes is defined as follows:

$$1.1 \cdot D_e \cdot Fr \leq x \leq 5.6 \cdot D_e \cdot Fr \tag{8.29}$$

Example 8.4

For the carbon dioxide jet of the previous example, find the Froude number and discuss the findings.

Solution

Substance	Molecular Weight	Temperature (K)	Density (kg/m^3)	Diameter (m)	Velocity (m/s)
Carbon dioxide	44	255	2.1	0.05	220

$$Fr = \frac{220}{\sqrt{9.81 \cdot 0.05 \cdot \frac{2.1 - 1.2}{1.2}}} = 362.72 \tag{8.30}$$

The linear distance from the outlet, where momentum flow is taken over by the buoyant flow calculated in the previous example, is 17 m. Equating, with the unknown coefficient K:

$$K \cdot D_e \cdot Fr = K \cdot 0.05 \cdot 367.72 = 17 \tag{8.31}$$

$$K = 0.92 \tag{8.32}$$

This value is very close to 1.1, that is the upper value of the momentum-driven range, according to Papanicolaou and List (1988).

8.6 Density and Thermal Positively Buoyant Plumes

Positively buoyant plumes are formed when either gas density is lower than air density or gas temperature is significantly greater than ambient temperature. These cases are very specific and have been analysed by Briggs. For positive buoyant gases, the following approach can be adopted.

8.6.1 Momentum-Dominated Positive Buoyant Plume

When initial momentum flux is significantly higher than positive buoyancy flux, the importance of the latter shall be neglected, and the study case will fall within the momentum dominated models.

8.6.2 Combination of Momentum and Positive Buoyant Plume

This is the most general scenario, which accounts both for initial momentum and buoyancy. Briggs (1984) has proposed the following general equation for the trajectory of a plume assumed as initially upward:

$$y_b = \left(19 \cdot \frac{\rho_e \cdot M_e}{\rho_a \cdot {u_w}^2} \cdot x - 4.2 \cdot \frac{B}{{u_w}^3} \cdot x^2\right)^{1/3} \tag{8.33}$$

where y_b is the height above the source:

$$M_e = {u_e}^2 \cdot \frac{{D_e}^2}{4} \tag{8.34}$$

$$B = g \cdot \frac{(\rho_e - \rho_a) \cdot u_e \cdot {D_e}^2}{4 \cdot \rho_a} \tag{8.35}$$

This equation is valid in general and can be used to assess the influence of the gas (positive or negative) buoyancy with respect to the momentum. In case of neutral buoyancy, the equation coincides with the trajectory related to momentum only, and it is mathematically similar to Patrick's equation.

Wilson and Angle (1979) have proposed the following very useful formula to calculate the maximum crosswind trajectory travelled by the momentum jet:

$$z_{cw} = \frac{4.8 \cdot}{u_w} \sqrt{\frac{\rho_e \cdot M_o}{\rho_a}} \tag{8.36}$$

Van den Bosch and Duijm (TNO, 2005) proposed the following formula for the plume radius at down-wind distances and for the concentration distribution of buoyant plumes:

$$R(x) = \frac{D_e}{2} \cdot \sqrt{\frac{\rho_e u_e}{\rho_a u_a}} + 0.41 \cdot [y(x) - h] \tag{8.37}$$

$$C(x) = C_o \cdot \left[\frac{D_e}{2 \cdot R(x)}\right]^2 \cdot \frac{u_e}{u_w} \tag{8.38}$$

where h is the height of the release point.

Example 8.5

Hydrogen is released at ambient temperature (20°C) and atmospheric pressure from grades upwards of 1.0 m/s from a 200 mm flange to the atmosphere, with a 1.0 m/s wind speed. Find the plume radius when concentration is reduced to 10% and its corresponding height.

Solution

Substance	Molecular Weight	Temperature (K)	Density (kg/m^3)	Diameter (m)	Velocity (m/s)
Hydrogen	2	293.16	0.08	0.2	1

$$M_e = {u_e}^2 \cdot \frac{{D_e}^2}{4} = 1 \cdot \frac{0.04}{4} = 0.01 \text{ m}^4/\text{s}^2 \tag{8.39}$$

$$B = g \cdot \frac{(\rho_e - \rho_a) \cdot u_e \cdot {D_e}^2}{4 \cdot \rho_a} = 9.81 \cdot \frac{(0.08 - 1.2) \cdot 1 \cdot 0.04}{4 \cdot 1.2} = -0.09156 \text{ m}^4/\text{s}^3 \tag{8.40}$$

We note that buoyancy affects the plume by an order of magnitude more than initial momentum.

$$\frac{C(x)}{C_o} = 0.1 = \left[\frac{0.2}{2 \cdot R(x)}\right]^2 \cdot 1 \rightarrow R(x) = \frac{0.2}{2 \cdot \sqrt{0.1}} = 0.32 \text{ m} \tag{8.41}$$

$$B = g \cdot \frac{(\rho_e - \rho_a) \cdot u_e \cdot {D_e}^2}{4 \cdot \rho_a} \tag{8.42}$$

$$C(x) = C_o \cdot \left[\frac{D_e}{2 \cdot R(x)}\right]^2 \cdot \frac{u_e}{u_w} \tag{8.43}$$

$$Y(x) = \frac{R(x) - \frac{D_e}{2} \cdot \sqrt{\frac{\rho_e u_e}{\rho_a u_a}}}{0.41} = \frac{0.32 - 0.1\sqrt{\frac{0.08}{1.2}}}{0.41} = 0.71 \text{ m} \tag{8.44}$$

$$R(x) = \frac{D_e}{2} \cdot \sqrt{\frac{\rho_e u_e}{\rho_a u_a}} + 0.41 \cdot [y(x) - h] \tag{8.45}$$

8.7 Thermal Plume Rise

The plume rise related to a temperature difference between gas and ambient temperature is proposed by Briggs (1972) as:

$$\Delta H = 1.6 \cdot \left[g \cdot \frac{{D_e}^2 \cdot u_e}{4} \cdot \left(\frac{T_s - T_a}{T_s}\right)\right]^{1/3} \cdot \frac{x^{2/3}}{u_w} \tag{8.46}$$

where all the terms are known and:

- T_s is the stack temperature.
- x is the downwind distance.
- the following parameter $\frac{T_s - T_a}{T_s}$ is the driving force of thermal plume

$$B = g \cdot \frac{(\rho_e - \rho_a) \cdot u_e \cdot D_e^2}{4 \cdot \rho_a} \tag{8.47}$$

The USEPA (1988) provides a method, based on the rate of sensible heat, to quantify the buoyancy property of very hot gases from a fire or a flare with the aim to calculate the plume trajectory. In the equation:

$$F_b = 1.66 \cdot 10^{-5} \cdot H_s \tag{8.48}$$

where:

- Fb is the flux term, m^4/s^3.
- Hs is the total heat release rate, cal/s.

8.8 Negatively Buoyant Plume (Dense Gas)

Dense gases with density greater, or apparently greater, than air may slump down as momentum is lost, until mixing with air promotes further passive dispersion. Due to their characteristics, the persistence of a dense cloud at ground level can be very critical in terms of toxic and/or explosive hazards.

Relatively simple rules can be used for practical calculations to determine the applicability of dense gas models and to adopt the most appropriate approach.

8.8.1 Onset of a Dense Gas Dominating Regime

Apparent or greater-than-air molecular weight can become significant after initial momentum is lost or is negligible. The distance from the source can be found using the following formula, already discussed for positive buoyancy releases:

$$x_{M-B} = \frac{\left(\frac{M_o}{\rho_o}\right)^{3/4}}{\left(\frac{F_o}{\rho_o}\right)^{1/2}} \tag{8.49}$$

very close to the Papanicolaou and List (1988) upper limit:

$$x_{M-B} = 1.1 \cdot D_e \cdot Fr \tag{8.50}$$

Beyond this distance, negative buoyancy becomes the driving property.

A very specific equation for the maximum initial rise of an upward-pointing dense plume has been given by Hoot et al. (1973):

$$\frac{\Delta y_{max}}{D_e} = 1.32 \cdot u_e \cdot \sqrt[3]{\frac{\rho_e^{\,2}}{D_e \cdot u_w \cdot \rho_a \cdot g \cdot (\rho_e - \rho_a)}} \tag{8.51}$$

The downwind distance at which the centreline of the dense plume strikes the ground can be calculated as Hoot et al. (1973):

$$x_g = \frac{u_e \cdot u_w \cdot \rho_e}{g \cdot (\rho_e - \rho_a)} + 0.56 \cdot \sqrt{\left(\frac{\Delta y_{max}}{D_e}\right) \cdot \left[\left(2 + \frac{h_s}{\Delta y_{max}}\right)^3 - 1\right] \cdot \frac{u_w^{\,3} \cdot \rho_e}{D_e \cdot u_e \cdot \rho_a \cdot g \cdot (\rho_e - \rho_a)}} \tag{8.52}$$

where h_s is the emission source height from grade. The corresponding maximum plume to initial concentration ratios at the point of maximum rise and where the centerline strikes the ground respectively, are:

$$\frac{C_{max-y}}{C_o} = 1.688 \cdot \left(\frac{u_e}{u_w}\right) \cdot \left(\frac{\Delta y_{max}}{D_e}\right)^{-1.85} \tag{8.53}$$

$$\frac{C_{max-x}}{C_o} = 2.42 \cdot \left(\frac{u_e}{u_w}\right) \cdot \left(\frac{h_s + \Delta y_{max}}{D_e}\right)^{-1.85} \tag{8.54}$$

Example 8.6

For the carbon dioxide of Example 8.3, find the maximum momentum-dominated plume rise assumed as upward directed.

Solution

Substance	Molecular Weight	Temperature (K)	Density (kg/m^3)	Diameter (m)	Velocity (m/s)
Carbon dioxide	44	255	2.1	0.05	220

The wind speed is perpendicular to the jet direction and its value is 5 m/s.

$$\frac{\Delta y_{max}}{D_e} = 1.32 \cdot u_e \cdot \sqrt[3]{\frac{\rho_e^{\,2}}{D_e \cdot u_w \cdot \rho_a \cdot g \cdot (\rho_e - \rho_a)}} = 1.32 \cdot 220 \cdot \sqrt[3]{\frac{2.1^2}{0.05 \cdot 5 \cdot 1.2 \cdot 9.81 \cdot 0.9}} = 344 \tag{8.55}$$

This value coincides with the value obtained in the Example 8.3.

8.8.2 Decision of Dense Gas Model Application

The application of dense gas models is decided on the basis of the following criteria (CCPS, 1999a).

1. The initial buoyancy:

$$g_o = g \cdot \left(\frac{\rho_e - \rho_a}{\rho_a} \right) \tag{8.56}$$

is significantly >0.

2. Continuous or discontinuous release

If:

$$\frac{u_w \cdot R_d}{x} > 2.5 \tag{8.57}$$

with:

- R_d = release duration
- x = distance to target
- u_w = wind speed

the dense gas release has to be considered continuous, if it is ≤ 0.6, it has to be considered discontinuous. For any intermediate value, both models for continuous and instantaneous release shall be used, and the minimum resulting concentration will be selected.

Eq. (8.57) is valid for dense gases only. For neutrally and positively buoyant gas categorization, guidance is provided at Section 8.10.4.

Example 8.7

Ammonia is released for 5 min in the atmosphere, where the average wind velocity is 5 m/s.

Solution

$$\frac{u_w \cdot R_d}{x} = \frac{5 \cdot 5 \cdot 60}{x} = 2.5 \rightarrow x = 600 \text{ m} \tag{8.58}$$

Source
Continuous
Instantaneous
600 m
2,500 m
x

$$\frac{u_w \cdot R_d}{x} = \frac{5 \cdot 5 \cdot 60}{x} = 0.6 \rightarrow x = 2500 \text{ m} \tag{8.59}$$

3. Dense model applicability

A characteristic source dimension is defined dependent on the type of release. For continuous releases, it is:

$$D_c = \sqrt{\frac{Q_o}{u_w}} \tag{8.60}$$

where Q_o is the initial flow rate, and, for instantaneous releases:

$$D_i = \sqrt[3]{V_o} \tag{8.61}$$

where V_o is the released volume. On this basis, a dense model shall apply if:

$$\sqrt[3]{\frac{g_o \cdot Q_o}{{u_w}^3 \cdot D_c}} \geq 0.15 \tag{8.62}$$

or:

$$\frac{\sqrt{g_o \cdot V}}{u_w \cdot D_i} \geq 0.20 \tag{8.63}$$

Should not these inequalities be satisfied, neutral or positive buoyant dispersion shall be adopted.

8.8.3 Britter-McQuaid Model

This model (Britter and McQuaid, 1988) is relatively simple, complete, and has undergone some validations (Hanna et al., 1990) and experimental calibrations (McQuaid, 1987). Figs 8.8 and 8.9 relate the concentrations to distances on the basis of release, buoyancy, and wind data, both for continuous and instantaneous release.

For the correct use of the method, the following aspects have to be clearly considered:

- weather stability doesn't affect the dense gas evolution significantly.
- wind speed, as for all calculations, has to be calculated as usual at 10 m from grade.
- the dashed box labelled *full scale data* region are relevant to data validated by field observations.
- the dashed vertical line labelled *passive limit* is the border line between passive and dense gas behaviour.
- the method is not intended to deal with jet releases; however, when significant momentum is lost, within the limits of validity of the model, it can be used, after accounting for the concentration dilution due to air entrainment.
- this model is not suitable for two-phase flow and should be used in rural areas for releases at ground level.
- effective concentration for non-isothermal releases (flashing LPG, LNG, refrigerated ammonia) is given by the formula:

$$C_{eff} = \frac{C_{non\text{-}isoth}}{C_{non\text{-}isoth} + (1 - C_{non\text{-}isoth}) \times (T_{amb}/T_o)} \tag{8.64}$$

where $C_{non\text{-}isoth}$ is the release concentration at temperature T_o other than the ambient temperature T_{amb}.

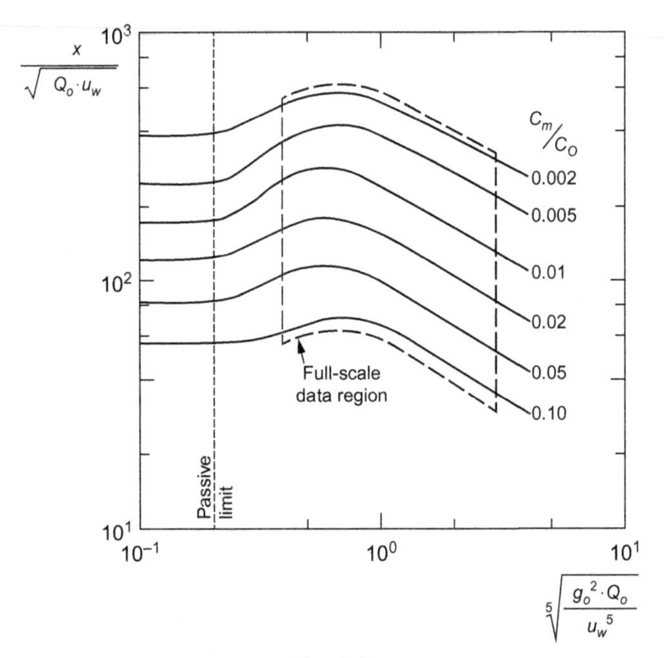

Fig. 8.8
Britter McQuaid chart for continuous release (Britter and McQuaid, 1988).

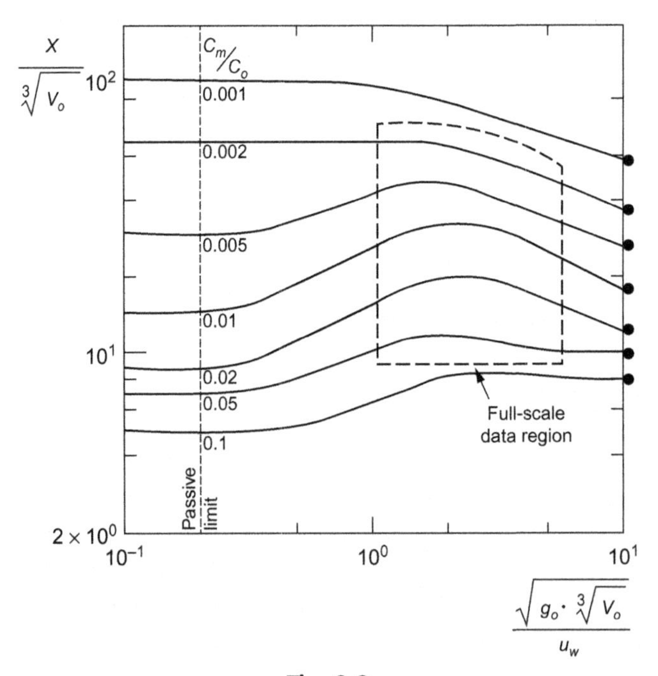

Fig. 8.9
Britter McQuaid chart for instantaneous release (Britter and McQuaid, 1988).

Example 8.8

LNG at normal boiling temperature evaporates from a 25 m^2 pool 5 cm high. If wind speed is 3 m/s, ambient temperature is 293.16 K, and LNG can be assumed as methane ($T_b = -161.5°$ C), find the distance at which the cloud methane concentration falls below the flammability range.

Solution

Substance	Molecular Weight	Nb Temperature (K)	Ambient (T K)	Pool Area (m^2)
LNG (methane)	16	111.66 K	293.16	0.2

Gas Constant R (m^3 atm K^{-1} $kmol^{-1}$)	Vapour Pressure (atm)	Liquid Density (kg/m^3)	Vapour density (kg/m^3)	LEL (%)
0.08205	1	425.6	1.76	25

The pool volume and mass is:

$$V = 25 \cdot 0.05 = 1.25 \text{ m}^3 \tag{8.65}$$

$$M = 1.25 \cdot 425.6 = 532 \text{ kg} \tag{8.66}$$

According to Matthiesen (1986), the evaporation rate is:

$$W = \frac{MW \cdot A \cdot Pv^o}{R \cdot T_L} \cdot 0.0083 \cdot \left(\frac{18}{MW}\right)^{\frac{1}{3}} = \frac{16 \cdot 25 \cdot 1}{0.08205 \cdot 111.66} \cdot 0.0083 \cdot \left(\frac{18}{16}\right)^{\frac{1}{3}} = 0.377 \text{ kg/s} \tag{8.67}$$

And the release time:

$$R_d = \frac{532}{0.377} = 1411 \text{ s} \tag{8.68}$$

The buoyancy is:

$$g_o = g \cdot \left(\frac{\rho_e - \rho_a}{\rho_a}\right) = 9.81 \cdot \frac{1.76 - 1.2}{1.2} = 4.29 \text{ m/s}^2 \tag{8.69}$$

Maximum distance for continuous release.

$$x_c \frac{u_w \cdot R_d}{2.5} = \frac{3 \cdot 1411}{2.5} = 1693.2 \text{ m} \tag{8.70}$$

Dense model application check

$$D_c = \sqrt{\frac{Q_o}{u_w}} = \sqrt{\frac{0.377}{1.76 \cdot 3}} = 0.071 \text{ m} \tag{8.71}$$

$$\sqrt[3]{\frac{g_o \cdot Q_o}{u_w{}^3 \cdot D_c}} = \sqrt{\frac{4.29 \cdot 0.377}{27 \cdot 1.76 \cdot 0.071}} = 0.69 \geq 0.15 \tag{8.72}$$

$$C_{eff} = \frac{C_{non\text{-}isoth}}{C_{non\text{-}isoth} + (1 - C_{non\text{-}isoth}) \times (T_{amb}/T_o)} = \frac{0.05}{0.05 + (1 - 0.05) \times (293.16/111.66)} = 0.019 \tag{8.73}$$

Effective concentration for non-isothermal release:

$$\frac{c_m}{c_o} = \frac{0.019}{1} = 0.019 \tag{8.74}$$

$$\left(\frac{g_o{}^2 Q_o}{u_w{}^5}\right)^{1/5} = \left(\frac{4.29^2 \times 0.377}{3^5 \cdot 1.76}\right)^{1/5} = 0.43 \tag{8.75}$$

$$\frac{x}{\sqrt{Q_o \cdot u_w}} = 160 \rightarrow x = 160 \cdot \sqrt{\frac{3 \cdot 0.377}{1.76}} = 128\,\text{m} \tag{8.76}$$

Example 8.9

For the propane horizontal flashing jet shown in the following table, develop a practical dispersion approach on the basis of a dense gas model.

Mass Flow Rate (kg/s)	T_f (K)	Vapour Quality at T_f (kg/kg)	Vapour Density at T_f (kg/m^3)	Liquid Density at T_f (kg/m^3)
10.8	231.16	0.3	2.5	581

Enthalpy of Liquid at T_f (J/kg)	Velocity at Exit (m/s)	Mass Flow Rate (kg/s)	Vapour Density at T_f (kg/m^3)	Liquid Density at T_f (kg/m^3)
−196,000	21.2	10.8	2.5	581

Solution

The LPG will flash, and the two phases will follow different routes. The vapour will travel according to the turbulent jet model, and the liquid will rain out in a bunded or unbunded area, and then will evaporate. In case of unbunded area, equation of Wu and Schroy (1979) can be adopted:

$$r(t) = \sqrt[5]{\frac{t^3 \cdot \rho_L \cdot Q \cdot \sin 2\alpha}{2 \cdot C^3 \cdot \mu_L \cdot \pi^2/6 \cdot g}} \tag{8.77}$$

In case of bunded area, this will coincide with the basin area (see Fig. 8.10).

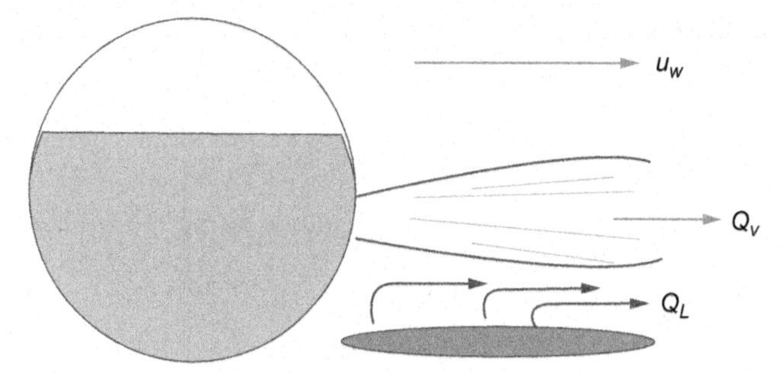

Fig. 8.10
Horizontal flashing jet.

On the basis of the assumption that vapour and liquid evaporating flow are both downwind oriented, an approximate conservative approach could be to consider an overall vapour travelling plume consisting of the sum of the two flows, where:

$$Q_v = \frac{0.3 \cdot 10.8}{2.5} = 1.296 \text{ kg/s} \tag{8.78}$$

This flow rate will be included in the model when jet momentum is lost. And Q_L is the liquid evaporating flow from the pool formed by the mass flow rate:

$$Q_L = 0.7 \cdot 10.8 = 7.56 \text{ kg/s} \tag{8.79}$$

8.8.4 Hoehne and Luce Model

This model (Hoehne and Luce, 1970) describes a plume trajectory, which is a function of the stack exit velocity and gas properties. It also predicts the dilution of the stack gases along the plume centreline. The basis equation of this model is:

$$\frac{y_s}{x_o} = 2.05 \cdot \left(\frac{x}{x_o}\right)^{0.28} \tag{8.80}$$

where:

- y_s is the vertical height above the stack.
- x is the horizontal downwind distance from the stack.
- x_o is

$$x_o = d_e \cdot \left(\frac{u_e}{u_w}\right) \cdot \sqrt{\frac{\rho_e}{\rho_a}} \tag{8.81}$$

where the subscript e refers to the exit conditions, w to wind, a to the ambient, u to the speed at the outlet, and ρ to the density. This equation is valid in the interval:

$$0 < \frac{x}{x_o} < 3 \tag{8.82}$$

And the centreline molar fraction is:

$$c = 4.95 \cdot \frac{u}{u_w} \cdot \frac{MW}{29} \cdot \left(\frac{s}{x_o}\right)^{-1.6} \tag{8.83}$$

where MW is the gas molar weight, and s is the distance along the plume axis. This equation is valid in the interval:

$$0 < \frac{S}{x_o} < 3 \tag{8.84}$$

8.9 Transition to Passive Dispersion

Passive dispersion is governed by the atmospheric turbulence and is the final outcome of any gas and vapour release. As passive dispersion prevails, the importance of released fluid properties decreases, since its fate is predominantly determined by air convective motion by wind and by weather characteristics. A large and comprehensive literature has been developed, which provides accurate and detailed prediction data on the Gaussian behaviour of passive plumes. However, due to the combined complex contribution of many key-drivers, such as buoyancy, initial momentum, and wind, the straightforward adoption of passive dispersion models in consequence assessment is often misleading and sometimes wrong, if it is not properly calibrated accounting for the multi-faceted behaviour of the released fluid from the release source. It is not infrequent to read reports where the simplistic adoption of passive dispersion models leads to significant gaps with respect to experimental results. A typical example is the prediction of carbon dioxide concentration profiles in the Carbon Capture and Storage (CCS) project. Even the effectiveness of the utilisation of validated software such as FRED, PHAST, and TNO EFFECT are strongly dependent on the user's full understanding and expertise about the fluid-thermodynamic-thermochemical frame of the input data and selected scenarios. This section summarises the dispersion model calculation methods, but aims mainly to provide guidance on some practical criteria to be followed to determine, in the various release scenarios, where and when passive dispersion may be assumed as the governing dispersion mechanism, and which thermodynamic and chemical data have to be assumed in the model input file. This is not a trivial aspect in consequence assessment.

8.9.1 Momentum-Jet to Passive-Plume Transition

This is the case of substances with molecular weight or apparent density which are comparable to air density, for example ethylene (molecular weight $\cong$ 28) and hydrogen sulphide (molecular weight $\cong$ 34) at atmospheric temperature, or methane at 161 K when ambient temperature is 293 K. For substances like these, buoyancy does not significantly affect the gas behaviour, as momentum and the distance from the outlet where this happens can be easily identified, applying one of the methods presented in the previous sections, along with the definition of the concentration profile. From this point onward passive dispersion will be assumed as the governing dispersion mechanism.

Example 8.10

Ethylene at 10 atm is released to the atmosphere from a pressure vessel through a 50 mm hole. In the exit plane, gas speed is 300 m/s, pressure is 6 atm, and temperature is 250 K. Wind speed is 5 m/s and ambient temperature is 273.16 K. Find:

- the distance from the outlet where momentum is lost and passive dispersion prevails.
- the concentration at this distance.

T_a (K)	T_e (K)	Air Density (kg/m³)	Ethylene Density at T_e and P_e (kg/m³)	Gas Speed at the Outlet Plane (m/s)	D_e (m)	u_w (m/s)
293.16	250	1.2	8.1	300	0.05	5

Solution

Applying Papanicolau and List method:

$$Fr = \frac{u}{\sqrt{g \cdot D_e \cdot \left(\frac{\rho_e}{\rho_a} - 1\right)}} \tag{8.85}$$

$$x_{M-B} = 5.6 \cdot D_e \cdot Fr = 1.1 \cdot 0.05 \cdot \frac{300}{\sqrt{9.81 \cdot 0.05 \cdot \frac{8.1 - 1.2}{1.2}}} = 9.83 \text{ m} \tag{8.86}$$

Applying Lee and Chu method:

$$M_o = A_e \cdot \rho_e \cdot {u_e}^2 = \pi \cdot \frac{0.05^2}{4} \cdot 8.1 \cdot 300^2 = 1430.6 \text{ kg} \cdot \text{m/s}^2 \tag{8.87}$$

$$F_o = A_e \cdot \frac{\rho_e - \rho_a}{\rho_a} \cdot u_e \cdot g \cdot \rho_e = \pi \cdot \frac{0.05^2}{4} \cdot \frac{8.1 - 1.2}{1.2} \cdot 300 \cdot 5 \cdot 9.81 = 166 \text{ kg} \cdot \text{m/s}^3 \tag{8.88}$$

$$x_{M-B} = \frac{\left(\frac{M_o}{\rho_e}\right)^{3/4}}{\left(\frac{F_o}{\rho_e}\right)^{1/2}} = \frac{\left(\frac{1430}{8.1}\right)^{3/4}}{\left(\frac{166}{8.1}\right)^{1/2}} = 10.7 \text{ m} \tag{8.89}$$

Passive dispersion becomes dominating at around 10 m from the source. At this distance, molar fraction can be calculated as (Benintendi, 2010):

$$y = \frac{1}{\frac{MW}{MW_a} \cdot \left[\left(\frac{x_{M-B}}{3.1 \cdot D_e}\right) - 1\right] + 1} = \frac{1}{\frac{28}{29} \cdot \left[\left(\frac{10}{3.1 \cdot 0.05}\right) - 1\right] + 1} = 0.015 = 1.5\% = 15000 \text{ ppmv} \tag{8.90}$$

8.9.2 Positively Buoyant to Passive-Plume Transition

Positively buoyant gases or vapours are, for example, hydrogen, methane, and ammonia. Their release falls into three categories:

1. Very low momentum releases, where buoyancy initially governs the flow model, and momentum is negligible.
2. Combination of momentum and positive buoyancy, where both are significant in the flow development.
3. Significant momentum flow resulting in a turbulent or laminar jet.

These scenarios have jointly been treated through Briggs and highly turbulent jet has been largely described in Chapter 4. When buoyancy governs, passive dispersion onset is attained at the end of the plume rise described by Briggs formula. When the flow is momentum-driven, in practical application, passive dispersion is assumed as significant because momentum is lost, for instance, by applying Lee and Chu equation. This is not true for negatively buoyant gases.

8.9.3 Negatively Buoyant Gas to Neutrally Buoyant Transition

Unlike lighter-than-air gases, buoyancy of dense gases significantly affects the outcome of a release when momentum is lost. Again, the possible scenarios are those identified for positively buoyant fluids, along with the relevant considerations:

1. Very low momentum releases, where buoyancy initially governs the flow model, and momentum is negligible.
2. Combination of momentum and positive buoyancy, where both are significant in the flow development.
3. Significant momentum flow resulting in a turbulent or laminar jet.

However, for dense gases, when momentum is lost, buoyancy becomes very important and drives the flow prior to the intervention of passive dispersion. Often, the most hazardous toxic or flammable scenarios are those corresponding to dense gas clouds.

Two methods can be used to locate the transition point from negatively to neutrally buoyant. The first one, very approximate and possibly over-conservative, is to assume a lower limit of the density ratio:

$$\frac{\rho_1 - \rho_a}{\rho_a} \tag{8.91}$$

where ρ_1 is the density of the substance diluted with air. CCPS (1996) suggests a limit for this parameter falling in the range 0.01–0.001.

A much more robust and technically justified criterion is given by Crowl and Louvar (2011), which suggest updating the limits of application of the dense model for continuous and instantaneous release, replacing the dense gas density ρ_o of initial buoyancy with the buoyancy of diluted mixture ρ_1:

$$\rho_1 = \rho_o \cdot \left(\frac{c_x}{c_o}\right) + \rho_a \cdot \left(1 - \frac{c_x}{c_o}\right) \tag{8.92}$$

and the buoyancy factor:

$$g_1 = g \cdot \left(\frac{\rho_1 - \rho_a}{\rho_a}\right) \tag{8.93}$$

The updated flow rate can be calculated from the initial flow rate Q_o by applying the law of mass conservation:

$$Q_o \cdot c_o = Q_x \cdot c_x \tag{8.94}$$

and:

$$D_1 = \sqrt{\frac{Q_x}{u_w}} \tag{8.95}$$

The updated volume:

$$V_o \cdot c_o = V_x \cdot c_x \tag{8.96}$$

and:

$$D_1 = \sqrt[3]{V_x} \tag{8.97}$$

The dense model application limit criteria become:

$$\left(\frac{c_x}{c_o}\right)^{1/6} \sqrt[3]{\frac{g_o \cdot Q_o}{u_w{}^3 \cdot D_c}} = 0.15 \tag{8.98}$$

and:

$$\left(\frac{c_x}{c_o}\right)^{1/3} \frac{\sqrt{g_o \cdot V_o}}{u_w \cdot \sqrt[3]{V_o}} = 0.20 \tag{8.99}$$

Example 8.11

2700 m^3/h of carbon dioxide are released for 20 min through a 100 mm hole at ambient temperature. Wind speed is 5 m/s. Find the distance from the source where passive dispersion becomes significant.

Solution

Assuming a continuous plume:

$$g_o = g \cdot \left(\frac{\rho_e - \rho_a}{\rho_a}\right) = 9.81 \cdot \frac{44 - 29}{29} = 5.07 \text{ m/s}^2 \tag{8.100}$$

$$D_C = \sqrt{\frac{Q_o}{u_w}} = \sqrt{\frac{0.83}{5}} = 0.39 \text{ m} \tag{8.101}$$

The molar fraction at the transition point is:

$$c_x = c_o \left(\cdot \frac{0.15}{\sqrt[3]{\frac{g_o \cdot Q_o}{u_w{}^3 \cdot D_c}}} \right)^6 = \left(\cdot \frac{0.15}{\sqrt[3]{\frac{5.07 \cdot 0.75}{125 \cdot 0.39}}} \right)^6 \cong 0.002 \tag{8.102}$$

So:

$$\left(\frac{g_o{}^2 Q_o}{u_w{}^5}\right)^{1/5} = \left(\frac{5.07^2 \times 2.23}{5^5}\right)^{1/5} = 0.36 \tag{8.103}$$

$$\frac{x}{\sqrt{Q_o \cdot u_w}} = 500 \rightarrow x = 500 \cdot \sqrt{5 \cdot 0.75} = 968 \text{ m} \tag{8.104}$$

$$\frac{u_w \cdot R_d}{x} = \frac{5 \cdot 20 \cdot 60}{968} = 6.2 > 2.5 \tag{8.105}$$

The assumption of continuous plume is correct.

8.10 Gaussian Dispersion

The Gaussian dispersion model has been broadly described in the literature (CCPS, 1999a). In this section the basic equations are presented, along with some practical guidance relative to the definition and selection of the relevant parameters.

8.10.1 Weather Input

Weather data should be as recent as possible and related to the site under investigation. Prevailing wind, wind rose, and weather stability classes are the essential information needed for consequence assessment and dispersion prediction.

Wind data are quoted at 10 m height. If meteorological data are not available, dispersion studies are typically carried out on the basis of the following combinations (Table 8.7):

Table 8.7 P-G class wind speed

P-G Class	Wind Speed (m/s)
D	5
F	1.5–2

Stability class D is, typically, by far the most frequent, while F is the second most frequent. In addition, class F presents a low turbulence degree, so the expected results can cover the worst case. Class D is for day time, class F for night time.

8.10.2 Averaging Time

According to Woodward (1998), passive dispersion data are taken with averaging times typically of 10 to 30 min. For flammable mass calculations, this time should be short, and typically not exceeding 10 s. To take into account the different averaging times, the following empirical formula is recommended for converting concentrations from one averaging time to another (Hanna et al., 1993):

$$\frac{c_t}{c_{600}} = \left(\frac{600}{t}\right)^{0.2} \tag{8.106}$$

where time is in seconds. Hanna claims that experimentally:

$$c_{max} = 2 \cdot c_{10\ \mathrm{min}} \tag{8.107}$$

where c_{max} is the maximum peak concentration in the plume. Solving for t, it is:

$$t = 18.75\ \mathrm{s} \tag{8.108}$$

This time should be adopted to carry out worst case predictions.

8.10.3 Release Time

Selection of release times is not univocal and can be made following different criteria. The most robust method is to identify the inventories of the various segments of the system and to carry out a transient analysis. This the most credible scenario, both in terms of release time, concentration profile, and identification of the mass amount emitted to the atmosphere.
A general formula useful for taking into account the pressure and inventory decay is provided by Spouge (DNV Technica, 1999):

$$W(t) = W_o \cdot \exp\left(-\frac{W_o}{M_o} \cdot t\right) \tag{8.109}$$

where:

- $W(t)$ is the mass flow rate.
- W_o is the initial flow rate.
- M_o is the inventory.

A prescriptive approach is provided by API Recommended Practice 581 (2016). For large leaks, including rupture, a release time of 3 min is adopted, as the midpoint of the range 1–5 s.

For small and relatively small leaks, the detection and isolation systems play a fundamental role in the determination of the release time. According to API Recommended Practice 581 (2016), the total leak duration L_d is the sum of the following:

- Time to detect the leak
- Time to analyse the incident and decide upon corrective action
- Time to complete appropriate corrective actions

Table 8.8 classifies the detection and isolation systems on the basis of level and degree of independency, specificity, automation, and actuation of the relevant components. Table 8.9 correlates detection and isolations codes, allowing for some possible reduction factors of releases. Table 8.10 provides maximum leak durations related to leak size and combinations of detection and isolation.

8.10.4 Decision on Instantaneous or Continuous Release

Categorisation of the release as instantaneous or continuous, according to Eq. (8.57), are strictly valid for dense gases. For neutrally or positively buoyant gases, it is recommended to adopt the method provided by the standard API Recommended Practice 581 (2016). The procedure is the following:

Step 1: Calculation of the release time for a given mass

The time to release 10,000 lbs (4536 kg) shall be calculated.

Table 8.8 Detection and isolation classification (API 581, 2016)

Type of Detection System	Detection Classification
Instrumentation designed specifically to detect material losses by changes in operating conditions (i.e. loss of pressure or flow) in the system	A
Suitably located detectors to determine when the material is present outside the pressure-containing envelope	B
Visual detection, cameras, or detectors with marginal coverage	C
Type of Isolation System	**Isolation Classification**
Isolation or shutdown systems activated directly from process instrumentation or detectors, with no operator intervention	A
Isolation or shutdown systems activated by operators in the control room or other suitable locations remote from the leak	B
Isolation dependent on manually operated valves	C

Table 8.9 Reduction factors of release rate or mass (API 581, 2016)

System Classifications		Release Magnitude Adjustment	Reduction Factor
Detection	Isolation		
A	A	Reduce release rate or mass by 25%	0.25
B	B	Reduce release rate or mass by 20%	0.20
A or B	C	Reduce release rate or mass by 10%	0.10
B	B	Reduce release rate or mass by 25%	0.15
C	C	No adjustment to release rate or mass	0.00

Table 8.10 Leak durations based on detection and isolation system (API 581, 2016)

Detection System Rating	Isolation System Rating	Maximum Leak Duration
A	A	20 min for 6.4 mm leaks 10 min for 25 mm leaks 5 min for 102 mm leaks
A	B	30 min for 6.4 mm leaks 20 min for 25 mm leaks 10 min 102 mm leaks
A	C	40 min for 6.4 mm leaks 30 min for 25 mm leaks 20 min 102 mm leaks
B	A or B	40 min for 6.4 mm leaks 30 min for 25 mm leaks 20 min 102 mm leaks
B	C	1 h for 6.4 mm leaks 30 min for 25 mm leaks 20 min 102 mm leaks
C	A, B or C	1 h for 6.4 mm leaks 40 min for 25 mm leaks 20 min 102 mm leaks

Step 2: Consideration of the diameter size

If the diameter size is 6.35 mm (0.25 in.) or less, then the release is continuous. Time calculated in Step 1 is to be evaluated.

Step 3: Mass

If time calculated in Step 1 is less than 3 min, then the release is instantaneous, otherwise it is continuous.

8.10.5 Pasquill-Gifford Model

Puff model

$$\langle C\rangle(x,y,z,t) = \frac{M}{(2\cdot\pi)^{1.5}\cdot\sigma_x\cdot\sigma_y\cdot\sigma_z}\cdot\exp\left[-\frac{1}{2}\cdot\left(\frac{y}{\sigma_y}\right)^2\right]\cdot \left\{\exp\left[-\frac{1}{2}\cdot\left(\frac{z-h}{\sigma_z}\right)^2\right]+\exp\left[-\frac{1}{2}\cdot\left(\frac{z+h}{\sigma_z}\right)^2\right]\right\} \quad (8.110)$$

where

- $\langle C\rangle$ is the time average concentration.
- M is the total mass of material released.
- σ_x, σ_y, σ_z are the dispersion coefficients in the x, y and z directions (length) (Table 8.11).
- x is downwind direction.
- y is the cross-wind direction.
- z is the above ground direction.
- h is the release height above the ground.

If the coordinate system is fixed at the release point, then the previous equation has to be multiplied by the factor:

$$\exp\left[-\frac{1}{2}\left(\frac{x-u_w\cdot t}{\sigma_x}\right)^2\right] \quad (8.111)$$

Plume Model

$$\langle C\rangle(x,y,z) = \frac{W}{2\cdot\pi\cdot\sigma_y\cdot\sigma_z\cdot u_w}\cdot\exp\left[-\frac{1}{2}\cdot\left(\frac{y}{\sigma_y}\right)^2\right]\cdot \left\{\exp\left[-\frac{1}{2}\cdot\left(\frac{z-h}{\sigma_z}\right)^2\right]+\exp\left[-\frac{1}{2}\cdot\left(\frac{z+h}{\sigma_z}\right)^2\right]\right\} \quad (8.112)$$

Table 8.11 Dispersion coefficients for puff model

P-G Class	σ_x or σ_y	σ_z
A	$0.18\cdot x^{0.92}$	$0.60\cdot x^{0.75}$
B	$0.14\cdot x^{0.92}$	$0.53\cdot x^{0.73}$
C	$0.10\cdot x^{0.92}$	$0.34\cdot x^{0.71}$
D	$0.06\cdot x^{0.92}$	$0.15\cdot x^{0.70}$
E	$0.04\cdot x^{0.92}$	$0.10\cdot x^{0.65}$
F	$0.02\cdot x^{0.92}$	$0.06\cdot x^{0.61}$

where:

- $\langle C \rangle$ is the time average concentration.
- W is the mass flow rate of the released material, and σ_x, σ_y, σ_z are the dispersion coefficients in the x, y and z directions (length).
- x is downwind direction.
- y is the cross-wind direction.
- z is the above ground direction.
- h is the release height above the ground.
- u_w is the wind speed.
- h is the release height above the ground + the plume rise.

Dispersion coefficients for plume model are shown in Table 8.12.

For rural areas, the downwind distance X, and the total isopleth area, for a given concentration of interest $\langle C \rangle$, mass flow rate W, and wind speed u_w can be found utilising the following equations (see Table 8.13) (CCPS, 1999b):

Table 8.12 Dispersion coefficients for plume model

P-G Class	σ_y	σ_z
	Rural Conditions	
A	$0.22 \cdot x \cdot (1+0.0001 \cdot x)^{-0.5}$	$0.2 \cdot x$
B	$0.16 \cdot x \cdot (1+0.0001 \cdot x)^{-0.5}$	$0.12 \cdot x$
C	$0.11 \cdot x \cdot (1+0.0001 \cdot x)^{-0.5}$	$0.08 \cdot x \cdot (1+0.0002 \cdot x)^{-0.5}$
D	$0.08 \cdot x \cdot (1+0.0001 \cdot x)^{-0.5}$	$0.06 \cdot x \cdot (1+0.0015 \cdot x)^{-0.5}$
E	$0.06 \cdot x \cdot (1+0.0001 \cdot x)^{-0.5}$	$0.03 \cdot x \cdot (1+0.0003 \cdot x)^{-1}$
F	$0.04 \cdot x \cdot (1+0.0001 \cdot x)^{-0.5}$	$0.016 \cdot x \cdot (1+0.0003 \cdot x)^{-1}$
	Urban Conditions	
A-B	$0.32 \cdot x \cdot (1+0.0004 \cdot x)^{-0.5}$	$0.24 \cdot x \cdot (1+0.001 \cdot x)^{-0.5}$
C	$0.22 \cdot x \cdot (1+0.0004 \cdot x)^{-0.5}$	$0.2 \cdot x$
D	$0.16 \cdot x \cdot (1+0.0004 \cdot x)^{-0.5}$	$0.14 \cdot x \cdot (1+0.003 \cdot x)^{-0.5}$
E-F	$0.11 \cdot x \cdot (1+0.0004 \cdot x)^{-0.5}$	$0.08 \cdot x \cdot (1+0.0015 \cdot x)^{-0.5}$

Table 8.13 Coefficients of Eqs (8.113), (8.114)

P-G Class		a_o	a_1	a_2	a_3
B	X/L	1.28868	0.037616	-0.0170972	0.00367183
D		2.00661	0.016541	0.000142451	0.0029
F		2.76837	0.0340247	0.0219798	0.00226116
		b_o	b_1	b_2	b_3
B	$(A/L)^2$	1.35167	0.0288667	-0.0287847	0.0056558
D		1.86243	0.0239251	-0.00704844	0.00503442
F		2.75493	0.0185086	0.0326708	0.00392425

$$\frac{X}{L^{\bullet}} = a_o + a_1 \cdot \ln(L^{\bullet}) + a_2 \cdot [\ln(L^{\bullet})]^2 + a_3 \cdot [\ln(L^{\bullet})]^3 \tag{8.113}$$

$$\frac{A}{(L^{\bullet})^2} = b_o + b_1 \cdot \ln(L^{\bullet}) + b_2 \cdot [\ln(L^{\bullet})]^2 + b_3 \cdot [\ln(L^{\bullet})]^3 \tag{8.114}$$

$$L^{\bullet} = \sqrt{\frac{W}{u_w \cdot \langle C \rangle}} \tag{8.115}$$

with:

Example 8.12

Ethylene at 10 atm is released to the atmosphere through a 50 mm hole. In the exit plane, gas speed is 300 m/s, pressure is 6 atm, and temperature is 250 K. Wind speed is 5 m/s and ambient temperature is 273.16 K. For this scenario, a momentum to buoyancy/wind transition distance has been calculated in the Example 8.10, which is around 10 m, where the average concentration is 15,000 ppmv. Find the maximum concentration behaviour at ground level, 10 m height (plume rise), and 20 m height.

T_a (K)	T_e (K)	Air Density (kg/m^3)	Ethylene Density at T_e and P_e (kg/m^3)	Gas Speed at the Outlet Plane (m/s)	D_e (m)	u_w (m/s)	W (kg/s)	C_e (kg/m^3)	$H_{M\text{-}B}$ (m)
293.16	250	1.2	8.1	300	0.05	5	4.77	$1.16 \cdot 10^{-3}$	10

Solution

See Fig. 8.11.

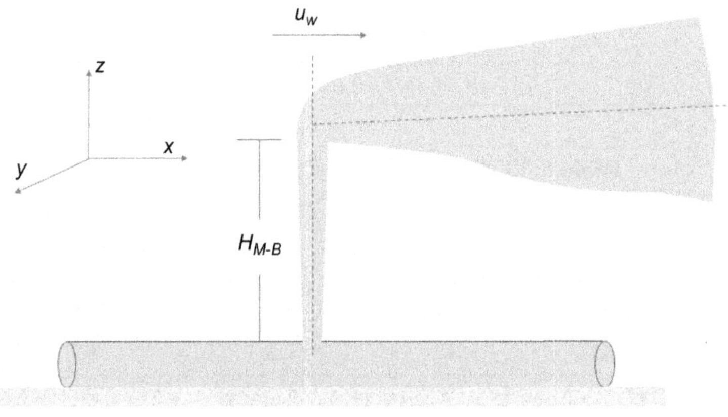

Fig. 8.11
Ethylene release of Example 8.12.

The wind speed is 5 m/s, so class D is selected. Maximum concentration at ground is at $z=0$ and $y=0$, so:

$$\langle C\rangle(x,0,0)=\frac{W}{2\cdot\pi\cdot\sigma_y\cdot\sigma_z\cdot u_w}\cdot\left\{2\cdot\exp\left[-\frac{1}{2}\cdot\left(\frac{h}{\sigma_z}\right)^2\right]\right\} \tag{8.116}$$

And, at 10 and 20 m height:

$$\langle C\rangle(x,0,10)=\frac{W}{2\cdot\pi\cdot\sigma_y\cdot\sigma_z\cdot u_w} \tag{8.117}$$

$$\langle C\rangle(x,0,20)=\frac{W}{2\cdot\pi\cdot\sigma_y\cdot\sigma_z\cdot u_w}\cdot\exp\left[-\frac{1}{2}\cdot\left(\frac{y}{\sigma_y}\right)^2\right]\cdot\left\{\exp\left[-\frac{1}{2}\cdot\left(\frac{10}{\sigma_z}\right)^2\right]+\exp\left[-\frac{1}{2}\cdot\left(\frac{30}{\sigma_z}\right)^2\right]\right\} \tag{8.118}$$

$\sigma_y=0.08\cdot x\cdot(1+0.0001\cdot x)^{-0.5}$	$\sigma_z=0.06\cdot x\cdot(1+0.0015\cdot x)^{-0.5}$

The downwind maximum concentration (centreline), respectively at, 10 m, 20 m, and grade are depicted in Figs 8.12 and 8.13.

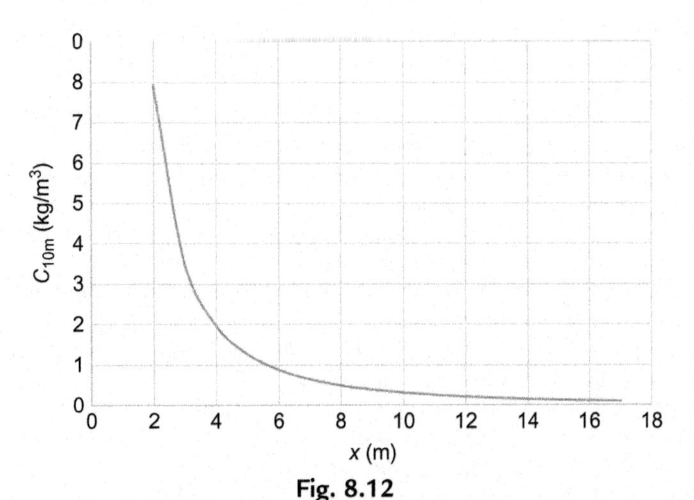

Fig. 8.12
Ethylene concentration decay at 10 m above grade.

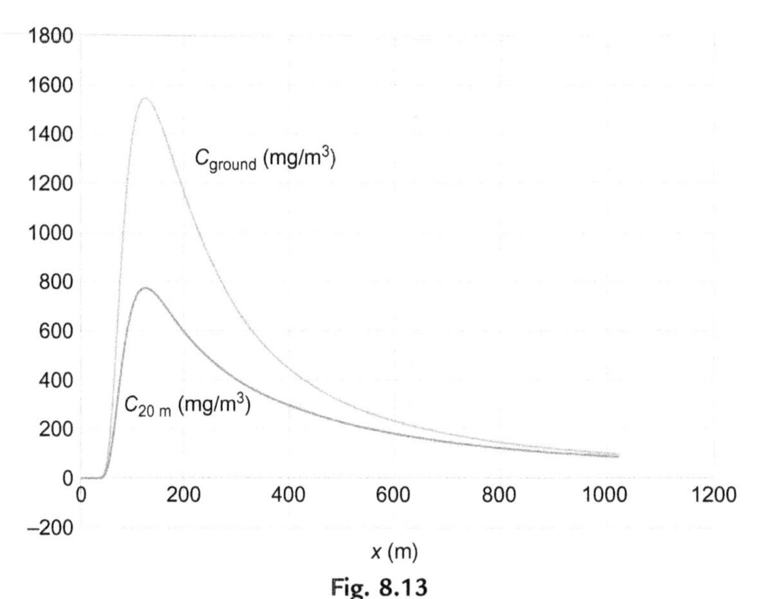

Fig. 8.13
Ethylene concentration decay at grade and at 20 m above grade.

References

API, 1996. A Guidance Manual for Modeling Hypothetical Accidental Releases to the Atmosphere. American Petroleum Institute. Publication No 4628, November.

API Standard 521, 2014. Pressure-Relieving and Depressuring Systems, sixth ed. American Petroleum Institute.

API Recommended Practice 581, 2016. Risk-Based Inspection Methodology, third ed. American Petroleum Institute.

Benintendi, R., 2010. Turbulent jet modelling for hazardous area classification. J. Loss Prev. Process Ind. 23, 373–378.

Briggs, G.A., 1972. Discussion: chimney plumes in neutral and stable surroundings. Atmos. Environ. 6, 507–510.

Briggs, G.A., 1984. Plume Rise and Buoyancy Effects. Atmospheric Science and Power Production. Report No. DOE/TIC-27601, pp. 327–336.

Britter, R.E., McQuaid, J., 1988. Workbook on the Dispersion of Dense Gases, Health and Safety Executive (Open Source). http://www.hse.gov.uk/research/crr_pdf/1988/crr88017.pdf.

CCPS, 1996. Guidelines for Use of Vapor Cloud Dispersion Models, second ed. American Institute of Chemical Engineers, Center for Chemical Process Safety.

CCPS, 1999a. Guidelines for Chemical Process Quantitative Risk Analysis, second ed. American Institute of Chemical Engineers, Center for Chemical Process Safety.

CCPS, 1999b. Guidelines for Consequence Analysis of Chemical Releases. American Institute of Chemical Engineers, Center for Chemical Process Safety.

Crowl, D.A., Louvar, J.F., 2011. Chemical Process Safety: Fundamentals with Applications, third ed. Prentice Hall International Series in Physical and Chemical Engineering.

Gifford, F.A., 1976. Turbulent diffusion-typing schemes: a review. Nucl. Saf. 17, 68–86.

Hanna, S.R., Briggs, G.A., Hosker, R.P., 1982. Handbook on Atmospheric Diffusion. Report No. DOE/TIC-11223. Department of Energy, Oak Ridge National Laboratory.

Hanna, S.R., Strimaitus, D.G., Chang, J., 1990. Results of Hazard Response Model Evaluation Using Desert Tortoise and Goldfish Data Eases., Vol. 1: Summary Report. American Petroleum Institute, Washington, DC.

Hanna, S.R., Strimaitus, D.G., Chang, J., 1993. Hazard Response Modeling Uncertainty (A Quantitative Method) Vol 11 - Evaluation of Commonly Used Hazardous Gas Dispersion Models, Environics Division Air Force Engineering & Services Center, Engineering & Services Laboratory.

Hoehne, V.O., Luce, R.G., 1970. The effect of velocity, temperature and molecular weight on flammability limits in wind-blown jets of hydrocarbon gases. In: 35th Mid-Year Meeting of the API Division of Refining, Houston, USA.

Hoot, T.G., Meroney, R.N., Peterka, J.A., 1973. Wind Tunnel Tests of Negatively Buoyant Plumes. Report No. CER73-74TGH-RNM-JAP-13. Colorado State Univ., Fort Collins, CO.

Lee, J.H.W., Chu, V.H., 2003. Turbulent Jets and Plumes: A Lagrangian Approach. Kluwer Academic Publisher, Dordrecht.

Matthiesen, R.C., 1986. Estimating chemical exposure levels in the workplace. Chem. Eng. Prog., April, 30.

McQuaid, J., 1987. Design of the thorney island release trials. J. Hazard. Mater. 16, 1–8.

Oke, T.R., 1978. Boundary Layer Climates. John Wiley and Sons, New York.

Oosthuizen, P.H., 1983. An experimental study of low reynolds number turbulent circular jet flow. In: American Society of Mechanical Engineers, Applied Mechanics, Bioengineering, and Fluids Engineering Conference, Houston, TX, June 20–22.

Panofsky, H.A., Dutton, J.A., 1984. Atmospheric Turbulence: Models and Methods for Engineering Applications. John Wiley and Sons, New York.

Papanicolaou, P.N., List, E.N., 1988. Investigations of round vertical turbulent buoyant jets. J. Fluid Mech. 195, 341–391.

Patrick, M.A., 1967. Experimental investigation of the mixing and penetration of a round turbulent jet injected perpendicularly into a transverse stream. Trans. Inst. Chem. Eng. 45, T16–T31.

Spouge, J., 1999. A Guide To Quantitative Risk Assessment for Offshore Installations, DNV Technica, CMPT.

USEPA, 1988. Screening Procedures for Estimating the Air Quality Impact of Stationary Sources. Report No. EPA-450/4-88-010. United States Environmental Protection Agency, Office of Air Quality Planning and Standards, Research Triangle Park, NC.

Van den Bosch, C.J.H., Duijm, N.J., 2005. In: Van den Bosch, C.J.H., Weterings, R.A.P.M. (Eds.), Methods for the Calculation of Physical Effects. Yellow Book.

Wilson, D.J., Angle, R.P., 1979. The Release and Dispersion of Gas from Pipeline Ruptures. Air Quality Control Branch, Pollution Control Division, Environmental Protection Services, Edmonton.

Woodward, J.L., 1998. Estimating the Flammable Mass of a Vapor Cloud: A CCPS Concept Book. American Institute of Chemical Engineers, New York.

Wu, J.M., Schroy, J.M., 1979. Emissions from Spills, 55 APCA and WPCF Joint Conference on Control of Specific (Toxic) Pollutants, Gainesville, Air Pollution Control Association, Florida Section, Feb. 13–16.

Further Reading

USEPA, 2004. Aermod: Description of Model Formulation Report No. EPA-454/R-03-004.

CHAPTER 9

Fire

Poca favilla gran fiamma seconda

(Dante Alighieri, Divine Comedy, Paradise I, 34)

9.1 Summary of Scenarios

Fire Scenario	Section
Pyrophoric materials	9.3
Ignition sources	9.5
Ignition probability	9.6
Fire scenarios	9.7
Flash fire (Eisenberg model)	9.7.1
Pool fires	9.7.2
Trench fires	9.7.3
Pool fires on water	9.7.4
Jet fires	9.79
Fire damage to people	9.8

9.2 Ignition Sources

Fire is a thermochemical phenomenon which requires the simultaneous availability of three chemical physical elements (Fig. 9.1).

Ignition sources provide the necessary energy amount to trigger and sustain combustion. Reaction rate of combustion is also affected by ignition process, possibly promoting fire to deflagration and to detonation, depending on the circumstances. When fuel is present in suitable proportions, and combination with oxygen, the attainment of the minimum ignition energy (MIE) threshold is the only conditional factor. Consequently, control of ignition sources, identification of their probability, and quantification of their occurrence are fundamental aspects in consequence and risk assessment.

Process Safety Calculations. https://doi.org/10.1016/B978-0-08-101228-4.00009-5

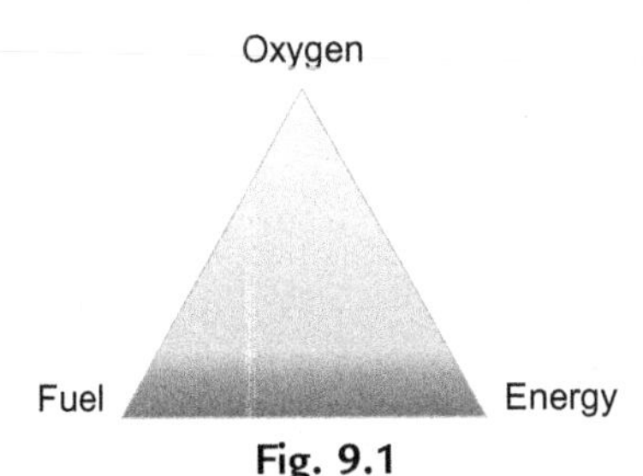

Fig. 9.1
Fire triangle for combustible gases, vapours, and dusts.

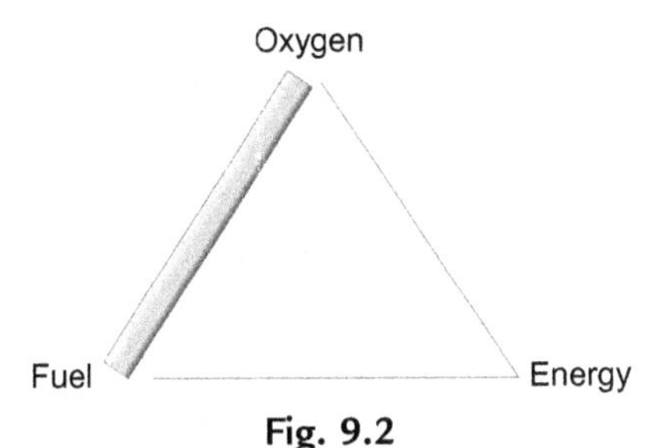

Fig. 9.2
Fire triangle for pyrophoric materials.

9.3 Pyrophoric Materials

Chemical–physical properties and general behaviour of pyrophoric materials have been discussed in Chapter 3. These materials don't need energy to undergo combustion, so separation from atmospheric oxygen is the only means to prevent ignition (Fig. 9.2).

9.4 Relevance and Effects of Ignition Sources

The relevance and the impact of ignition sources have to be evaluated against their effectiveness to trigger a combustion reaction and to sustain it. From the thermodynamic point of view, this consists of comparing their energy potential to the minimum energy required for ignition. The latter is the Minimum Ignition Energy, that is specific for any gases and dusts and has been discussed in Chapter 3. It is worth noticing that alkanes and unsaturated hydrocarbons have a very low MIE in oxygen, ranging between 0.001 and 0.1 mJ. MIE of light hydrocarbons, hybrid mixtures, and reactive dusts fall between 0.1 and 10 mJ. Combustible dusts have a MIE range of 0.01–10 J. An useful exercise is to quantify the ignition energy associated to a specific source, and this is particularly important for static electricity, due to the foreseeable serious impact this may have on design configuration and materials selection. Also, minimum ignition energy of dust depends on their size, as shown in the example included in Table 9.1 (BS 5958-1, 1991).

Table 9.1 Minimum ignition energy of dusts

Particle Size (μm)	Minimum Ignition Energy (mJ)
710–1680	>5000
355–709	250–500
180–354	50–250
105–179	<10
53–104	<10
5	<10

9.5 Analysis of Ignition Sources

Systematic and adequate analysis of the potential ignition sources is a very important but quite often missed step in process safety. The state of the art of fire risk assessment and the international best practices require to identify all ignition sources potentially existing in a process segment and to assess their likelihood and magnitude. This section illustrates this approach on the basis of some international standards. Notably, the systematic ignition source list included in the BS EN 1127-1 has been followed (Table 9.2).

9.5.1 Calculation Schemes for Static Electricity

See Table 9.3.

9.5.2 Streaming Current and Charge Balance

The following equation (Britton, 1999) can be used to calculate the maximum streaming current, which indicates transfer flow of electrons from one surface to another by a flowing of a fluid or a solid in a metal or glass pipe:

$$I_s = k \cdot u_p^2 \cdot D_p^2 \cdot \left[1 - \left(\frac{L_p}{u_p \cdot \tau}\right)\right] \tag{9.1}$$

where:

L_p, pipe length, m.
I_s is the streaming current, μA.
u_p is the liquid or solid velocity, m/s.
D_p is the pipe diameter, m.
k is a constant $= 10^{-5}$ A s^2/m^4.

Table 9.2 Ignition sources according to BS EN 1127-1

Ignition Source	Description	Recommendations
Hot surfaces	A hot surface can act as an ignition source and a dust layer or a combustible solid in contact with a hot surface and ignited by the hot surface can also act as an ignition source	The temperatures of all equipment, protective systems and components surfaces that can come into contact with explosive atmospheres should not exceed 80% of the auto ignition temperature of the combustible gas or liquid in °C, and in case of dust clouds, 2/3 of the minimum ignition temperature in °C of the dust cloud concerned (BS EN 1127-1, 2011). Guidance can be collected from API RP 2016 (2013)
Flames and hot gases	Flames are associated with combustion reactions at temperatures of more than 1000°C	Restrictions. Specific devices
Mechanically generated sparks	As a result of friction, impact or abrasion processes such as grinding, particles can become separated from solid materials and become hot because of the energy used in the separation process. If these particles consist of oxidisable substances, for example, iron or steel, they can undergo an oxidation process, thus reaching even higher temperatures. These particles (sparks) can ignite combustible gases and vapours and certain dust/air-mixtures (especially metal dust/air mixtures). Impacts involving rust and light metals (e.g., aluminium and magnesium) and their alloys can initiate a thermite reaction which can cause ignition of explosive atmospheres. The light metals titanium and zirconium can also form incentive sparks under impact or friction against any sufficiently hard material, even in the absence of rust	ATEX rated electrical and non-electrical equipment. Material selection (magnesium, titanium, zirconium) and construction specifications (EN, 13463-1)
Electrical apparatus	Electrical equipment and instruments	ATEX rated electrical equipment
Stray electric currents, cathodic corrosion protection	Return currents in power generating, systems especially in the vicinity of electric railways and large welding systems, when, for example, conductive electrical system components, such as rails and cable sheathing laid underground, lower the resistance of this return current path; a result of a short-circuit or of a short-circuit to earth because of faults in the electrical installations; which are a result of magnetic induction and a result of lightning. If parts	Compliance with API 2003 (2015)

Table 9.2 Ignition sources according to BS EN 1127-1—cont'd

Ignition Source	Description	Recommendations
	of a system able to carry stray currents are disconnected, connected or bridged even in the case of slight potential differences an explosive atmosphere can be ignited as a result of electric sparks and/or arcs. Moreover, ignition can also occur due to the heating up of these current paths When impressed current cathodic corrosion protection is used, the above-mentioned ignition risks are also possible. However, if sacrificial anodes are used, ignition risks due to electric sparks are unlikely, unless the anodes are aluminium or magnesium	
Static electricity	Charge build-up and release generated by: - the contact and separation of solids, e.g. the movement of conveyor belts, plastics film, etc. over rollers, the movement of a person - the flow of liquids or powders, and the production of sprays - induction phenomena, i.e. objects reach high potential or become charged due to being in an electric field	Compliance with NFPA 77 - PD CLC/TR 60079-32-1:2015 - CLC/TR 50404
Lightning	Electrical discharge of atmospheric origin between cloud and earth consisting of one or more strokes	Compliance with BS EN 62305-1 (2011)
Radio frequency electromagnetic waves (10^4 Hz–$3 \cdot 10^{11}$ Hz)	All conductive parts located in the radiation field function as receiving aerials. If the field is powerful enough, and if the receiving aerial is sufficiently large, these conductive parts can cause ignition in explosive atmospheres. The received radio-frequency power can, for example, make thin wires glow or generate sparks during the contact or interruption of conductive parts. The energy picked up by the receiving aerial, which can lead to ignition, depends mainly on the distance between the transmitter and the receiving aerial, as well as on the dimensions of the receiving aerial at any particular wavelength and RF power	Comply with distances and spacing as stated by BS EN 1127-1 (2011)
Radio frequency electromagnetic waves ($3 \cdot 10^{11}$ 10^4 Hz–$3 \cdot 10^{15}$ Hz)	Radiation in this spectral range can—especially when focused—become a source of ignition through absorption by explosive atmospheres or solid surfaces Sunlight, for example, can trigger an ignition if objects cause a convergence of the radiation	Compliance with BS EN 1127-1 (2011)

Continued

Table 9.2 Ignition sources according to BS EN 1127-1—cont'd

Ignition Source	Description	Recommendations
	(e.g. bottles acting as lenses, concentrating reflectors). Under certain conditions, the radiation of intense light sources (continuous or flashing) is so intensively absorbed by dust particles that these particles become sources of ignition for explosive atmospheres or for dust deposits	
Ionising radiations	Ionising radiation generated, for example, by X-ray tubes and radioactive substances can ignite explosive atmospheres (especially explosive atmospheres with dust particles) as a result of energy absorption. Moreover, the radioactive source itself can heat up, owing to internal absorption of radiation energy, to such an extent that the minimum ignition temperature of the surrounding explosive atmosphere is exceeded. Ionising radiation can cause chemical decomposition or other reactions which can lead to the generation of highly reactive radicals or unstable chemical compounds. This can cause ignition. Such radiation can also create an explosive atmosphere by decomposition (e.g. a mixture of oxygen and hydrogen by radiolysis of water)	Compliance with BS EN 1127-1 (2011)
Ultrasonics	In the use of ultrasonic sound waves, a large proportion of the energy emitted by the electroacoustic transducer is absorbed by solid or liquid substances. As a result, the substance exposed to ultrasonics warms up so that, in extreme cases, ignition may be induced	Compliance with BS EN 1127-1 (2011)
Adiabatic compression and shock waves	In the case of adiabatic or nearly adiabatic compression, and in shock waves, such high temperatures can occur that explosive atmospheres (and deposited dust) can be ignited. The temperature increase depends mainly on the pressure ratio, not on the pressure difference. In pressure lines of air compressors, and in containers connected to these lines, explosions can occur as a result of the compression ignition of lubricating oil mists. Shock waves are generated, for example, during the sudden venting of high-pressure gases into pipelines. In this process the shock waves are propagated into regions of lower pressure faster than the speed of sound. When they are diffracted or reflected by pipe bends, constrictions, connection flanges, closed valves, etc., very high temperatures can occur	Processes causing adiabatic compression shall be avoided, if potentially hazardous

Table 9.2 Ignition sources according to BS EN 1127-1—cont'd

Ignition Source	Description	Recommendations
Exothermic reactions, including self-ignition of dusts	Exothermic reactions can act as an ignition source when the rate of heat generation exceeds the rate of heat loss to the surroundings. Many chemical reactions are exothermic. Whether a reaction can reach a high temperature is dependent, among other parameters, on the volume/surface ratio of the reacting system, the ambient temperature and the residence time. These high temperatures can lead to ignition of explosive atmospheres and also the initiation of smouldering and/or burning Materials that are not capable of self-sustained combustion or smouldering in dust layers may still be capable to dust explosions when dispersed in air Such reactions include those of pyrophoric substances with air, alkali metals with water, self-ignition of combustible dusts, self-heating of feed-stuffs induced by biological processes, the decomposition of organic peroxides, or polymerisation reactions. Catalysts can also induce energy-producing reactions (e.g. hydrogen/air atmospheres and platinum). Some chemical reactions (e.g. pyrolysis and biological processes) can also lead to the production of flammable substances, which in turn can form an explosive atmosphere with the surrounding air Violent reactions resulting in ignition can occur in some combinations of construction materials with chemicals (e.g. copper with acetylene, heavy metals with hydrogen peroxide). Some combinations of substances, especially when finely dispersed, (e.g. aluminium/rust or sugar/chlorate) react violently when exposed to impact or friction (see 5.3 of the standard). For protective measures against ignition hazards due to chemical reactions, see 6.4.14 of the standard. Hazards can also arise from chemical reactions due to thermal instability, high heat of reaction and/or rapid gas evolution. These hazards are not considered in this standard.	- Inerting - stabilisation - improvement of heat dissipation, e.g. by dividing the substances into smaller portions - limiting temperature and pressure - storage at lowered temperatures - limiting residence times Construction materials which react hazardously with the substances being handled shall be avoided

Table 9.3 Calculation scheme for static electricity

Item	Units	Definition	Description	References
Dielectric constant, ε_r	Unitless		–	–
Permittivity constant, ε_o	s/ohm cm $Coulomb^2/N\ m^2$ Coulomb/V m	8.85×10^{-14} 8.85×10^{-12} 8.85×10^{-12}	–	
Pipe diameter, D_p	m	–	–	
Pipe length, L_p	m	–	–	
Specific conductivity, γ_c	mho/cm	–	–	
Solid or fluid velocity in pipe, u_p	m/s	–	–	
Relaxation time, τ	s	$\tau = \frac{\varepsilon_r \cdot \varepsilon_o}{\gamma_c}$	Time required for a charge to dissipate (ε_r is solid or liquid relative dielectric constant)	
Capacitance, C	farad		Electrical capacitance of generic items (containers, persons, cars)	
Capacitance for a sphere, C_s	farad	$C_s = 4 \cdot \pi \cdot \varepsilon_o \cdot \varepsilon_r$	(ε_r is air relative dielectric constant = 1)	
Voltage, V	volts	$V = R \cdot I_s$	–	
Electric charge, Q	coulomb	$V = R \cdot I_s$	–	
Resistance, R	ohm		–	
Energy, J	joules	$J = \frac{C \cdot V^2}{2}$ $J = \frac{Q \cdot V}{2}$ $J = \frac{Q^2}{2 \cdot C}$	–	
Streaming current, I_s	A	$I_s = k \cdot u_p^2 \cdot D_p^2 \cdot \left[1 - \left(\frac{L_p}{u_p \cdot \tau}\right)\right]$	Current transferred between two surfaces due to the flow of a solid or a fluid in a pipe	(Britton, 1999; CCPS, 1995)
k	$A\ s^2/m^4$	10^{-5}	Streaming current constant	
Specific charge, q	$\mu Coulomb/m^3$	$q = 5 \cdot u_p$	Specific electric charge density for an infinitely long pipe (low conductivity liquids, in particular saturated hydrocarbon liquids)	CLC/TR 50404:2003

If some solid or liquid streams are discharged into a containment, and others are carried away from the containment, the general balance of charge is given by the formula:

$$\frac{dQ(t)}{dt} = \sum I_{S_{in}} - \sum I_{S_{out}} - \frac{Q(t)}{\tau} \tag{9.2}$$

where the first term of the second member of the equation represents the incoming current, the second one the leaving current, and the third one is the charges lost through the relaxation (Tables 9.4–9.8).

Table 9.4 Electrical properties of some liquid materials

Liquid Material	Electrical Conductance, µmhos/cm (Temperature, °C)	Dielectric Constant (Temperature, °C)
Acetaldehyde	$1.7 \cdot 10^{-6}$ (15)	21.8 (10), 21 (18)
Acetone	$2.0 \cdot 10^{-8}$ (18)	21 (20), 20.7 (25), 17.6 (56)
Ammonia	$1.3 \cdot 10^{-7}$(−79)	
Aniline	$2.4 \cdot 10^{-8}$ (25)	7.06 (20), 5.93 (70)
Benzene	$7.6 \cdot 10^{-8}$	2.292 (15), 2.283 (20), 2.274 (25)
Chlorine	$<1.0 \cdot 10^{-16}$ (−70)	
Cyclohexane	$<210^{-14}$	
Ethyl acetate	$<1.0 \cdot 10^{-9}$ (25)	6.081 (20), 5.30 (77)
Ethyl alcohol	$1.35 \cdot 10^{-9}$ (25)	25.3 (20), 20.21 (55)
Ethylamine	$4.0 \cdot 10^{-7}$ (0)	8.7 (0), 6.94 (10)
Ethyl ether	$<4.0 \cdot 10^{-13}$ (25)	
Heptane	$<1.0 \cdot 10^{-13}$	1.921 (20), 1.85 (70)
Hexane	$<1.0 \cdot 10^{-18}$ (18)	1.904 (15), 1.89 (20)
Kerosene	$<1.7 \cdot 10^{-8}$ (25)	
Methyl alcohol	$4.4 \cdot 10^{-7}$ (18)	41.8 (−20), 33 (20)
Methyl ethyl ketone	$1.0 \cdot 10^{-7}$ (25)	
Nonane	$1.7 \cdot 10^{-8}$ (25)	1.972 (20), 1.85 (110)
Pentane	$<2.0 \cdot 10^{-10}$ (19.5)	2.011 (−90), 1.837 (20)
Petroleum	$3.0 \cdot 10^{-13}$	
n-Propyl alcohol	$5.0 \cdot 10^{-8}$ (18), $2.0 \cdot 10^{-8}$ (25)	20.8 (20), 20.33 (25)
iso-Propyl alcohol	$3.5 \cdot 10^{-6}$ (25)	20.18 (20), 18.3 (25), 16.2 (40)
Pyridine	$5.3 \cdot 10^{-8}$ (18)	13.26 (20), 12.3 (25), 9.4 (116)
Sulphur	$1.0 \cdot 10^{-12}$ (115), $5.0 \cdot 10^{-12}$ (130), $1.2 \cdot 10^{-7}$ (440)	
Toluene	$<1.0 \cdot 10^{-14}$	2.385 (20), 2.364 (30)
Xylene	$<1.0 \cdot 10^{-15}$	o 2.562 (20), 2.54 (30) m 2.359 (20). 2.35 (30) P 2.273 (20), 2.222 (50)
Water	$4.0 \cdot 10^{-8}$ (18)	80.4

Lange's handbook of chemistry (Dean, 1999)

Table 9.5 Dielectric constant of air

Gas	Dielectric Constant
Air	1

Table 9.6 Electrical properties of some powders

Powder	Relaxation Time (s)	Electrical Conductance (mhos/cm)
Grain	10^{-4}–10^{-2}	10^{-9}–10^{-11}
Sugar	1	10^{-13}
Milk	1–100	10^{-13}–10^{-15}
Nylon	2–100	10^{-15}
Polythylene	>10,000	$>10^{-17}$

Table 9.7 Electrical capacitances of equipment (*CLC/TR 50404:2003*)

Item	Capacitance (pF)
Small items (scoop, hose, nozzle, can)	10–20
Small containers (bucket, 50 L drum)	10–100
Medium containers (250–500 L)	50–300
Major plant items surrounded by earthed structure	100–1000
Human body	100–300

Table 9.8 Mass charge density for operation with powders (*CLC/TR 50404:2003*)

Operation	Mass Charge Density (μC/kg)
Sieving	10^{-3}–10^{-5}
Purging	10^{-1}–10^{-3}
Scroll feed transfer	1–10^{-2}
Grinding	1–10^{-1}
Micronising	10^{-2}–10^{-1}
Pneumatic conveying	10^{-3}–10^{-1}

Example 9.1

340,000 kg/h of sugar are supplied to a 10 m diameter unearthed spherical tank until the tank, initially empty, has been completely filled. Sugar is fed in a 100 mm diameter pipe. Find the accumulated charge, the stored energy, and evaluate if this is sufficient to ignite the powder.

Solution

ε_r, air	ε_o (C/V m)	ρ (kg/m^3)	D (m)	W (kg/h)	MIE (mJ)	τ (s)
1	8.85×10^{-12}	800	10	340,000	10	1

Volume of sphere:

$$V_s = \frac{4}{3} \cdot \pi \cdot 5^3 = 524 \, \text{m}^3 \tag{9.3}$$

Filling time:

$$t_f = \frac{V_s \cdot \rho_s}{W} = \frac{524 \cdot 800}{340,000} = 1.23 \, \text{h} = 4439 \, \text{s} \tag{9.4}$$

Powder velocity:

$$u_p = \frac{W}{\rho_s \cdot \pi \cdot \frac{D_p^2}{4}} = \frac{10,000}{3600 \cdot 800 \cdot 3.14 \cdot \frac{0.01}{4}} = 14.3 \, \text{m/s} \tag{9.5}$$

Assuming infinite length pipe, streaming current:

$$I_s = k \cdot u_p^2 \cdot D_p^2 = 10^{-5} \cdot 14.3^2 \cdot 0.01 = 2 \cdot 10^{-5}\,\text{A} \tag{9.6}$$

Accumulated charge in the sphere is obtained integrating Eq. (9.2) with $\sum I_{S_{out}} = 0$:

$$Q(t_f) = I_s \cdot \tau \cdot \left(1 - e^{-\frac{t_f}{\tau}}\right) \cong I_s \cdot \tau = 2 \cdot 1 \cdot 10^{-5} = 2 \cdot 10^{-5}\,\text{C} \tag{9.7}$$

The capacitance of the sphere is:

$$C_s = 4 \cdot \pi \cdot \varepsilon_o \cdot \varepsilon_r \cdot r = 4 \cdot \pi \cdot 1 \cdot 8.85 \cdot 10^{-12} \cdot 5 = 5.56 \cdot 10^{-10}\,\text{F} \tag{9.8}$$

The energy stored is:

$$J = \frac{Q^2}{2 \cdot C} = \frac{(8.18 \cdot 10^{-6})^2}{2 \cdot 5.56 \cdot 10^{-10}} \cdot 1000 = 360 > 10\,\text{mJ} \tag{9.9}$$

The energy stored is sufficient to ignite the powder.

Example 9.2

5 m^3/h of acetone are fed by gravity to a reactor by means of a 50 mm diameter 5 m long non-conductive pipe. The pipe is connected to the reactor by means of a flanges couple, the downstream one of which is earthed along with the reactor. A capacitance of 20 μF can be assumed for this equivalent capacitor (Eichel, 1967). Assess the ignition hazard (Fig. 9.3).

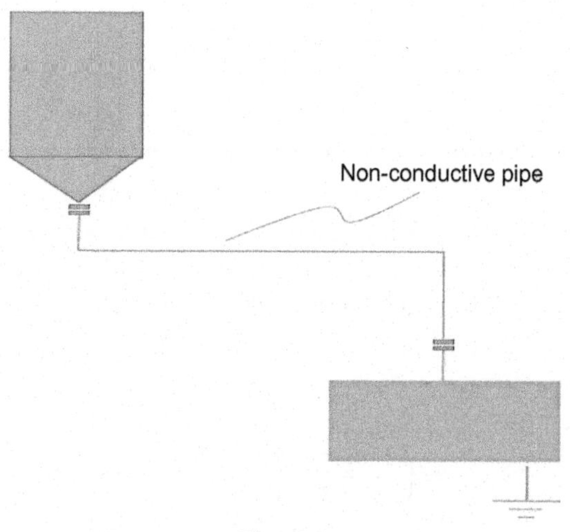

Fig. 9.3
Plant configuration of Example 9.3.

Solution

ε_o (C/V m)	γ_c (μmhos/m)	ε_r	Flow Rate (m^3/h)	MIE (mJ)	L (m)	D_p (m)
8.85×10^{-12}	$2.0\ 10^{-6}$	21	5	0.19	5	0.05

Acetone velocity:

$$u_p = \frac{5}{3600 \cdot 3.14 \cdot \frac{0.025}{4}} = 0.07\,\text{m/s} \tag{9.10}$$

Streaming current:

$$I_s = k \cdot u_p^2 \cdot D_p^2 \cdot \left[1 - \exp\left(\frac{L_p}{u_p \cdot \tau}\right)\right] = 10^{-5} \cdot 0.07^2 \cdot 0.05^2 \cdot (1-0) = 1.225^{-10}\,\text{A} \tag{9.11}$$

Electrical resistance associated to acetone in the pipe:

$$R = \frac{1}{\gamma_c} \cdot L \cdot \frac{4}{\pi \cdot D_p^2} = 5 \cdot 10^5 \cdot 5 \cdot \frac{4}{3.14 \cdot 0.05^2} = 1.27 \cdot 10^9\,\text{ohms} \tag{9.12}$$

Voltage drop down to the pipe:

$$V = R \cdot I_s = 1.27 \cdot 10^9 \cdot 1.25 \cdot 10^{-10} = 0.16\text{V} \tag{9.13}$$

The equivalent capacitor consisting of the two flanges, of which the upstream one is charged and the downstream one earthed, can be associated with the following energy:

$$J = \frac{C \cdot V^2}{2} = \frac{20 \cdot 10^{-12} \cdot 0.16^2}{2} = 2.56 \cdot 10^{-13}\,\text{J} \tag{9.14}$$

The ignition hazard is negligible.

9.5.3 *Adiabatic Compression*

Example 9.3

A stoichiometric mixture of hydrogen and air is compressed from atmospheric pressure and 20°C. Find the minimum compression ratio which can ignite the mixture, assuming that ideal gas law is applicable. Air and hydrogen have the same specific heat, ratio ≅1.4.

Solution

Hydrogen AIT in air is 400°C. Adiabatic compression:

$$\frac{P_1}{P_o} = \left(\frac{273.16 + 400}{293.16}\right)^{\frac{1.4}{0.4}} = 18.35 \tag{9.15}$$

It must be noted that this is an approximate result. Beyond 10 atm, non-ideal behaviour of gases should be accounted for.

9.5.4 Cool Flame

In presence of some contaminants, ethane and higher hydrocarbons can undergo a two-staged ignition, producing the so called cool flame. It is a flame burning at a much lower temperature than the normal one (Stull, 1976).

Partial combustion of ethane cool flame occurs according to the following reaction:

$$C_2H_6 + O_2 \rightarrow CH_3CHO + H_2O \tag{9.16}$$

with a reaction heat of 77,300 kcal/kmol. Acetaldehyde will then be burnt in a further stage:

$$CH_3CHO + 3.5O_2 \rightarrow 2CO_2 + 3H_2O \tag{9.17}$$

with a reaction heat of 263,900 kcal/kmol.

9.6 Ignition Probability

9.6.1 Generic Ignition Probabilities

For screening purposes, or for very simplified and well identified scenarios, the following probabilities can be used in the QRA studies.

Cox et al. (1990)

Cox et al. (1990) provided the following ignition probabilities (Table 9.9).

These values can be used also for offshore releases (Spouge, 1999)

Example 9.4

10,000 Nm^3/h of methane are released to the atmosphere. Find the probability for the gas to be ignited and to explode, according to Cox data.

Solution

The flow rate is given in standard units, so, using the molar volume 22.414 Nm^3/kmol, the released mass flow rate can be calculated:

$$W = \frac{10.000}{22.414 \cdot 3600} \cdot 16 = 2\,kg/s \tag{9.18}$$

So, methane will have a probability of 7% to be ignited and $0.07 \times 0.03 = 0.21\%$ to explode.

Table 9.9 Ignition probabilities (Cox et al., 1990)

Release Rate Category	Release Rate (kg/s)	Gas Leak	Oil Leak	Explosion After Ignition
Minor	<1	0.01	0.01	0.01
Major	1–50	0.07	0.03	0.03
Massive	>50	0.30	0.08	0.08

Table 9.10 Ignition probabilities (Purple book, 2005)

Source		Substance		
Continuous	Instantaneous	K_1-Liquid	Gas Low Reactivity	Gas Medium/Low Reactivity
Minor	<1	0.065	0.02	0.02
Major	1–50	0.065	0.04	0.05
Massive	>50	0.065	0.09	0.07

Uijt de Haag et al. (TNO, Purple book, 2005)

The Purple Book provides the following ignition probabilities, differentiated by liquid and gas reactivity (Table 9.10).

K_1-liquids: flammable liquid having a flash point less than 21°C and a vapour pressure at 50°C less than 1.35 bar (pure substances) or 1.5 bar (mixtures) (Table 9.11).

Substances not included in this table should be assumed as highly reactive.

Probabilities of direct ignition for transport units are given in Table 9.12.

EGIG (European Gas Pipeline Incident Data Group, 2015)

The probabilities of failure of pipelines are given by EGIG (if ignition probabilities apply to pipelines) (Table 9.13):

9.6.2 Ignition Delays Models

Ignition probabilities presented in the previous sections are referred to as overall ignition probabilities, since a flammable cloud may be ignited either at the release section (direct

Table 9.11 Reactivity of substances

Reactivity of Substances (Purple book)		
Low	Medium	High
Ammonia	1-Butene	Acetylene
Bromomethane	1,3-Butadiene	Benzene
Carbon monoxide chloroethane	Acetaldehyde	Carbon disulphide
	Acetonitrile	Ethylene oxide
Methane	Butane	Formaldehyde
	Chloroethene	Hydrogen sulphide
	Dimethylamine	Methyl acrylate
	Ethane	Methyl formate
	Ethene	Naphta, solvent
	Formic acid	Vinyl acetate
	Propane	
	Propene	

Table 9.12 Probability of direct ignition for transport units (TNO)

Source	Probability of Direct Ignition
Road tanker continuous	0.1
Road tanker instantaneous	0.4
Tank wagon continuous	0.1
Tank wagon instantaneous	0.8

Table 9.13 Failure probability of pipelines (EGIG, 2015)

Size of Leak	Ignition Percentage (%)
Pinhole crack	4.40
Hole	2.30
Rupture (all diameters)	13.9
Rupture $\leq$16 in.	10.3
Rupture $\geq$16 in.	32.0

ignition) or after a certain time, depending on several factors (delayed ignition). For example, ignition of a maximum size cloud could be the peak scenario, or some distant ignition sources could be considered more effective. In all cases. It is:

$$P_{ign} = P_{direct} + P_d \tag{9.19}$$

According to the OGP standard 434-6 (2010), an immediate ignition probability of 0.001 might be assumed as being constant and independent of the release rate, as per the look-up UKOOA correlations (see next section). The method assumes that the immediate ignition probability is 0.001 and is independent of the release rate. Delayed ignition probabilities might be obtained by subtracting 0.001 from the total ignition probability, e.g. an ignition probability value of 0.004 obtained from the look-up correlations can be considered as an immediate ignition probability of 0.001 and a delayed ignition probability of 0.003.

Uijt de Haag and Ale (TNO, Purple book, 2005) propose that the probability of delayed ignition caused by an ignition source is given by the following formula:

$$P_d(t) = P_{present} \cdot \left(1 - e^{-\omega \cdot t}\right) \tag{9.20}$$

where:

$P(t)$ is the probability of ignition in the interval $0 \rightarrow t$.
$P_{present}$ is the probability corresponding to the cloud.
ω is the ignition effectiveness (s^{-1}).
t is time (s).

The Purple Book provides the probability of ignition for a time interval of one minute per some sources. From these data, the ignition effectiveness can be calculated (Table 9.14).

Table 9.14 Probability of ignition in 1 min (Purple book, 2005)

Source	Probability of Ignition in 1 min
Point source	
Motor vehicle	0.40
Flare	1.00
Outdoor furnace	0.90
Indoor furnace	0.45
Outdoor boiler	0.45
Indoor boiler	0.23
Ship	0.50
Ship transporting flammable material	0.30
Fishing vessel	0.20
Pleasure craft	0.10
Diesel train	0.40
Electric train	0.80
Line source	
Transmission line	0.20 × 100 m
Road	See note[a]
Railway	See note[a]
Aerial sources	
Chemical plant	0.90 × site
Oil refinery	0.90 × site
Heavy industry	0.70 × site
Light industrial warehousing	As for population
Population sources	
Residential	0.01 per person
Employment force	0.01 per person

[a]Note

$$P(t) = d \cdot (1 - e^{-\omega \cdot t})\_if\_d \leq 1$$

$$P(t) = \left(1 - e^{-d \cdot \omega \cdot t}\right)\_if\_d \geq 1$$

$d = N \cdot E/u_v$.
N is the number of vehicles/h, E is the length of a road or railway section (km).
u_v is the average velocity of a vehicle (km/h).

Note:
The probability of ignition for a grid cell in a residential area in the interval 0 to t, $P(t)$, is given by:

$$P(t) = (1 - e^{-n \cdot \omega \cdot t})$$

where n is the average number of people present in the grid cell.

Example 9.5

Define the formula for the ignition probability of a cloud by an outdoor boiler and calculate the delayed ignition probability after 3 min.

Solution

The boiler will be assumed to be a permanently present ignition source, so we can write (Fig. 9.4):

$$0.45 = P_{present} \cdot \left(1 - e^{-\omega \cdot 60}\right) \tag{9.21}$$

Solving for ω:

$$\omega \cong 0.01 \tag{9.22}$$

$$P_d(t) = \left(1 - e^{-0.01 \cdot t}\right) \tag{9.23}$$

$$P_d(180) = \left(1 - e^{-0.01 \cdot 180}\right) = 0.834 \tag{9.24}$$

Delayed ignition probabilities for offshore QRA studies have been provided by Spouge (1999) and have been included in Table 9.15.

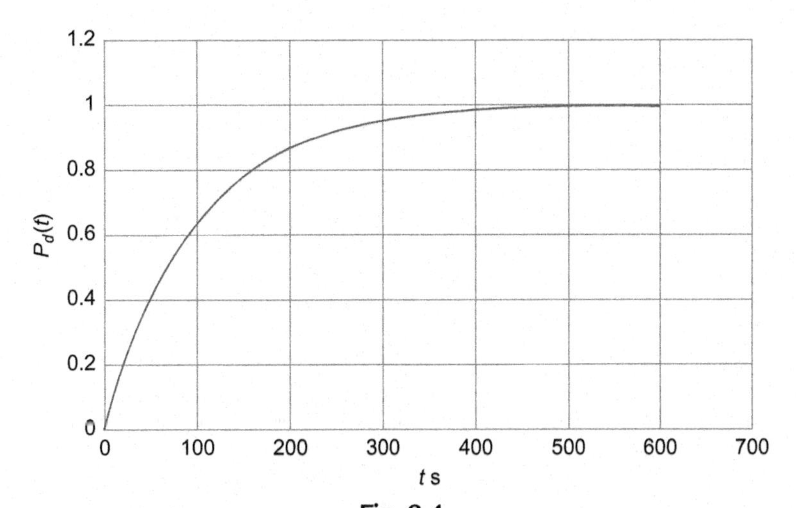

Fig. 9.4
Ignition probability with time for Example 9.5.

Table 9.15 Ignition probability for off-shore plants (Spouge, 1999)

Release	Early Ignition (<5 min)	Delayed Ignition (5–60 min)	Very Delayed Ignition (5–60 min)
Shallow gas blowout	0.07	0.11	0.07
Deep blowout	0.09	0.11	0.16
Deep wheel release	0.03	0.11	0.16

9.6.3 UKOOA Ignition Model

The UKOOA ignition model (Energy Institute, 2006) has deeply improved and made the prediction of ignition probability specific, and is strongly recommended in the execution of a detailed QRA. However, it is dependent on a large number of parameters, so its adoption is not always easy. Several practical approaches have been proposed to simplify the utilisation of the method, keeping its reliability high. Bain et al. (2006) have proposed a method which has been presented in this paragraph. Tabulated couples of values relating mass flows (kg/s) to ignition probabilities (P_{ign}) corresponding to 4–7 points of the relative curves have been listed in Table 9.16. A specific ignition probability can be calculated as follows:

Table 9.16 UKOOA ignition probability model scenarios

Case No	Case Description
1	Pipe liquid industrial
2	Pipe liquid rural
3	Pipe gas LPG industrial
4	Pipe gas LPG rural
5	Small gas plant LPG
6	Small plant liquid
7	Small plant liquid bund
8	Large plant gas LPG
9	Large plant liquid
10	Large plant liquid bund
11	Large plant confined gas LPG
12	Tank liquid 300 × 300 m bund
13	Tank liquid 100 × 100 m bund
14	Tank gas LPG storage plant
15	Tank gas LPG storage industrial
16	Tank gas LPG storage rural
17	Offshore process liquid
18	Offshore process liquid NUI
19	Offshore process gas open deck NUI
20	Offshore process gas typical
21	Offshore process gas large module
22	Offshore process gas–Congested mechanical vented module
23	Offshore riser
24	Offshore FPSO gas
25	Offshore FPSO gas wall
26	Offshore FPSO liquid
27	Offshore engulfed blowout riser
28	Cox, Lees, Ang – Gas
29	Cox, Lees, Ang – Liquid
30	Tank liquid diesel, fuel, oil

$$P_{ign} = e^{\ln P_n + \frac{(\ln Q - \ln Q_n)\cdot(\ln P_{n+1} - \ln P_n)}{\ln Q_{n+1} - \ln Q_n}} \tag{9.25}$$

where:

P_{ign} is the ignition probability corresponding to the release rate Q (unknown intermediate point between the tabulated couples of values).
P_n is the ignition probability corresponding to the release rate Q_n (tabulated).
P_{n+1} is the ignition probability corresponding to the release rate Q_{n+1} (tabulated).

According to this formula, ignition probabilities related to any release rate can be calculated (Table 9.17).

Example 9.6

On the basis of the tabulated values, draw up the diagram related to ignition probability for case No 8 (Large Plant Gas LPG) and find the ignition probability corresponding to $Q = 190$ kg/s.

Solution

Setting the values of case n. 8 the following diagram will be obtained (Fig. 9.5).

190 kg/s falls between 100 and 260 kg/s, with $P_{ign} = 0.25$ and $P_{ign} = 0.65$, respectively. Solving:

$$P_{ign}(190) = e^{\ln 0.250 + \frac{(\ln 190 - \ln 100)\cdot(\ln 0.65 - \ln 0.25)}{\ln 260 - \ln 100}} = 0.475 \tag{9.26}$$

9.7 Fire Scenarios

A mixture of air or oxygen and a flammable substance within the flammability limits, if properly energised, will undergo a combustion reaction, resulting in fire or explosion. The difference between these two phenomena is essentially related to the energy release rate which is related, in turn, to a significant pressure rise, depending on some factors, such as congestion and turbulence. In process safety calculations, fire is generally associated to thermal effects only, whereas explosion is also accompanied by significant mechanical effects. It has to be taken into account that a fire event with a flame speed travelling slower than the corresponding sound speed is always termed as a deflagration.

9.7.1 Flash Fire

Flash fire can be defined as the combustion of a flammable gas or vapour and air mixture in which the flame propagates through that mixture in a manner such that negligible or no damaging overpressure is generated. (CCPS, 2005). Typically, a flash fire occurs if there is a release of a flammable gas or vapour so that a cloud is formed, and a delayed ignition follows. The incident of Finxborough, according to the witnesses, was possibly initiated from the ignition of cyclohexane at the old hydrogen plant. Far from the reactor train and, prior to the

Table 9.17 UKOOA ignition probabilities for scenarios included in Table 9.16

No	Point 1		Point 2		Point 3		Point 4		Point 5		Point 6		Point 7	
	Q (kg/s)	P_{ign}	Q (kg/s)	P_{ign}	Q (kg/s)	P_{ign}	Q (kg/s)	P_{ign}	Q (kg/s)	P_{ign}	Q (kg/s)	P_{ign}	Q (kg/s)	P_{ign}
1	0.01	0.00100	0.03493	0.00100	0.100	0.00180	70.000	0.07000	100,000	0.07000				
2	0.01	0.00100	0.03787	0.00100	0.100	0.00180	0.300	0.00350	70.000	0.07000	70.000	0.00700	100,000	0.00700
3	0.01	0.00100	0.08791	0.00100	0.100	0.00110	1000	1.00000	100,000	1.00000				
4	0.01	0.00100	0.04799	0.00100	0.100	0.00110	10.000	0.00200	1000	0.08000	23408.54	1.00000	100,000	1.00000
5	0.01	0.00100	0.07654	0.00100	0.100	0.00110	1.000	0.00250	3.000	0.01400	498.991	0.60000	100,000	0.60000
6	0.01	0.00100	0.07548	0.00100	0.100	0.00110	1.000	0.00240	100.000	0.10000	100,000	0.100000		
7	0.01	0.00100	0.07548	0.00100	0.100	0.00110	1.000	0.00240	8.053	0.01300	100.000	0.01300	100,000	0.01300
8	0.01	0.00100	0.07654	0.00100	0.100	0.00110	1.000	0.00250	100.000	0.25000	260.000	0.65000	100,000	0.065000
9	0.01	0.00100	0.07654	0.00100	0.100	0.00110	1.000	0.00250	100.000	0.12000	109.990	0.13000	100,000	0.13000
10	0.01	0.00100	0.07548	0.00100	0.100	0.00110	1.000	0.00240	42.492	0.05000	100.000	0.05000	100,000	0.05000
11	0.01	0.00100	0.07654	0.00100	0.100	0.00110	1.000	0.00250	70.000	0.43000	325.028	0.70000	100,000	0.70000
12	0.01	0.00100	0.05250	0.00100	0.100	0.00105	1.000	0.00125	7.000	0.00270	519.617	0.12000	100,000	0.12000
13	0.01	0.00100	0.05250	0.00100	0.100	0.00105	1.000	0.00125	7.000	0.00270	49.035	0.01500	100,000	0.01500
14	0.01	0.00104	0.00160	0.00100	0.100	0.00110	1.000	0.00116	100.000	0.96000	102.838	1.00000	100,000	1.00000
15	0.01	0.00104	0.00160	0.00100	0.100	0.00110	1.000	0.00116	100.000	0.22700	988.106	1.00000	100,000	1.00000
16	0.01	0.00104	0.00160	0.00100	0.100	0.00110	1.000	0.00116	10.000	0.01540	52,551.53	0.50000	100,000	0.50000
17	0.01	0.00100	0.07882	0.00100	0.100	0.00110	100.000	0.01750	100,000	0.01750				
18	0.01	0.00100	0.07882	0.00100	0.100	0.00110	24.731	0.01000	100.000	0.01000	100,000	0.01000		
19	0.01	0.00101	0.00803	0.00100	0.100	0.00110	1.000	0.00120	30.000	0.02400	31.423	0.02500	100,000	0.02500
20	0.01	0.00100	0.08833	0.00100	0.100	0.00110	3.000	0.01500	10.000	0.02400	37.008	0.04000	100,000	0.04000
21	0.01	0.00100	0.08933	0.00100	0.100	0.00110	5.000	0.03000	30.000	0.05000	100,000	0.05000		
22	0.01	0.00100	0.09194	0.00100	0.100	0.00110	1.000	0.01500	50.000	0.03500	92.624	0.04000	100,000	0.04000
23	0.01	0.00100	0.08340	0.00100	0.100	0.00110	30.000	0.0200	38.267	0.02500	100,000	0.02500		
24	0.01	0.00100	0.08628	0.00100	0.100	0.00110	1.000	0.00130	50.000	0.15000	100,000	0.15000		
25	0.01	0.00100	0.08393	0.00100	0.100	0.00110	0.300	0.00200	10.000	0.15000	100,000	0.15000		
26	0.01	0.00100	0.08160	0.00100	0.100	0.00110	100.000	0.02800	100,000	0.02800				
27	0.01	0.00100	0.08642	0.00100	0.100	0.00110	100.000	0.10000	100,000	0.10000				
28	0.01	0.00081	0.50000	0.01000	100.0000	0.30000	100,000	0.30000						
29	0.01	0.00215	0.50000	0.01000	100.0000	0.08000	100,000	0.08000						
30	0.01	0.00100	0.10000	0.00100	1.00	0.00103	7.000	0.0117	25.551	0.0240	100,000	0.00240		

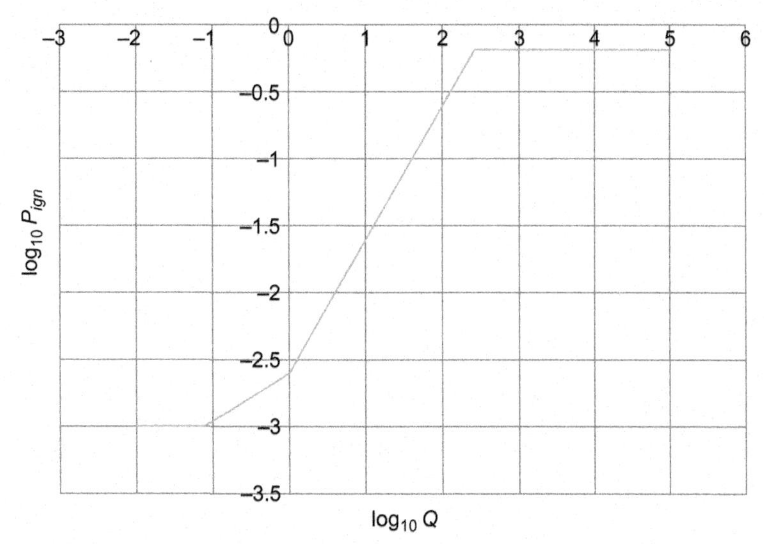

Fig. 9.5
Ignition probability of Example 9.6.

explosion, a flash fire was observed (Mannan, 2012). Two phase or flashing liquid can also produce vapour, which evaporates and is ignited. Effects of a flash fire consist of potential damage to people and to structures, due the high temperature and to oxygen depletion. No mechanical impact is associated.

During a flash fire, a turbulent flame spreads through the cloud, propagating both in downwind and upwind. The behaviour of a well-established flash fire flame is both as a diffusion and a pre-mixed flame.

Modelling

Raj and Emmons (1975) presented a computational model, consisting of turbulent flame propagating at constant speed. For screening purposes, the Hajek and Ludwig formula can be used, for more accurate calculations, the following method proposed by Eisenberg et al. (1975) is recommended for its completeness and relatively simple approach.

Hajek and Ludwig method (1960)

This method is recommended by the standard API 521 (2014) to determine the flame radiation to a point of interest. It is a point-source model valid for laminar and turbulent jet flames assuming the flame to be located in the flame centre.

$$Q_H = \frac{4 \cdot \pi \cdot E \cdot x}{\tau \cdot X_R} \tag{9.27}$$

where:

Q_H is heat release, kW.
E is the radiant heat intensity, kW/m^2.
x is the minimum distance from the epicentre of the flame to the object being considered.
X_R is the fraction of heat radiated.
τ is the heat transmissivity.

Eisenberg method (1975)

(1) Input data

LFL	UFL	Adiabatic Flame Temperature (T_{ad})	Ambient Temperature (T_a)	Density of Hot Gas Layer (Air at 1500°C) (ρ_{air})	Dispersion Coefficient (m)	Released Mass (kg)	Stefan–Boltzmann Constant
kg/m^3	kg/m^3	K	K	kg/m^3	σ_x, σ_y, σ_z	M	$5.6703 \cdot 10^{-8}$ (W/m^{2}K^4)

(2) Volume V_r and area A_r of radiation

$$V_r = \frac{2 \cdot \pi}{3} \cdot \sigma_x \cdot \sigma_y \cdot \sigma_z \cdot \left(R_L{}^3 - R_U{}^3\right) \tag{9.28}$$

$$A_r = \frac{2 \cdot \pi}{3} \cdot \left(\sigma_x^2 + \sigma_y^2 + \sigma_z^2\right) \cdot \left(R_L^2 + R_U^2\right) \tag{9.29}$$

and the unitless functions of LFL and UEL:

$$R_L = \sqrt{2 \cdot \ln\left[\frac{M}{(2 \cdot \pi)^{1.5} \cdot \sigma_x \cdot \sigma_y \cdot \sigma_z \cdot \mathrm{LFL}}\right]} \tag{9.30}$$

$$R_U = \sqrt{2 \cdot \ln\left[\frac{M}{(2 \cdot \pi)^{1.5} \cdot \sigma_x \cdot \sigma_y \cdot \sigma_z \cdot \mathrm{UFL}}\right]} \tag{9.31}$$

It can be recognised that V_r represents a half ellipsoid. In these formulas the coefficients of dispersion are those for puff release.

(3) Effective duration of the flash fire

The effective duration of the flash fire is given by the following formula:

$$t_{FF} = \frac{3}{2 \cdot \alpha \cdot T_a^3} \cdot \left[\tan^{-1}\left(\frac{\beta+1}{2}\right) - \tan^{-1}\beta - \frac{1}{2} \cdot \ln\left(\frac{\beta+1}{\beta+3}\right) \right] \tag{9.32}$$

with:

$$\alpha = \frac{A_r \cdot \sigma}{\rho_{air} \cdot V_r} \tag{9.33}$$

$$\beta = \frac{T_{ad}}{T_a} \tag{9.34}$$

(4) Radiation to a target

$$Q_H = \varepsilon \cdot \sigma \cdot \left(\frac{T_{ad} + T_a}{2}\right)^4 \cdot F \cdot \tau \tag{9.35}$$

with:

ε is the emissivity, assumed equal to 1.
σ is the Stefan–Boltzmann constant.
τ is the atmospheric transmissivity.

$$\tau = P_o \cdot (P_w \cdot x)^{-0.09} \tag{9.36}$$

P_o is 2.02 $(N/m^2)^{0.09} \cdot m^{0.09}$.
P_w is partial water vapour pressure of water in air at a relative humidity RH, in N/m^2:

$$P_w = 1013.25 \cdot RH \cdot \exp\left(14.4114 - \frac{5238}{T_a}\right) \tag{9.37}$$

F is the view factor between the flame and the target.

The selection of the view factor can be reasonably made on the basis of two very simple assumptions:

(1) Flame shape is a half ellipsoid with (Fig. 9.6):

- A semi axis $\sigma_x \cdot \sqrt[3]{(R_L^3 - R_U^3)}$
- An axis $\sigma_y \cdot \sqrt[3]{(R_L^3 - R_U^3)}$
- An axis $\sigma_z \cdot \sqrt[3]{(R_L^3 - R_U^3)}$

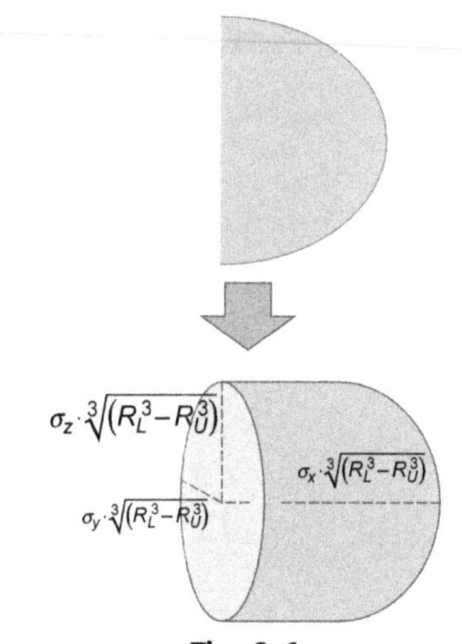

Fig. 9.6
Flame shape for Eisenberg method.

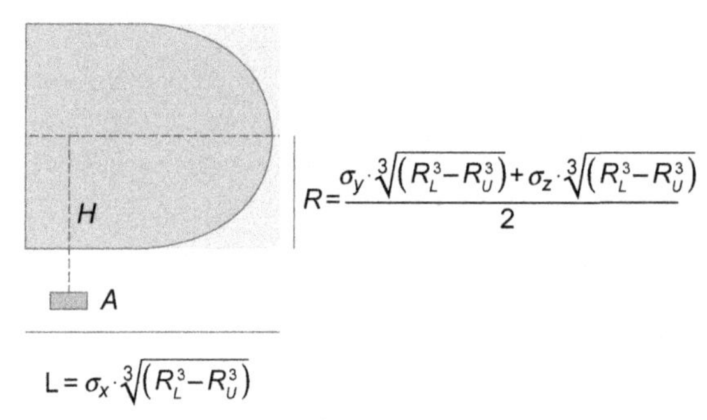

Fig. 9.7
Flame shape for Eisenberg method—Case 1.

(2) Two cases are possible:
   (a) The target A is located somewhere along a segment parallel to the semi-axis at a distance H from the semi-axis: in this case the radiator will be assumed as a cylinder of radius R and height L (Fig. 9.7).
   The view factor can be found in the next diagram (Treccani) (Fig. 9.8).
   (b) The target A is located somewhere on a radial line on the concave side of the of semi-ellipsoid (Fig. 9.9).
   The view factor can be calculated according to the very simple formula given by Siegel and Howell (1981), assuming a spherical radiator:

$$F=\left[\frac{R}{R+H}\right]^2 \tag{9.38}$$

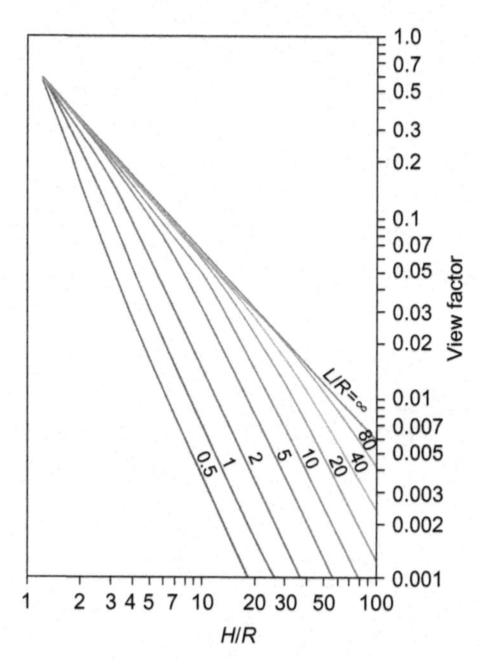

Fig. 9.8
View factor for Eisenberg method—Case 1 (Treccani).

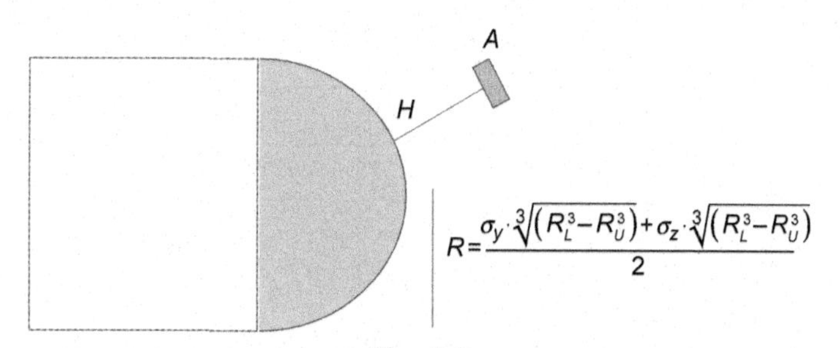

Fig. 9.9
Flame shape for Eisenberg method—Case 2

Example 9.7

10,000 kg of methane are released to the atmosphere, forming a horizontal semi-elliptical cloud, which is ignited 100 m from the source. Find the overall heat radiated to a target located on the approximately semi-spherical side of the cloud at 50 m from the surface. The ambient temperature is 20°C, humidity is 50%, and wind speed is 5 m/s.

Solution

LFL (kg/m^3)	UFL (kg/m^3)	T_a (K)	Density of Hot Gas Layer (Air at 1500°C), ρ_{air}	Released Mass (kg)	Stefan–Boltzmann Constant (W/m^2K^4)	Adiabatic Flame Temperature, T_{ad} (°C)
0.033	0.099	293.16	kg/m^3	10,000	$5.6703 \cdot 10^{-8}$	1953

Weather category $D/5$ is adopted (Table 9.18).

$\sigma_x = \sigma_y = 4.15$ m, $\sigma_z = 3.77$ m

$$R_L = \sqrt{2 \cdot \ln\left[\frac{M}{(2\cdot\pi)^{1.5}\cdot\sigma_x\cdot\sigma_y\cdot\sigma_z\cdot \text{LFL}}\right]} = \sqrt{2\cdot\ln\left[\frac{10{,}000}{(2\cdot\pi)^{1.5}\cdot 4.15^2\cdot 3.77\cdot 0.033}\right]} = 3.37 \quad (9.39)$$

$$R_U = \sqrt{2 \cdot \ln\left[\frac{M}{(2\cdot\pi)^{1.5}\cdot\sigma_x\cdot\sigma_y\cdot\sigma_z\cdot \text{UFL}}\right]} = \sqrt{2\cdot\ln\left[\frac{10{,}000}{(2\cdot\pi)^{1.5}\cdot 4.15^2\cdot 3.77\cdot 0.1}\right]} = 3.03 \quad (9.40)$$

$$V_r = \frac{2\cdot\pi}{3}\cdot\sigma_x\cdot\sigma_y\cdot\sigma_z\cdot\left(R_L^3 - R_U^3\right) = \frac{2\cdot\pi}{3}\cdot 4.15^2\cdot 3.77\cdot\left(3.37^3 - 3.03^3\right) = 1421\,\text{m}^3 \quad (9.41)$$

$$A_r = \frac{2\cdot\pi}{3}\cdot\left(\sigma_x^2+\sigma_y^2+\sigma_z^2\right)\cdot\left(R_L^2+R_U^2\right) = \frac{2\cdot\pi}{3}\cdot\left(4.15^2+4.15^2+3.77^2\right)\cdot\left(3.37^2+3.03^2\right) = 2092\,\text{m}^2 \quad (9.42)$$

Radiation to target:

$$R = \frac{\sigma_y\cdot\sqrt[3]{\left(R_L^3-R_U^3\right)}+\sigma_z\cdot\sqrt[3]{\left(R_L^3-R_U^3\right)}}{2} = \frac{4.15\cdot\sqrt[3]{\left(3.37^3-3.03^3\right)}+3.77\cdot\sqrt[3]{\left(3.37^3-3.03^3\right)}}{2} = 8.66\,\text{m} \quad (9.43)$$

$$F = \left[\frac{R}{R+H}\right]^2 = \frac{8.66}{58.66} = 0.02 \quad (9.44)$$

$$P_w = 1013.25\cdot RH\cdot\exp\left(14.4114 - \frac{5238}{T_a}\right) = 1013.25\cdot 50\cdot\exp\left(14.4114-\frac{5238}{293.26}\right) = 1599\,\text{Pa} \quad (9.45)$$

$$\tau = P_o\cdot(P_w\cdot x)^{-0.09} = 2.02\cdot(1{,}599\cdot 50)^{-0.09} = 0.73 \quad (9.46)$$

$$Q_H = \varepsilon\cdot\sigma\cdot\left(\frac{T_{ad}+T_a}{2}\right)^4\cdot F\cdot\tau = 1\cdot 5.6703\cdot 10^{-8}\cdot\left(\frac{2226.16+293.16}{2}\right)^4\cdot 0.02\cdot 0.73 = 2084\ \text{W/m}^2 \quad (9.47)$$

Table 9.18 Dispersion coefficients for Example 9.7

P–G Class	σ_x or σ_y	σ_z
D	$0.06\cdot x^{0.92}$	$0.15\cdot x^{0.70}$

A general evaluation of the fire duration can be carried out adopting the laminar flame speed, which is realistic because no particular pressure increase is associated to the fire. Assuming for methane 0.34 m/s (Bradley, 1969) and calculating the time required for the flame to cross the semi axis:

$$\frac{\sigma_x \cdot \sqrt[3]{(R_L^3 - R_U^3)}}{u_b} = \frac{9.07}{0.34} = 26.7\,\text{s} \tag{9.48}$$

9.7.2 Pool Fires With Round or Equivalent Basis

Pool fires can occur with combustible liquid, so with crude oil, oil fractions, LPG, and LNG. Methanol presents a particular fire behaviour. In this section, pool fires on land will be analysed. Notably, pool fires with circular or equivalent basis will be analysed. As a rule of thumb, the sentence *equivalent basis* refers to as rectangular-shaped basis with aspect ratio not very different from 1, approximately up to 2.5, according to Moorhouse (1982). Trench fires, with an aspect ratio up to 30, have been presented in Section 9.7.3.

Modelling

Input data

Pool fire main parameters

- burning rate
- pool size
- flame geometry, including height, tilt, and drag
- flame surface emitted power
- geometric view factor with respect to the receiving source
- atmospheric transmissivity
- soot and smoke
- received thermal flux

Burning rate and burning velocity

The specific burning rate can be calculated according to the two following formulas (Burgess and Zabetakis, 1961):

$$W_b = W_\infty \cdot [1 - \exp(-k \cdot \beta \cdot D)] \tag{9.49}$$

with:

- W_b = burning rate, kg/m^2 s
- W_∞ = burning rate at $D \rightarrow \infty$
- k = flame absorption extinction coefficient, m^{-1}
- β = mean beam length correction

The parameters are included in Table 9.19 (Babrauskas, 1983)

For those compounds which are not included in this table, the following general equation can be used: (Burgess and Hertzberg, 1974):

$$W_b = C \cdot \frac{\Delta H_c}{\Delta H_v + c_p \cdot (T_b - T_a)} \tag{9.50}$$

where:

$C = 0.001$ kg/m^2 s
ΔH_c = heat of combustion at its normal boiling point, J/kg

Table 9.19 Parameters for burning rate (Babrauskas, 1983)

Material	W_∞ (kg/m^2 s)	$k \times \beta$ (m^{-1})	k (m^{-1})	T_{ref} (K)
		Cryogenics		
Liquid hydrogen	0.169	6.1	–	1600
LNG (mostly CH_4)	0.078	1.1	0.5	1500
LPG (mostly C_3H_8)	0.099	1.4	0.4	–
		Alcohols		
Methanol	0.015	Independent of diameter	–	1300
Ethanol	0.015	Independent of diameter	0.4	1490
		Pure organic fuels		
Butane	0.078	2.7	–	–
Benzene	0.085	2.7	4.0	1460
Hexane	0.074	1.9	–	1300
Heptane	0.101	1.1	–	–
Xylene	0.09	1.4	–	–
Acetone	0.041	1.9	0.9	–
Dioxane	0.018	5.4	–	–
Dyethylether	0.085	0.7	–	–
		Petroleum products		
Benzine	0.048	3.8	–	–
Gasoline	0.055	2.1	2.0	1450
Kerosene	0.039	3.5	2.6	1480
JP-4	0.051	3.6	–	1120
JP-5	0.054	1.6	0.5	1250
Transformer oil (HC)	0.039	0.7	–	1500
Fuel oil (heavy)	0.035	1.7	–	–
Crude oil	0.022 (rel. density = 0.83) 0.045 (rel. density = 0.88)	2.8	–	–
		Solids		
Polymethylmetacrylate	0.02	3.3	–	1260

ΔH_v = heat of vaporisation at its normal boiling point, J/kg
c_p = specific heat of liquid, J/kg K
T_b, normal boiling temperature, K
T_a, ambient temperature, K

Effect of wind on burning rate Babrauskas (1983) has shown that the effect of wind on burning rate is significant, due to the diffusive rate limiting step of the combustion reaction. Notably, he has proposed the following expression for windy condition:

$$W_{bw} = W_b \times \left(1 + 0.15 \cdot \frac{u_w}{D_p}\right) \tag{9.51}$$

where u_w is wind velocity at 10 m above grade, and D_p is pool diameter.

The burning velocity u_b, m/s, is defined as:

$$u_b = \frac{W_b}{\rho_L} \tag{9.52}$$

where ρ_L is the liquid density, kg/m^3.

Example 9.8

Heptane spilled into a 5 m diameter circular basin is ignited. The ambient temperature is 20°C. Find the mass burning rate, the burning velocity, and analyse the effect of wind.

Solution

T_a (K)	T_b (K)	ΔH_c (kJ/kg K)	ΔH_v (kJ/kg K)	c_p (J/kg K)	W_∞ (kg/m s^2)	D (m)	$k \times \beta$ (m^{-1})	ρ_L (kg/m^3)
293.16	371.53	44,558	318	2.24	0.101	5	1.1	680

Burgess and Zabetakis formula

$$W_b = W_\infty \cdot [1 - \exp(-k \cdot \beta \cdot D)] = 0.101 \cdot [1 - \exp(-1.1 \cdot 5)] = 0.1006\,\text{kg/m}^2\text{s} \tag{9.53}$$

Burgess and Hertzberg formula

$$W_b = 0.001 \cdot \frac{44,558}{318 + 2.24 \cdot (371.53 - 293.16)} = 0.090\,\text{kg/m}^2\text{s} \tag{9.54}$$

It is worth noting that the Burgess and Zabetakis formula is sensitive for large diameter.

The burning velocity u_b is:

$$u_b = \frac{W_b}{\rho_L} = \frac{0.1006}{680} \times 1000 = 0.15\,\text{mm/s} \tag{9.55}$$

The wind effect is shown in Fig. 9.10. It has to be considered that this approach is valid as long as oxygen transfer is the governing mechanism.

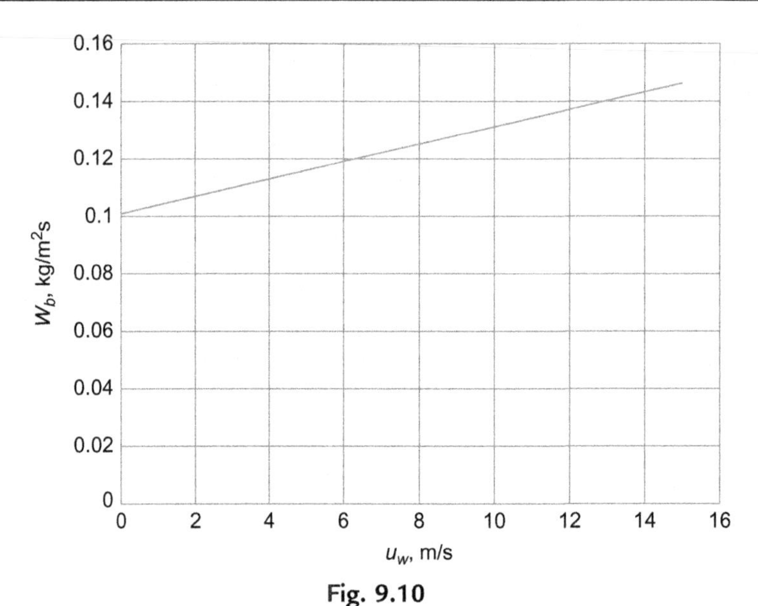

Fig. 9.10
Effect of wind speed on burning rate for Example 9.8.

Instantaneous and continuous spill

This differentiation is made by Mudan (1984) on the basis of the following parameter:

$$\tau_{cr} = \frac{t_s \cdot u_b}{\sqrt[3]{V_L}} = 0.002 \tag{9.56}$$

where:

τ_{cr} is a non-dimensional time.
t_s is the spill duration, s.
V_L is the volume of spilled liquid, m^3.
u_b is the burning velocity, m/s.

$$u_b = \frac{W_b}{\rho_L} \tag{9.57}$$

ρ_L is the liquid density, kg/m^3.

Spills with non-dimensional times less than 0.002 are classified as continuous.

Transient pool formation

In addition to the Wu and Schroy formula presented in Chapter 7, the following specific expressions can be used for continuous and instantaneous spill on land and on water (Shaw and Briscoe, 1978):

Instantaneous spill on land:

$$\frac{d}{2} = \left[\sqrt{\frac{8 \cdot g \cdot V_o}{\pi}} \cdot t + \frac{{d_o}^2}{4} \right] \tag{9.58}$$

Continuous spill on land:

$$\frac{d}{2} = \frac{\sqrt{2}}{3} \cdot \sqrt[4]{\frac{8 \cdot g \cdot Q_b}{\pi}} \cdot t^{3/4} \tag{9.59}$$

Instantaneous spill on water:

$$\frac{d}{2} = \left[\left(\frac{8 \cdot g \cdot (\rho_w - \rho_L) \cdot V_o}{\pi \cdot \rho_w} \right)^{1/2} \cdot t + \frac{{d_o}^2}{4} \right]^{1/2} \tag{9.60}$$

Continuous spill on water:

$$\frac{d}{2} = \frac{\sqrt{2}}{3} \cdot \sqrt[4]{\frac{8 \cdot g \cdot (\rho_w - \rho_L) \cdot Q_b}{\pi \cdot \rho_w}} \cdot t^{3/4} \tag{9.61}$$

with:

d, pool radius, m
g, acceleration of gravity, m/s^2
V_o, volume of instantaneous spill, m^3
t, time after initial spill, s
d_o, initial diameter of contained liquid, m
Q_b, continuous liquid spill rate, m^3/s
ρ_w, water density, kg/m^3
ρ_L, liquid density, lower than water density, kg/m^3

According to U.S. EPA (1996), the worst case assumption for the height of an unconfined pool is 1 cm.

Transient unconfined pool fire diameter

A situation can occur where spillage and burning compete. This happens when combustible liquid catches on fire while it is continuously released on an unconfined domain. Two different rates can be identified, notably the release rate and the burning rate. A simple balance can be written to find the steady state diameter of the pool, which will exist when the burning rate is equal to the spill rate (Mudan, 1984):

$$D_s = 2 \cdot \sqrt{\frac{Q_L}{\pi \cdot u_b}} \tag{9.62}$$

where:

D_s is the steady state diameter, m.
Q_L is liquid spill rate, m^3/s.
u_b is the burning velocity, m/s.

and time to reach equilibrium, s:

$$t_{eq} = 0.564 \cdot \frac{D_s}{\sqrt[3]{\pi \cdot u_b \cdot D_s}} \tag{9.63}$$

For an instantaneous release, the spreading diameter with the time t on a smooth horizontal surface is given by the expression (Mudan, 1984):

$$D_p(t) = D_m \cdot \sqrt{\frac{\sqrt{3}}{2} \cdot \left(\frac{t}{t_m}\right) \cdot \left[1 + \left(\frac{2}{\sqrt{3}} - 1\right) \cdot \left(\frac{t}{t_m}\right)^2\right]} \tag{9.64}$$

where D_m is the maximum diameter, m:

$$D_m = 2 \cdot \sqrt[8]{\frac{{V_L}^3 \cdot g}{u_b}} \tag{9.65}$$

and t_m is time to reach the maximum diameter:

$$t_m = 0.6743 \cdot \sqrt[4]{\frac{V_L}{g \cdot {u_b}^2}} \tag{9.66}$$

$D_p(t)$, pool diameter at time t, m
D_m maximum diameter, m
t_m, time to reach the maximum diameter, s
V_L, volume of liquid spilled, m^3
u_b, the kinematic burning rate, m/s

Mudan recommends the following formula for the average pool diameter of an instantaneous liquid release:

$$D_{av} = 0.683 \cdot D_m \tag{9.67}$$

Unconfined pool fire of a liquid flowing from a storage tank

In accordance with the Stevin–Torricelli law, the volumetric flow rate is given by the following expression:

$$Q_L(t) = \frac{\pi \cdot d^2}{4} \cdot C_D \cdot \sqrt{2 \cdot g \cdot h(t)} \tag{9.68}$$

The emptying time has been calculated in Chapter 7:

$$h(t) = h_o \cdot \left[1 - \sqrt{\frac{g}{2 \cdot h_o}} \cdot C_D \cdot \left(\frac{d}{D_t}\right)^2 \cdot t\right]^2 \tag{9.69}$$

where h_o is the initial level.

Example 9.9

Heptane is released on land from a 10 m diameter nitrogen blanketed atmospheric tank through a 100 mm diameter hole located 12 m below the initial liquid level. Assuming atmospheric pressure in the nitrogen head space, analyse the pool formation if this is ignited (Fig. 9.11).

Solution

W_b (kg/m s^2)	h_o (m)	d_e (m)	C_D	D_t (m)	ρ_L (kg/m^3)
0.101	5	0.1	0.61	10	680

In accordance with Stevin–Torricelli law, the volumetric flow rate is given by the following expression:

$$Q_L(t) = \frac{\pi \cdot d^2}{4} \cdot C_D \cdot \sqrt{2 \cdot g \cdot h(t)} \tag{9.70}$$

According to the EPA (1996), the worst case assumption for the height of an unconfined pool is 1 cm. A transient mass balance is:

$$0.01 \cdot \rho_L \cdot \frac{\pi}{4} \cdot \frac{dD_p(t)^2}{dt} = \frac{\pi \cdot d^2}{4} \cdot C_D \cdot \sqrt{2 \cdot g \cdot h_o} \cdot \left(1 - \sqrt{\frac{g}{2 \cdot h_o}} \cdot C_D \cdot \left(\frac{d}{D_t}\right)^2 \cdot t\right) - W_b \cdot \frac{\pi \cdot D_p(t)^2}{4} \tag{9.71}$$

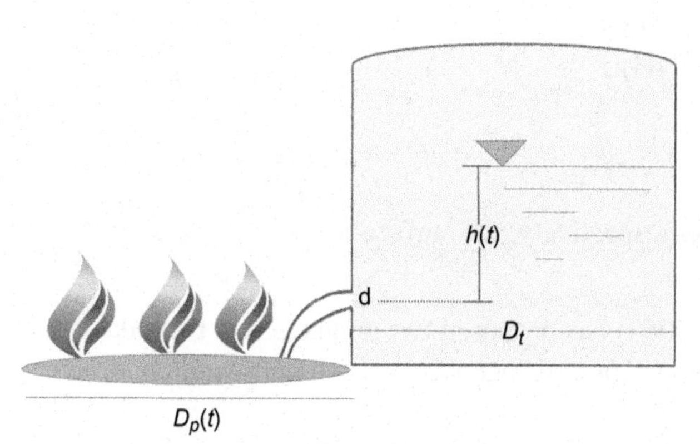

Fig. 9.11
Heptane storage tank of Example 9.9.

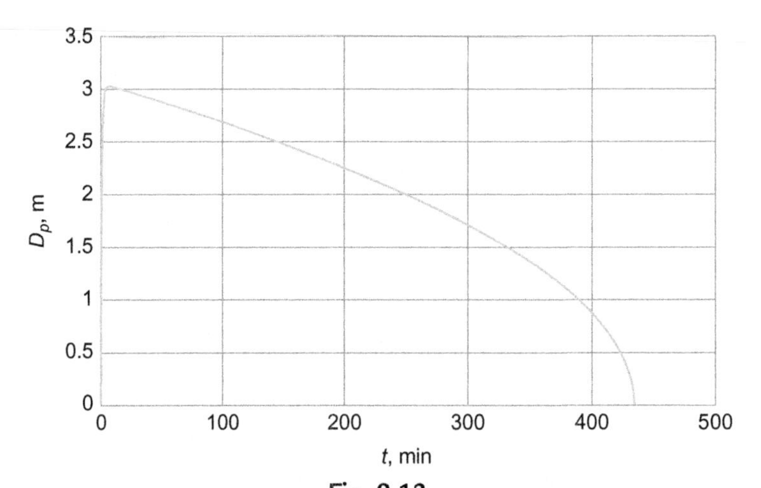

Fig. 9.12
Evolution of pool fire diameter of Example 9.9.

Including the numerical values:

$$5.34 \cdot \frac{dD_p(t)^2}{dt} = 0.74 \cdot \left(1 - 3.9 \times 10^{-5} \times t\right) - 0.079 \cdot D_p(t)^2 \tag{9.72}$$

This is a form of D'Alembert equation, the integral of which is:

$$D_p(t) = \sqrt{9.39 \cdot [1 - \exp(-0.0147 \times t)] - 3.6 \times 10^{-4} \cdot t} \tag{9.73}$$

In the following diagrams the behaviour of D_p against the time has been depicted (Fig. 9.12).

Pool fire on land: the flame

Flame can be modelled as a cylinder which may be tilted if wind is present. In addition, a significant flame drag is expected, resulting in the extension of the flame base downwind of the pool edge. The flame diameter of a confined pool can be calculated as (Uijt de Haag and Ale, Yellow Book, 2005):

$$D_{eq} = 4 \cdot \frac{A_p}{S_p} \tag{9.74}$$

where A_p and S_p are, respectively, the surface area of the pool (m^2) and the perimeter (m).

Flame size Thomas (1963) has proposed two fundamental formulas for the calculation of the flame height to diameter ratio respectively in still and wind conditions:

$$\frac{L_f}{D_p} = 42 \cdot \left(\frac{W_b}{\rho_{air} \cdot \sqrt{g \cdot D_p}}\right)^{0.61} \tag{9.75}$$

where:

D_p, pool diameter, m
ρ_{air}, air density, kg/m^3
W_b, mass burning rate, kg/m^2 s
L_f, flame height, m

and:

$$\frac{L_f}{D_p} = 55 \cdot \left(\frac{W_{bw}}{\rho_{air} \cdot \sqrt{g \cdot D_p}} \right)^{0.67} \times (u^o) \qquad (9.76)$$

with:

D_p, pool diameter, m
ρ_{air}, air density, kg/m^3
W_{bw}, mass burning rate in windy condition, kg/m^2 s
L_f, flame height, m
u^o, unitless velocity

$$u^o = \frac{u_w}{u_c} = \frac{u_w}{\sqrt[3]{g \cdot W_{bw} \cdot \frac{D_p}{\rho_v}}} \qquad (9.77)$$

where u_c is called critical velocity (m/s), and ρ_v is the vapour density at the normal boiling temperature (kg/m^3).

Example 9.10

A 50%-mass mixture of heptane and octane is released and catches on fire in a 10 m diameter circular basin. The ambient temperature is 20°C. Find the flame height if wind speed is 3 m/s.

Solution

T_a (K)	T_b (K)	ΔH_c (kJ/kg K)	ΔH_v (kJ/kg K)	c_p (J/kg K)	W_∞ (kg/m s^2)	D (m)	$k \times \beta$ (m^{-1})	ρ_L (kg/m^3)	ρ_v (kg/m^3)
Heptane									
293.16	371.53	44,558	318	2.24	0.101	10	1.1	680	3.2
Octane									
293.16	398.75	44,510	298	2.15	Unknown	10	Unknown	700	3.48

Burgess and Zabetakis formula
This formula can be used for heptane only:

$$W_b = W_\infty \cdot [1 - \exp(-k \cdot \beta \cdot D)] = 0.101 \cdot [1 - \exp(-1.1 \cdot 10)] \cong 0.101\,\text{kg/m}^2\text{s} \qquad (9.78)$$

Burgess and Hertzberg formula
For n-octane:

$$W_b = 0.001 \cdot \frac{44,510}{298 + 2.15 \cdot (398.75 - 293.16)} = 0.085\,\text{kg/m}^2\text{s} \qquad (9.79)$$

Being octane and heptane mass percentages the same:

$$W_{b7-8} = \frac{0.101 + 0.85}{2} = 0.0925\,\text{kg/m}^2\text{s} \tag{9.80}$$

$$\rho_{L7-8} = \frac{680 + 700}{2} = 690\,\text{kg/m}^3 \tag{9.81}$$

$$\rho_{v7-8} = \frac{3.2 + 3.48}{2} = 3.34\,\text{kg/m}^3 \tag{9.82}$$

And, due to the effect of wind:

$$W_{bw7-8} = W_{b7-8} \times \left(1 + 0.15 \cdot \frac{3}{10}\right) = 0.097\,\text{kg/m}^2\text{s} \tag{9.83}$$

The burning velocity u_b is:

$$u_b = \frac{0.0925}{690} \times 1000 = 0.000134\,\text{m/s} \tag{9.84}$$

$$u^o = \frac{u_w}{\sqrt[3]{g \cdot W_{bw7-8} \cdot \dfrac{D_p}{\rho_{v7-8}}}} = \frac{3}{\sqrt[3]{9.81 \cdot 0.0925 \cdot \dfrac{10}{3.34}}} = 1.395 \tag{9.85}$$

$$L_f = 55 \cdot D_p \cdot \left(\frac{W_{bw7-8}}{\rho_{air} \cdot \sqrt{g \cdot D_p}}\right)^{0.67} \times (u^o) = 55 \cdot 10 \cdot \left(\frac{0.097}{1.2 \cdot \sqrt{9.81 \cdot 10}}\right)^{0.67} \times 1.395 = 30.7\text{ m} \tag{9.86}$$

Height of a highly emissive clear flame

This height can be calculated according to Rew's formula:

$$L_f = \min \left[11.404 \cdot D_p \cdot \left(\frac{W_b}{\rho_{air} \cdot \sqrt{g \cdot D_p}}\right)^{1.13} \cdot \left(\frac{u_w}{\dfrac{g \cdot W_b \cdot D_p}{\rho_{air}}}\right)^{0.179} \cdot \left(\frac{H}{C}\right)^{2.49} \right] \tag{9.87}$$

Flame tilt

According to Mudan and Croce (1986), flame tilt angle θ due to wind is defined by (Fig. 9.13):

$$\cos\theta = \begin{cases} 1\_\text{for}\_u^o \leq 1 \\ \dfrac{1}{\sqrt{u^o}}\ \text{for}\_u^o \geq 1 \end{cases} \tag{9.88}$$

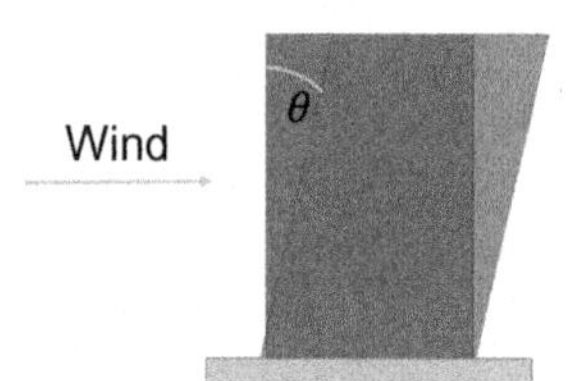

Fig. 9.13
Flame tilt angle.

Example 9.11

For the previous example, find the tilt angle θ.

Solution

$$u^o = 1.395 > 1 \rightarrow \cos\theta = \frac{1}{\sqrt{u^o}} = \frac{1}{\sqrt{1.395}} = 0.847 \tag{9.89}$$

$$\theta = 32° \tag{9.90}$$

Flame elongation (drag)
Wind will cause an elongation of the flame basis shape which, if round-shaped with no wind, will become elliptical. For a cylindrical flame the Moorhouse correlation is used:

$$D_d = 1.5 \cdot Fr_{nd}^{0.069} = 1.5 \cdot D_o \cdot \left(\frac{u_w^2}{g \cdot D_o}\right)^{0.069} \tag{9.91}$$

where Fr_{nd} is the non-densimetric Froude number.

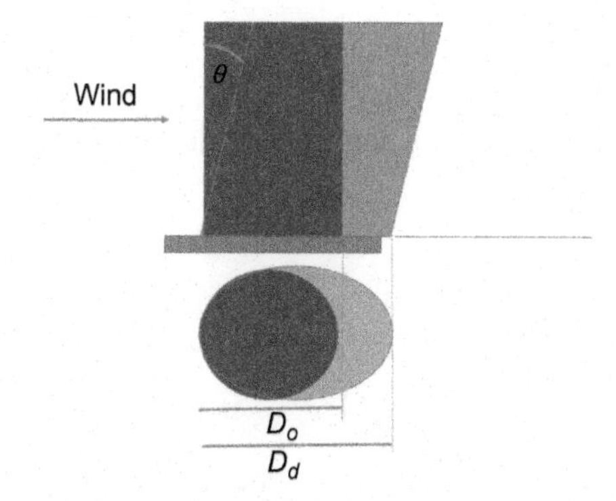

Example 9.12

Find the elongated diameter for the 50%-mass mixture of heptane and octane of the previous examples.

Solution

$$D_d = 1.5 \cdot D_o \cdot \left(\frac{u_w^2}{g \cdot D_o}\right)^{0.069} = 1.5 \cdot 10 \cdot \left(\frac{9}{98.1}\right)^{0.069} = 12.72\,\text{m} \tag{9.92}$$

9.7.3 Trench Fires

Trench LNG fires of pool with aspect ratios up to 30 have been studied by Mudan (1984), which have found that sensitivity to wind is much greater and have developed specific formulas to identify fire shape and size. Although these formulas are possibly the only

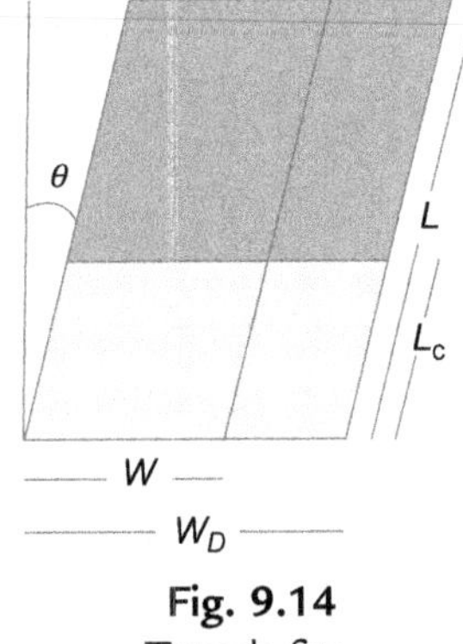

Fig. 9.14
Trench fire.

ones available at a validated level, their specific reference to LNG fires should properly be accounted for.

Flame height

With reference to the flame cross section, perpendicular to the long axis of the trench, the following formulas apply (Fig. 9.14).

$$L=\begin{cases} 2.2\cdot W \text{ if } \dfrac{u_W}{2\cdot\sqrt{g\cdot W}}\geq 0.25 \\ 0.88\cdot W\cdot\left(\dfrac{u_W}{2\cdot\sqrt{g\cdot W}}\right)^{-0.65} \text{ if } 0.1\leq\dfrac{u_W}{2\cdot\sqrt{g\cdot W}}\leq 0.25 \\ 4.0\cdot W \text{ if } \dfrac{u_W}{2\cdot\sqrt{g\cdot W}}\leq 0.1 \end{cases} \tag{9.93}$$

where:

L is the flame height.
W is the flame width in downwind direction.

All other terms are known.

Flame elongation (drag)

$$L_{f-d}=\begin{cases} 3.5\cdot W \text{ if } \dfrac{u_W}{2\cdot\sqrt{g\cdot W}}\geq 0.25 \\ 23.3\cdot W\cdot\left(\dfrac{u_W}{2\cdot\sqrt{g\cdot W}}\right)^{1.37} \text{ if } 0.1\leq\dfrac{u_W}{2\cdot\sqrt{g\cdot W}}\leq 0.25 \\ 1.0\cdot W \text{ if } \dfrac{u_W}{2\cdot\sqrt{g\cdot W}}\leq 0.1 \end{cases} \tag{9.94}$$

where $L_{f\text{-}d}$ is flame drag.

Flame tilt

$$\theta=\begin{cases}55.94^\circ \ \text{if}\ \dfrac{u_w}{2\cdot\sqrt{g\cdot W}}\geq 0.25\\[2ex] ar\cos\left[0.36\cdot\left(\dfrac{u_w}{2\cdot\sqrt{g\cdot W}}\right)^{-0.32}\right] \ \text{if}\ 0.042\leq\dfrac{u_w}{2\cdot\sqrt{g\cdot W}}\leq 0.25\\[2ex] 0^\circ\cdot\text{if}\ \dfrac{u_w}{2\cdot\sqrt{g\cdot W}}\leq 0.042\end{cases} \tag{9.95}$$

where θ is the tilt angle.

Height of a highly emissive clear flame

This height can be calculated according to the Rew's formula:

$$L_C=\min\left[11.404\cdot D_p\cdot\left(\frac{W_b}{\rho_{air}\cdot\sqrt{g\cdot D_p}}\right)^{1.13}\cdot\left(\frac{u_w}{\dfrac{g\cdot W_b\cdot D_p}{\rho_{air}}}\right)^{0.179}\cdot\left(\frac{H}{C}\right)^{2.49}\right],[L\cdot\cos\theta] \tag{9.96}$$

9.7.4 Pool Fire Diameter on Water (Sea Surface)

This is the case when combustible liquids are spilled onto the sea surface from off-shore installations or tankers. The calculation approach included in this section is related to various ignition times.

Diameter—Immediate ignition

Formulas valid for smooth solid surfaces may be used, provided the acceleration of gravity g is replaced by (Sintef, 2003):

$$g'=g\cdot\left(1-\frac{\rho_L}{\rho_{sw}}\right) \tag{9.97}$$

where:

ρ_L is the density of the liquid spill.
ρ_{sw} is the seawater density.

Diameter—Delayed ignition

Continuous spill

Sintef (2003) recommends to treat this case as if the ignition were immediate.

Discontinuous spill

Sintef (2003) distinguished three regimes and as many diameters:

Gravity and inertial force driven regime

$$D_1 = 2.28 \cdot \left[\left(\frac{\rho_{sw} - \rho_L}{\rho_{sw}} \right) \cdot g \cdot V_L \cdot t^2 \right]^{1/4} \tag{9.98}$$

with:

V_L is the volume instantaneously released.
t is time, s.

Large slicks and viscous forces driven regime

$$D_2 = 1.96 \cdot \left[\left(\frac{\rho_{sw} - \rho_L}{\rho_{sw}} \right) \cdot \frac{g \cdot V_L{}^2 \cdot t^{3/2}}{\nu_{sw}} \right]^{1/6} \tag{9.99}$$

where ν_{sw} is seawater kinematic viscosity $\cong 1.31 \times 10^{-6}$ m^2/s.

Surface tension driven regime This regime can be considered as governing the spreading process when the slick has sufficiently thinned, and surface tension becomes predominant.

$$D_2 = 3.2 \cdot \left[\left(\frac{\sigma_w^2 \cdot t^3}{\rho_{sw}^2 \cdot \nu_{sw}} \right) \cdot \right]^{1/4} \tag{9.100}$$

where σ_w is the interphasial tension (0.005–0.02 N/m).

Flame length

Defining Q^o as:

$$Q^o = \frac{W_b \cdot \pi \cdot \frac{D_p^2}{4} \cdot \Delta H_c}{\rho_{air} \cdot c_{pair} \cdot T_a \cdot D_p^2 \cdot \sqrt{g \cdot D_p}} \tag{9.101}$$

where ΔH_c and all of the terms are known, it is (Cetegen et al., 1984):

$$L_f = D_p \cdot 3.3 \cdot Q_o{}^{2/3} \tag{9.102}$$

This result is valid for large fires, 10–40 m, on the sea surface.

Maximum ignition delay for discontinuous spill

According to Sintef, 2003:

$$t_d\max = 0.0975 \cdot V_L^{0.46} \tag{9.103}$$

This time shall apply to each regime.

9.7.5 Heat Transmission in Pool Fires

An accurate and realistic model is the called two-zone model, which accounts for the luminous and the smoky parts of the solid cylindrical flame. This model is strongly recommended for the calculations.

The surface emissive power

Fuels for liquid pool fires are assumed to burn with either a luminous or a smoky flame. The flame Surface Emissive Power, SEP, a fundamental factor of the radiative heat definition, will be calculated according to the following alternative equations.

Luminous fires

In general, hydrocarbons lighter than pentane burn with a luminous flame, and heavier hydrocarbons burn with a smoky flame. For luminous flames (Mudan and Croce, 1988) (Table 9.20):

$$SEP_{av} = SEP_{\max} \cdot \left(1 - e^{-\frac{D_p}{L_s}}\right) \tag{9.104}$$

where:

SEP_{av} is the average Surface Emissive Power, W/m^2.
$SEP_{\max}$ is maximum Surface Emissive Power for luminous part of flame, W (100% in this case).
D_p the pool fire diameter, m.
L_s is emissive power characteristic length, m.

Sooty fires

For smoky/sooty flames the following general equation applies (Mudan and Croce, 1988; Sintef, 2003):

Table 9.20 Emissive data for some hydrocarbons

Material	Flame Type	SEP_{max} (W/m^2)	L_s (m)	Reference
LNG	Luminous	220,000	2.5	Cook et al. (1990)
Propane	Luminous	160,000	2.75	DNV
Kerosene	Sooty	140,000	8.33	DNV

$$SEP_{av} = SEP_{\max} \cdot e^{-\frac{D_p}{L_s}} + SEP_{soot} \cdot \left(1 - e^{-\frac{D_p}{L_s}}\right) \quad (9.105)$$

where:

SEP_{av} is the Surface Emissive Power, W/m^2.
$SEP_{\max}$ is maximum Surface Emissive Power for the luminous part of flame, 140 kW and 150 kW according to Sintef (2003).
SEP_{soot} is maximum Surface Emissive Power for the smoky part of the flame, 20 kW according to DNV.
D_p is the pool fire diameter, m.
L_s is the emissive power characteristic length, 8.33 for smoky flames m.

General fires

DNV suggests considering, as general fire, any fire cases for which experimental data are not available, for instance, white phosphorous. In this case (Moorhouse, 1982):

$$SEP_{av} = \frac{\chi_R \cdot W_b \cdot \Delta H_c}{\left(1 + 4 \cdot \frac{L_f}{D_p}\right)} \quad (9.106)$$

where χ_R is the fraction of total energy converted to radiation (between 0.3 and 0.4 for hydrocarbons). Some fractions are shown in Table 9.21 (Mudan and Croce, 1988).

The heat transfer equation

The heat transfer equation from a source to a target is the following:

$$Q_H = F \cdot SEP_{av} \cdot \tau \quad (9.107)$$

where:

F is the overall view factor.
SEP_{av} is the surface emissive power.
τ is the transmissivity.

Table 9.21 Fraction of energy converted to radiation (Mudan and Croce, 1988)

Fuel	χ_R
Hydrogen	0.20
Methane	0.20
Ethylene	0.25
Propane	0.30
Butane	0.30
Pentane and higher	0.40

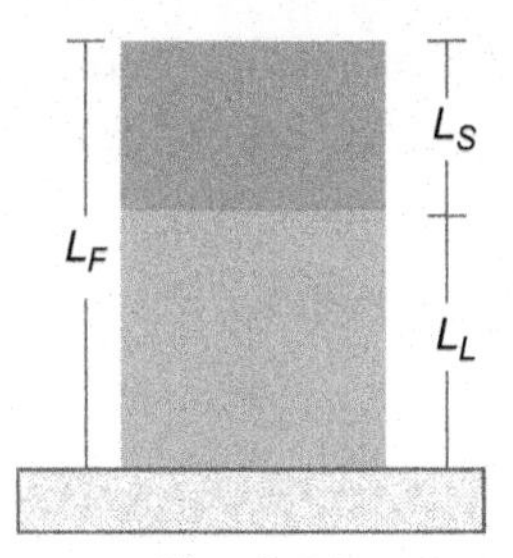

Fig. 9.15
Solid flame model.

View factors for solid flame model

Hydrocarbons heavier than pentane show the upper part of the flame, which is sooty, whereas the lower part behaves as luminous flame. Accordingly, for these cases, the average surface emissive power accounts for the different fractions of the flame. This affects the calculation of the heat transfer. Mudan and Croce (1986) have experimentally shown that the height of the luminous fraction (where smoke generation starts) L_L is given by the formula (Fig. 9.15):

$$L_L = L_F \cdot e^{-0.12 \cdot D_p} \tag{9.108}$$

The circumstances affect the selection of the view factors, as the average surface effective power is different for the two fractions.

The most general equation can be written as follows:

$$Q_H = F_s \cdot SEP_{soot} \cdot \tau_s + F_L \cdot SEP_{\max} \cdot \tau_L \tag{9.109}$$

This is the most rigorous equation for heat flux received by a target.

Depending on the accuracy, screening and accurate methods can be found in the literature.

Screening methods

Three simple screening methods are presented for the calculation of heat radiation to a target.

The Shokri and Beyler simple correlation is as follows:

$$Q_H = 15.5 \cdot \left(\frac{L}{D_p}\right)^{-1.59} \tag{9.110}$$

where Q_H is, as usual, the heat radiated (kW/m^2), and L is the distance between the centre of the pool fire to the edge of a vertical target located at ground level.

Point source model is widely used to predict the thermal radiation field. It assumes that the flame is a point source located at the centre of the real flame (Uijt de Haag and Ale, TNO).

$$Q_H = \frac{\chi_R \cdot W_b \cdot \frac{D_p^2}{4} \cdot \Delta H_c}{4 \cdot \pi \cdot R^2} \tag{9.111}$$

where all the terms are known, and R is the distance between the target and the centre of the flame.

Hurley (2016) recommends to apply a safety factor of 2 when this method is used (Fig. 9.16).

View factors for solid flame model

The solid flame model assumes a cylindrical flame and accounts for the luminous and smoky fractions of it. Being vertical F_v and horizontal F_v view factors vectorial entities, the overall view factor is always:

$$F = \sqrt{F_v^2 + F_h^2} \tag{9.112}$$

Should the average surface emissive power be used:

$$Q_H = F \cdot SEP_{av} \cdot \tau \tag{9.113}$$

the following, much more accurate, view factors are to be used for no-wind heat radiation towards a vertical and horizontal target at ground level (Hurley, 2016) (Fig. 9.17).

$$\pi \cdot F_v = \frac{(B - 1/S)}{\sqrt{B^2 - 1}} \cdot \tan^{-1}\left[\sqrt{\frac{(B+1)\cdot(S-1)}{(B-1)\cdot(S+1)}}\right] - \frac{(A - 1/S)}{\sqrt{A^2 - 1}} \cdot \tan^{-1}\left[\sqrt{\frac{(A+1)\cdot(S-1)}{(A-1)\cdot(S+1)}}\right] \tag{9.114}$$

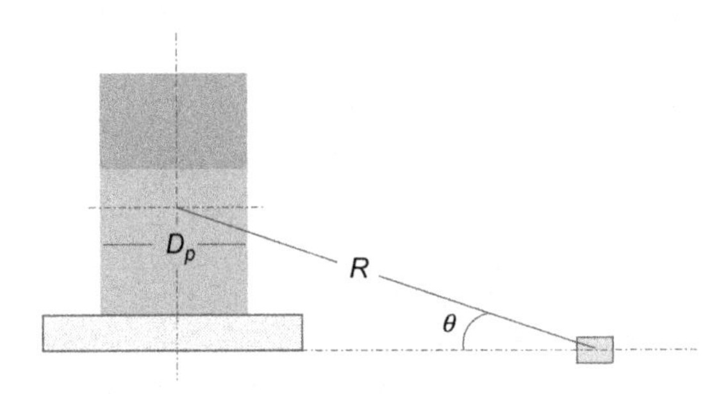

Fig. 9.16
Point source model.

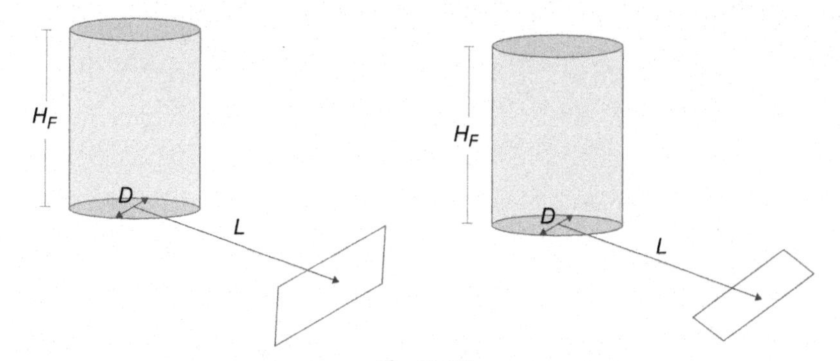

Fig. 9.17
View factors for solid flame model. Target at ground.

$$\pi \cdot F_h = \frac{1}{S} \cdot \tan^{-1}\left[\frac{h}{\sqrt{S^2-1}}\right] - \frac{h}{S} \cdot \tan^{-1}\left[\sqrt{\frac{(S-1)}{(S+1)}}\right] + \frac{A \cdot h}{S \cdot \sqrt{A^2-1}} \cdot \tan^{-1}\left[\sqrt{\frac{(A+1)\cdot(S-1)}{(A-1)\cdot(S+1)}}\right] \quad (9.115)$$

with:

$$A = \frac{h^2+S^2+1}{2 \cdot S} \quad (9.116)$$

$$B = \frac{S^2+1}{2 \cdot S} \quad (9.117)$$

$$S = \frac{2 \cdot L}{D} \quad (9.118)$$

$$h = \frac{2 \cdot H}{D} \quad (9.119)$$

When the luminous and smoky fractions of the flame are accounted for, the previous view factors can be used for the associated configuration, taking into account the respective locations (Fig. 9.18).

The configuration shown in this figure depicts the case of a target which is not at ground level. In this case, two vertical view factors have to be calculated, notably two vertical view factors if the target is vertical, or two horizontal view factors if the target is horizontal (the tilted case can be approached vectorially). In the previous equations, should the target be vertical, a F_{v1} calculations including H_{f1} will be calculated, and then a F_{v2} calculation including H_{f2} will be calculated. Finally:

$$F_v = F_{v_1} + F_{v_2} \quad (9.120)$$

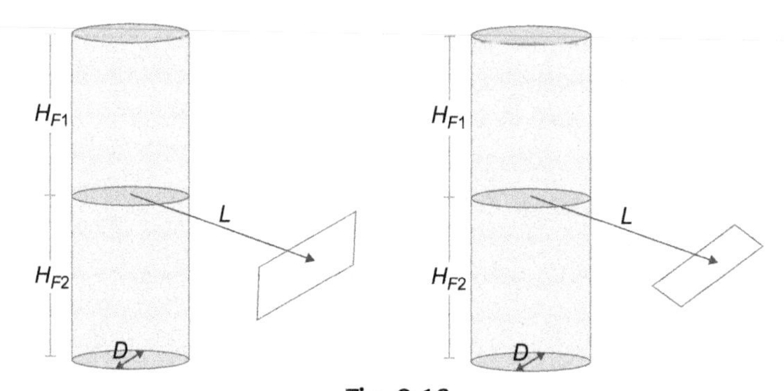

Fig. 9.18
View factors for solid flame model. Target at elevation.

The same approach should be adopted in the case of horizontal target. Just the vertical or the horizontal view factors are needed if the target is not tilted.

The set of view factors for a generic radiation receiving element located at ground in no-wind and wind conditions is as follows (Hurley, 2016).

No-wind condition

$$\pi \cdot F_v = \frac{1}{S} \cdot \tan^{-1}\left[\frac{h}{\sqrt{S^2-1}}\right] - \frac{h}{S} \cdot \tan^{-1}\left[\sqrt{\frac{(S-1)}{(S+1)}}\right] + \frac{A \cdot h}{S \cdot \sqrt{A^2-1}} \cdot \tan^{-1}\left[\sqrt{\frac{(A+1)\cdot(S-1)}{(A-1)\cdot(S+1)}}\right] \tag{9.121}$$

$$\pi \cdot F_v = \frac{(B-1/S)}{\sqrt{B^2-1}} \cdot \tan^{-1}\left[\sqrt{\frac{(B+1)\cdot(S-1)}{(B-1)\cdot(S+1)}}\right] - \frac{(A-1/S)}{\sqrt{A^2-1}} \cdot \tan^{-1}\left[\sqrt{\frac{(A+1)\cdot(S-1)}{(A-1)\cdot(S+1)}}\right] \tag{9.122}$$

with:

$$A = \frac{h^2+S^2+1}{2 \cdot S} \tag{9.123}$$

$$B = \frac{S^2+1}{2 \cdot S} \tag{9.124}$$

$$S = \frac{2 \cdot L}{D} \tag{9.125}$$

$$h = \frac{2 \cdot H}{D} \tag{9.126}$$

Wind condition

$$\pi \cdot F_v = -\frac{a \cdot \cos\theta}{b - a \cdot \sin\theta} \cdot \tan^{-1}\left(\sqrt{\frac{b-1}{b+1}}\right) + \frac{a \cdot \cos\theta}{b - a \cdot \sin\theta}$$

$$\times \frac{a^2 + (b+1)^2 - 2 \cdot b \cdot (1 + a \cdot \sin\theta)}{\sqrt{a^2 + (b+1)^2 - 2 \cdot a \cdot (b+1)} \cdot \sqrt{a^2 + (b-1)^2 - 2 \cdot a \cdot (b-1) \cdot \sin\theta}}$$

$$\times \tan^{-1}\left(\sqrt{\frac{a^2 + (b+1)^2 - 2 \cdot a \cdot (b+1) \cdot \sin\theta}{a^2 + (b-1)^2 - 2 \cdot a \cdot (b-1) \cdot \sin\theta}} \cdot \sqrt{\frac{b-1}{b+1}}\right)$$

$$+ \frac{\cos\theta}{\sqrt{1 + (b^2 - 1) \cdot \cos\theta^2}}$$

$$\times \left\{ \tan^{-1}\left[\frac{a \cdot b - (b^2 - 1) \cdot \sin\theta}{\sqrt{(b^2 - 1)} \cdot \sqrt{1 + (b^2 - 1) \cdot \cos\theta^2}}\right] + \tan^{-1}\left[\frac{(b^2 - 1) \cdot \sin\theta}{\sqrt{(b^2 - 1)} \cdot \sqrt{1 + (b^2 - 1) \cdot \cos\theta^2}}\right] \right\} \tag{9.127}$$

And:

$$\pi \cdot F_h = \tan^{-1}\left(\sqrt{\frac{b+1}{b-1}}\right) + \frac{\sin\theta}{\sqrt{1 + (b^2 - 1) \cdot \cos\theta^2}}$$

$$\times \left[\tan^{-1} \frac{a \cdot b - (b^2 - 1) \cdot \sin\theta}{\sqrt{(b^2 - 1)} \cdot \sqrt{1 + (b^2 - 1) \cdot \cos\theta^2}} + \tan^{-1} \frac{(b^2 - 1) \cdot \sin\theta}{\sqrt{(b^2 - 1)} \cdot \sqrt{1 + (b^2 - 1) \cdot \cos\theta^2}} \right]$$

$$- \frac{a^2 + (b+1)^2 - 2 \cdot (b + 1 + a \cdot b \cdot \sin\theta)}{\sqrt{a^2 + (b+1)^2 - 2 \cdot a \cdot (b+1) \cdot \sin\theta} \cdot \sqrt{a^2 + (b-1)^2 - 2 \cdot a \cdot (b-1) \cdot \sin\theta}}$$

$$\times \tan^{-1}\left(\sqrt{\frac{a^2 + (b+1)^2 - 2 \cdot a \cdot (b+1) \cdot \sin\theta}{a^2 + (b-1)^2 - 2 \cdot a \cdot (b-1) \cdot \sin\theta}} \cdot \sqrt{\frac{b+1}{b-1}}\right) \tag{9.128}$$

being R the pool fire radius, H the height, L the distance between the centre of the flame at the ground and the target θ, the smaller angle of the flame vertical axis with the ground. It is:

$$a = \frac{H}{R} \tag{9.129}$$

$$b = \frac{L}{R} \tag{9.130}$$

Graphical view factors

The following view factor is provided by Treccani (Figs. 9.19 and 9.20):

The following diagram, including view factors for various tilted angles, has been obtained from Mudan and Croce (1988). L is the target distance from the axis to pool radius ratio (Fig. 9.21).

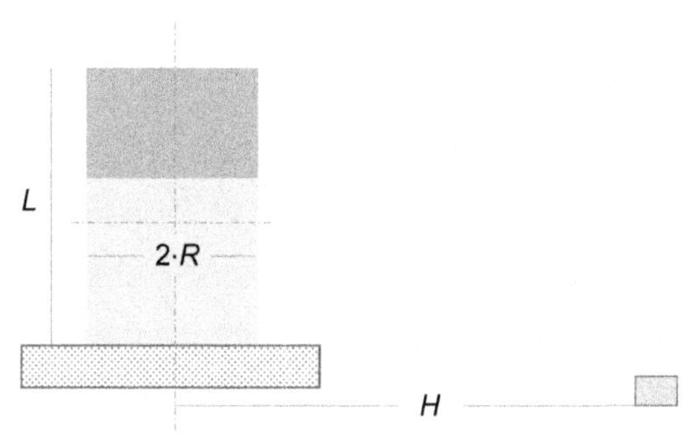

Fig. 9.19
Notation for graphical view factor (Treccani).

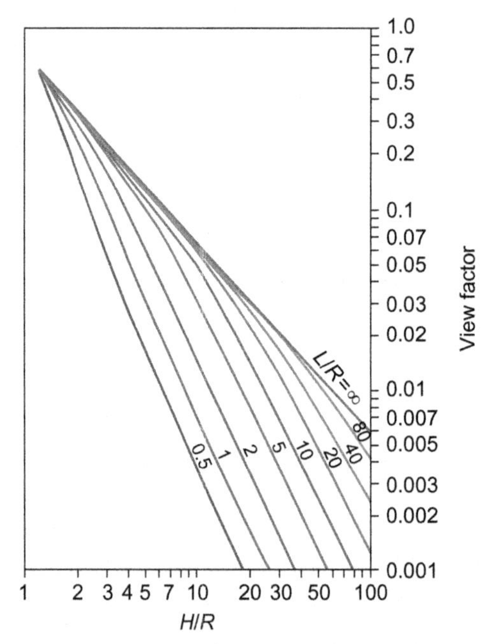

Fig. 9.20
Graphical view factor (Treccani).

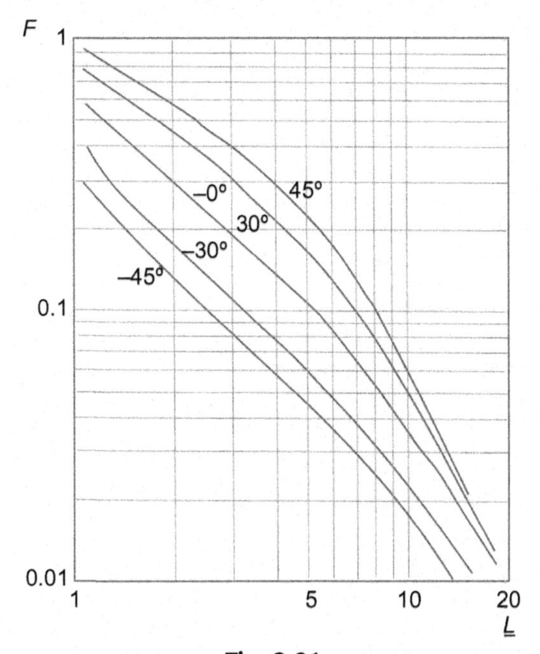

Fig. 9.21
Graphical view factor at various angles (Treccani).

Example 9.13

A 50%-mass mixture of heptane and octane is released and catches on fire into a 10 m diameter circular basin. The ambient temperature is 20°C, and the relative humidity is 50%. Find the heat radiated to a vertical target 50 m far from the fire edge.

Solution

T_a (K)	T_b (K)	ΔH_c (kJ/kg K)	ΔH_v (kJ/kg K)	c_p (J/kg K)	W_∞ (kg/m s^2)	D (m)	$k \times \beta$ (m^{-1})	ρ_L (kg/m^3)	ρ_v (kg/m^3)
Heptane									
293.16	371.53	44,558	318	2.24	0.101	10	1.1	680	3.2
Octane									
293.16	398.75	44,510	298	2.15	Unknown	10	Unknown	700	3.48

For this example, the flame height, the tilted angle, and the flame elongation have already been calculated.

- $D_d = 12.72$ m
- $\theta = 32°$
- $L_f = 30.7$ m

This example can be approached in two ways.

One-zone model

According to this model, assuming conservatively that the pool fire diameter corresponds to the elongated flame basis, the average surface emissive power for the entire solid flame is:

$$SEP_{av} = SEP_{max} \cdot e^{-\frac{D_p}{L_s}} + SEP_{soot} \cdot \left(1 - e^{-\frac{D_p}{L_s}}\right) = 150{,}000 \cdot e^{-\frac{12.72}{8.33}} + 20{,}000 \cdot \left(1 - e^{-\frac{12.72}{8.33}}\right)$$
$$= 32{,}577 + 15.656 = 48{,}233\,\text{W/m}^2 \tag{9.131}$$

The transmissivity is calculated as follows:

$$\tau = P_o \cdot (P_w \cdot x)^{-0.09} \tag{9.132}$$

P_o is 2.02 $(\text{N/m}^2)^{0.09} \cdot \text{m}^{0.09}$.

P_w is the partial water vapour pressure of water in air at a relative humidity RH, in N/m^2:

$$P_w = 1013.25 \cdot RH \cdot \exp\left(14.4114 - \frac{5238}{T_a}\right) \tag{9.133}$$

$$P_w = 1013.25 \cdot RH \cdot \exp\left(14.4114 - \frac{5238}{T_a}\right) = 1013.25 \cdot 50 \cdot \exp\left(14.4114 - \frac{5238}{293.16}\right)$$
$$= 1599\,\text{Pa} \tag{9.134}$$

$$\tau = P_o \cdot (P_w \cdot x)^{-0.09} = 2.02 \cdot (1599 \cdot 50)^{-0.09} = 0.73 \tag{9.135}$$

The target is vertical, so just the vertical view factor is to be calculated by the following equation:

$$\pi \cdot F_v = -\frac{a \cdot \cos\theta}{b - a \cdot \sin\theta} \cdot \tan^{-1}\left(\sqrt{\frac{b-1}{b+1}}\right) + \frac{a \cdot \cos\theta}{b - a \cdot \sin\theta}$$
$$\times \frac{a^2 + (b+1)^2 - 2 \cdot b \cdot (1 + a \cdot \sin\theta)}{\sqrt{a^2 + (b+1)^2 - 2 \cdot a \cdot (b+1)} \cdot \sqrt{a^2 + (b-1)^2 - 2 \cdot a \cdot (b-1) \cdot \sin\theta}}$$
$$\times \tan^{-1}\left(\sqrt{\frac{a^2 + (b+1)^2 - 2 \cdot a \cdot (b+1) \cdot \sin\theta}{a^2 + (b-1)^2 - 2 \cdot a \cdot (b-1) \cdot \sin\theta}} \cdot \sqrt{\frac{b-1}{b+1}}\right)$$
$$+ \frac{\cos\theta}{\sqrt{1 + (b^2 - 1) \cdot \cos\theta^2}}$$
$$\times \left\{\tan^{-1}\left[\frac{a \cdot b - (b^2 - 1) \cdot \sin\theta}{\sqrt{(b^2 - 1)} \cdot \sqrt{1 + (b^2 - 1) \cdot \cos\theta^2}}\right] + \tan^{-1}\left[\frac{(b^2 - 1) \cdot \sin\theta}{\sqrt{(b^2 - 1)} \cdot \sqrt{1 + (b^2 - 1) \cdot \cos\theta^2}}\right]\right\} \tag{9.136}$$

with:

$a = H/R = 4.82$
$b = L/R = 8.86$
$\sin\theta = 0.53$
$\cos\theta = 0.848$

$$\pi \cdot F_v = -\underbrace{\frac{a\cdot\cos\theta}{b-a\cdot\sin\theta}}_{1}\cdot\underbrace{\tan^{-1}\left(\sqrt{\frac{b-1}{b+1}}\right)}_{2}+\underbrace{\frac{a\cdot\cos\theta}{b-a\cdot\sin\theta}}_{3}\times\underbrace{\frac{a^2+(b+1)^2-2\cdot b\cdot(1+a\cdot\sin\theta)}{\sqrt{a^2+(b+1)^2-2\cdot a\cdot(b+1)}\cdot\sqrt{a^2+(b-1)^2-2\cdot a\cdot(b-1)\cdot\sin\theta}}}_{4}\times\underbrace{\tan^{-1}\times\left(\sqrt{\frac{a^2+(b+1)^2-2\cdot a\cdot(b+1)\cdot\sin\theta}{a^2+(b-1)^2-2\cdot a\cdot(b-1)\cdot\sin\theta}}\cdot\sqrt{\frac{b-1}{b+1}}\right)}_{5}+\underbrace{\frac{\cos\theta}{\sqrt{1+(b^2-1)\cdot\cos\theta^2}}}_{6}\times\left\{\underbrace{\tan^{-1}\left[\frac{a\cdot b-(b^2-1)\cdot\sin\theta}{\sqrt{(b^2-1)}\cdot\sqrt{1+(b^2-1)\cdot\cos\theta^2}}\right]}_{7}+\underbrace{\tan^{-1}\left[\frac{(b^2-1)\cdot\sin\theta}{\sqrt{(b^2-1)}\cdot\sqrt{1+(b^2-1)\cdot\cos\theta^2}}\right]}_{8}\right\} \tag{9.137}$$

And the eight factors of the equation are:

1	2	3	4	5	6	7	8
−0.649	0.729	0649	1.666	0.840	0.554	0.112	0.0254

Solving:

$$F_v = 0.16 \tag{9.138}$$

$$Q_H = F \cdot SEP_{av} \cdot \tau = 0.131 \times 48.233 \cdot 0.73 \cong 4.63\,\text{kW/m}^2 \tag{9.139}$$

Two-zones model

The two zone model states that the overall radiated heat is the summation of the contributions of the smoky and luminous part of the flame (Fig. 9.22).

Applying the Shokri and Beyler correlation:

$$H_L = H \cdot e^{-0.12 \cdot D_p} = 30.7 \cdot e^{-0.12 \cdot 12.72} = 6.67\,\text{m} \tag{9.140}$$

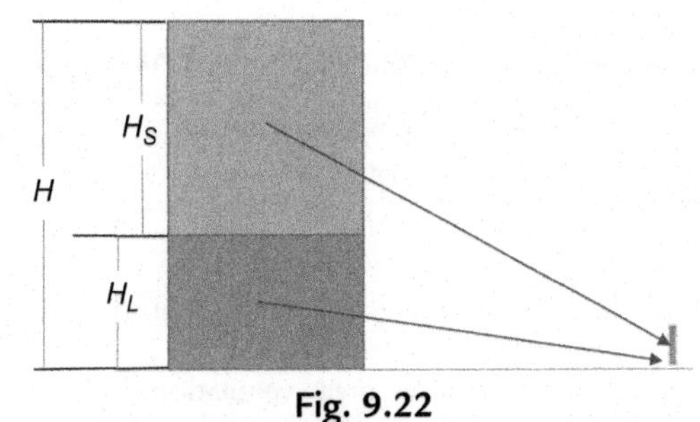

Fig. 9.22
Two zone model for Example 9.13.

and:

$$H_S = H - H_L = 30.7 - 6.67 = 24.03\,\text{m} \tag{9.141}$$

Two view factors are to be calculated, one relating to the luminous part and the other to the smoky part of the flame. The latter can be calculated as:

$$F_{v-S} = F_v - F_{v-L} \tag{9.142}$$

where the subscripts S and L are, respectively, related to the smoky and luminous parts of the flame. Applying the vertical view factor equation with H and H_L, its $F_{v-L} = 0.009$ and $F_{v-S} = 0.15$ Being:

$$Q_H = F_s \cdot SEP_{soot} \cdot \tau_s + F_L \cdot SEP_{\max} \cdot \tau_L \tag{9.143}$$

it is:

$$Q_H = F_s \cdot SEP_{soot} \cdot \tau_s + F_L \cdot SEP_{\max} \cdot \tau_L = 0.15 \cdot 20 + 150 \cdot 0.009 = 4.35\,\text{kW} \tag{9.144}$$

9.7.6 Recommended Incident Heat Fluxes

See Table 9.22.

9.7.7 Practical Data on Pool Fires

Heat fluxes in hydrocarbons pool fire (FABIG, 2010)

See Table 9.23.

Pool fire of liquefied natural gas

The flame height proposed by FABIG (2010) to be calculated according to the following formulas, based on the parameter Q^*, defined as:

Table 9.22 Recommended incident heat fluxes

Local Peak Heat Load (kW/m^2)	Global Average Heat Load (kW/m^2)	Reference
150	100	Scandpower (2004)
190	–	API 521 (2014)
150 (Open or closed area—fuel controlled) 200 (Enclosed area—ventilation controlled)	100 (Open or closed area—fuel controlled) 130 (Enclosed area—ventilation controlled)	Norsok S-001 (2008)

Table 9.23 Heat fluxes in hydrocarbon pool fires (FABIG, 2010)

Pool Size (m)	Fuel	Heat Flux (kW/m^2)
0.5 × 9.5	Kerosene	150
4.0 × 7.0	Kerosene	94–112
9.0 × 18	JP-5	80–164
35 m^2	LNG	250–300
12.5 × 15.5	LPG	230–270

$$Q^* = \frac{m_b \cdot \Delta H_c}{\rho_a \cdot c_p \cdot T_a \cdot D_p{}^2 \cdot \sqrt{g \cdot D_p}} \tag{9.145}$$

where:

m_b is the burning rate, kg/s.
ΔH_c is the combustion heat of methane, 50,000,000 J/kg.
ρ_a is the density of air, 1.2 kg/m^3.
T_a is the ambient temperature, 288 K.
c_p is the specific heat of air, 1000/kg K.
D_p is the pool diameter, m.
g is the acceleration due to gravity, 9.81 m/s^2.

For intermittent fires:

$$0.23 < Q^* < 1.9 \rightarrow \frac{L_f}{D_p} = 2.6 \cdot Q^{*0.66} \tag{9.146}$$

For mass fires:

$$0.23 > Q \rightarrow \frac{L_f}{D_p} = 18.8 \cdot Q^{*2} \tag{9.147}$$

The specific burning rate W_b can be assumed as 0.14 kg/m^2 s on land and 0.22 kg/m^2 s on water (FABIG, 2010).

Pool fire of methanol

Methanol combustion is very particular because, even if this alcohol burns with a non-luminous flame, no soot is produced. Operating data are summarised in Table 9.24 (FABIG, 2010):

Practical data of pool fire on sea water

The following flame size and emissive data, relevant to a pool fire of a C_7 material, have been predicted by DNV (Spouge, 1999) on the basis of FLARE program findings with ambient temperature of 20°C, 70% of relative humidity, and a wind speed of 3 m/s (Tables 9.25 and 9.26).

Table 9.24 Methanol pool fire data (FABIG, 2010)

Methanol Pool Fire Data	Data
Typical pool diameter (m)	5
Flame length (m)	Equal to pool diameter
Mass burning rate (kg m^2 s)	0.03
Fraction of heat radiated	0.15
CO level and smoke concentration	Negligible
Radiative flux (kW/m^2)	35
Convective flux (kW/m^2)	0
Flame temperature (K)	1250
Flame emissivity	0.25

Table 9.25 Data for pool fires on water (Spouge, 1999)

Release Rate (kg/s)	Pool Diameter (m)	Flame Length (m)	Tilt Angle (°)
0.4	2.5	8	58
1.7	5	13	53
7	10	21	48
42	25	39	39
60	30	44	37
167	50	63	29
427	80	88	19
699	100	102	11

Table 9.26 Data for pool fires on water (Spouge, 1999)

Release Rate (kg/s)	Pool Diameter (m)	Flame Length (m)	Distance to 5 kW/m^2	Distance to 12.5 kW/m^2
0.4	2.5	109	12	6.3
1.7	5	86	19	9.4
7	10	56	28	48
42	25	26	30	39
60	30	23	32	37
167	50	20	43	29
427	80	20	64	19
699	100	102	77	11

The follow data have been produced by SINTEF (2003) and FABIG (2010) and consist of typical values of relevance to pool fires on the sea surface (Tables 9.27 and 9.28).

9.7.8 Leak Rate Categories for Fire Risk Assessment (FABIG, 2010)

The following data summarise the ranges of the leak rate to be adopted in QRA and CFD studies with the size categorisation (Table 9.29).

Table 9.27 Data for pool fires on sea water (SINTEF, 2003)

Pool Fire Data	Value	Reference
Minimum ignitable thickness for fresh crude oil	1 mm	SINTEF
Minimum ignitable thickness for aged crude oil	3–5 mm	
Minimum ignitable thickness for fuel oil	10 mm	
Minimum heat flux to induce a flammable mixture above the oil stick	5 kW over 1 m^2	
Burning rate for crude oil	0.055 kg/s m^2	
Burning rate for condensate	0.100 kg/s m^2	
Flame tilt (circular hydrocarbon fires) (Mudan and Croce)	$\cos\theta = 1.1 \cdot D_p/{u_w}^{0.5}$	
Flame temperature	1300°C	
Heat flux onto a cold steel target	200 kW/m^2	
Peak heat flux onto a target enveloped in flames	400 KW/m^2	

Table 9.28 Data for pool fires on sea water (FABIG, 2010)

Variable	Pool <9 m	Pool >9 m
Flame length (m)	$L/D = 2.5 \cdot Q^{*0.66}$	$L/D = 2.5 \cdot Q^{*0.66}$
Burning rate (kg m^2 s)	0.042	As on land for large fires
Minimum thickness before fire extinction (mm)	3–5	0.5 (condensate) 1 (light crude) 1–3 (heavy crude)
SEP (kW/m^2)	53	–
Radiative flux (kW/m^2)	110	230
Flame temperature (K)	1200	1460
Flame emissivity	0.9	0.9

9.7.9 Jet Fire

Scenario

Jet Fire or Jet Flame can be defined as a fire type resulting from combustion of pressurised release of gas and/or liquid (CCPS, 2005). Thermo-fluid-dynamics of jet fires can be framed within the jet theory (Chapter 4) with the additional thermal and kinetic effects. Generally speaking, jet fires are expected when flammable gases, vapours, mixtures, and, in specific cases, flashing liquids and non-flashing liquids are released with a significant momentum. High pressure liquid/gas mixtures can undergo jet fires thanks to the gas content, and to the mechanical energy, which are effective in atomising the liquid in very small droplets vaporised by radiation from the flame. Jet fires can take place also with highly pressurised liquids, when

Table 9.29 Leak rate categories for fire risk assessment (FABIG, 2010)

		Small			Medium			Large						
QRA	Range	0.1–2			2–16			>16						
	kg/s	0.5			5			50						
CFD coarse	Range	0.1–1		1.2	2–8		8–16	16–32	32–64	64–256		>256		
	kg/s	0.5		1.25	5		10	25	50	100		500		
CFD fine	Range	0.1–0.5	0.5–1	1.2	2.4	4–8	8–16	16–32	32–64	64–128	128–256	256–512	512–1024	1024–2048
	kg/s	0.3	0.75	1.5	2	6	12	24	48	96	192	384	768	1536

atomisation energy provided by the release dynamics is sufficient to produce appropriate droplet size, and the temperature sustains combustion. Many scenarios can be practically met by a process safety analyst, some of which present some tricky interpretation aspects. For these reasons, some simple, and relatively easy, applicable rules have been provided here.

Fluid-dynamics of jet fire

Gas and vapour jet fires

Releases of gases or vapours with upstream pressures greater than the choked pressure shall be considered a turbulent jet, which, if ignited, will behave like jet fires. It is worth stressing that, for a given upstream pressure P_o, as long as downstream pressure will be reduced, a certain downstream pressure will be attained, in correspondence of which a further reduction will not result in any further flow increase P_a. The outlet speed will be the sound speed at the local temperature, and any increase of the upstream pressure will just result in a pressure drop increase. The upstream pressure is the critical pressure, and it is given by the expression:

$$P_{chocked} = P_a \cdot \left(\frac{\gamma + 1}{2}\right)^{\frac{\gamma}{\gamma - 1}} \qquad (9.148)$$

Being $P_a = 1$ atm, it is easy to calculate the choked pressures for the most frequent combustible gas, and for carbon dioxide, on the basis of the specific heats provided by CCPS (1995) (Table 9.30).

All pressures are below 2.0 atm and above 1.7, so 2 atm can be assumed as the minimum pressure when jet fire is expected. When upstream pressure is other than choked pressure, a specific evaluation is needed. As addressed in the jet theory (Chapter 4), a Reynolds number greater than 10,000 is representative of a turbulent jet (API 521, 2014). Significantly lower

Table 9.30 Choked pressure for some materials (FABIG, 2010)

Material	Γ	P_{choked} (atm)
Acetylene	1.30	1.832416
Ammonia	1.32	1.844545
Butane	1.11	1.716479
Carbon dioxide	1.4	1.892929
Ethane	1.22	1.78376
Ethylene	1.22	1.78376
Hydrogen	1.41	1.898963
Hydrogen sulphide	1.3	1.832416
Methane	1.32	1.844545
Methyl chloride	1.2	1.771561
Natural gas	1.27	1.814196
Propane	1.5	1.953125
Propene	1.14	1.734874
Sulphur dioxide	1.26	1.808116

Reynolds numbers define jet flows characterised by a lower air entrainment. These jets with lower momentum may behave as laminar flows, air entrainment can be reduced intensely, and wind will affect the shape and dispersion of the mixture. A cloud can be generated, and a transition from potential jet to flash fire has to be considered. For example, as confirmed by Woodward and Pitbaldo (2010), LNG jet fire is likely at high pressures.

Liquid jet fires

Jet fires of combustible liquids do not have to be ruled out on the assumption that liquid is the predominant phase. In practical applications, three general categories of liquids can undergo a jet fire:

(1) Flashing liquids
High momentum flashing liquids, such as LPG, are known to easily undergo jet fires. Gómez-Mares et al. (2008) have reported that 61% out of 84 jet fire incidents that occurred since 1961 involved LPG. The high velocity vapour fraction is immediately available for ignition and produces a significant spray effect on the liquid, which is atomised in very small ignitable droplets.

(2) Pure liquids
Practical and general methods to establish the occurrence of fire in pure liquid turbulent jets are not available, despite the huge coverage of the scientific and technical literature and the very complete understanding of the phenomena underpinning liquid droplet formation and fragmentation (f.i. Uijt de Haag and Ale, Yellow Book, 2005). As a matter of fact, the process safety engineer needs some practical data, and a general approach, in order to decide whether or not a liquid jet should be included in the jet fire hazard analysis. The main parameters affecting the ignition of a liquid jet are:
- The droplet size which, in turn, is related to the surface to volume ratio, which governs the air transfer to the droplet surface.
- The liquid vapour pressure and, therefore, the bulk temperature
- The minimum ignition energy for the combustible liquid

These three parameters are in reality inter-related to many other variables which affect the overall highly turbulent jet ignition process.
As a general rule, the following practical approach can be followed:
The SMD (Sauter Mean Diameter) of the ignitable liquid particles falls in the range 10–20 mm. This has been largely proved theoretically and experimentally. The application of the Locally Homogenous Flow (LHF) to spray combustion by numerous researchers has shown that there exist striking similarities between the structure of flames fuelled with gases and with well-atomised sprays having a maximum drop number densities for $10 < \text{SMD} < 20\ \mu m$ (Kuo and Acharya, 2012). This has been confirmed by many studies. Many further researches have analysed the atomisation pressures required by the system to achieve the necessary droplet fragmentation. Typical values for

hydrocarbons fall in the range 4–10 bar, as per Liu et al. (2015). The same authors have related the heptane particle sizes to the Minimum Ignition Energy, and have shown that in the range $10 < SMD < 20\ \mu m$, MIE lies in the range 0.5–2 mJ. For hydrocarbons, a rule of thumb can be adopted:

- Depending on molecular weight, presence of gases, temperature, and vapour pressure, jet fire can be expected between 2 bar and 10 bar, with 4–5 bars as average.
- Spouge (1999) states that typical conditions are pressures over 2 bars for condensate and 7 bar for light crude oil provided the mass flow is lower than 10 kg/s.
- For a generic liquid, the following evaluation procedure could be followed:

(a) The liquid jet Reynolds number is calculated as:

$$Re_L = \frac{\rho_L \cdot u_L \cdot D_e}{\mu_L} \tag{9.149}$$

where all the parameters are related to the liquid downstream of the outlet, and D_e is the diameter.

(b) The liquid jet Weber number is calculated as:

$$We_L = \frac{2 \cdot \rho_L \cdot u_L^2 \cdot D_e}{\sigma_L} \tag{9.150}$$

where σ_L is the surface tension between liquid and vapour.

(c) The droplet size is calculated as (Uijt de Haag and Ale, Yellow Book, 2005):

$$d_L = 3.78 \cdot \sqrt{1 + 3 \cdot \frac{\sqrt{We_L}}{Re_L}} \tag{9.151}$$

If $We_L < 10^6 \times Re_L^{-0.45}$ and T_e (temperature at the outlet) is lower than the normal boiling temperature. If not:

$$d_L = 15 \cdot \frac{\sigma_L}{u_L^2 \cdot \rho_a} \tag{9.152}$$

where ρ_a is air density.

Example 9.14

Methanol is released from a 10 barg vessel through a 100 mm hole. Analyse the case and find whether or not a jet fire is likely. Temperature of liquid is 20°C at ambient temperature.

Solution

T_a (K)	T_b (K)	P_o (barg)	D_e (m)	ρ_L (kg/m^3)	μ (Pa/s)	σ (N/m)	MIE (mJ)	ρ_a (kg/m^3)
293.16	337.8	10	0.1	790	0.000594	0.0227	0.14	1.2

The Torricellian velocity corresponding to this head is:

$$h_L = \frac{101{,}972}{790} = 129\,\text{m} \tag{9.153}$$

$$u_L = C_D \cdot \sqrt{2 \cdot g \cdot h_o} = 0.61 \cdot \sqrt{2 \cdot 9.81 \cdot 129} = 50.3\,\text{m/s} \tag{9.154}$$

$$Re_L = \frac{\rho_L \cdot u_L \cdot D_e}{\mu_L} = \frac{790 \cdot 50.3 \cdot 0.1}{0.000594} = 6{,}649{,}831.65 \tag{9.155}$$

$$We_L = \frac{2 \cdot \rho_L \cdot u_L^2 \cdot D_e}{\sigma_L} = \frac{790 \cdot 50.3^2 \cdot 0.1}{0.0227} = 8{,}805{,}159 \tag{9.156}$$

Being $We_L > 10^6 \times Re_L^{-0.45} = 850.7$, it is:

$$d_L = 15 \cdot \frac{0.0227}{50.3^2 \cdot 1.2} = 7.4\,\mu\text{m} \tag{9.157}$$

The spray droplet diameter is below 20 μm, so jet fire has to be considered.

Flame stability

Flame onset and its stability depend on many factors, such as:

- Nature of the fuel: methane and hydrogen show very different flame properties
- Hole size
- Characteristics of the surroundings

Jet fires from small holes are not likely or unstable. The basic concept underpinning this evidence is the quenching distance, that is the minimum size of a burner supporting a flame (Bradley, 1969). This is the principle upon which the flame arrestors are based. By increasing the gas speed, a specific value is reached at which the flame is blown off: this is the blow out velocity, which is, in general, proportional to the laminar burning velocity and for vertical natural gas flames is given by the expression (Chamberlain, 1995):

$$u_{bo} = S_u \cdot 0.0028 \cdot Re(H)^{1.1} \cdot \left(\frac{\rho_a}{\rho_g}\right)^{1.5} \tag{9.158}$$

where S_u is the laminar burning velocity, ρ_a and ρ_g are, respectively, air and expanded gas densities. Re(H) is the Reynolds calculation accounting for the distance to stoichiometric concentration H:

$$H = \left[\left(\frac{4 \cdot \theta}{X}\right) Y \cdot \sqrt{\frac{\rho_g}{\rho_a}} + 5.8\right] \cdot D_e \tag{9.159}$$

In this formula:

- X is the fuel mass fraction in the stoichiometric mixture, 0.055 for methane, 0.06 for propane.

- Y is the fuel mass fraction at the hole.
- D_e is the hole diameter.

The quenching distance is defined as:

$$d_q = 8 \cdot \frac{\alpha}{u_{bo}} \tag{9.160}$$

It can be concluded that the greater the laminar burning velocity, the smaller the quenching distance. This background is very useful to understand the flame behaviour. When a burner flame is started, the first part of the flame is blue, which indicates the fuel oxygen mixture is oversaturated (Fig. 9.23).

We know from jet theory that air is entrained, and that mixture leaves the flammability region at a certain distance from the outlet, which is around 100–120 diameters for light hydrocarbon turbulent jets. This is the reason why the central and last part of the flame is yellow, due to the air to fuel stoichiometric ratio. By increasing the flame speed, it will lift-off and will possibly blow out, according to Eq. (9.165). As anticipated, this is related to the flame's unstable behaviour. In the following approximate diagram, retrofitted on the basis of data provided by FABIG (2010), the flame instability of natural gas is shown. It is seen that a jet fire cannot be supported within a large region of diameters and pressures (Fig. 9.24).

According to the Chamberlain's equation, the flame stability band depends on the laminar burning velocity. It is reasonable to expect that this zone will be different for hydrogen and methane (Bradley, 1969) (Fig. 9.25; Table 9.31).

As discussed by Kalghatgi (1981), depending on momentum, flame stability is increased by cross-wind, and the tendency to lift-off is reduced. This is a very important aspect to be taken

Fig. 9.23
Flame colour-temperature zones.

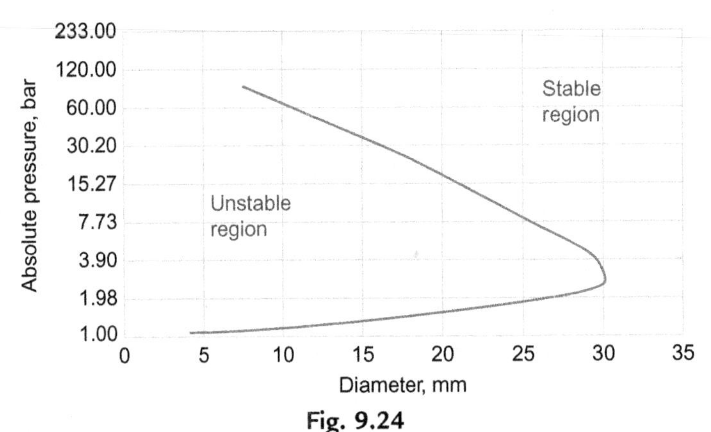

Fig. 9.24
Natural gas flame stability regions (FABIG, 2010).

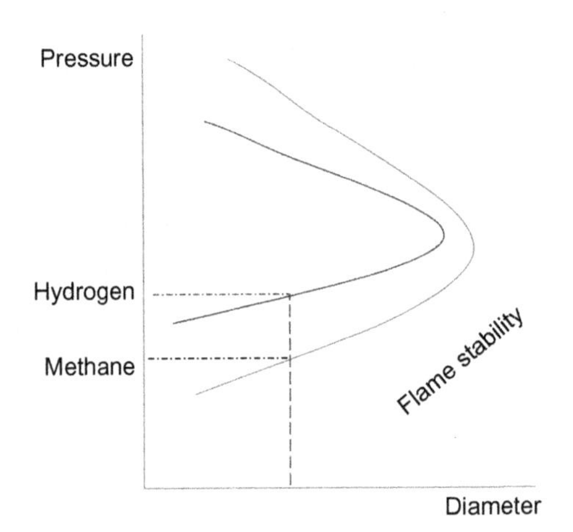

Fig. 9.25
Gas flame stability regions for hydrogen and methane.

Table 9.31 Laminar burning velocity for methane and hydrogen (Bradley, 1969)

Material	S_u (m/s)
Methane	0.30–0.37
Hydrogen	1.80

into account, especially for laminar and laminar-to-turbulent flame jets, for which the wind effectiveness in increasing air entrainment is very high.

Finally, it must be noticed that impinging jets increase the likelihood of their ignition due to the modification of the concentrations profile around the axis with respect to the free jet.

Table 9.32 Effective radiation temperature (SINTEF, 2003)

Material	Effective Radiation Temperature (°C)
Methane	1016
Ethane	1317
Ethylene	1449
Propane	1288
Isobutane	1281
Normal butane	1339
Propylene	1217
Isobutylene	1136

Flame temperature

Jet fires exhibit a higher temperature than the buoyancy controlled diffusion flames due to the very high air entrainment effect, assumed as 4–5 times the stoichiometric (SINTEF, 2003). Effective flame radiation temperatures are shown in Table 9.32 (SINTEF, 2003).

The Norwegian Fire Research Laboratory has detected an experimental maximum temperature of 1380°C. A significant gradient temperature exists along the axis and in the cross-section (Odgaard, 1983):

- The zone of maximum mean axial temperature is located at 60% of the flame length.
- The maximum temperature is located off-axis.
- Maximum jet temperature *should be* 1500°C.

Modelling

No-wind scenario

For jet fire with negligible wind, the method presented by Hawthorne et al. (1949) can be used.

Flame length The visible flame length, downstream of the lift-off distance, is given by the expression:

$$L_f = D_e \cdot \frac{5.3}{C_{fr}} \cdot \left[\frac{T_{ad}}{C_{rp} \cdot T_o} \cdot \sqrt{C_{fr} + (1 - C_{fr}) \cdot \frac{MW_{air}}{MW_f}} \right] \quad (9.161)$$

where:

D_e is the orifice diameter, m.
C_{fr} is the fuel to reactants molar ratio.
C_{rp} is the reactants to products molar ratio.
T_{ad} is the adiabatic flame temperature, K.
T_o is the temperature before the release, K.
WM_{air} and WM_f are the molecular weights of air and fuel, respectively.

This equation is approximated as:

$$L_f = D_e \cdot \frac{15}{C_{fr}} \cdot \sqrt{\frac{MW_{air}}{MW_f}} \tag{9.162}$$

For choked flows that are the most likely, D_e should be replaced by (McCaffrey and Evans, 1986):

$$D_{e1} = D_e \cdot \frac{1}{\sqrt{M_a}} \cdot \left[\frac{2+(\gamma-1)\cdot M_a^2}{\gamma+1}\right]^{\frac{\gamma+1}{4\cdot(\gamma-1)}} \tag{9.163}$$

with Mach number of an expanding jet M_A (McCaffrey and Evans, 1986):

$$M_a = \sqrt{\frac{2\cdot\left(\frac{P_o}{P_a}\right)^{\frac{\gamma}{\gamma-1}} - 2}{(\gamma-1)}} \tag{9.164}$$

where:

γ is the adiabatic exponent.
P_o is the upstream static pressure.
P_a is the atmospheric pressure.

Flame lift-off distance According to Kent (1968), the lift-off distance is:

$$L_{LO} = 4\cdot\pi\cdot D_e \tag{9.165}$$

And the effective exit velocity is M_a times the sound speed at the stagnation temperature (McCaffrey and Evans, 1986):

$$u_e = M_a \cdot \sqrt{\frac{\gamma\cdot\frac{R}{MW_f}\cdot T_o}{1+\frac{\gamma-1}{2}\cdot M_a^2}} \tag{9.166}$$

with:

T_o, the upstream stagnation temperature, K
R, the gas constant for the specific gas

Baron (1954), has proposed the following formulas for the jet fire plume diameter:

$$D_{JF}(x) = 0.29\cdot x\cdot\sqrt{\ln\left(\frac{L_f}{x}\right)} \tag{9.167}$$

Being x the coordinate of the jet axis and:

$$D_{JF}\,\mathrm{max} = 0.12 \cdot L_f \tag{9.168}$$

the maximum diameter which is attained at $x = 0.6 \cdot L_f$.

Example 9.15

Release of hydrogen in calm air from 50 mm at a relative pressure of 9 atm and 20°C results in a jet fire. Find the flame length from the outlet and the behaviour of the flame diameter (Fig. 9.26).

Solution

T_a (K)	T_{ad} (K)	D_e (m)	MW_{air}	MW_{H2}	γ	P_o (atm)
293.16	2318.16	0.05	29	2	1.41	0.14

$$2H_2 + O_2 \rightarrow 2H_2O \tag{9.169}$$

$$C_{fr} = \frac{2}{3} = 0.67 \tag{9.170}$$

$$C_{rp} = \frac{3}{2} = 1.5 \tag{9.171}$$

$$M_a = \sqrt{\frac{2 \cdot \left(\frac{P_o}{P_a}\right)^{\frac{\gamma-1}{\gamma}} - 2}{\gamma - 1}} = \sqrt{\frac{2 \cdot \left(\frac{10}{1}\right)^{\frac{0.41}{1.41}} - 2}{0.41}} = 1.95 \tag{9.172}$$

$$D_{e1} = D_e \cdot \frac{1}{\sqrt{M_a}} \cdot \left[\frac{2 + (\gamma - 1) \cdot M_a^2}{\gamma + 1}\right]^{\frac{\gamma+1}{4 \cdot (\gamma-1)}} = 0.05 \cdot \frac{1}{\sqrt{1.95}} \cdot \left[\frac{2 + 0.41 \cdot 1.95^2}{2.41}\right]^{\frac{2.41}{4 \cdot 0.41}}$$

$$= 0.05 \cdot 1.26 = 0.063\,\mathrm{m} \tag{9.173}$$

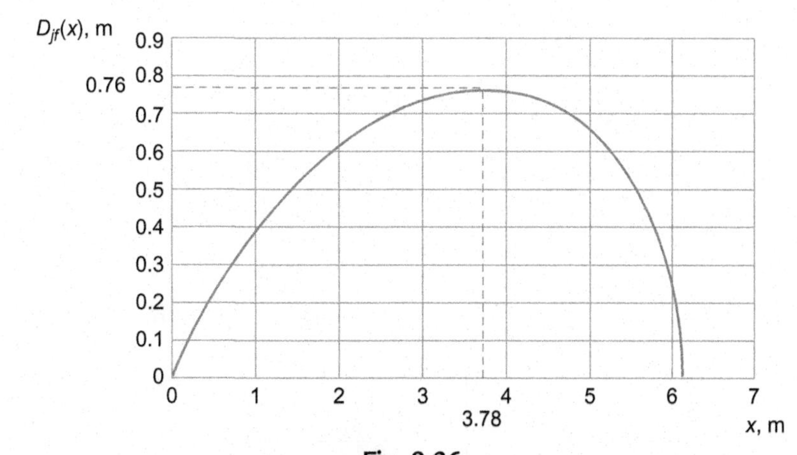

Fig. 9.26
Jet fire plume diameter for the Example 9.16.

$$L_f = D_{el} \cdot \frac{5.3}{C_{fr}} \cdot \left[\frac{T_{ad}}{C_{rp} \cdot T_o} \cdot \sqrt{C_{fr} + (1 - C_{fr}) \cdot \frac{MW_{air}}{MW_f}} \right]$$

$$= 0.063 \cdot \frac{5.3}{0.67} \cdot \left[\frac{2318.16}{1.5 \cdot 293.16} \cdot \sqrt{0.67 + (1 - 0.67) \cdot \frac{29}{2}} \right] = 6.13\,\text{m} \quad (9.174)$$

$$L_{LO} = 4 \cdot \pi \cdot D_{e_1} = 4 \cdot \pi \cdot 0.063 = 0.79\,\text{m} \quad (9.175)$$

$$L_{flame} = L_f + L_{LO} = 6.13 + 0.79 = 6.92\,\text{mm} \quad (9.176)$$

$$D_{JF}(x) = 0.29 \cdot x \cdot \sqrt{\ln\left(\frac{6.13}{x}\right)} \quad (9.177)$$

Wind scenario

Several models are available to describe jet flames in wind conditions, such as Brzustowski and Chamberlain's model.

API 521 Model Graphical API 521 method is recommended to evaluate the distortion of flame due to the wind in vertical upward jet fires.

Example 9.16

For ethane at 5 atm gauge the exiting at choked condition 273.16 K, and find the distortion due to a wind speed of 10 m/s. The hole size is 100 mm.

Solution

The following diagrams, calculated on the basis of the data provided by API 521, show the behaviour of the ratio of the distorted flame Δx and Δy to the length flame L_f against the wind to release speed ratio. Data for calm conditions have to be calculated according to the Hawthorne method.

$$M_a = \sqrt{\frac{2 \cdot \left(\frac{P_o}{P_a}\right)^{\frac{\gamma-1}{\gamma}} - 2}{\gamma - 1}} = \sqrt{\frac{2 \cdot \left(\frac{6}{1}\right)^{\frac{0.18}{1.18}} - 2}{0.18}} = 1.82 \quad (9.178)$$

$$D_{e1} = D_e \cdot \frac{1}{\sqrt{M_a}} \cdot \left[\frac{2 + (\gamma - 1) \cdot M_a^2}{\gamma + 1}\right]^{\frac{\gamma+1}{4 \cdot (\gamma-1)}} = 0.1 \cdot \frac{1}{\sqrt{1.82}} \cdot \left[\frac{2 + 0.18 \cdot 1.82^2}{2.18}\right]^3 = 0.125\,\text{m} \quad (9.179)$$

$$L_f = D_{e1} \cdot \frac{15}{C_{fr}} \cdot \sqrt{\frac{MW_{air}}{MW_f}} = 0.125 \cdot \frac{15}{0.23} \cdot \sqrt{\frac{29}{30}} = 8\,\text{m} \quad (9.180)$$

$$u_e = M_a \cdot \sqrt{\frac{\gamma \cdot R \cdot T_o}{1 + \frac{\gamma - 1}{2} \cdot M_a^2}} = 1.82 \cdot \sqrt{\frac{1.18 \cdot \frac{8,314}{30} \cdot 273.16}{1 + \frac{0.18}{2} \cdot 1.82^2}} = 262.3\,\text{m/s} \quad (9.181)$$

From the diagrams (Figs. 9.27 and 9.28):

$$\frac{u_w}{u_e} = \frac{20}{262.3} = 0.038 \rightarrow \frac{\Delta x}{L_f} = 0.6 \rightarrow \Delta x = 4.8\,\text{m} \tag{9.182}$$

$$\frac{u_w}{u_e} = \frac{20}{262.3} = 0.038 \rightarrow \frac{\Delta y}{L_f} = 0.63 \rightarrow \Delta y = 5\,\text{m} \tag{9.183}$$

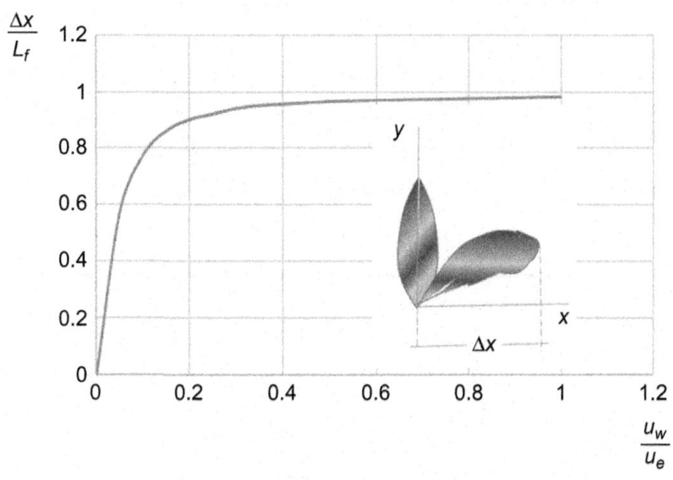

Fig. 9.27
Distortion distance ratio (API 521).

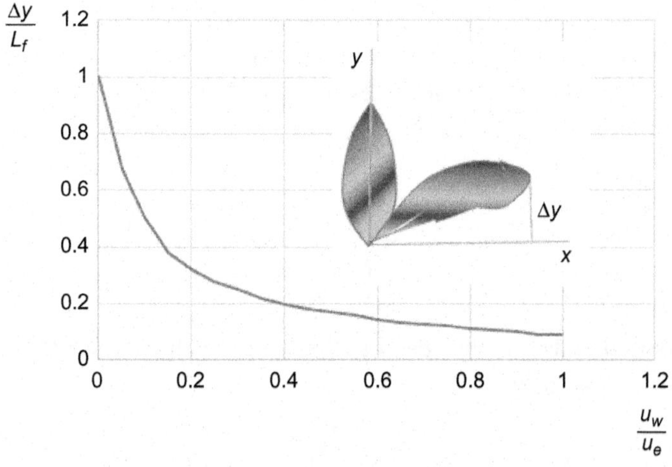

Fig. 9.28
Distortion distance ratio (API 521).

This model should necessarily account for the enhancement of the air entrainment promoted by wind, and it should result in a shorter flame. Chamberlain has provided the following equation for the calculation of the flame length modified by wind speeds greater than 2 m/s:

$$L_{fw} = L_f \cdot \left(0.51 \cdot e^{-0.4 \cdot u_w} + 0.49\right) \cdot [1 - 0.00607 \cdot (\theta - 90°)] \tag{9.184}$$

This equation is valid for tilt angles with a vertical less than 45°.

Example 9.17

Find the flame length for wind conditions for ethane, which was analysed in the previous example.

Solution

Being the tilt angle very close to 45°, Chamberlain's equation can be applied:

$$L_{fw} = 8 \cdot \left(0.51 \cdot e^{-0.4 \cdot 5} + 0.49\right) \cdot [1 - 0.00607 \cdot (45° - 90°)] = 5.7\,\text{m} \tag{9.185}$$

Kalghatgi solid flame model

This model is relatively simple and is considered particularly accurate (SINTEF, 2003) both in terms of flame size and characterisation and, consequently, in term of the calculation of the flame radiation. Its validity and calibration have been analysed and implemented on the basis of experimental results. It's a solid flame model, whereas the surface emissive power calculated by means of the Brzustowski will be based on a source point model, which inevitably is much less accurate. Kalghatgi (1983) assumed the flame as a frustum (truncated cone) (Fig. 9.29).

The model presents the following equations for the calculation of some geometrical parameters of the flame, over the range $0.02 < R < 0.25$.

Wind to discharge speed ratio

$$R = \frac{u_w}{u_e} \tag{9.186}$$

where u_w is wind speed and u_e is the speed at the exit, which can be calculated applying the following known equations:

$$M_a = \sqrt{\frac{2 \cdot \left(\frac{P_o}{P_a}\right)^{\frac{\gamma - 1}{\gamma}} - 2}{\gamma - 1}} \tag{9.187}$$

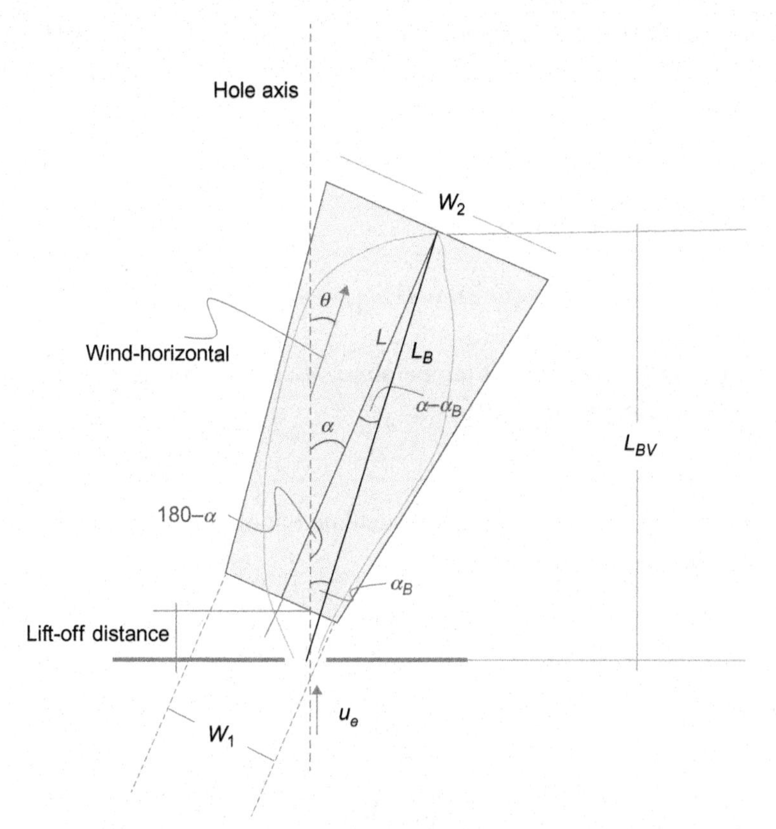

Fig. 9.29
Kalghatgi solid flame model.

$$u_e = M_a \cdot \sqrt{\frac{\gamma \cdot \frac{R}{MW_f} \cdot T_o}{1 + \frac{\gamma - 1}{2} \cdot M_a^2}} \tag{9.188}$$

Expanding jet data

$$T_e = T_o \cdot \left(\frac{P_{air}}{P_o}\right)^{\frac{\gamma-1}{\gamma}} \tag{9.189}$$

$$P_e = P_o \cdot \left(\frac{2}{\gamma + 1}\right)^{\frac{\gamma}{\gamma-1}} \tag{9.190}$$

$$\rho_e = \frac{P_e \cdot WM_f}{R \cdot T_e} \tag{9.191}$$

where subscript e refers to the exit plane, o to the stagnation condition, WM_F is the fuel molecular weight, and R is the gas constant expressed as 8314 J/kmol K. All variables are given according to the international unit (mks) system.

Source diameter

The source diameter D_s is defined as:

$$D_S = D_e \cdot \sqrt{\frac{\rho_f}{\rho_{air}}} \tag{9.192}$$

where ρ_f is the fuel density at ambient temperature. It corresponds to the approximation of Thring and Newby (Benintendi, 2010).

Mass flow rate

The most realistic mass flow rate should be calculated adopting the jet characteristics at the outlet, so the sound speed has been used:

$$W_e = \pi \cdot \frac{De^2}{4} \cdot \pi \cdot \rho_e \cdot u_s \tag{9.193}$$

where u_s is the sound speed (choked flow) at the outlet.

$$u_s = \sqrt{\gamma \cdot \frac{R}{MW_f} \cdot T_e} \tag{9.194}$$

These formulas don't take into account the coefficient of discharge C_D. It has to be considered in the practical calculations what has been stated in Chapter 4.

Vertical flame length

$$L_{BV} = D_S \cdot \left(6 + \frac{2.35}{R} + 20 \cdot R\right) \tag{9.195}$$

Calculation of angle α_B

(angle between the hole axis and the line joining the hole centre to downstream flame tip.)

$$\alpha_B = 94 - \frac{1.6}{R} - 35 \cdot R \tag{9.196}$$

Calculation of angle α

(angle between the hole axis and the frustum axis)

$$\alpha = 94 - \frac{1.1}{R} - 30 \cdot R \tag{9.197}$$

Calculation of L_B

$$L_B = \frac{L_{BV}}{\cos \alpha_B} \tag{9.198}$$

Visible flame length

Applying trigonometric Euler 'sin formula to the triangle, including angles α_B, $\alpha - \alpha_B$, and $180 - \alpha$ (Fig. 9.30):

$$L = \frac{L_{BV} \cdot \sin \alpha_B}{\cos \alpha_B \cdot \sin(180 - \alpha)} \tag{9.199}$$

This is the visible flame length.

Lift-off distance

The lift-off distance can be calculated applying Carnot's cosine formula to the triangle, including angles α_B, $\alpha - \alpha_B$, and $180 - \alpha$ (Fig. 9.31):

$$L_{LO} = \sqrt{L_B^2 + L^2 - 2 \cdot L_B \cdot L \cdot \cos(\alpha - \alpha_B)} \tag{9.200}$$

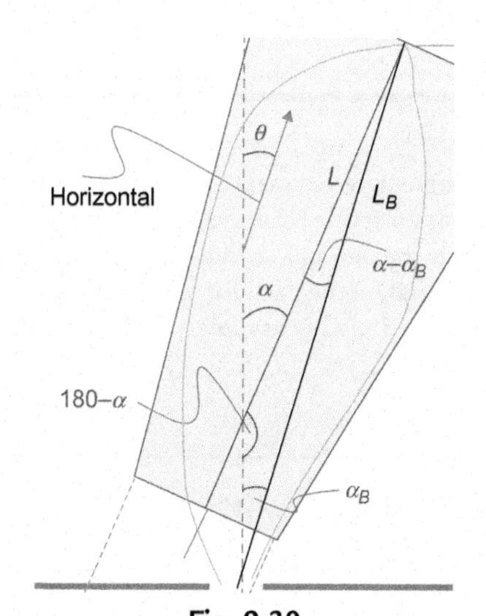

Fig. 9.30
Angles of Kalghatgi solid flame model.

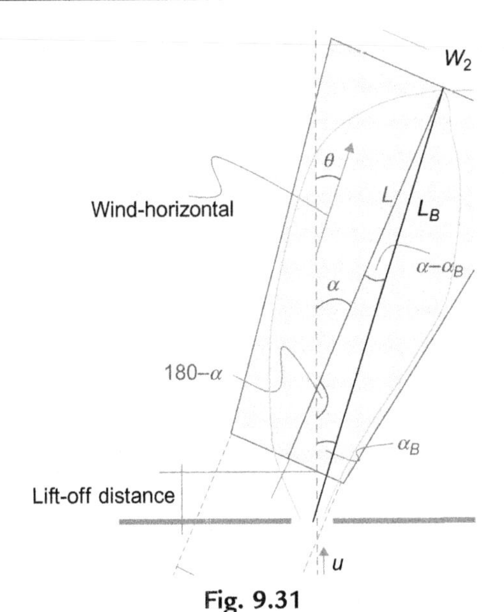

Fig. 9.31
Kalghatgi solid flame model—Lift-off distance.

Flame widths

$$W_1 = D_S \cdot \left(49 - \frac{0.22}{R} - 380 \cdot R + 950 \cdot R^2\right) \tag{9.201}$$

$$W_2 = D_S \cdot \left(80 - \frac{0.57}{R} - 570 \cdot R + 1470 \cdot R^2\right) \tag{9.202}$$

Example 9.18

LPG is stored in a sphere at ambient temperature (20°C). Pressurised liquid consists of 75% propane and 25% *n*-butane (mass fraction). Liquid is released through an 80 mm hole, the axis of which forms a 30° angle with the vertical, and a jet flame is seen. The hole is located on the gas-phase part of the shell. Analyse the case for wind speeds of 4/ms and 15 m/s. No liquid rain-out is assumed, and all emitted LPG is assumed to be ignited (Fig. 9.32).

Solution

Propane

T_a (K)	T_{NB} (K)	D_e (m)	MW_{air}	$MW_{propane}$	γ	x_m	Vapour pressure $P_{p°}$(20°C) (MPa)
293.16	231	0.08	29	44.1	1.13	0.75	0.83

***n*-Butane**

T_a (K)	T_{NB} (K)	D_e (m)	MW_{air}	$MW_{n\text{-}butane}$	γ	x_m	Vapour pressure $P_b°$(20°C) (MPa)
293.16	273	0.08	29	58.1	1.18	0.25	0.208

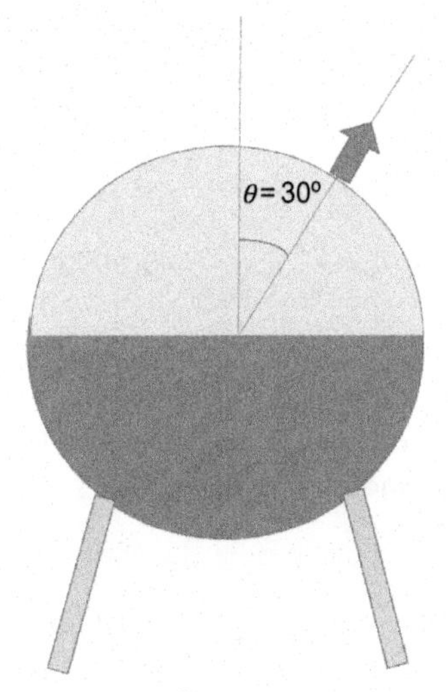

Fig. 9.32
LPG sphere of Example 9.19.

The average specific heat ratio is:

$$\gamma = x_{mp} \cdot \gamma_p + x_{mb} \cdot \gamma_b = 0.75 \cdot 1.13 + 0.25 \cdot 1.18 \cong 1.14 \tag{9.203}$$

The molar fractions are promptly calculated assuming 1 kg of liquid mixture:

$$x_p = \frac{\frac{750}{44.1}}{\frac{750}{44.1} + \frac{250}{58.1}} \cong 0.8 \tag{9.204}$$

$$x_b = \frac{\frac{250}{58.1}}{\frac{750}{44.1} + \frac{250}{58.1}} \cong 0.2 \tag{9.205}$$

The average molecular weight is:

$$WM = x_p \cdot WM_p + x_b \cdot WM_b = 0.8 \cdot 44.1 + 0.2 \cdot 58.1 \cong 47\,\text{g/mol} \tag{9.206}$$

Storage pressure

The storage pressure can be calculated adopting Raoult's equation:

$$P_o = P_p \cdot x_p + P_b \cdot x_b = 0.83 \cdot 0.8 + 0.208 \cdot 0.2 \cong 0.7\,\text{MPa} \tag{9.207}$$

$$M_a = \sqrt{\frac{2 \cdot \left(\frac{P_o}{P_a}\right)^{\frac{\gamma-1}{\gamma}} - 2}{\gamma - 1}} = \sqrt{\frac{2 \cdot \left(\frac{0.70}{0.10}\right)^{\frac{0.14}{1.14}} - 2}{0.14}} = 1.96 \tag{9.208}$$

$$u_e = M_a \cdot \sqrt{\frac{\gamma \cdot \dfrac{R}{MW_f} \cdot T_o}{1 + \dfrac{\gamma - 1}{2} \cdot M_a^2}} = 1.96 \cdot \sqrt{\frac{1.14 \cdot \dfrac{8314}{47} \cdot 293.16}{1 + \dfrac{0.14}{2} \cdot 1.96^2}} = 423\,\text{m/s} \tag{9.209}$$

$$R_1 = \frac{u_w}{u_e} = \frac{4}{423} = 0.0095 \tag{9.210}$$

$$R_2 = \frac{u_w}{u_e} = \frac{15}{423} = 0.035 \tag{9.211}$$

For the first scenario $R < 0.02$. According to Kalghatgi and Beyler (2016), this case has to be considered not affected by wind.

Expanding jet data

$$T_e = T_o \cdot \left(\frac{P_{air}}{P_o}\right)^{\frac{\gamma - 1}{\gamma}} = 293.16 \cdot \left(\frac{0.1}{0.7}\right)^{\frac{0.14}{1.14}} = 230.7\,\text{K} \tag{9.212}$$

$$P_e = P_o \cdot \left(\frac{2}{\gamma + 1}\right)^{\frac{\gamma}{\gamma - 1}} = 0.7 \cdot \left(\frac{2}{2.14}\right)^{\frac{1.14}{0.14}} = 0.4\,\text{MPa} \tag{9.213}$$

$$\rho_e = \frac{P_e \cdot WM_f}{R \cdot T_e} = \frac{400{,}000 \cdot 47}{8314 \cdot 230.7} = 9.8\,\text{kg/m}^3 \tag{9.214}$$

Source diameter

The source diameter D_s is defined as:

$$D_S = D_e \cdot \sqrt{\frac{\rho_f}{\rho_{air}}} = 0.08 \cdot \sqrt{\frac{1.94}{1.2}} = 0.1\,\text{m} \tag{9.215}$$

Mass flow rate

$$u_s = \sqrt{1.14 \cdot \frac{8{,}314}{47} \cdot 230.7} = 215.7\,\text{m/s} \tag{9.216}$$

$$W_e = \pi \cdot \frac{0.08^2}{4} \cdot 9.8 \cdot 215.7 = 10.62\,\text{kg/s} \tag{9.217}$$

Vertical flame length

$$L_{BV} = D_S \cdot \left(6 + \frac{2.35}{R_2} + 20 \cdot R_2\right) = 0.1 \cdot \left(6 + \frac{2.35}{0.035} + 20 \cdot 0.035\right) = 7.3\,\text{m} \tag{9.218}$$

Calculation of angle α_B

(angle between the hole axis and the line joining the hole centre to downstream flame tip.)

$$\alpha_B = 94 - \frac{1.6}{R_2} - 35 \cdot R_2 = 94 - \frac{1.6}{0.035} - 35 \cdot 0.035 = 47° \tag{9.219}$$

Calculation of angle α

(angle between the hole axis and the frustum axis)

$$\alpha = 94 - \frac{1.1}{R_2} - 30 \cdot R_2 = 94 - \frac{1.1}{0.035} - 30 \cdot 0.035 = 61.5° \tag{9.220}$$

Calculation of L_B

$$L_B = \frac{L_{BV}}{\cos\alpha_B} = \frac{7.3}{0.68} = 10.7\,\text{m} \tag{9.221}$$

Visible flame length

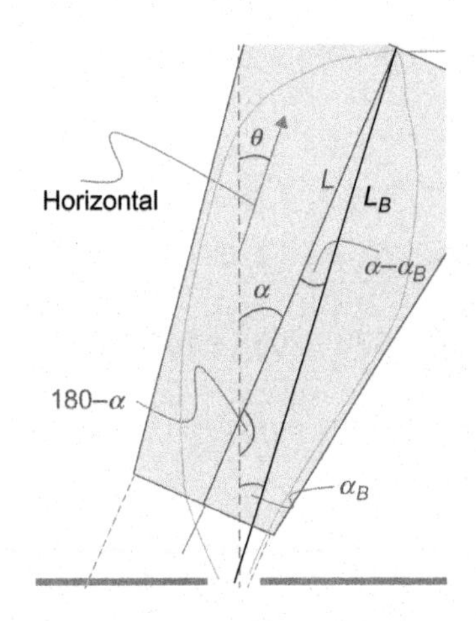

$$L = \frac{L_{BV} \cdot \sin\alpha_B}{\cos\alpha_B \cdot \sin(180 - \alpha)} = \frac{7.3 \cdot 0.73}{0.68 \cdot 0.88} = 8.9\,\text{m} \tag{9.222}$$

Lift-off distance

The lift-off distance can be calculated applying Carnot's cosine formula to the triangle including angles α_B, $\alpha - \alpha_B$, and $180 - \alpha$:

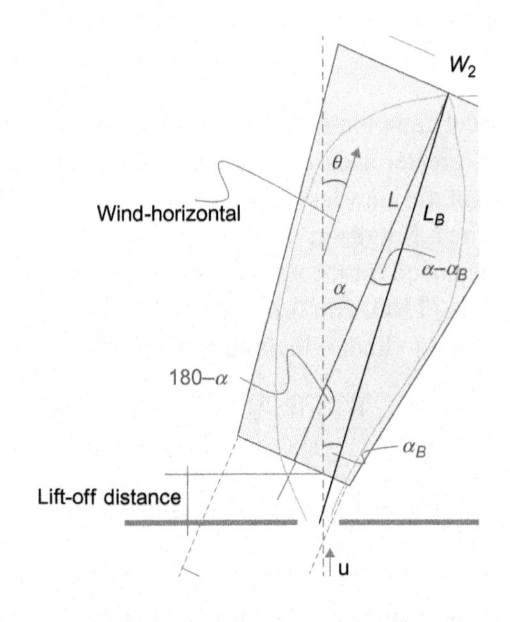

$$L_{LO} = \sqrt{L_B^2 + L^2 - 2 \cdot L_B \cdot L \cdot \cos(\alpha - \alpha_B)} = \sqrt{10.7^2 + 8.9^2 - 2 \cdot 10.7 \cdot 8.9 \cdot 0.97} = 3.0\,\mathrm{m} \tag{9.223}$$

Flame widths

$$W_1 = D_S \cdot \left(49 - \frac{0.22}{R_2} - 380 \cdot R_2 + 950 \cdot R_2^2\right) = 0.1 \cdot \left(49 - \frac{0.22}{0.035} - 380 \cdot 0.035 + 950 \cdot 0.035^2\right) = 3.0\,\mathrm{m} \tag{9.224}$$

$$W_2 = D_S \cdot \left(80 - \frac{0.57}{R} - 570 \cdot R + 1470 \cdot R^2\right) = 0.1 \cdot \left(80 - \frac{0.57}{0.035} - 570 \cdot 0.035 + 1470 \cdot 0.035^2\right) = 4.6\,\mathrm{m} \tag{9.225}$$

The overall distance of the flame from the outlet is:

$$L_f = L + L_{LO} = 8.9 + 3 = 11.9\,\mathrm{m} \tag{9.226}$$

Heat transfer

The heat transfer equation from a source to a target is the same presented for pool fire:

$$Q_H = F \cdot SEP_{av} \cdot \tau \tag{9.227}$$

where:

F is the overall view factor.
SEP_{av} is the surface emissive power.
τ is the transmissivity.

Kalghatgi's solid flame model equation is recommended instead of the point source model because of its simplicity, its greater accuracy for the flame length prediction and for the much more robust definition of the view factors for close targets, exactly those presented for the solid flame model of pool fires. Kalghatgi's model consists of a truncated cone (frustum), whereas the view factors developed for the pool fires are based on a cylinder. In order to cover the gap, Uijt de Haag and Ale (TNO, 2005) provide the following formula related to the surface area A of the equivalent cylinder, including end discs:

$$A = \frac{\pi}{2} \cdot \left(\frac{W_1 + W_2}{2}\right)^2 + \pi \cdot L \cdot \frac{W_1 + W_2}{2} \tag{9.228}$$

The Surface Emissive Power, SEP_{av}, is given by the general formula:

$$SEP_{av} = \varepsilon \cdot \sigma \cdot T_R \tag{9.229}$$

where T_R is the effective flame radiation temperature, and ε is the emissivity. The former is tabulated for some gases but not for all gases. The emissivity, in turn, depends on other variables, notably the extinction coefficient and the mean beam length of the fire plume. Uijt de Haag and Ale (TNO, 2005) have proposed a generalised method which is consistent with Kalghatgi's solid flame model. This method consists of the following steps:

Step 1—Calculation of mass flow rate

$$u_S = \sqrt{\gamma \cdot \frac{R}{MW_f} \cdot T_e} \tag{9.230}$$

$$W_e = \pi \cdot \frac{De^2}{4} \cdot \pi \cdot \rho_e \cdot u_s \tag{9.231}$$

Step 2—Calculation of the net heat released per unit time

$$Q_H = W_e \cdot \Delta H_c \tag{9.232}$$

where:

Q_H is the radiated heat power, W.
W_e is the initial mass flow rate at the exit, kg/s.
ΔH_c is the combustion heat (lower heating value), J/kg.

Step 3—Calculation of heat radiated fraction

$$X_R = 0.21 \cdot \exp(-0.00323 \cdot u_e) + 0.11 \tag{9.233}$$

Step 4—Calculation of the SEP_{av}

$$SEP_{AV} = \frac{Q_H \cdot X_R}{0.25 \cdot (W_1^2 + W_2^2) + 0.5 \cdot (W_1 + W_2) \cdot \sqrt{\left(L^2 + \left(\frac{W_2 - W_1}{2}\right)^2\right)}} \tag{9.234}$$

Example 9.19

Find the average surface emissive power for LPG discussed in the previous example.

Solution

x_m	Propane—Enthalpy of Combustion (J/kg)	*n*-Butane—Enthalpy of Combustion (J/kg)	x_m
0.75	46,434,318	45,815,862	0.25

Enthalpy of combustion of the mixture:

$$\Delta H_c = 0.75 \cdot 46{,}434{,}318 + 0.25 \cdot 45{,}815{,}862 = 46{,}279{,}704 \text{J/kg} \tag{9.235}$$

Sound speed (speed at the outlet):

$$u_s = \sqrt{1.14 \cdot \frac{8314}{47} \cdot 230.7} = 215.7\,\mathrm{m/s} \tag{9.236}$$

Speed of the expanding jet:

$$u_e = M_a \cdot \sqrt{\frac{\gamma \cdot \frac{R}{MW_f} \cdot T_o}{1 + \frac{\gamma - 1}{2} \cdot M_a^2}} = 1.96 \cdot \sqrt{\frac{1.14 \cdot \frac{8314}{47} \cdot 293.16}{1 + \frac{0.14}{2} \cdot 1.96^2}} = 423\,\mathrm{m/s} \tag{9.237}$$

Mass flow rate:

$$W_e = \pi \cdot \frac{0.08^2}{4} \cdot 9.8 \cdot 215.7 = 10.62\,\mathrm{kg/s} \tag{9.238}$$

Area of the equivalent cylinder:

$$W_1 = 3.0\,\mathrm{m} \tag{9.239}$$

$$W_2 = 4.6\,\mathrm{m} \tag{9.240}$$

$$L = 8.9\,\mathrm{m} \tag{9.241}$$

$$A = \frac{\pi}{2} \cdot \left(\frac{W_1 + W_2}{2}\right)^2 + \pi \cdot L \cdot \frac{W_1 + W_2}{2} = \frac{\pi}{2} \cdot 3.8^2 + \pi \cdot 8.9 \cdot 3.8 = 129\,\mathrm{m}^2 \tag{9.242}$$

Radiate heat fraction:

$$X_R = 0.21 \cdot \exp(-0.00323 \cdot u_e) + 0.11 = 0.21 \cdot \exp(-0.00323 \cdot 423) + 0.11 = 0.116 \tag{9.243}$$

$$Q_H = W_e \cdot \Delta H_c = 10.62 \cdot 46{,}279{,}704 = 491{,}490{,}456.5\,\mathrm{W} \simeq 491{,}490\,\mathrm{kW} \tag{9.244}$$

$$SEP_{AV} = \frac{Q_H \cdot X_R}{0.25 \cdot (W_1^2 + W_2^2) + 0.5 \cdot (W_1 + W_2) \cdot \sqrt{\left(L^2 + \left(\frac{W_2 - W_1}{2}\right)^2\right)}} = \frac{491{,}490 \cdot 0.116}{129} = 442\,\mathrm{KW/m^2} \tag{9.245}$$

The order of magnitude of this value is consistent with those included in the literature. SINTEF (2003) reports an effective emissive power of 383 kW/m^2 for *n*-butane jet flames.

Recommended incident heat fluxes

See Table 9.33.

Table 9.33 Recommended incident heat fluxes

Local Peak Heat Load (kW/m^2)	Global Average Heat Load (kW/m^2)	Reference
350 (Leak rates > 2.0 kg/s) 250 (Leak rate > 0.1 kg/s)	100 (Leak rates > 2.0 kg/s) 0 (Leak rate > 0.1 kg/s)	Scandpower (2004)
300	–	API 521 (2014)
250 (Initial flux)	–	Norsok S-001 (2008)

9.8 *Fire Damage to People*

The Eisenberg, Lynch, and Breeding vulnerability model is based on probit functions included in Chapter 6. Tables 9.34 and 9.35, provided by the standard API 521, correlate specific thermal levels to time taken to feel pain and to the shielding and protective clothing requirements.

Table 9.34 Time to pain (API 521)

Radiation (kW/m^2)	Time to Pain (s)
1.740	60
2.330	40
2.900	30
4.730	16
6.940	9.0
11.67	4.0
19.87	2

Table 9.35 Thermal radiation thresholds (API 521)

Permissible Design Level (kW/m^2)	Conditions
9.46	Maximum radiant intensity where urgent emergency action is required. Being this level >6.31 kW/m^2, shielding is required
6.31	Maximum radiant intensity where urgent emergency action lasting >30 s is required without shielding or protective cloth
4.73	Maximum radiant intensity where urgent emergency action lasting 2–3 min is required without shielding or protective cloth
1.58	Maximum radiant intensity where urgent emergency action lasting 2–3 min is required without shielding or protective cloth

References

API 2003, 2015. Protection Against Ignitions Arising out of Static, Lightning, and Stray Currents, eighth ed.

API Standard 521, 2014. Pressure-relieving and Depressuring Systems, sixth ed. January.

Babrauskas, V., 1983. Estimating Large Pool Fire Burning Rates. Fire. Technol 1, 251–261.

Bain, B., Celnik, M., Korneliussen, G., 2006. Practical application of the UKOOA ignition model. In: Hazards XXIII. Sympoisum Series no. 158, IChemE.

Baron, T., 1954. Reactions in turbulent free jets. The turbulent diffusion flame. Chem. Eng. Prog. 50(N2).

Benintendi, R., 2010. Turbulent jet modelling for hazardous area classification. J. Loss Prev. Process Ind. 23, 373–378.

Beyler, C., 2016. Fire Hazard Calculations for Large, Open Hydrocarbon Fires in SFPE Handbook of Fire Protection Engineering, fifth ed. Springer.

Bradley, J.N., 1969. Flame and Combustion Phenomena. Barnes & Noble.

Britton, L.G., 1999. Avoiding Static Ignition Hazards in Chemical Operations. Center of Chemical Process Safety of the American Institute of Chemical Engineers.

BS 5958-1, 1991. Code of Practice for Control of Undesirable Static Electricity—Part 1: General Considerations.

BS EN 1127-1, 2011. Explosive Atmospheres—Explosion Prevention and Protection. Part 1: Basic Concepts and Methodology.

BS EN 13463-1, 2009. Non-electrical Equipment for Use in Potentially Explosive Atmospheres, Part 1: Basic Method and Requirements.

BS EN 62305-1, 2011. Protection Against Lightning. Part 1: General Principles.

Burgess, D., Hertzberg, M., 1974. Radiation from pool flames. In: Afgan, N.H., Beer, J.M. (Eds.), Heat Transfer in Flames.

Burgess, D.S., Zabetakis, M.G., 1961. Research on the Hazards Associated with the Production and Handling of liquid Hydrogen. R.I. 5707, Bureau of Mine, Pittsburgh, PA.

CCPS, 1995. Guidelines for Safe Storage and Handling of Reactive Materials. AIChE, Wiley.

CCPS, 2005. Combined Glossary of Terms.

Cetegen, B.M., Zukoski, E.E., Kubota, T., 1984. Entrainment in the near and far field of fire plumes. J. Combust. Sci. Technol. 39 (1–6), 305–331.

Chamberlain, G.A., 1995. An experimental study of water deluge on compartment fires. In: Modeling and Mitigating the Consequences of Accidental Releases of Hazardous Material. AIChE, pp. 763–776.

Cook, J., Bahrami, Z., Whitehouse, R.J., 1990. A comprehensive program for calculation of flame radiation levels. J. Loss Prev. 3.

Cox, A., Lees, F., Ang, M., 1990. Classification of Hazardous Locations. Report of the Inter-Institutional Group on the Classification of Hazardous Locations, IChemE.

Dean, J.A., 1999. Lange's Handbook of Chemistry, 15th ed. McGraw-Hill Inc.

EGIG, 2015. 9th Report of the European Gas Pipeline Incident Data Group https://www.egig.eu/uploads/bestanden/ba6dfd62-4044-4a4d-933c-07bf56b82383.

Eichel, F.G., 1967. Electrostatics, Chemical Engineering. McGraw-Hill Inc. March 13.

Eisenberg, N.A., Lynch, C.J., Breeding, R.J., Eisenberg, N.A., Lynch, C.J., Breeding, R.J., 1975. Vulnerability Model. A Simulation System for Assessing Damage Resulting from Marine Spills. National technology information service report, AD-A015-245, Springfield, MA, (1975).

Energy Institute, 2006. Model Development and Look-up Correlations. IP Research Report. IP. Energy Institute, HSE, UKOOA, London. ISBN 978-0-85293-454-8.

FABIG, 2010. Fire Loading and Structural Response. SCI Knowledge.

Gómez-Mares, M., Zárate, L., Casal, J., 2008. Jet Fires and the Domino Effect. Fire Saf. J. 43 (8), 583–588.

Hajek, J.D., Ludwig, E.E., 1960. How to Design Safe Flare Stacks, Part 1, Petro/Chem Engineer, 1960, Volume 32, Number 6, pp. C31–C38; Part 2, Petro/Chem Engineer, Volume 32, Number 7, pp. C44–C51.

Hawthorne, W.R., Weddell, D.S., Hottel, H.C., 1949. Mixing and Combustion in Turbulent Gas Jets. In: 3rd International Symposium on Combustion. Combustion Institute, Pittsburgh, PA, pp. 266–288.
Hurley, M.J., 2016. SFPE Handbook of Fire Protection Engineering, fifth ed. Springer.
Kalghatgi, G.T., 1981. Blow-Out Stability of Gaseous Jet Diffusion Flames. Part I: In Still Air. Shell Research Ltd., Thornton Research Centre, Chester.
Kalghatgi, G.T., 1983. The visible flame shape and size of a turbulent hydrocarbon jet diffusion flame in a crosswind. Combust. Flame 52, 91–103.
Kent, G.R., 1968. Find radiation effects on flares. Hydrocarb. Process. 47(6).
Kuo, K.K., Acharya, Ragini, 2012. Fundamentals of Turbulent and Multiphase Combustion. John Wiley & Sons Inc.
Liu, X., Zhang, Q., Wang, Y., 2015. Influence of particle size on the explosion parameters in two-phase vapor–liquid n-hexane/air mixtures. Process Saf. Environ. Prot. 95, 184–194. IChemE.
Mannan, S. (Ed.), 2012. Lees' Loss Prevention in the Process Industries, fourth ed. Butterworth-Heinemann.
McCaffrey, B.J., Evans, D.D., 1986. Very Large Methane Jet Diffusion Flames. Center for Fire Research National Bureau of Standards, Gaithersburg, MD.
Moorhouse, J., 1982. Scaling criteria for pool fires derived from large scale experiments. Inst. Chem. Eng. Symp. 71, 165–179.
Mudan, K.S., 1984. Thermal radiation hazards from hydrocarbon pool fires. Prog. Energy Combust. Sci. 10, 59–80.
Mudan, K.S., Croce, P.A., 1986. Calculating impacts for large open hydrocarbon fires. Fire Saf. J. 11, 99–112.
Mudan, K.S., Croce, P.A., 1988. Handbook of Fire Protection Engineering. Society of Fire Protection Engineers. Chapters 2–4.
NFPA 77, 2014. Recommended Practice on Static Electricity.
Norsok S-001, 2008. Technical Safety.
Odgaard, E., 1983. Characteristics of Hydrocarbon Fires in the Open-Air, Veritas. Report No. 82-0704, Det Norske Veritas Research Division, Norway. January.
OGP, 2010. Ignition probabilities, Risk Assessment Data Directory Report No. 434 – 6 March.
PD CLC/TR 60079-32-1, 2015. Explosive Atmospheres. Electrostatic Hazards, Guidance.
Raj, P.P.K., Emmons, H.W., 1975. On the Burning of a Large Flammable VaporCloud. In: Paper Presented at the Joint Technical Meeting of the Western and Central States Section of the Combustion Institute, San Antonio, TX.
Scandpower, 2004. Guidelines for the Protection of Pressurised Systems Exposed to Fire.
Shaw, P., Briscoe, F., 1978. Evaporation from spills of hazardous liquids. UKAEA, SRD R100, In: Marshall, V.C. (1987). (Ed.), Major Chemical Hazards. Ellis Horwood Ltd, Risley, p. 333. ISBN: 085312969X.
Siegel, R., Howell, J.R., 1981. Thermal Radiation Heat Transfer. McGraw-Hill Book Corp.
Sintef, 2003. Handbook for Fire Calculations and Fire Risk Assessment in the Process Industry. Norwegian Fire Research Laboratory.
Spouge, J., 1999. A Guide to Quantitative Risk Assessment for Offshore Installations. DNV Technica, CMPT.
Stull, D.R., 1976. Fundamentals for Fire and Explosion. Corporate Safety and Loss Dpt, The Dow Chemical Company, Midland.
Thomas, P.H., 1963. The size of flames from natural fires. In: Ninth Symposium (International) on Combustion. U.S. Combustion Institute, New York, pp. 844–859.
Treccani, Fires. http://www.treccani.it/portale/opencms/handle404?exporturi=/export/sites/default/Portale/sito/altre_aree/Tecnologia_e_Scienze_applicate/enciclopedia/inglese/inglese_vol_5/449_468_ing.pdf&%5D.
U.S. EPA, 1996. Guidance on the Application of Refined Dispersion Models to Hazardous/Toxic Air Pollutant Releases, EPA 454/R-93-002.
Uijt de Haag, P.A.M., Ale, B.J.M., 2005. Guideline for Quantitative Risk Assessment, CPR 18E, Purple Book. RVIM.
Woodward, J.L., Pitbaldo, R., 2010. LNG risk based safety: modeling and consequence analysis. In: Zukoski, E. (Ed.), Proceedings of the First International Symposium on Fire Safety Science. Wiley/Hemisphere Press, New York. 1984.

Further Reading

API Recommended Practice 2216, 2003. Ignition Risk of Hydrocarbon Liquids and Vapors by Hot Surfaces in the Open Air, third ed. December.

Brzustowski, T.A., 1976. Flaring in the energy industry. Prog. Energy Combust. 2, 129–141.

Croce, P.A., Mudan, K.S., Moorhouse, J., 1984. Thermal Radiation From LNG Trench Fires—Volume I. Report # GR 84/0151.1, Gas Research Institute, Chicago, IL. September.

Esposito, C., Mazzei, A., Mazzei, N., 2006. Valutazione dei Rischi Derivati dall'Elettricità Statica: Individuazione e Valutazione delle Misure di Prevenzione e Protezione. http://www.mednat.org/bioelettr/elettricita_statica-danni.pdf.

Rew, C., Deaves, D., 1999. Fire spread and flame length in ventilated tunnels, a model used in channel tunnel assessments. In: Proceedings of the 1st International Conference on Tunnel Fires and Escape From Tunnels, Lyon, France, 5–7 May, pp. 397–406.

Scandpower Risk Management AS, 2006. Blowout and Well Release Frequencies—Based on SINTEF Offshore Blowout Database, Report No. 90.005.001/R2.

Shokri, M., Beyler, C.L., 1989. Radiation from large pool fires. J. Fire. Prot. Eng. National Fire Protection Association.

CHAPTER 10

Explosions

The explosion was of the equivalent force to that of some 15-45 tons TNT

(Flixborough Court of Inquiry)

10.1 Summary of Scenarios

Explosion Scenario	Section
Confined explosion	10.6
Deflagration	10.6.1
Detonation	10.6.2
Detonation in pipes	10.7
Semiconfined explosion	10.9
Unconfined vapour cloud explosion (UVCE)	10.10
Multienergy method (MEM)	10.10.1
Baker-Strehlow-Tang method (BST)	10.10.2
Congestion_assessment_method (CAM)	10.10.3
Vessel burst	10.11
BLEVE	10.12
Rapid phase transition (RPT)	10.13
Thermal runaway	10.14
Pressure piling	10.16
Drag loads	10.17
Loads on buildings	10.18
Blast resistant buildings in the QRA framework	10.19
Hazardous area classification	10.20
Inerting	10.21

10.2 The Explosion

The definition of the term *explosion* is potentially related to a broad range of scenarios:

- The release of nonreacting pressurised gas stored in a vessel resulting in a significant overpressure.

Process Safety Calculations. https://doi.org/10.1016/B978-0-08-101228-4.00010-1

- The confined or unconfined production of energy by a chemical reaction resulting in a significant pressure rise.
- The violent expansion of a liquid due to the rapid transfer of a significant heat amount resulting in an intense pressure rise.
- The violent expansion of a pressurised liquid stored at a temperature greater than its normal boiling temperature due to a full loss of containment.

All explosive scenarios are associated with a significant pressure rise which can take place internally or externally to the system. The former case is classified as a confined explosion, the latter as an unconfined explosion. An intermediate scenario is the semiconfined explosion. A confined explosion is a chemical or physical phenomenon occurring isochorically (constant volume) inside a containment. In a semiconfined explosion, the potential energy is released, partially isochorically, with the containment still in place, whereas the remainder is released after the enclosure rupture. A semi confined-explosion is also that occurring in partially enclosed system, such as ducts or open pipes.

According to the definitions included in Chapter 5, it is important to note:

- A combustive explosion with flame speed smaller than the local sound velocity is always a deflagration.
- A deflagration is not always an explosion, for example, a flash fire is a deflagration with no significant pressure effect.
- According to the CCPS, a shock wave is a transient change in the gas density, pressure, and velocity of the air surrounding an explosion point. The initial change can be either discontinuous or gradual. A discontinuous change is referred to as a shock wave, and a gradual change is known as a pressure wave.

The consequences of an explosive phenomenon, depending on the circumstances, can be:

- Containment collapse
- Rapid gas expansion
- Heat, flame, light generation
- Pressure wave
- Shock wave
- Containment rupture
- Fragments/debris projection

The following flow chart identifies the various explosion scenarios. Even if not explicitly indicated, many confined scenarios can occur also as unconfined scenarios (Fig. 10.1).

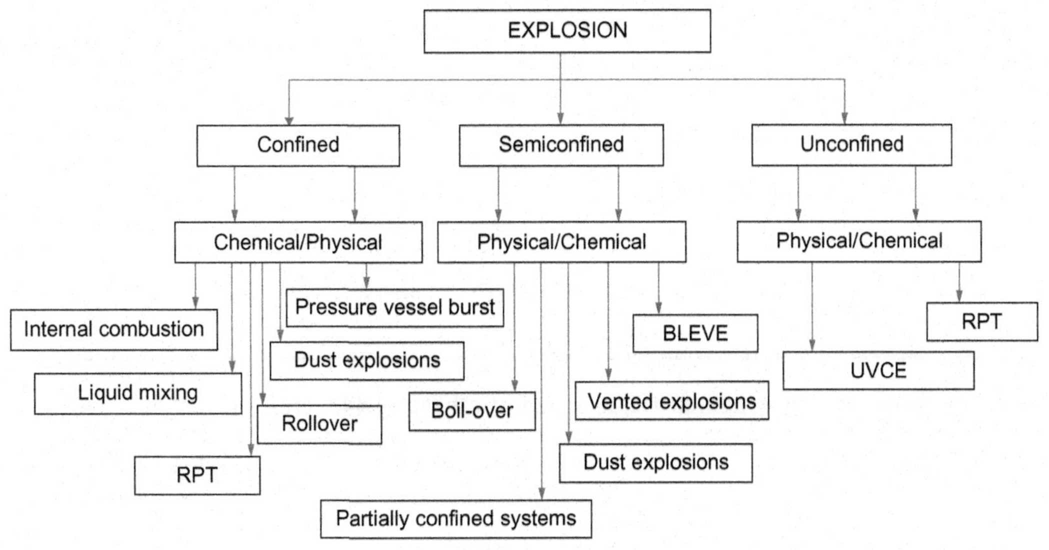

Fig. 10.1
Explosion scenarios.

10.3 Laminar Burning Velocity

The laminar burning velocity is one of the fundamental parameters in confined and unconfined assessment and has a role in the definition of the mechanisms describing the evolution of an explosion scenario. Data for selected gases have been tabulated in Chapter 3 (Bradley, 1969). An estimation of the burning velocity can be done according to Britton (2000):

$$S_u = 1666.1 + 34.228 \cdot \left(\frac{\Delta H_c}{\chi_{o\text{-}f}}\right) + 0.18039 \cdot \left(\frac{\Delta H_c}{\chi_{o\text{-}f}}\right)^2 \tag{10.1}$$

In this formula:

- S_u is the laminar burning velocity (cm/s).
- ΔH_c is the fuel heat combustion (kcal/mole).
- $\chi_{o\text{-}f}$ is the oxygen to fuel ratio.

Example 10.1

Find the laminar burning velocity for ethane. The standard enthalpy of combustion is 1,428,640 kJoule/kmole.

Solution

The ethane reaction of combustion is:

$$C_2H_6 + \frac{7}{2} \cdot O_2 = 2CO_2 + 3 \cdot H_2O \tag{10.2}$$

$$\chi_{\text{o-f}} = \frac{7}{2} \tag{10.3}$$

$$S_u = 1666.1 - 34.228 \cdot \left(\frac{1{,}428{,}460 \cdot 2}{7 \cdot 4186}\right) + 0.18039 \cdot \left(\frac{1{,}428{,}460 \cdot 2}{7 \cdot 4186}\right)^2 = 43.7\,\text{cm/s} \tag{10.4}$$

According to Bradley, the laminar flame speed of ethane is 44.5 cm/s.

Laminar burning velocity of mixtures follows a Le Chatelier-like law (NFPA 67, 2016).

10.4 Flame Speed and Turbulent Velocity

The speed of flame front is given by the following formula:

$$S_f = S_u + S_p \tag{10.5}$$

where S_p is the flow velocity induced in the unburned gas by the rate of production of the combustion products.

Estimates of the turbulent velocity can be carried out by adopting expressions available in the scientific literature which include the Lewis or Markstein numbers. An approximate but very simple formula has been set up by Abdel-Gayed and Bradley (1977).

$$u_t = 3.788 \cdot S_u \cdot \left(\sqrt{R_e} - 2500\right)^{0.238} \tag{10.6}$$

where R_e is the Reynolds number of the unburned gas. This approximate expression can be adopted provided $Re > 12{,}500$ and an S_u much smaller than the mean turbulent velocity square root.

10.5 Explosion With Oxidizers Other Than Oxygen

It is important to consider that the reaction leading to an explosion is an oxidation-reduction reaction, so an oxidiser is required. Oxygen is not the only oxidiser, because fluorine and chlorine can easily lead to an explosion, the former being the most reactive. Bromine's tendency toward explosion is low, and iodine does not promote explosion.

Table 10.1 (Cardillo, n.d.) presents flammable ranges of hydrogen, methane, ethane, and ethylene with various oxidizers:

Table 10.1 Flammability ranges of fuels with different oxidisers

Oxidiser	Air LEL-UEL (%)	O_2 LEL-UEL (%)	Cl_2 LEL-UEL (%)	N_2O LEL-UEL (%)	NO LEL-UEL (%)
Fuel					
Methane	5.0–15	5.1–61	56–70	4.3–22	8.6–21
Ethane	3.0–12	3.0–66	6.1–58	–	–
Ethylene	2.7–36	2.9–80	–	1.9–40	–
Hydrogen	4.0–75	4.0-94	4.0–89	3.0–84	6.6–66

10.6 Confined Explosion: Deflagration and Detonation

10.6.1 Deflagration

A confined explosion occurs in an unvented vessel and theoretically can be a detonation or a deflagration involving all or part of the gas volume. For practical calculations, it is assumed that the explosive regions both for detonation and deflagration are the same, depending on the different mechanism, the ignition energy, and the geometrical considerations. Deflagration can be associated with flame speeds ranging from the laminar speed, whose order of magnitude is 0.5–1 to 500–1000 m s^{-1}. Correspondingly, the peak pressure will range from a few mbar to several bar.

The maximum theoretical pressure following the ignition of a hydrocarbons-air mixture can be evaluated applying the ideal gas law:

$$P_2 = P_1 \cdot \frac{n_2 \cdot T_2}{n_1 \cdot T_1} \tag{10.7}$$

where P, n, and T are respectively pressure, number of kmoles, and temperature of the system in the initial state 1 and 2, before and after the explosion. Temperature ratio in the previous formula is approximately equal to the adiabatic flame temperature to initial (ambient) temperature ratio. This is the reason why the maximum theoretical hydrocarbons' pressure, with initial pressure and temperature between 1 and 40 bar and 0–300°C, is (deflagration):

$$\frac{P_2}{P_1} = 8 \tag{10.8}$$

for hydrocarbon-air mixtures and:

$$\frac{P_2}{P_1} = 16 \tag{10.9}$$

for hydrocarbon-oxygen mixtures.

Shepherd's model

Shepherd et al. (1997) developed a relatively simple and sufficiently accurate model on the basis of the energy conservation law applied to a constant volume adiabatic system. If E is the constant energy, $P(t)$ the average pressure at time t, V the volume, γ the specific heats ratio, $M_{uf}(t)$ the unburned (fuel) gas at time t, and ΔH_c the standard enthalpy of combustion, the following formula applies:

$$E = \frac{P(t)\cdot V}{\gamma - 1} + M_{uf}(t)\cdot \Delta H_c \tag{10.10}$$

At time zero:

$$E = \frac{P_o\cdot V}{\gamma - 1} + M_o\cdot \Delta H_c \tag{10.11}$$

where M_o is the initial fuel mass, and at the end of the explosion:

$$E = \frac{P_{max}\cdot V}{\gamma - 1} \tag{10.12}$$

Equating the terms:

$$\frac{P_{max}\cdot V}{\gamma - 1} = \frac{P_o\cdot V}{\gamma - 1} + M_o\cdot \Delta H_c \tag{10.13}$$

and

$$P_{max} = P_o + \frac{(\gamma - 1)\cdot M_o\cdot \Delta H_c}{V} \tag{10.14}$$

Example 10.2

Find the maximum overpressure for ethane at atmospheric pressure and 0°C in air and oxygen. The standard enthalpy of combustion is 47,621,333 J/kg.

Solution

Air

The ethane reaction of combustion is:

$$C_2H_6 + \frac{7}{2}\cdot O_2 + \frac{79}{21}\cdot\frac{7}{2}\cdot N_2 = 2CO_2 + 3\cdot H_2O + \frac{79}{21}\cdot\frac{7}{2}\cdot N_2 \tag{10.15}$$

In the equation:

$$P_{max} = P_o + \frac{(\gamma - 1)\cdot M_o\cdot \Delta H_c}{V} \tag{10.16}$$

M_o/V is the ethane stoichiometric concentration at 1 atm and 273.16 K. The ethane molar fraction is:

$$\frac{1}{1+\frac{7}{2}+\frac{79}{21}\cdot\frac{7}{2}}=0.057 \tag{10.17}$$

At the given temperature, 1 kmole occupies 22.414 m^3 (molar volume) so ethane concentration is:

$$\frac{0.057\cdot 30}{22.414}=0.076\,\mathrm{kg/m^3} \tag{10.18}$$

$$P_{max}=P_o+\frac{(\gamma-1)\cdot M_o\cdot \Delta H_c}{V}=101{,}325+(1.18-1)\cdot 0.076\cdot 47{,}621{,}333=752{,}785\,\mathrm{Pa} \tag{10.19}$$

Oxygen

The ethane reaction of combustion is:

$$C_2H_6+\frac{7}{2}\cdot O_2=2CO_2+3\cdot H_2O \tag{10.20}$$

In the equation:

$$P_{max}=P_o+\frac{(\gamma-1)\cdot M_o\cdot \Delta H_c}{V} \tag{10.21}$$

The ethane molar fraction is:

$$\frac{1}{1+\frac{7}{2}}=0.22 \tag{10.22}$$

Ethane concentration is:

$$\frac{0.22\cdot 30}{22.414}=0.297\,\mathrm{kg/m^3} \tag{10.23}$$

$$P_{max}=P_o+\frac{(\gamma-1)\cdot M_o\cdot \Delta H_c}{V}=101{,}325+(1.18-1)\cdot 0.297\cdot 47{,}621{,}333=2{,}647{,}161\,\mathrm{Pa} \tag{10.24}$$

The maximum pressure values of selected materials, obtained by Bartknecht (1993) in a 5 L sphere and 10 mJ ignition energy, have been included in Table 10.2 (Bartknecht, 1993).

Table 10.2 Maximum explosion pressure in a 5 L sphere (10 mJ MIE) (Bartknecht, 1993)

Material	P_{max} (bar abs)
Acetylene	10.6
Ammonia	5.4
Butane	8.0
Carbon disulphide	6.4
Dyethyl ether	8.1
Dimethyl formamide	8.4
Ethane	7.8
Ethyl alcohol	7.0
Ethyl benzene	7.4
Hydrogen	6.8
Hydrogen sulphide	7.4
Isopropanol	7.8
Methane	7.1
Methanol	7.5
Methylene chloride	5.0
Neopentane	7.8
Octanol	6.7
Pentane	7.8
Propane	7.9
South African crude oil	6.8–7.6
Toluene	7.8

For acetylene, maximum pressure is as high as 10 bar.

Practical data

For practical calculations and conventionally designed vessels the assumable bursting range is the following:

$$\frac{P_b}{P_1} = 4\_to\_5 \tag{10.25}$$

Design considerations for such vessels are the following:

- 5 atm could be assumed as the design pressure for those vessels to prevent rupture but not deformation.
- 10 atm is to be assumed as the design pressure to prevent rupture and deformation.

These considerations are valid in case of deflagrative explosion only.

Effect of initial temperature on maximum pressure is counter-intuitive but very particular and important, especially in cryogenic process safety, notably when LNG or LPG are managed. Specifically, reduction in the initial temperature results in increase of the maximum pressure. The following diagram shows the behaviour of maximum pressure rise with an inverse of temperature for methane (Fenning, 1926) (Fig. 10.2).

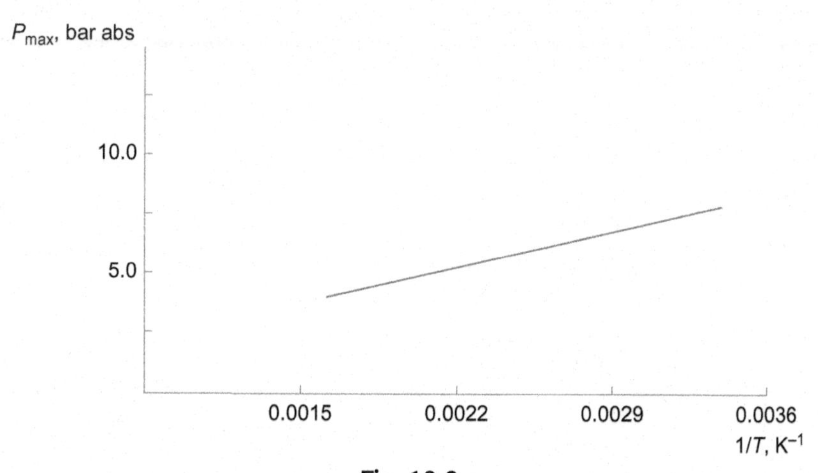

Fig. 10.2
Maximum pressure of methane with inverse of temperature.

Deflagration of a gas pocket

Bodurtha (1980) considers the interesting case of a gas pocket exploding within a larger enclosure. It is reasonable to assume that maximum pressure is attained when the fuel oxygen concentration ratio is stoichiometric. In the absence of experimental data, this assumption can be made. A flammable gas volume v_G is assumed to be supplied into a volume V_V vessel. Indicating with C_nH_m, a generic hydrocarbon, the reaction stoichiometry can be written as:

$$C_nH_m + \left(n+\frac{m}{4}\right)\cdot O_2 + \left(n+\frac{m}{4}\right)\cdot\frac{79}{21}\cdot N_2 = n\cdot CO_2 + \frac{m}{2}\cdot H_2O + \left(n+\frac{m}{4}\right)\cdot\frac{79}{21}\cdot N_2 \quad (10.26)$$

The stoichiometric overall (air+fuel) volume needed for combustion of v_G is:

$$V_s = \frac{v_G}{y_{st}} \quad (10.27)$$

where y_{st} is the fuel stoichiometric molar fraction:

$$y_{st} = \frac{1}{1+\left(n+\frac{m}{4}\right)\cdot\left(1+\frac{79}{21}\right)} \quad (10.28)$$

Upon explosion, the final vessel pressure will be:

$$P_2 = 8\cdot P_1\cdot\frac{V_s}{V_V} = 8\cdot P_1\cdot\frac{v_G\cdot\left[1+\left(n+\frac{m}{4}\right)\cdot\left(1+\frac{79}{21}\right)\right]}{V_V} \quad (10.29)$$

Example 10.3

0.01 m^3 of propane are supplied into a 10 m^3 vessel and ignited. Find the theoretical maximum final pressure.

Solution

V_V (m^3)	v_G (m^3)	n	m	P_1 (atm)
10	0.1	3	8	1

$$P_2 = 8 \cdot P_1 \cdot \frac{v_G \cdot \left[1 + \left(n + \frac{m}{4}\right) \cdot \left(1 + \frac{79}{21}\right)\right]}{V_V} = 8 \cdot 1 \cdot \frac{0.1 \cdot \left[1 + \left(n + \frac{m}{4}\right) \cdot \left(1 + \frac{79}{21}\right)\right]}{10} = 1.98\,\text{atm} \tag{10.30}$$

Cube root law

The maximum pressure attained by a gas in a vessel does not significantly depend on its volume, provided some specific parameters such as ignition location, shape, concentration, and location remain the same. This is due to the fact that maximum pressure is an indicator of the specific intrinsic properties of the fuel, such as its flame adiabatic temperature and combustion enthalpy, and these characteristics are not theoretically affected by volume size. Conversely, the rate at which the explosion progresses is significantly affected by the volume size. This is due to the circumstance that the pressure rise rate depends on the thermo-kinetic and is related to the time needed for the spherical flame to cover the distance existing between the epicentre of the explosion and the vessel wall. This has been represented on the following diagram which shows the pressure behaviour with the time (Fig. 10.3).

From the mathematical point of view, this is expressed as the cube root law (CRL), which is valid for explosive gases and dusts:

$$k_G = \left(\frac{dP}{dt}\right)_{\max} \times \sqrt[\frac{1}{3}]{V} \tag{10.31}$$

From the physical point of view, this expression has a very clear meaning. Being the laminar burning rate constant, if the volume increases (in reality if the radius of an ideal spherical vessel increases), the time needed for the flame to reach the vessel increases proportionally, and this is detectable from the system pressure rise rate. The CRL states that this opposite tendency results in the constant value of k_G. This constant gives an indication of the violence of the explosion and has a practical importance in the sizing of explosion venting devices. It must be noticed that the CRL and k_G, which is experimentally determined, are valid provided the following conditions apply:

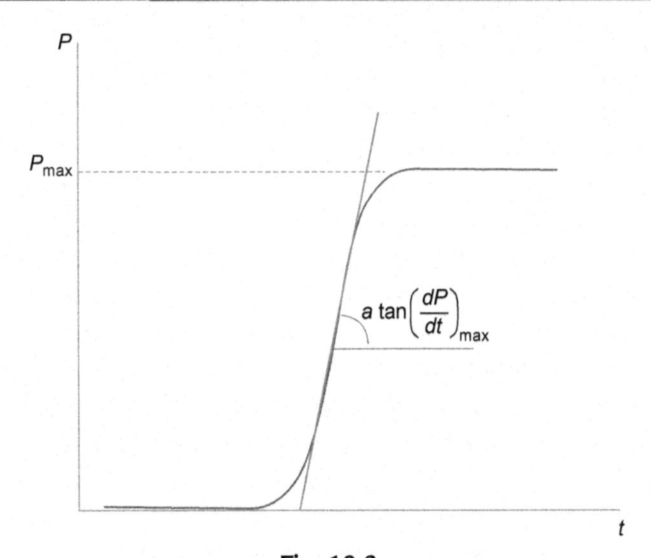

Fig. 10.3
Pressure rise in confined explosion.

- the gas-air concentration is the same
- same vessel shape
- same turbulence degree
- same ignition source

Explosion constant k_G for methane, propane, and hydrogen have been provided in Table 10.3 (Bartknecht, 1981). The values are referred to quiescent conditions and ignition energy = 10 mJ.

Bartknecht (1981) reports the CRL data for chlorine hydrogen explosion (7 L vessel, 10 mJ) (Table 10.4).

Table 10.3 Maximum pressure for light hydrocarbons—Ignition energy = 10 mJ (Bartknecht, 1981)

Material	P_{max} (bar abs)
Methane	55
Propane	75
Hydrogen	550

Table 10.4 Maximum pressure and explosion constant for chlorine-hydrogen explosion (Bartknecht, 1981)

Condition	P_{max} (bar)	k_G (bar m/s)
No turbulence	8.5	1250
Turbulence	9.4	2500

10.6.2 Detonation

Detonation occurs with a propagation of the combustion reaction to the unreacted material, whose speed is greater than the speed of sound in the unreacted material. The very high overpressure propagation is referred to as a shock wave. This is defined as a transient change in the gas density, pressure, and velocity of the air surrounding an explosion point. The initial change can be either discontinuous or gradual. A discontinuous change is referred to as a shock wave, whereas a gradual change is known as a pressure wave (CCPS, 1999). Flammability detonation ranges are typically assumed as the same as for the deflagration, even if the detonability ranges are generally narrower than deflagration ranges (Bodurtha, 1980). Flames propagating at speed of 2000–3000 m/s exceed the sound speed. Formation of a detonation (shock) wave can be produced in a laboratory through a series of pressure pulses which travel at the local sound speeds according to the increasing temperature. These small pulses can coalesce, and the resulting pressure pulse can travel at a velocity that is greater than the sound speed of the undisturbed gas. The system can be described in Fig. 10.4.

The variables are related to the initial and final conditions across the shock transition plane. The detonation velocity $u_{C\text{–}J}$ is given by the expression (Bradley, 1969):

$$u_{C-J} = v_1 \cdot \sqrt{\frac{P_2 - P_1}{v_1 - v_2}} \tag{10.32}$$

The stable detonation is described by the so-called Chapman Jouget (C-J) state. C-J detonation velocity and pressure for some gases at $P_1 = 1$ atm and $T_1 = 25°C$ have been included in Table 10.5 (Baker et al., 1983).

Similar data are provided by Lewis and von Elbe (1961), along with some useful parameters related to detonation with pure oxygen (Table 10.6).

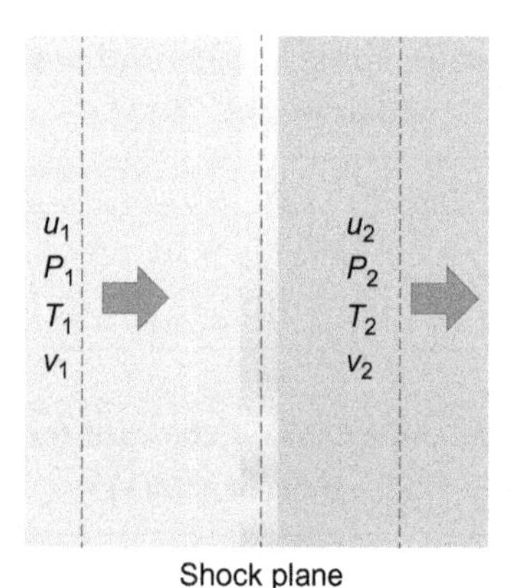

Fig. 10.4
Detonation shock wave.

Table 10.5 C-J detonation data (Baker et al., 1983)

Fuel	% Fuel	Pressure (MPA)	Temperature (K)	Velocity (m/s)
Methane	9.48	1.742	2784	1802
Ethylene	6.53	1.863	2929	1822
Hydrogen	29.52	1.584	2951	1968
Ethylene oxide	7.73	1.963	2949	1831
Acetylene	7.73	1.939	3114	1864
Propane	4.02	1.863	2840	1804

Table 10.6 C-J detonation data (Lewis and von Elbe, 1961)

Reactants	Detonability Limits LEL-UEL (%)	Velocity (m/s)	Temperature (K)	Pressure (atm)
		Fuel + oxygen		
Hydrogen	15–90	2840	3600	18.8
Carbon monoxide	38–90	1790	3500	18.6
Acetylene	3.5–92	2400	4200	44.0
Ammonia	25.4–75	2400	–	–
Propane	3.1–37	2500	–	–
		Fuel + air		
Hydrogen	18.2–59	1900	2950	15.6
Acetylene	4.2–50	1900	3100	19.0
Methane	–	1800	2736	17.2
Propane	–	1800	2823	18.3

Conditions for detonation

Under certain conditions, detonation of gas-air mixtures is possible in pipes and unlikely in vessels. The minimum aspect (height to diameter) ratio of a pipe or vessel for the detonation to be possible is around 10. For piping, a minimum critical diameter can conservatively be assumed to be 12 mm. When detonation occurs, peak pressures are of the order of 30 times of the initial absolute pressure and the stresses in the piping or vessels exceed the UTS. To properly manage the detonation hazard, the following useful rules can be considered (Englund, 1990):

1. Detonation in a vessel is prevented ensuring the minimum aspect ratio (height or length to diameter ratio).
2. Equipment designed to withstand 3.5 MPa are generally adequate to contain a detonation of a fuel in air.
3. Dished heads survive detonations much better than flat heads.
4. Splitting a line bends angle into two bends with a smaller angle reduces the deflagration to detonation hazard.
5. Flame arrestors can arrest detonations.

10.7 Detonation in Pipes

Detonation in pipelines is much more likely than in vessels, depending on several factors. Bartknecht (1981) differentiates the ignition of an explodable mixture at the open end of a pipeline, which is closed at the other end, from that at the closed end of a pipeline which is open at the other end. Pipelines which are closed at one end show the strongest explosion effects when ignition starts at the closed end. Furthermore, as a general rule, the explosion velocity will rise with increasing pipe diameter. The explosion velocity of the former case considered by Bartknecht is reported as:

$$u_{\text{exp}} = R_{\text{f}} \cdot S_{\text{u}} \tag{10.33}$$

where S_{u} is the laminar burning velocity and R_{f} is the flame surface area to the pipe cross section ratio. For pipes, R_{f} can be estimated as:

$$R_{\text{f}} = 2 \cdot \left[1 + 0.7 \cdot \left(\frac{x}{D_{\text{p}} \cdot \sqrt{0.785}} - 1.3 \right) \right] \tag{10.34}$$

where x is the distance and D_{p} is the pipe diameter.

NFPA 67 (2016) provides the detonation limits for some gas/mixtures at 300 K and 1 atm with respect to the inner pipe diameter (Table 10.7).

Table 10.7 Detonation limits in pipes (NFPA 67, 2016)

<table>
<tr><th>Gas</th><th>Pipe Inner Diameter (mm)</th><th>Detonation LEL (%)</th><th>Detonation UEL (%)</th></tr>
<tr><td>Methane</td><td rowspan="5">25</td><td>–</td><td>–</td></tr>
<tr><td>Ethylene</td><td>4.7</td><td>12.3</td></tr>
<tr><td>Hydrogen</td><td>19.0</td><td>58.0</td></tr>
<tr><td>Acetylene</td><td>4.0</td><td>–</td></tr>
<tr><td>Propane</td><td>3.3</td><td>5.4</td></tr>
<tr><td>Methane</td><td rowspan="5">50</td><td>–</td><td>–</td></tr>
<tr><td>Ethylene</td><td>4.0</td><td>14.9</td></tr>
<tr><td>Hydrogen</td><td>17.0</td><td>59.0</td></tr>
<tr><td>Acetylene</td><td>3.4</td><td>–</td></tr>
<tr><td>Propane</td><td>3.1</td><td>6.3</td></tr>
<tr><td>Methane</td><td rowspan="5">100</td><td>9.5</td><td>9.5</td></tr>
<tr><td>Ethylene</td><td>3.4</td><td>17.4</td></tr>
<tr><td>Hydrogen</td><td>15.0</td><td>61.0</td></tr>
<tr><td>Acetylene</td><td>2.9</td><td>7.5</td></tr>
<tr><td>Propane</td><td></td><td></td></tr>
</table>

10.7.1 Length to Detonation

Bartknecht (1981) reports that the length of pipe needed for acceleration to detonation velocity is shortened considerably by decreasing pipe diameter. The tendency to reach detonation velocity decreases with increasing pipe diameter. Table 10.8 (Bartknecht) summarises the pipe length for acceleration to detonation for different gas/air mixtures.

10.7.2 Deflagration to Detonation Transition

The standard NFPA 67 (2016) suggests as a common practice a pipe length of approximately 100 pipe diameters for the transition from deflagration to detonation. A stable detonation C-J velocity and pressure are about 20 bar and velocity 1800 m/s for most fuels.

10.8 Reflected Shock Wave and Pressure Temporal-Spatial Distribution

As it happens for deflagration, when the detonation reaches a wall, a reflected shock wave is created (Chapter 5), so the moving gas immediately behind the shock wave is brought to rest. Pfreim's universal equation (1941), which is independent of the initial pressure, can be used to find the ratio of the reflected overpressure to the incident overpressure:

$$\frac{P_R}{P_I} = 5 \cdot \gamma + 1 + \sqrt{\frac{25}{4} \cdot \gamma + 1} \qquad (10.35)$$

and, being that the specific heat ratio γ for all real gas at high temperature is approximately equal to 1.3, it can be assumed (Craven and Greig, 1968):

$$\frac{P_R}{P_I} \cong 2.5 \qquad (10.36)$$

This value is not very different from the value related to the deflagration scenario.

Table 10.8 Pipe length for acceleration to detonation

	Pipe Length (m)		
	Pipeline Diameter (mm)		
Fuel	100	200	400
Methane	12.5	18.5	>30
Hydrogen	7.5	12.5	12.5
Propane	12.5	17.5	22.5

Karnesky's equation (Karnesky et al., 2010) has been validated for the calculation of the spatial and temporal distribution of the overpressure along the axis (The subscript 1 denotes the pre-detonation region and the subscript 3 denotes the post-expansion region.):

$$P(x,t) \cong \begin{cases} P_1 & \text{if\_}u_{C-J} < \frac{x}{t} < \infty \\ P_3 \cdot \left[1 - \frac{\gamma-1}{\gamma+1} \cdot \left(1 - \frac{x}{c_3 \cdot t}\right)\right]^{\frac{2\cdot\gamma}{\gamma-1}} & \text{if\_}c_3 < \frac{x}{t} < u_{C-J} \\ P_3 & \text{if\_}0 < \frac{x}{t} < c_3 \end{cases} \tag{10.37}$$

where

- P_1 is the pre-detonation pressure.
- P_3 is the post-expansion pressure.
- $P_{C\text{-}J}$ is the C-J pressure.
- x is the pipe axis.
- $u_{C\text{-}J}$ is the detonation velocity.
- $c_{C\text{-}J}$ is sound speed at C-J state.
- γ is the specific heat ratio.

$$c_3 = \frac{\gamma+1}{2} \cdot c_{C-J} - \frac{\gamma-1}{2} \cdot u_{C-J} \tag{10.38}$$

$$P_3 = P_{C-J} \cdot \left(\frac{c_3}{c_{C-J}}\right)^{\frac{2\cdot\gamma}{\gamma-1}} \tag{10.39}$$

10.8.1 Maximum Overpressure in Pipes

Fig. 10.5 recalculated on the basis of data included in the EN, 14994:2007, gives an approximate quantitative understanding of the maximum overpressure attained in a smooth, straight pipe closed at one end, and vented at the other, for three pipe categories. *L/D* is the length to diameter ratio. It is valid for gases with a fundamental burning velocity of less than 0.6 m/s and for flow velocities of 2 m/s or less (Fig. 10.5).

10.9 Semiconfined Explosion

For practical calculations, four representative cases have been considered for semiconfined explosions.

1. Semiconfined explosions in enclosures with some degree of confinement, usually inside the buildings.
2. Explosions in blast-resistant vessels and pipes with free openings not fitted with venting panels.

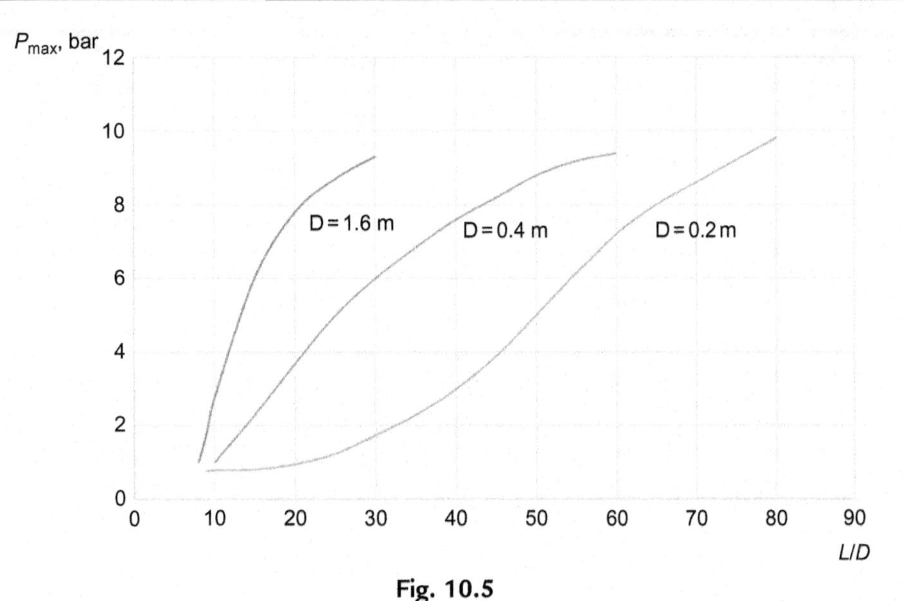

Fig. 10.5
Maximum overpressure in smooth straight pipes closed at one end and vented at the other.

3. Explosions in blast-resistant structures protected with venting panels.
4. Explosions in nonblast-resistant vessels which rupture other than BLEVE.

Case 1 is assumed to refer to not particularly congested enclosures. Congested scenarios have been dealt with within unconfined explosion scenarios with congested areas, such as modules and plant units.

10.9.1 Semiconfined Explosions in Enclosures With Some Degree of Confinement

This is the case of buildings with some openings. A simple method to determine the peak pressure has been described by Weibull (1968). This is based on the assumption that the influence of the area of the vent openings is not significant, being that the overpressure is dependent on the fuel mass to chamber volume ratio only. It is given by the formula:

$$P_{\text{peak}} = 22.5 \cdot \left(\frac{M}{V}\right)^{0.72} \tag{10.40}$$

where P_{peak} is in bar, M is in kg, and V in m^3.

Notably, Weibull reports maximum pressures for vented chambers of various shapes having a single vent, with a range of vent areas of $(A/V^{2/3}) < 0.02\ 1\ 5$. These maximum quasi-static pressures are shown by Weibull to be independent of the vent area ratio and to be a function of charge-to-volume ratio (M/V) up to 0.3 12 lb/ft^2.

Example 10.4

An enclosure whose volume is 15 m^3 is filled with methane. Determine the maximum value of all possible peak pressures if temperature is 293.16 K, assuming the building is blast-resistant.

Solution

Depending on the peak pressure on the mass amount, the worst case will be the one related to the methane Upper Explosion Limit (UEL).

V (m^3)	UEL (%)
15	15

The amount of methane can be easily calculated, considering that the overall molar density is:

$$\rho_m = \frac{1}{22.414} \cdot \frac{273.16}{293.16} = 0.0415\,\text{kmoles/m}^3 \tag{10.41}$$

So, the amount of methane is:

$$M = 0.0415 \cdot 0.15 \cdot 16 \cdot 15 = 1.5\,\text{kg} \tag{10.42}$$

$$P_{\text{peak-MAX}} = 22.5 \cdot \left(\frac{1.5}{15}\right)^{0.72} \cong 4.3\,\text{bar} \tag{10.43}$$

Simmonds and Cubbage (1960) have studied the pressure rise following the explosions in industrial ovens and have detected two pressure peaks, P_1 and P_2, the former being related to the immediate relief of the combustion gases, and the latter being related to the progress of combustion, resulting in a significant increase of the pressure drop and in a second peak pressure (Fig. 10.6).

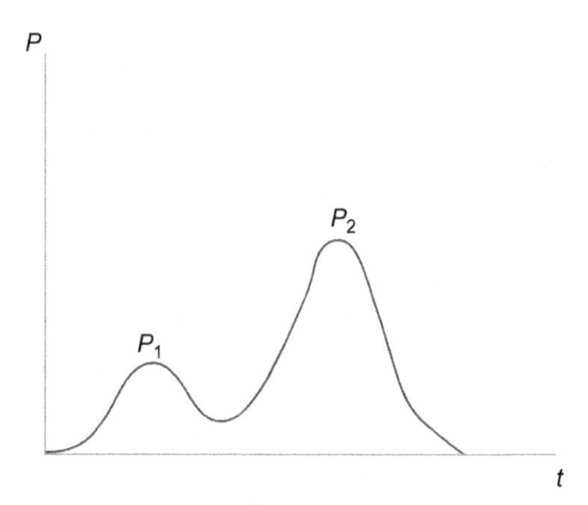

Fig. 10.6
Pressure peaks in semiconfined explosion.

The two peaks can be calculated according to the formulas:

$$P_1 = \frac{(437 \cdot K \cdot W + 2817) \cdot S_u}{\sqrt[3]{V}} \tag{10.44}$$

$$P_2 = 5800 \cdot S_u \cdot K \tag{10.45}$$

Pressure are in Pa (gauge), W is the weight per unit area of an existing vent, and K is the vent coefficient, defined as:

$$K = \frac{A_s}{A_v} \tag{10.46}$$

where A_v is the opening (vent) area, A_s is the area of the vessel side where the opening lies.

Explosions in blast-resistant vessels and pipes with free openings not fitted with venting panels

This is the case of a vessel fitted with some openings or interconnected to another enclosure (Fig. 10.7).

In a study dealing with pressure piling of interconnected vessels, Singh (1994) states that the maximum pressure in the primary larger compartment is close to the value attained in a single (closed) vessel. Unless a deeper analysis is carried out, the burst pressure should be assumed as the burst pressure of a single closed vessel.

Explosions in blast-resistant structures protected with venting panels

Venting devices are provided to prevent the explosion pressure from exceeding a specific design pressure. The following diagram (Fig. 10.8) describes the principle of explosion protection of venting devices. In an unvented system the overpressure would reach its

Fig. 10.7
Semiconfined explosion in containments not fitted with venting panels.

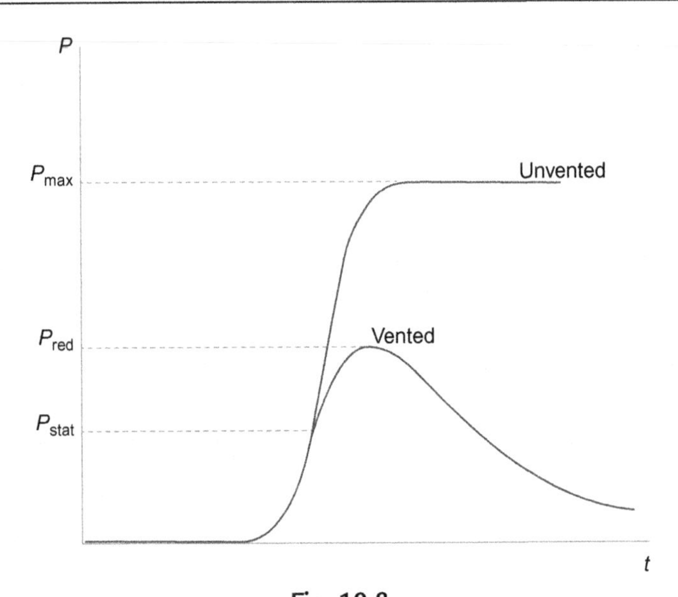

Fig. 10.8
Explosion in containments fitted with venting panels.

maximum value P_{max}. In a vented system pressure will initially follow the same trend until the vent static activation pressure P_{stat} is reached. Afterwards, the pressure rise curve will grow less than the unvented rise until a reduced pressure P_{red} is reached and then will decline. P_{red} is the designed maximum pressure for the protected system. Design and calculation formulas are provided by international standards dealing with gas and dust explosions, such as NFPA 68 and EN, 14994:2007. Some equations are included in these sections.

Venting of elongated enclosures vented at each end (EN, 14994:2007)

$$P_{red} = P_{stat} + \frac{0.023 \cdot S_u^2 \cdot K \cdot W \cdot \sqrt[3]{\frac{L}{D}}}{\sqrt[3]{V}} \tag{10.47}$$

where

- P_{red} and P_{stat} are given in bar.
- S_u is the gas burning velocity, m/s.
- K is the vent coefficient, vessel cross sectional area to total area of vents ratio.
- W is the vent device weight unit area, kg/m^2.
- V is the enclosure volume, m^3.
- L is the enclosure length, m.
- D is the enclosure diameter, m.

This equation is valid for:

- atmospheric conditions
- S_u comparable to methane and propane
- $V < 200\ m^3$
- $P_{stat} < 0.1$ bar
- $P_{red} < 1$ bar
- $2 < L/D < 10$

Venting of elongated enclosures vented along the enclosure (EN, 14994:2007):

For methane:

$$P_{red} = P_{stat} + 0.07 \cdot K \cdot d \tag{10.48}$$

For propane:

$$P_{red} = P_{stat} + 0.085 \cdot K \cdot d \tag{10.49}$$

where d is the ratio of the maximum possible distance that can exist between a potential ignition source and the nearest vent to the diameter of the enclosure.

10.10 Unconfined Vapour Cloud Explosion

Unconfined vapour cloud explosions (UVCEs) are among the most serious scenarios in QRA consequence assessment, due to the potential huge and large impact on people and assets. The very wide number of variables and factors, along with some inherent difficulties in the selection and calibration of the most appropriate dispersion scenarios, and of the most realistic explosion models, make this phenomenon complex and affected by a significant degree of uncertainty. In addition, the process safety engineer is inevitably involved in dealing with the consequences of the explosions on buildings, equipment, and stick-built structures, and this increases the complexity of the overall scenario. This section aims to provide some practical, and hopefully comprehensive, guidelines to properly approach the subject within some specific scenarios that are typically met in the calculations.

Four main models can be found in the literature for the calculation of the UVCE:

- The TNT model
- The MultiEnergy (ME/TNO) model
- The Baker-Strehlow-Tang (BST) model
- Congestion Assessment Model (CAM) model

It is probably correct to state that the degree of approximation, uncertainty, and subjectivity decreases from the first to the last of these models, being the three latter substantially comparable in terms of accuracy. The choice has been made not to deal with the TNT model in this book, due to the very high level of inaccuracy and subjectivity, and due to the circumstance that is not generally accepted in the process safety calculations. However, due to its simplicity, it remains a very good and quick screening method. Many literature sources present a full description of the TNT model. The MULTIENERGY (MEM) and the BST models have been presented and benchmarked with the aim to provide the user with some criteria to increase their confidence in the calculations. Finally, a general overview of the CAM model has been provided.

10.10.1 Multienergy Model

The MEM (van den Bosch and Duijm, 2005) has been developed by TNO, and is based on blast charts showing the behaviour of scaled pressure and scaled duration against a scaled distance. It can be considered an accurate model with some significant weaknesses. Notably, a high degree of subjectivity is included, even if some useful guidance has been provided to make the choice of the blast curves less uncertain. This inevitably results in an overconservatism.

A very strong quantitative support to reduce the subjective approach has been provided through the GAME project, where independent correlations were developed, including all the physical factors affecting explosion. This has been effective in relating the blast curves to the physics of explosive phenomena. This approach has been presented here with the aim to provide a complete guidance on the application of the method to real calculation cases.

Summary of the MEM

The MEM blast charts relate the scaled pressure P_s and the scaled duration t_s to a scaled distance R_s.

In Fig. 10.9:

$$P_s = \frac{\Delta P}{P_a}(R_s) \tag{10.50}$$

and

$$R_s = R \cdot \sqrt[3]{\frac{P_a}{E}} \tag{10.51}$$

where

- ΔP is the overpressure, Pa.
- R is the distance of the target to centre of explosion, m.
- P_a is the atmospheric pressure, Pa.
- E is the blast energy, J.

The energy blast to be included in this equation is the combustion heat associated to the flammable cloud, which has to be calculated.

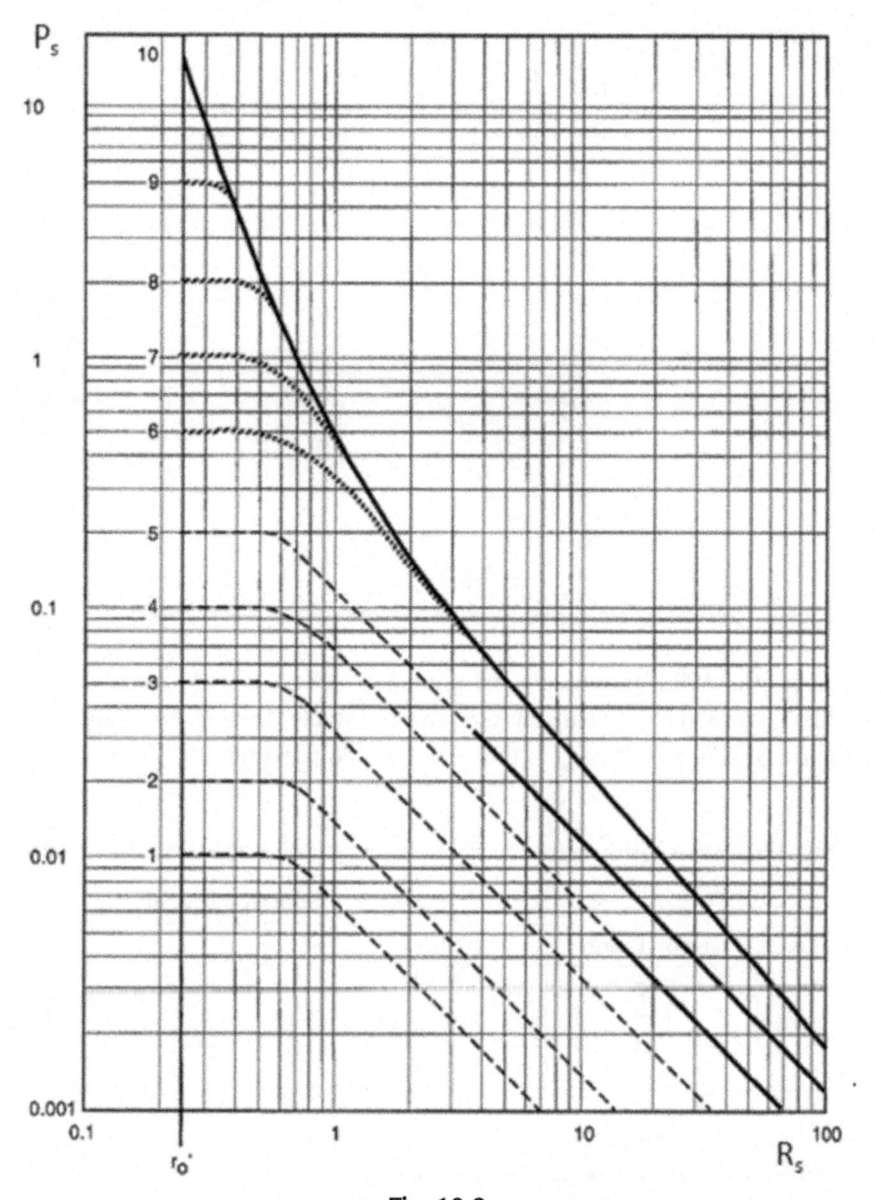

Fig. 10.9
Scaled pressure versus scaled distance (Van den Bosch and Duijm, 2005).

In Fig. 10.10:

$$t_s = u_s \cdot t_p \cdot \sqrt[3]{\frac{P_a}{E}} \tag{10.52}$$

and t_p is the pressure wave duration. As presented, the method would require selecting a blast curve and then relating distances to overpressures and durations. This implies some

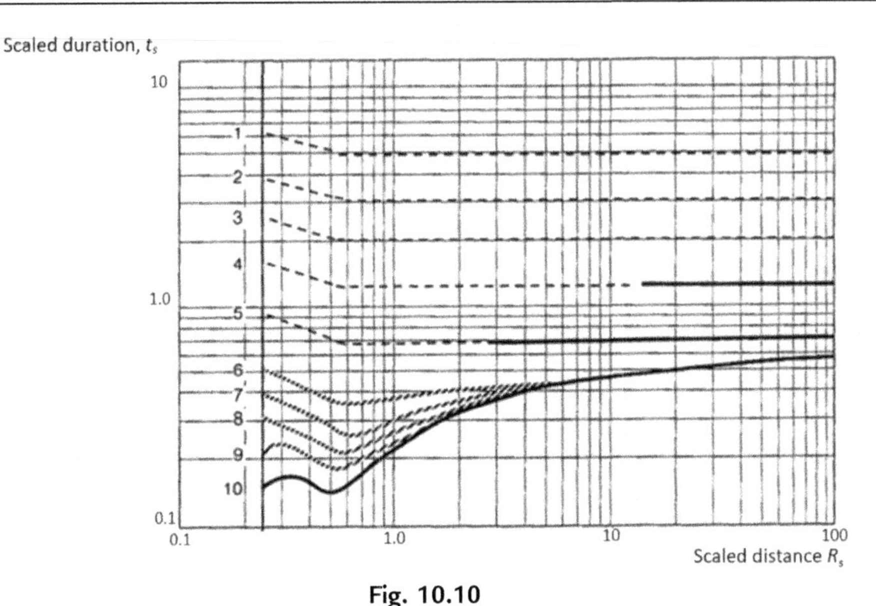

Fig. 10.10
Scaled duration versus scaled distance (van den Bosch and Duijm, 2005).

subjectivity. Kinsella (1993) provided some guidance on how to reduce the range of possible choices, on the basis of flame expansion modality, obstacle density, and ignition energy, as shown in Tables 10.9 and 10.10. P_{so} is the peak on-side pressure.

Table 10.9 MEM blast strengths related to flame, energy, and congestion data (Kinsella, 1993)

		Obstacle Density		
Flame Expansion	**Ignition Energy**	**High**	**Low**	**None**
2D	High	7–10	5–7	4–6
	Low	5–7	3–5	1–2
3D	High	7–10	4–6	4–5
	Low	4–5	2–3	1

Table 10.10 MEM blast strengths related to peak side-on pressure (Kinsella, 1993)

Blast Strength	P_{so} (bar)
10	>10
7	1.0
6	0.5
5	0.2
4	0.1
3	0.05
2	0.02
1	0.01

The same author has specified:

Obstructions
High: Closely packed obstacles within gas clouds giving an overall volume blockage fraction (i.e. the ratio of the volume of the obstructed area occupied by the obstacles to the total volume of the obstructed area itself) in excess of 30% and with spacing between obstacles less than 3 m.

Low: Obstacles in gas cloud but with overall blockage fraction less than 30% and/or spacing between obstacles larger than 3 m.

None: No obstacles within gas cloud.

Parallel plane confinement
Yes: Gas clouds, or parts of them, are confined by walls/barriers on two or three sides.

No: Gas cloud is not confined, other than by the ground.

Ignition strength
High: The ignition source is, for instance, a confined vented explosion. This may be due to the ignition of part of the cloud by a low energy source, for example, inside a building.

Low: The ignition source is a spark, flame, hot surface, etc.

Although these specifications have reduced the uncertainty in the blast curve choice, nevertheless, some subjectivity still affects the real cases. For these reasons, the following systematic procedure, based on the GAME project findings, is considered much more accurate and reliable.

Procedure to use the MEM model

The following flow chart summarises the steps to be followed for a conservative approach to the multienergy model, implemented with the contribution of the GAME project correlations (Fig. 10.11).

Step 1—Definition of the release and the dispersion modality

The definition of the release will be done on the basis of the source models described in Chapter 7.

Step 2—Calculation of release duration

The release duration can be calculated following the criteria included in Chapter 8, dealing with:

- Classification of continuous or instantaneous release
- Transient analysis of the release from isolated plant segments
- Adoption of prescriptive times, based on the standard API 581

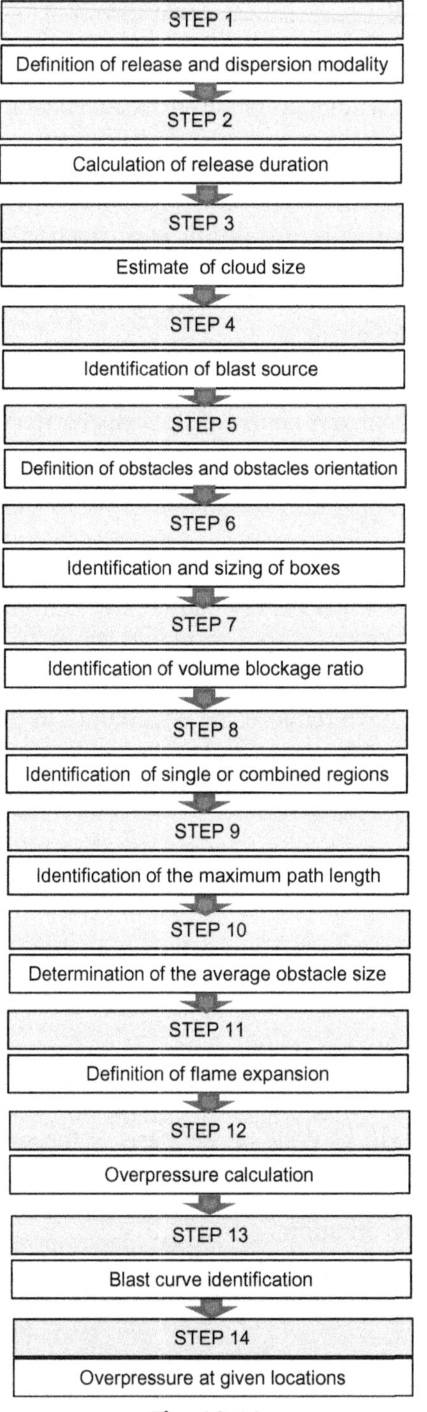

Fig. 10.11
Procedure to use the MEM.

Step 3—Estimation of the cloud size

Three main scenarios can be identified for which the flammable mass has to be found:

1. Passive dispersion—Instantaneous release
2. Passive dispersion—Continuous release
3. Dense gases (Continuous/instantaneous)

The first two scenarios can be properly approached through van Buijtenen's models. For dense gas, a simplified method has been presented. All these scenarios have to take into account also the behaviour of the released gas downstream of the outlet, until the buoyancy-driven regime dominates.

Instantaneous release: van Buijtenen's model

The occurrence of instantaneous or continuous releases for passive or neutral dispersion can be assessed adopting the presented method provided by API RP 581 (2016). If the release is judged as instantaneous, the following expression can be adopted to find the maximum flammable mass to the total released gas mass ratio (van Buijtenen, 1980), which depends on the UEL to LEL ratio only:

$$R_m = \mathrm{erf}\left[\sqrt{v_2}\right] - \mathrm{erf}\left[\sqrt{v_1}\right] - \frac{2}{\sqrt{\pi}} \cdot \exp(-v_2) \cdot \sqrt{v_2} + \frac{2}{\sqrt{\pi}} \cdot \exp(-v_1) \cdot \sqrt{v_1} \qquad (10.53)$$

with:

$$v_2 = \frac{v^2 \cdot \ln(v)}{v^2 - 1}, \quad v_1 = \frac{\ln(v)}{v^2 - 1} \qquad (10.54)$$

and

$$v = \frac{C_{UEL}}{C_{LEL}} \qquad (10.55)$$

where c_{UEL} and c_{LEL} are the upper and the lower explosion limits, respectively (formally given in kg/m^3 units, even if the ratio is independent of the units).

Continuous release: van Buijtenen's model

For continuous releases, the following equations apply to find the maximum R_m ratio:

Maximum released amount for a plume at ground level

The total amount in the explosive region (kg) is:

$$M_e = W_o \cdot \left(\frac{W_o}{a \cdot c\pi \cdot u_w}\right)^{\frac{1}{b+d}} \cdot \frac{b+d}{(b+d+1) \cdot u_w} \cdot \left[\left(\frac{1}{C_{LFL}}\right)^{\frac{1}{b+d}} - \left(\frac{1}{C_{UFL}}\right)^{\frac{1}{b+d}}\right] \qquad (10.56)$$

Maximum release amount for a free from the ground plume

$$M_e = W_o \cdot \left(\frac{W_o}{2 \cdot a \cdot c\pi \cdot u_w}\right)^{\frac{1}{b+d}} \cdot \frac{b+d}{(b+d+1) \cdot u_w} \cdot \left[\left(\frac{1}{C_{LFL}}\right)^{\frac{1}{b+d}} - \left(\frac{1}{C_{UFL}}\right)^{\frac{1}{b+d}}\right] \quad (10.57)$$

where

- W_o is the release mass flow rate, kg/s.
- a, b, c and d are constants depending on the weather stability class (see Table 10.11).
- M_e is the flammable amount in the cloud, kg.
- C_{UEL} and C_{LEL} are the gas explosion upper and lower limits, kg/m^3.

Table 10.11 Coefficients of van Buijtenen's model

Stability	a	b			
B	0.371	0.866			
C	0.209	0.897			
D	0.128	0.905			
E	0.098	0.902			
F	0.065	0.902			
	Distance (m)				
	1–10	10–100	100–1000	1000–20,000	20,000–100,000
	Values for c				
B	0.15	0.1202	0.0371	0.054	0.0644
C	0.15	0.0963	0.0992	0.099	0.1221
D	0.15	0.0939	0.2066	0.924	1.0935
E	0.15	0.0864	0.1975	2.344	2.5144
F	0.15	0.0880	0.0984	6.528	2.8745
	Values for d				
B	0.8846	0.9807	1.1530	1.0997	1.0843
C	0.7533	0.9456	0.9289	0.9255	0.9031
D	0.6368	0.8403	0.7338	0.5474	0.5229
E	0.5643	0.8037	0.6965	0.4026	0.3756
F	0.4771	0.7085	0.7210	0.2593	0.3010
	Values for e				
B	0	0	3.1914	2.5397	0
C	0	0	0.2444	1.7383	0
D	0	0	−1.3659	−9.0641	0
E	0	0	−1.1644	−16.3186	0
F	0	0	−0.3231	−25.1583	0

Example 10.5

Hydrogen at 30 bar is released through a 100 mm at ambient temperature (20°C). Find the flammable mass, assuming a 1 min release duration and constant choked conditions.

This example does not account for the initial very high momentum, which has to be considered in a real case. The Gaussian model is strictly valid provided that the initial momentum is low. If not, air entrainment will significantly reduce the concentration in a very short distance, and then passive dispersion will prevail (look at the Lee's and Chu's formula in Chapter 8 to establish the momentum to the buoyancy distance), and then the P-G model will be run considering the different initial point of the passive dispersion.

Solution

γ (cp/cv)	Ambient Temperature (K)	P_o (bar)	R (J/kmole K)	MW (kg/kmole)	Release Time (min)	C_D
1.41	293.16	30	8314	2	1	0.85

$$\begin{aligned} W_c &= P_o \cdot A \cdot C_D \cdot \sqrt{\frac{MW \cdot \gamma}{R \cdot T_o} \cdot \left[\frac{2}{\gamma+1}\right]^{(\gamma+1)/(\gamma-1)}} \\ &= 3{,}000{,}000 \cdot \frac{\pi \cdot 0.1^2}{4} \cdot 0.85 \cdot \sqrt{\frac{2 \cdot 1.41}{8{,}314 \cdot 293.16} \cdot \left[\frac{2}{1.41+1}\right]^{(1.41+1)/(1.4-1)}} \\ &= 12.45\,\text{kg/s} \end{aligned} \tag{10.58}$$

747 kg/373.5 kmoles will be released.

$$v_2 = \frac{18.75^2 \cdot \ln(18.75)}{18.75^2 - 1} = 2.94, \quad v_1 = \frac{\ln(18.75)}{18.75^2 - 1} = 0.0084 \tag{10.59}$$

The erfunction is tabulated as shown in Table 10.12.

$$R_m = \text{erf}\left[\sqrt{v_2}\right] - \text{erf}\left[\sqrt{v_1}\right] - \frac{2}{\sqrt{\pi}} \cdot \exp(-v_2) \cdot \sqrt{v_2} + \frac{2}{\sqrt{\pi}} \cdot \exp(-v_1) \cdot \sqrt{v_1} \tag{10.60}$$

$$R_m = \text{erf}[1.71] - \text{erf}[0.092] - \frac{2}{\sqrt{\pi}} \cdot \exp(-2.94) \cdot 1.71 + \frac{2}{\sqrt{\pi}} \cdot \exp(-0.0084) \cdot 0.092 \tag{10.61}$$

$$R_m = 0.984 - 0.100 - 0.102 + 0.103 = 0.885 \tag{10.62}$$

Combustion reaction:

$$2H_2 + O_2 + 79/21 N_2 \rightarrow 2H_2O + 79/21 N_2 \tag{10.63}$$

Hydrogen stoichiometric molar fraction:

$$y_{H_2} = \frac{2}{2 + 1 + \frac{79}{21}} = 0.296 \tag{10.64}$$

Table 10.12 Values of erfunction

x	erf(x)	x	erf(x)
0.00	0.0000000	1.30	0.9340079
0.05	0.0563720	1.40	0.9522851
0.10	0.1124629	1.50	0.9661051
0.15	0.1679960	1.60	0.9763484
0.20	0.2227026	1.70	0.9837905
0.25	0.2763264	1.80	0.9890905
0.30	0.3286268	1.90	0.9927904
0.35	0.3793821	2.00	0.9953223
0.40	0.4283924	2.10	0.9970205
0.45	0.4754817	2.20	0.9981372
0.50	0.5204999	2.30	0.9988568
0.55	0.5633234	2.40	0.9993115
0.60	0.6038561	2.50	0.9995930
0.65	0.6420293	2.60	0.9997640
0.70	0.6778012	2.70	0.9998657
0.75	0.7111556	2.80	0.9999250
0.80	0.7421010	2.90	0.9999589
0.85	0.7706681	3.0	0.9999779
0.90	0.7969082	3.10	0.9999884
0.95	0.8208908	3.20	0.9999940
1.00	0.8427008	3.30	0.9999969
1.10	0.8802051	3.40	0.9999985
1.20	0.9103140	3.50	0.9999993

Hydrogen stoichiometric mass concentration:

$$x_{H_2} = \frac{1}{22.414} \cdot \frac{273.16}{293.16} \cdot 0.296 = 0.0123\,\text{kg/m}^3 \tag{10.65}$$

According to Hess et al. (1973) at the exit plane:

$$T_e = T_o \cdot \left(\frac{2}{\gamma + 1}\right) = 293.16 \cdot \left(\frac{1.41}{2.41}\right) = 171.5\ \text{K} \tag{10.66}$$

The cloud will have a temperature dependent on the thermal balance between air and hydrogen. The final cloud temperature T_{cl} can be found equating:

$$x_{H_2} \cdot c_{p_{H_2}} \cdot (T_{cl} - T_e) = (1 - x_{H_2}) \cdot c_{p_{air}} \cdot (T_{air} - T_{cl}) \tag{10.67}$$

with

- $c_{p\text{-}H2} = 28$ kJ/kmole.
- $c_{p\text{-}air} = 29$ kJ/kmole.

$$0.286 \cdot 28 \cdot (T_{cl} - 171.5) = (1 - 0.286) \cdot 29 \cdot (293.16 - T_{cl}) \tag{10.68}$$

Solving:

$$T_{cl} = 259.24\text{K} \tag{10.69}$$

The molar mass released in 1 min is 373.5 kmoles.

The flammable mass is:

$$R_m \cdot m_{mol} \cdot MW = 0.885 \cdot 373.5 \cdot 2 = 661.1\,\text{kg} \tag{10.70}$$

Example 10.6

Hydrogen at 30 bar is released at a rural site, 10 m above grade, through a 100 mm hole at ambient temperature (20°C). The release is continuous. Find the flammable mass, assuming choked conditions. The P-G weather category is *D*/5.

This example does not account for the initial very high momentum, which has to be considered in a real case. The Gaussian model is strictly valid provided that the initial momentum is low. If not, air entrainment will significantly reduce the concentration in a very short distance, and then passive dispersion will prevail (look at the Lee and Chu's formula in Chapter 8 to establish the momentum to buoyancy distance), and then the P-G model will be run considering different initial point of the passive dispersion.

Solution

γ (cp/cv)	Ambient Temperature (K)	P_o (bar)	R (J/kmole K)	MW (kg/kmole)	Release Time (min)	C_D
1.41	293.16	30	8314	2	10	0.85

$$\begin{aligned} W_c &= P_o \cdot A \cdot C_D \cdot \sqrt{\frac{MW \cdot \gamma}{R \cdot T_o} \cdot \left[\frac{2}{\gamma+1}\right]^{(\gamma+1)/(\gamma-1)}} \\ &= 3{,}000{,}000 \cdot \frac{\pi \cdot 0.1^2}{4} \cdot 0.85 \cdot \sqrt{\frac{2 \cdot 1.41}{8314 \cdot 293.16} \cdot \left[\frac{2}{1.41+1}\right]^{(1.41+1)/(1.4-1)}} \\ &= 12.45\,\text{kg/s} \end{aligned} \tag{10.71}$$

For rural areas, the downwind distance X, for a given concentration of interest $\langle C \rangle$, mass flow rate W_c, and wind speed u_w can be found utilising the following equations (CCPS, 1999):

$$\frac{X}{L^{\bullet}} = a_o + a_1 \cdot \ln(L^{\bullet}) + a_2 \cdot [\ln(L^{\bullet})]^2 + a_3 \cdot [\ln(L^{\bullet})]^3 \tag{10.72}$$

with:

$$L^{\bullet} = \sqrt{\frac{W_c}{u_w \cdot \langle C \rangle}} \tag{10.73}$$

P-G Class	a_o	a_1	a_2	a_3
B	1.28868	0.037616	−0.0170972	0.00367183
$X/L^{\bullet}$				
D	**2.00661**	**0.016541**	**0.000142451**	**0.0029**
F	2.76837	0.0340247	0.0219798	0.00226116

So, introducing the LEL at around 20°C:

$$\langle C_{LEL} \rangle = \frac{0.04 \cdot 2}{22.414 \cdot \frac{293.16}{273.16}} = \frac{0.08}{24} = 0.0033\,\text{kg/m}^3 \tag{10.74}$$

$$L^{\bullet} = \sqrt{\frac{W_c}{u_w \cdot \langle C_{LEL} \rangle}} = \sqrt{\frac{12.45}{4 \cdot 0.0033}} = 30.7\,\text{m} \tag{10.75}$$

$$X = 30.7 \cdot \left\{2.0061 + 0.016541 \cdot \ln(30.7) + 0.000142451 \cdot [\ln(30.7)]^2 + 0.0029 \cdot [\ln(30.7)]^3\right\} \cong 67\,\text{m} \tag{10.76}$$

On the basis of this datum, the coefficients of van Buijtenen for $D/5$ are:

Adopting:

a
0.128
b
0.905
Distance (m)
10–100
Values for c
0.0939
Values for d
0.8403

$$M_e = W_o \cdot \left(\frac{W_o}{a \cdot c\pi \cdot u_w}\right)^{\frac{1}{b+d}} \cdot \frac{b+d}{(b+d+1) \cdot u_w} \cdot \left[\left(\frac{1}{C_{LFL}}\right)^{\frac{1}{b+d}} - \left(\frac{1}{C_{UFL}}\right)^{\frac{1}{b+d}}\right] \tag{10.77}$$

It will be:

$$M_e = 12.45 \cdot \left(\frac{12.45}{0.128 \cdot 0.0929 \cdot \pi \cdot 5}\right)^{\frac{1}{0.905+0.8403}} \cdot \frac{0.905 + 0.8403}{(0.905 + 0.8403 + 1) \cdot 5} \cdot \left[\left(\frac{1}{0.0033}\right)^{\frac{1}{0.905+0.8403}} - \left(\frac{1}{0.062}\right)^{\frac{1}{0.905+0.8403}}\right] = 377\,\text{kg} \tag{10.78}$$

Dense gases

For dense gases, the models presented by Britter and McQuaid (1988) can be adopted. They are in line with the method presented in Chapter 8. A simple empirical formula has been proposed by Harvey (1979), which has been tested, comparing calculated values to experimental data:

$$R_c = 30 \cdot \sqrt[3]{M_e} \tag{10.79}$$

where R_c is the radius of the cloud (m), and M_e is the gas mass. This equation is based on the assumption that the cloud radius to height ratio is 5. Mannan (2012) reports that a very good accordance with the cyclohexane cloud size of Flixborough incident has been found.

Step 4—Identification of blast sources

Examples:

- Process equipment
- Space between parallel planes
- Multistorey buildings
- Buildings
- Space in tube-like structures
- Tunnels, bridge, corridors
- Sewage systems

It is worth noting that stairs, pipes, and stick-built structures don't significantly contribute to pressure rise. However they have to be considered in the explosion effects assessment (Blast loads or explosion wind).

Step 5—Definition of obstacles and obstacles orientation

Take the *smallest dimension* oriented in a plane perpendicular to the flame propagation direction to be D_1, then (Fig. 10.12):

- D_1 is H_c or D_c for a cylinder.
- D_1 is the smallest of the three sides for a box.
- D_1 is the diameter of a sphere.

Take the obstacle dimension *parallel to the flame propagation* direction to be D_2 (Fig. 10.13).

> *An obstacle belongs to an obstructed region if the distance from its centre to the centre of any obstacle in the obstructed region is smaller than 10 times* D_1*, or 1.5 times* D_2*, of the obstacle under consideration in the obstructed region (*D_1 *and* D_2 *belonging to any obstacle in the obstructed region). If the distance between the outer boundary of the obstructed region and the outer boundary of the obstacle is larger than 25 m, then the obstacle does not belong to that obstructed region.*
>
> ***(Van den Bosch and Duijm, Yellow Book, 2005)***

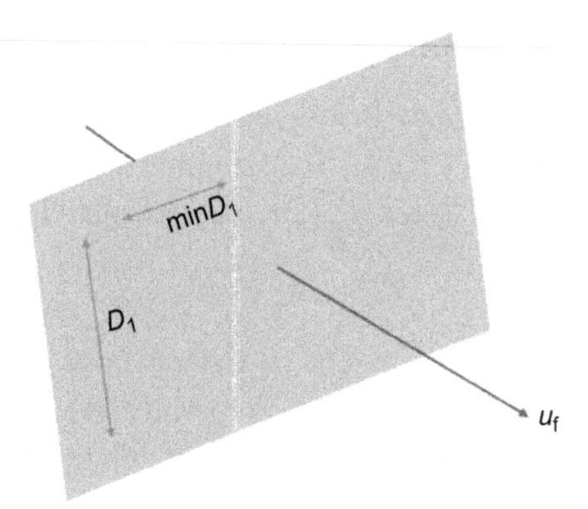

Fig. 10.12
Definition of obstacles and obstacles orientation.

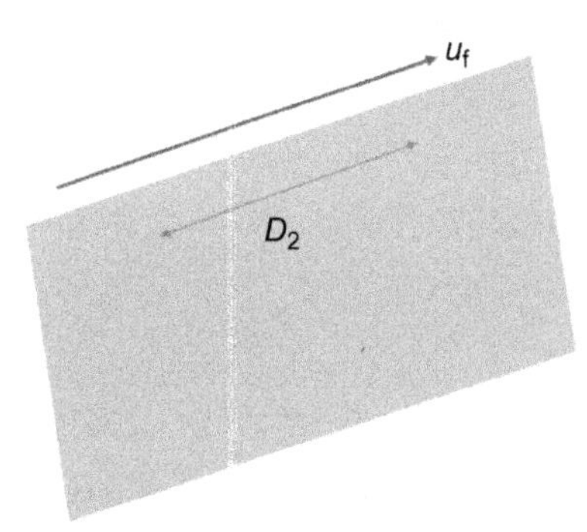

Fig. 10.13
Definition of obstacles and obstacles orientation.

Example 10.7

The following example consists of a unit including the following main items (Fig. 10.14 and 10.15; Table 10.13):

- Two liquid hydrocarbon storage tanks
- Two hydrogen compressors
- Three reactors fed with hydrogen and hydrocarbons
- Seven heat-exchangers
- Two pumps
- Pipe rack
- Substation
- Field Auxiliary Room (FAR)

Ignition of hydrogen is assumed to take place between two reactors. Criteria to select ignition locations are the following:

- Worst case(s)
- High degree of congestion

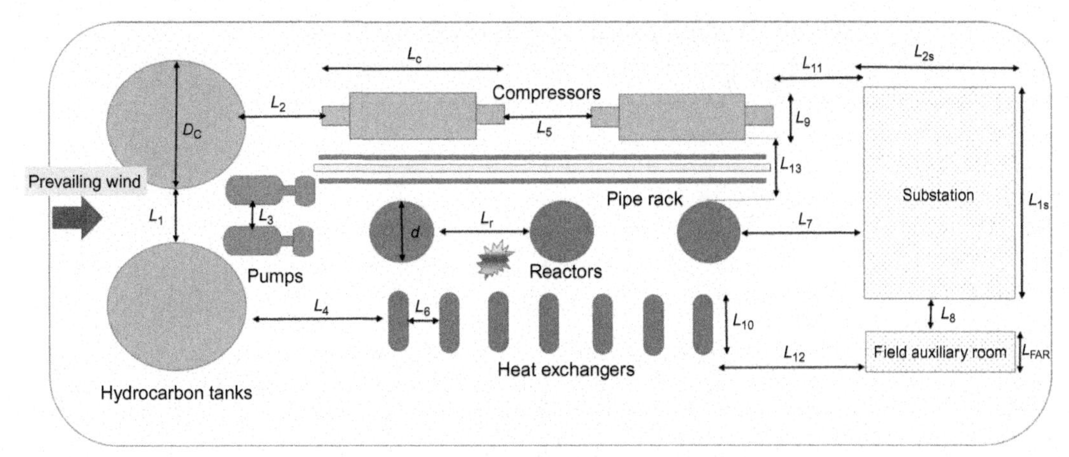

Fig. 10.14
Lay out of the unit of Example 10.7.

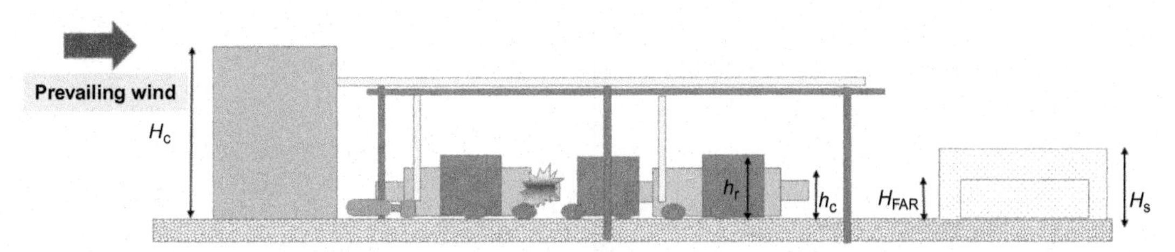

Fig. 10.15
Side view of the unit of Example 10.7.

- Potential presence of flammable gas/vapour
- Likelihood of presence of ignition sources
- Choice consistent with prevailing wind direction

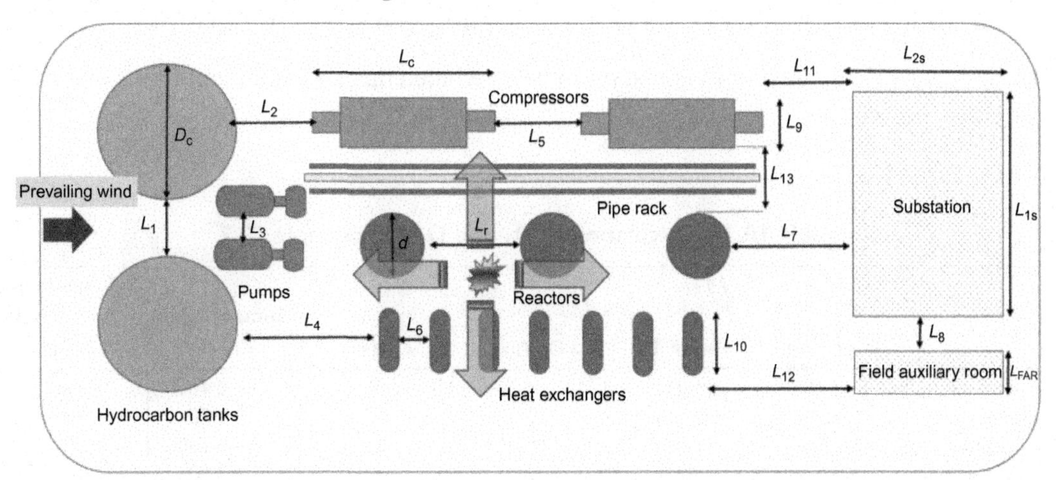

Table 10.13 Sizes, heights and distances for Example 10.7

Length	Size (m)
D_c	10
d	4.0
H_c	15
L_{1s}	18
L_{2s}	10
H_s	7.0
H_{FAR}	3.0
h_c	5.0
h_r	7.0
L_r	7.0
L_c	10.0
L_{FAR}	5.0
L_1	6.0
L_2	7.0
L_3	3.0
L_4	8.0
L_5	7.0
L_6	3.0
L_7	8.0
L_8	4.0
L_9	6.0
L_{10}	6.0
L_{11}	7.0
L_{12}	8.0
L_{13}	6.0

Definition of D_1 and D_2

The calculation of D_1 and D_2 for some equipment has been shown in Table 10.14.

Repeating the calculation for all items, it can be concluded that all equipment forms one single obstructed region.

Table 10.14 Calculation of D_1 and D_2 for Example 10.7

Item	Flame From	Flame To	D_1 (m)	D_2 (m)	Dimensional Check	Belongs to an Obstructed Zone
Reactor	Reactor	Reactor	4	7	$11 < 40$	Yes
Substation	Reactor	Substation	7	10	$15 < 70$	Yes
Tank	Reactor	Tank	10	10	$15 < 100$	Yes
Compressor	Reactor	Compressor	5	6	$12 < 50$	Yes

Step 6—Identification and sizing of boxes

Boxes containing obstructed regions have to be defined, according to the criterion to minimise noncongested or free space zones. Some height allowances may be considered for pipeworks and ancillary equipment.

Example 10.8

Define and size the boxes for the items included in the previous example (Fig. 10.16 and Table 10.15).

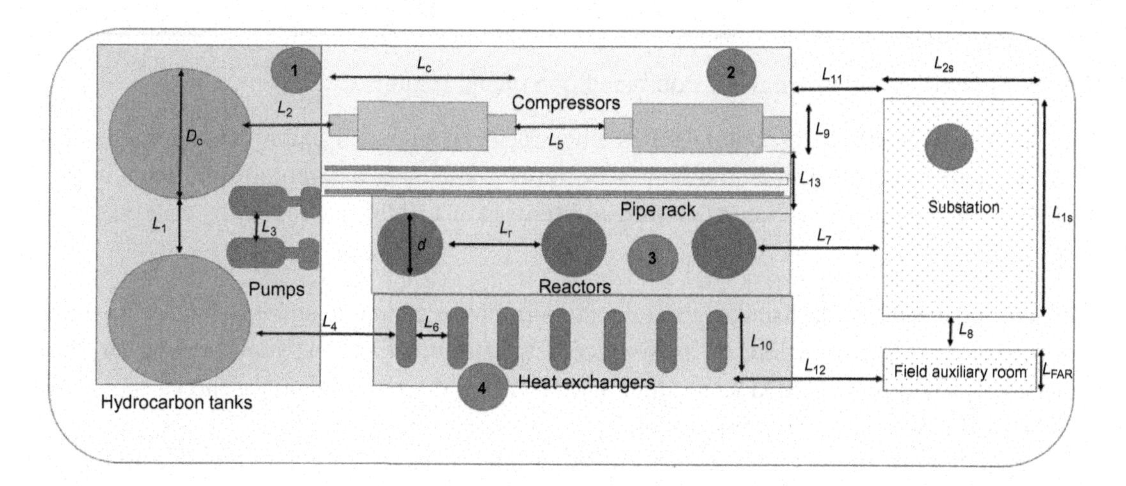

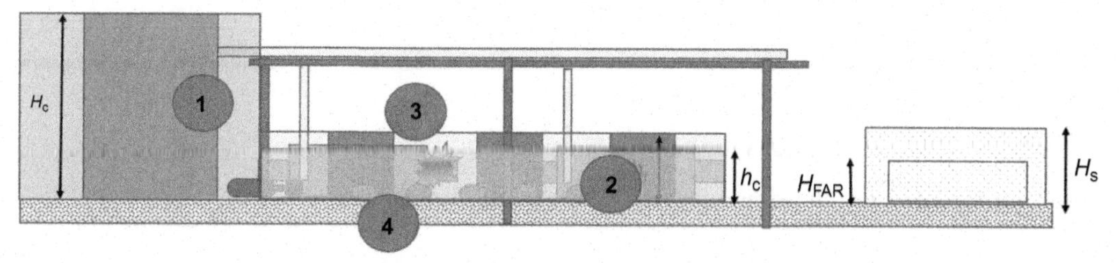

Fig. 10.16
Identification of boxes for Example 10.8.

Table 10.15 Calculation of volume of boxes for Example 10.8

Description	Box	Calculation	Volume (m^3)
Tanks area	1	$36 \times 22 \times 20$	15,840
Compressors area	2	$30 \times 20 \times 6$	3600
Reactors area	3	$26 \times 8 \times 7$	1456
Heat exchangers	4	$26 \times 8 \times 2$	416
	TOTAL	–	21,312

Step 7—Identification of volume blockage ratio (VBR)

The VBR is defined as the ratio of the volume occupied by congestion elements such as pipes, beams, plates, etc. to the volume of the portion of the plant under consideration (CCPS, 2005).

Example 10.9

Calculate the VBR for the previous example.

Solution

Pumps have not been included in the VBR calculation (Table 10.16).

Step 8—Identification of single or combined obstructed regions

The obstructed regions (OR) contained in the boxes can be managed as individual regions or as a combined region. This differentiation is very important and has a potential significant impact on the outcome of the explosion. According to Mercx et al. (1998), the following criteria can be used:

- The critical separation distance around a potential blast source is equal to half of its linear dimension in each direction; if the distance between potential blast sources is larger, the sources should be modelled as separate blasts; if not, the two potential sources should be modelled as a single superimposed blast.
- If separate blast sources are modelled together, their effects in the far-field is cumulative, and the combustion energy will be summed.

Example 10.10

Identify the obstructed regions in the previous example.

Applying Mercx's rule, all boxes will form a single OR. Builidings will be kept separated.

Table 10.16 Calculation of VBR for Example 10.9

Description	No. Items	Item Volume (m^3)	Total	Box	Calculation	Volume (m^3)	VBR
Tanks	2	1177.5	2355	1	$36 \times 22 \times 20$	15,840	0.149
Compressors	2	300	600	2	$30 \times 20 \times 6$	3600	0.167
Reactors	3	88	264	3	$26 \times 8 \times 7$	1456	0.180
Heat exchangers	7	10	70	4	$26 \times 8 \times 2$	416	0.168
			3289	TOTAL	–	21,312	0.154

Step 9—Identification of the maximum flame path length L_p (MFPL)

A conservative recommended approach to MFPL calculation is presented by Mercx et al. (1998) and consists of taking it as ½ the maximum length of a congested area. This is specifically valid for the ignition location assumed in the centre of the overall obstructed region. Should this be assumed elsewhere, for example at the upwind edge, L_p would be different.

Example 10.11

Calculate the MFPL ratio for the previous example.

Solution
The ½ of the longer size of the overall obstructed region is around 52 m.

Step 10—Determination of the average obstacle size

The recommended average obstacle size D, among those defined by Mercx et al. (1998), is given by the hydraulic mean diameter:

$$D_{hym} = 4 \cdot \frac{\sum_1^n V_i}{\sum_1^n A_i} \tag{10.80}$$

This value is based on the concept of hydraulic diameter, which is defined as four times the ratio of the summed volumes to the summed surface areas of an object distribution.

Example 10.12

Calculate the average obstacle size for the previous example.

Solution
See Table 10.17.

Table 10.17 Calculation of average obstacle size for Example 10.12

Description	Volume (m^3)	Surface Area (m^2)	D_{hym} (m)	No. Items	Items D_{hym} (m)
Tanks	1177.5	628	7.5	2	15
Compressors	300	280	4.3	2	8.6
Reactors	88	113	3.1	3	9.3
Heat exchangers	10	31	1.3	7	9.1
TOTAL				14	42

Step 11—Definition of flame expansion

The degrees of freedom of flame expansion has been categorised by Baker et al. (1994) and have been depicted in the next chapter dealing with the BST model.

Mercx et al. (1998) have proposed the following criteria to delimitate the 2-D from the 3-D flame expansion. Being *L*, *W*, and *H*, the length, the width, and the height of the obstructed region respectively, flame expansion will be 2-d if a confining plane (other than the ground) is present and:

- $L > 5 \cdot H$
- $W > 5\ H$
- the dimensions of the confining plane are of about the same size as *L* and *H*.

Example 10.13

Identify the flame expansion modality for the previous example.

Solution
The flame expansion will be 3-D.

Step 12—Overpressure calculation

Maximum overpressures for 2-D and 3-D confinement according to the findings of the GAME project are as follows.

2-D Confinement

$$\frac{\Delta P_{max}}{P_a} = 3.38 \cdot \left(\frac{VBR \cdot L_f}{D}\right)^{2.25} \cdot {S_u}^{2.7} \cdot D^{0.7} \tag{10.81}$$

3-D Confinement

$$\frac{\Delta P_{max}}{P_a} = 0.84 \cdot \left(\frac{VBR \cdot L_f}{D}\right)^{2.75} \cdot {S_u}^{2.7} \cdot D^{0.7} \tag{10.82}$$

with:

- P_a is the atmospheric pressure.
- ΔP_{max} is the side on pressure, same unit as P_o, occurring at the explosion centre.
- VBR is the volume blockage ratio.
- D is the obstacle size.
- S_u is laminar burning velocity.
- L_f is the length available for the flame travel.

Example 10.14

Calculate the overpressure for the previous example.

Solution
3-D Confinement

$$\frac{\Delta P}{P_a} = 0.84 \cdot \left(\frac{\text{VBR} \cdot L_f}{D}\right)^{2.25} \cdot S_u{}^{2.7} \cdot D^{0.7} = 0.84 \cdot \left(\frac{0.154 \cdot 18.8}{42}\right)^{2.25} \cdot 3.25.7^{2.7} \cdot 42^{0.7} \tag{10.83}$$
$$= 0.18\,\text{atm}$$

The overpressure at the epicentre of the explosion will be 0.18 atm.

Step 13—Blast curve identification

The knowledge of the maximum overpressure enables one to identify the applicable blast curve, removing the subjectivity of the selection, that is one the limits of the MEM.

Example 10.15

Identify the blast curve for the previous example. For hydrogen the maximum laminar burning velocity of 3.25 m/s will be adopted (Van den Bosch and Duijm, 2005).

Entering $P_s = 0.18$, a blast curve very close to 5 will be found. We will adopt this curve. On this basis, impulse and duration curves will be defined. A different ignition location, such as at the obstructed region edge, would result in 1.14 atm, a much higher overpressure (Fig. 10.17).

An interpolation can be carried out as recently confirmed also by Boot and van der Voort (2013) to define a more accurate curve.

Step 14—Overpressure at given locations

The calculation of an overpressure at a given location requires the full definition of the function scaled peak side-on pressure against the scaled distance:

$$\frac{\Delta P}{P_o}(R_s) \tag{10.84}$$

$$R_s = R \cdot \sqrt[3]{\frac{P_a}{E}} \tag{10.85}$$

- r is the distance to centre of explosion, m.
- Pa is the atmospheric pressure, Pa.
- E is the energy blast, J.

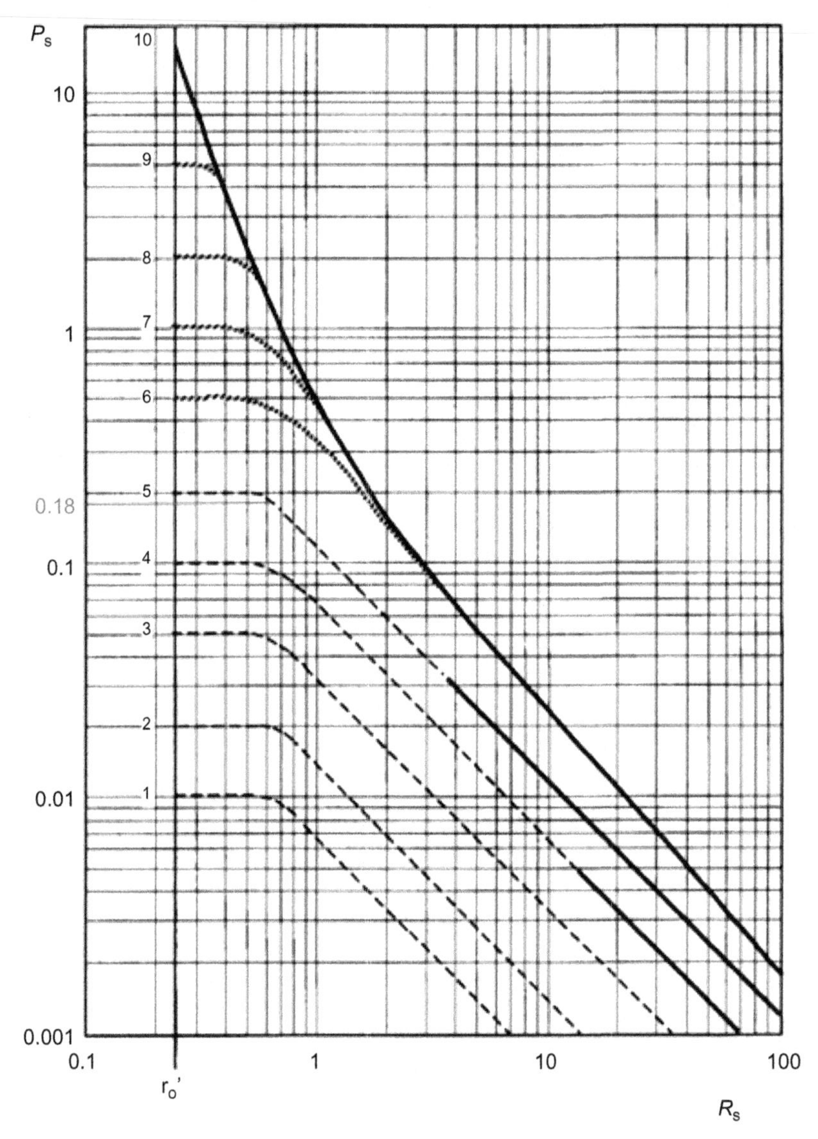

Fig. 10.17
Identification of blast curve for example 10.15.

The energy blast to be included in this equation is the combustion enthalpy associated to the flammable cloud calculated at Step 3. If this is larger than the obstructed region free volume, it will be included in the calculation.

Example 10.16

Find the overpressure and the duration of the pressure wave at the substation of the previous example. For hydrogen, a lower heating value of 10 MJ/m^3 can be assumed.

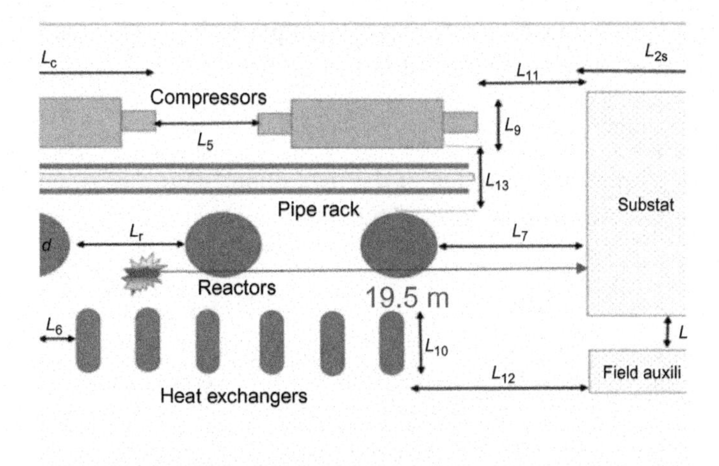

Fig. 10.18
Distance of substation from explosion epicentre for Example 10.16.

Solution
In this example, the calculated flammable cloud size has been found to be larger than the free volume of the obstructed region (Fig. 10.18). The latter, 27,819 m^3, will be used for the calculation. According to the reaction stoichiometry, the molar ratio of flammable mixture to pure hydrogen is 10.5 to 1. Pure hydrogen will be 2649 m ^{3}.

$$t_s = u_s \cdot t_p \cdot \sqrt[3]{\frac{P_a}{E}} \tag{10.86}$$

$$R_s = R \cdot \sqrt[3]{\frac{P_a}{E}} = 19.5 \cdot \sqrt[3]{\frac{101,325}{10,000,000 \cdot 2649}} = 0.3 \tag{10.87}$$

The overpressure is still very close to 0.18 atm. The scaled positive duration is 0.88.

$$t_s = u_s \cdot t_p \cdot \sqrt[3]{\frac{P_a}{E}} \tag{10.88}$$

$$t_p = 0.88 \cdot \frac{\sqrt[3]{\frac{10,000,000 \cdot 2649}{101,325}}}{340} = 0.165\,\text{s} = 165\ \text{ms} \tag{10.89}$$

10.10.2 Baker-Strehlow-Tang Model

The original Baker-Strehlow-Tang method was published in 1994 (Baker et al., 1994). As with the MEM, some blast curves are defined. However, the selection is based on flame speed. A further release of the model was issued in 1999 (Tang and Baker, 1999). In 2005, Pierorazio et al. conducted a more accurate experimental campaign and integrated the model with further data. This latest version has been presented in this section.

Summary of the BST method

The BST methodology requires selection of the maximum flame speed, on the basis of the combined effects of the affecting factors, which are:

- congestion
- fuel reactivity
- confinement

Congestion

Pierorazio et al. (2005) carried out experimental tests using specific congestion patterns. Table 10.18 summarises their statements.

A rule of thumb for congestion consists of the following (CCPS, 2010):

- If it is easy to walk through an area relatively unimpeded, and if there are only two layers of obstacles, it can be considered low congestion.
- Medium congestion is when, walking through an area of medium congestion, it is cumbersome to do so and often requires taking an indirect path.
- High congestion is when it is not possible to walk through an area because there is not sufficient space to pass obstacles and successive layers prevent transit through the area. Moreover, repeated layers of closely spaced obstacles block line of sight from one edge of the congested zone to the opposite side.

Potential Explosion Sites (PES)

Baker et al. (1997) define a PES as a congested or confined area where a flammable cloud is introduced and ignited. This concept is very important in order to determine whether or not a plant area has to be approached as a single or a multiple PES. In the former case, flame does not decelerate, and only one blast wave has to be considered. In the second case, separation is such that two different zones with two separate blast waves have to be considered. This has a significant impact on the layout.

Table 10.18 Congestion levels for BST method (Pierorazio et al., 2005)

Level	Area Blockage Ratio Per Plane	Number of Obstacle Layers	Pitch
Low	Less than 10%	One or two layers	>8D
Medium	Between 10% and 40%	Two to three layers	4D–8D
High	Greater than 40%	Three or more closely spaced layers	<4D

D: diameter of obstacles.

Some criteria are:

- Zones with less than 4.6 m of empty space between them are to be considered as a single PES.
- For cases where the edges of the respective zones are filled with repeated obstacles that encroach uniformly to the zone borders, a minimum of 9.1 m between zones is generally recommended to consider the zones to form separate PESs.
- Combination of blast waves from multiple sources can be approached by using the greater peak pressure, summing their impulses, and calculating the duration on the basis of the triangular assumption.

Reactivity (Zeeuwen and Wiekema, 1978)
Reactivity is defined on the basis of the laminar burning velocity (Table 10.19).

For mixtures, Baker et al. (1997) have proposed to calculate the burning velocity of mixtures by means of the Le Chatelier principle.

Confinement
Degrees of confinements, as defined by Pierorazio et al. (2005), is simplified in the following sketches. An additional category 2.5-D has been included, which has to be considered an average between 2-D and 3-D (Fig. 10.19).

BST flame speed correlations
The joint effect of congestion, reactivity, and confinement has been correlated by Pierorazio et al. (2005), resulting in the flame speed Mach numbers included in Table 10.20.

It is worth noting:

- Pierorazio et. al (2015) have deleted from their classification Confinement 1-D, because the maximum flame speed achieved in real one-dimensional expansion conditions (i.e. pipe) is a function also of the length-to-diameter ratio, in addition to the other factors. So researchers considered the representation of this case through the use of a single number an over-simplification. For 1-D confinement, a more complete analysis is recommended (Bartknecht, 1981, 1993).
- If DDT is attained, a Mach number equal to 5.2 has to be adopted.

Table 10.19 Reactivity of fuels (Zeeuwen and Wiekema, 1978)

Low Reactivity	Medium Reactivity	High Reactivity
$S_u < 45$ cm/s Methane and carbon monoxide	45 cm/s $< S_u <$ 75 cm/s All other fuels	$S_u > 75$ cm/s Hydrogen Acetylene Ethylene Ethylene oxide Propylene oxide

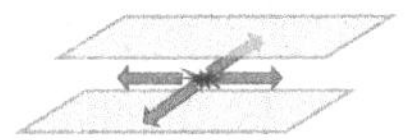

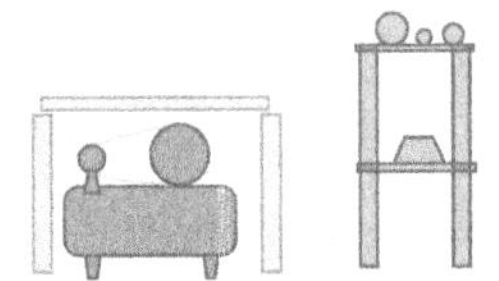

Fig. 10.19
Definition of confinements (Pierorazio et al., 2005).

Table 10.20 Flame speed correlations (Pierorazio et al., 2005)

		Congestion		
Confinement	**Reactivity**	**High**	**Medium**	**Low**
1-D	High	DDT	DDT	DDT
	Medium	2.27	1.77	1.03
	Low	2.27	1.03	0.294
2-D	High	DDT	DDT	DDT
	Medium	1.6	0.66	0.47
	Low	0.66	0.47	0.079
2.5-D	High	DDT	DDT	DDT
	Medium	1.0	0.55	0.29
	Low	0.50	0.35	0.053
3-D	High	DDT	DDT	0.36
	Medium	0.50	0.44	0.11
	Low	0.34	0.23	0.026

DDT: deflagration to detonation transition.

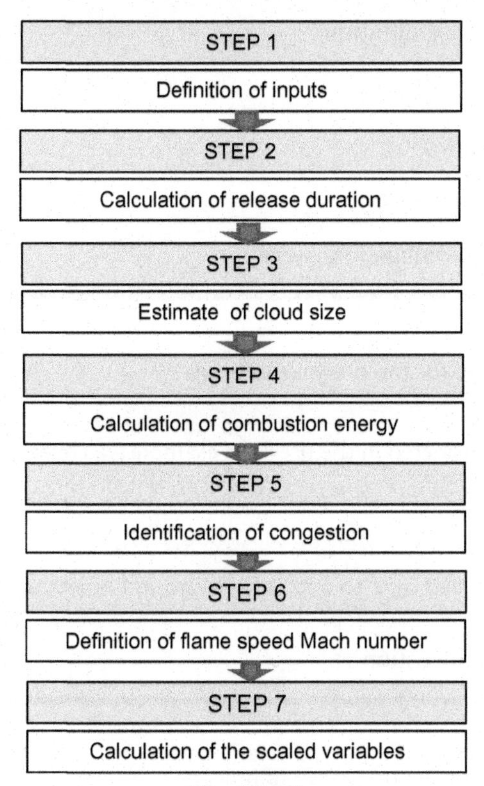

Fig. 10.20
Flow chart of BST method.

Procedure to use the BST model

The following flow chart summarises the steps to be followed for the application of the BST method (Fig. 10.20).

Step 1—Definition of inputs

Inputs to be defined for BST model are:

1. Identification of geometrical features
2. Identification of substance reactivity

Example 10.17

Define the input data for the BST application to the example described for the MEM model.

Solution
All data are available. Fuel is hydrogen. The flame expansion will be 3-D.

Step 2—Calculation of rclcase duration

(see MEM method)

Step 3—Estimate of cloud size

(see MEM method)

Step 4—Calculation of the combustion energy

Example 10.18

Find the combustion energy for the previous example.

Solution

The combustion energy considered in the previous example was related just to the fuel filling the unit free space, 2,649 m^3:

$$E = 2649\,m^3 \times 10{,}000{,}000 J/m^3 \tag{10.90}$$

This value has to be multiplied by 2 to account for ground reflection.

Step 5—Identification of congestion

Example 10.19

Find the congestion class for the previous example.

Solution

Even if there are not just two layers, it is very easy to walk through the unit, so low congestion is much more reasonable than medium.

Step 6—Definition of flame speed Mach number

Example 10.20

Find the flame speed Mach number for the previous example.

Solution

Entering the flame speed table, Mach number $M_f = 0.36$ (Fig. 10.21).

Step 7—Calculate the scaled variables

As for the MEM method, scaled variables are related to scaled distance on the basis of blast curves parameterised in terms of flame speed Mach numbers.

In the following diagram (Fig. 10.22):

$$P_s = \frac{\Delta P}{P_o}(R_s) \tag{10.91}$$

Confinement	Reactivity	Congestion		
		High	Medium	Low
1-D	High	DDT	DDT	DDT
	Medium	2.27	1.77	1.03
	Low	2.27	1.03	0.294
2-D	High	DDT	DDT	DDT
	Medium	1.6	0.66	0.47
	Low	0.66	0.47	0.079
2.5-D	High	DDT	DDT	DDT
	Medium	1.0	0.55	0.29
	Low	0.50	0.35	0.053
3-D	High	DDT	DDT	0.36
	Medium	0.50	0.44	0.11
	Low	0.34	0.23	0.026

Fig. 10.21
Flame speed correlation for Example 10.20.

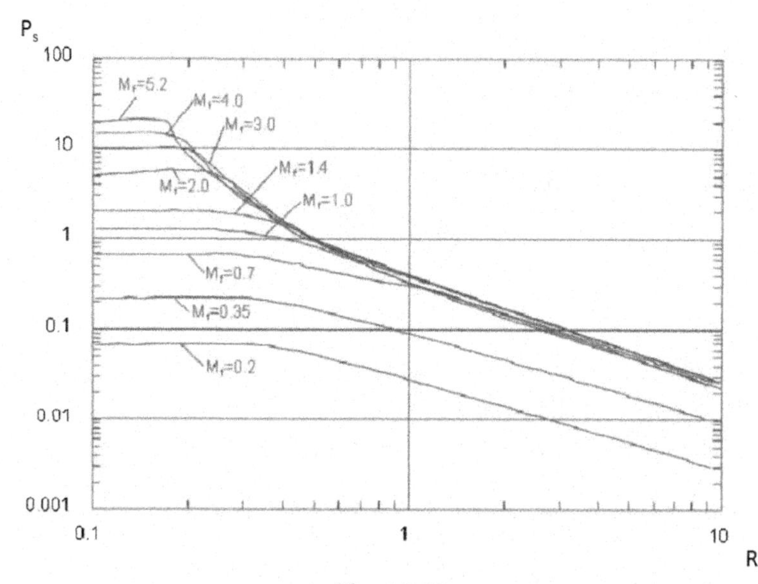

Fig. 10.22
Calculation of scaled overpressure.

$$R_s = R \cdot \sqrt[3]{\frac{P_a}{E}} \qquad (10.92)$$

where

- ΔP is the overpressure, Pa.
- R is the distance of the target to centre of explosion, m.

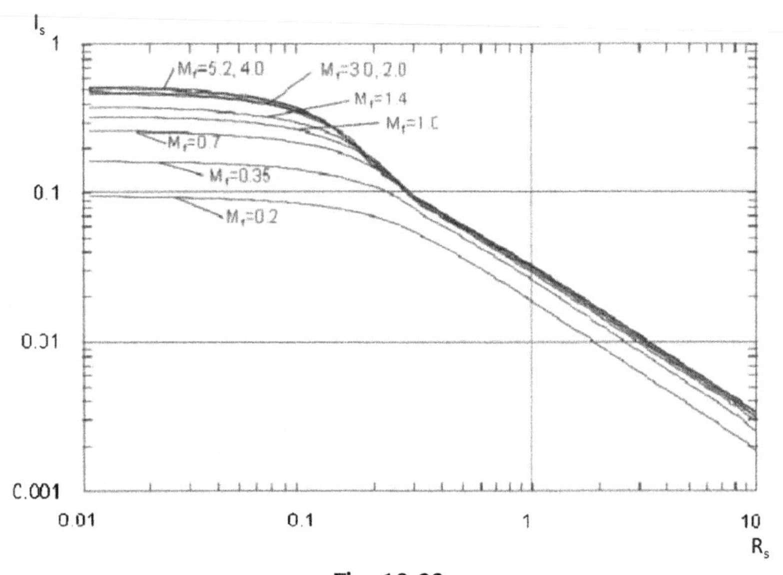

Fig. 10.23
Calculation of scaled impulse.

- P_a is the atmospheric pressure, Pa.
- E is the blast, J.

In the following diagram (Fig. 10.23):

$$I_s = \frac{I \cdot u_s}{\sqrt[3]{P_a^{\ 2} \cdot E}} \tag{10.93}$$

- I_s is the scaled impulse.
- I is the specific impulse for the positive phase.
- u_s is the acoustic velocity at ambient conditions.

Example 10.21

Find the maximum peak pressure for the previous example.

Solution
Entering the scaled pressure diagram, a value very close to 0.18 is found. This is the same result as the one found applying the MEM method (Fig. 10.24).

10.10.3 Congestion Assessment Method

This method is based on the analysis carried out by Cates (1991) and Puttock (1995, 2001) and has been implemented by Shell within the software FRED.

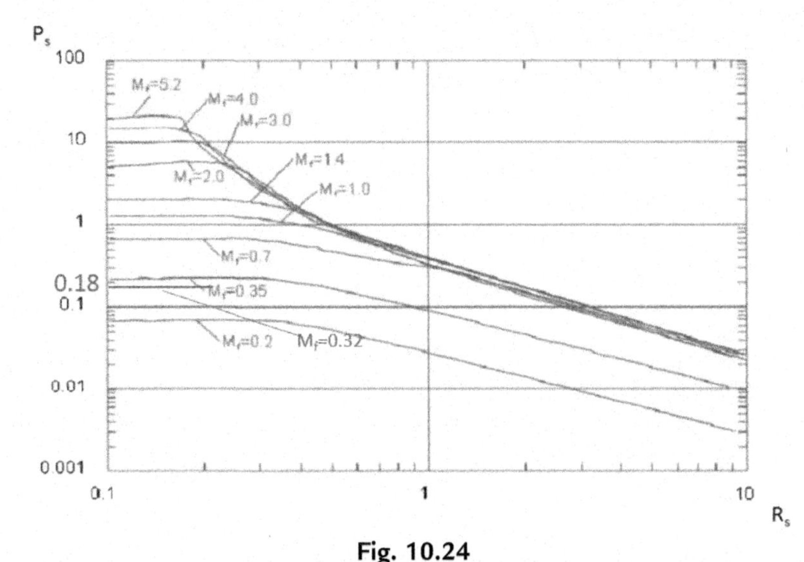

Fig. 10.24
Calculation of scaled overpressure for Example 10.21.

Summary of the CAM method

The CAM utilises two fuel parameters to characterise its explosive behaviour:

- Laminar burning velocity, S_u
- Expansion ratio, that is the unburned fuel to burned fuel ratio, V_o

Fuel Factor

The mentioned parameters are combined to give a fuel factor, F, which is (CCPS, 2010):

$$F = \frac{\left[S_u \cdot (V_o - 1)^{2.71}\right]_{\text{fuel}}}{\left[S_u \cdot (V_o - 1)^{2.71}\right]_{\text{propane}}} \tag{10.94}$$

where S_u is the laminar burning velocity, m/s, and V_o is thermodynamically defined as:

$$V_o = \frac{T_{ad}}{T_f} \cdot \frac{n_p}{n_r} \tag{10.95}$$

with:

- T_{ad}, fuel flame adiabatic temperature, K
- T_f, initial temperature, K
- n_p, n_r, number of combustion products and reactants kilomoles

The Fuel Factor for propane is 1. The expansion factors for some hydrocarbons and for hydrogen are included in Table 10.21 (Dugdale, 1985).

Table 10.21 Expansion factors for some hydrocarbons (Dugdale, 1985)

Fuel	Expansion Factor (V_o)
Hydrogen	6.9
Methane	7.5
Ethane	7.7
Propane	7.9
n-Butane	7.9
Pentane	8.1
Hexane	8.1
Acetylene	8.7
Ethylene	7.8
Propylene	7.8
Benzene	8.1
Cyclohexane	8.1

Table 10.22 Fuel factors for some hydrocarbons (Puttock, 1995)

Fuel	Fuel Factors
Methane	0.6
Propane	1.0
Toluene	0.7
Pentane	1.0
n-Butane	1.0
Ethanol	1.5
Acetone	1.0
Methanol	1.0
Propylene	1.5
Benzene	1.0
Cyclohexane	1.0
Butadiene	2.0

Puttock (1995) has reported the following fuel factors (Table 10.22):

Blockage ratio
The blockage ratio has experimentally been determined by Puttock as:

$$B_r = \exp(6.44 \cdot B) \tag{10.96}$$

where B is the blockage of rows of obstacles.

Other affecting parameters
Other parameters affecting the peak overpressure are:

- d, obstacle diameter, m
- N, the number of obstacles row in each direction, counting from the centre
- R_{pd}, the pitch-to-diameter ratio

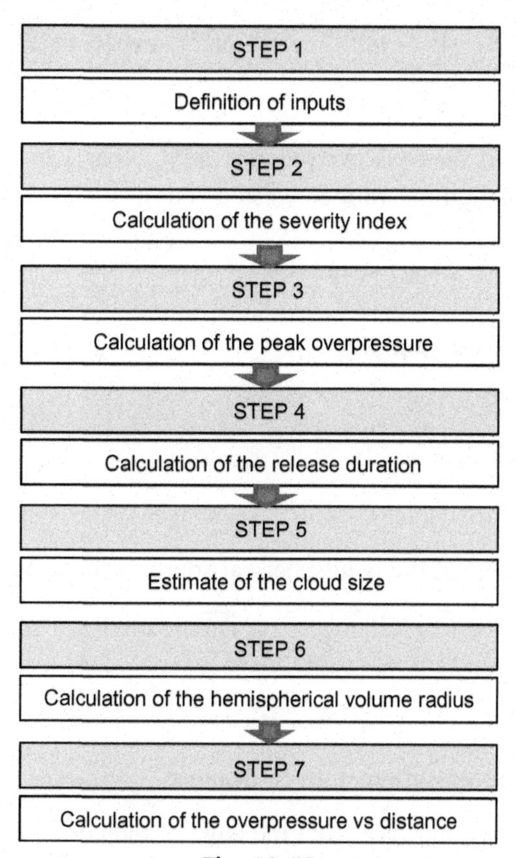

Fig. 10.25
Flow chart for CAM method.

Procedure to use the CAM model

The following flow chart summarises the steps to be followed for the application of the CAM model (Fig. 10.25).

Step 1—Definition of inputs

Inputs to be defined for BST model are:

1. Identification of geometrical features
2. Identification of substance expansion factor
3. Identification of substance fuel factor

Step 2—Calculate the severity index

A severity index S has been introduced by Puttock to account for the unrealistic behaviour of the peak overpressure at high values.

$$S = 0.000039 \cdot [S_u \cdot (V_o - 1)]^{2.71} \cdot d^{0.55} \cdot N^{2.54} \cdot \exp(6.44 \cdot B) \cdot R_{pd}^{0.55} \tag{10.97}$$

Step 3—Calculate the peak overpressure

Once S has been determined, the peak overpressure ΔP_{max} (bar) can be calculated by means of the following implicit equation (no-roof congested zone):

$$S = \Delta P_{max} \cdot \exp\left(0.4 \cdot \frac{\Delta P_{max}}{V_o^{1.08} - 1 - \Delta P_{max}}\right) \tag{10.98}$$

Step 4—Calculation of release duration

(see MEM method)

Step 5—Estimate of cloud size

(see MEM method)

Step 6—Calculate the radius of the hemispherical source volume V

$$R_o = \sqrt[3]{\frac{3 \cdot V}{2 \cdot \pi}} \tag{10.99}$$

where units are metres.

Step 7—Calculate the overpressure at a given distance

If r is the distance of a given receptor from the edge of the congestion area, the following parameters are defined:

$$R = R_o + r \tag{10.100}$$

$$\log_{10} P_1 = 0.08 \cdot l_r^4 - 0.592 \cdot l_r^3 + 1.63 \cdot l_r^2 - 3.28 \cdot l_r + 1.39 \tag{10.101}$$

with:

$$l_r = \log_{10}\left(\frac{R}{R_o}\right) + 0.2 - 0.02 \cdot \Delta P_{max} \tag{10.102}$$

It will be:

$$P(R) = \min\left(\frac{R_o}{R} \cdot \Delta P_{max}, P_1\right) \tag{10.103}$$

10.11 Vessels Burst

Effects of a vessel burst can be assessed, considering the equipment, as a sphere filled with gas. A practical method is to use Baker-Tang curves (Tang et al, 1996). It has to be considered:

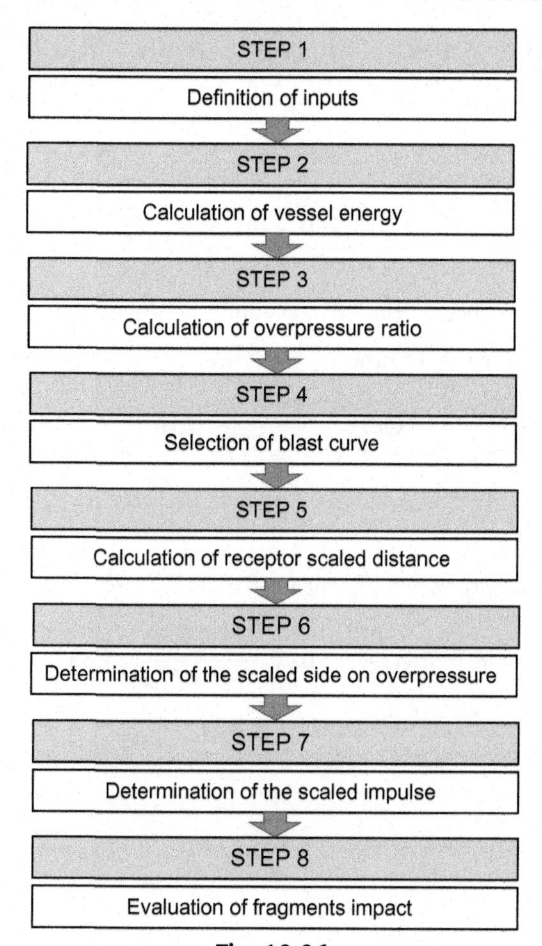

Fig. 10.26
Flow chart for calculation of vessel bursts.

- Effects of fragments are not accounted for.
- Kinetic energy is not considered.
- The model is based on ideal gas law.

The following flow chart can be followed to assess the peak side-on overpressure and impulse for a given receptor. Equations for the evaluation of fragment impact have been added (Fig. 10.26).

Example 10.22

A 15 m^3 air receiver at 10 atm absolute and 20°C bursts. Find the wave overpressure and impulse for a receptor far 20 m from the receiver. Ambient temperature is 20°C.

Solution

Step 1—Input data

Volume (m^3)	P_1 (Pa)	T (K)	P_o (Pa)	R (m)	T_{amb} (K)	g	[1]a_o (m/s)
15	1,013,250	293.16	101,325	20	293.16	1.4	340

[1]a_o is the pressurised gas speed ratio.

Step 2—Calculation of vessel energy

$$E_v = \frac{2 \cdot (P_1 - P_o)}{\gamma - 1} = \frac{2 \cdot (1{,}013{,}250 - 101{,}325)}{1.4 - 1} = 4{,}559{,}625\text{J} \tag{10.104}$$

In this equation, factor 2 has been included to account for ground reflection for the BST method.

Step 3—Calculation of overpressure ratio

$$\text{OP}_\text{R} = \frac{1{,}013{,}250}{101{,}325} = 10 \tag{10.105}$$

Step 4—Selection of blast curve

See Fig. 10.27.

Step 5—Calculation of receptor scale distance

$$R_S = R \cdot \sqrt[3]{\frac{P_o}{E_v}} = 20 \cdot \sqrt[3]{\frac{101{,}325}{4{,}559{,}625}} = 5.63 \tag{10.106}$$

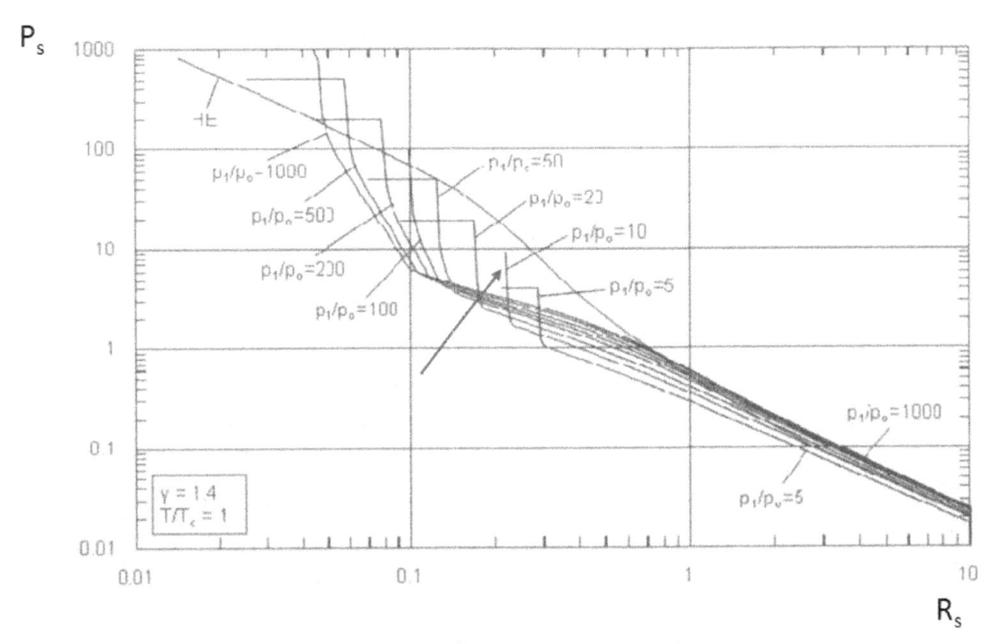

Fig. 10.27
Selection of blast curve for Example 10.22.

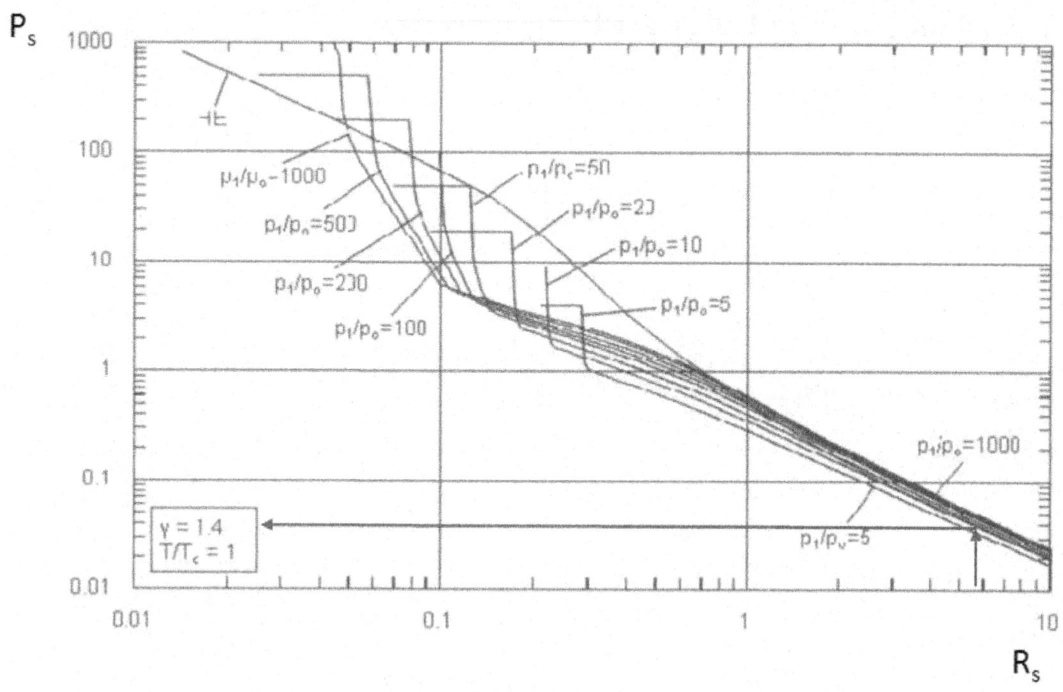

Fig. 10.28
Selection of peak side-on pressure for Example 10.22.

Step 6—Calculation of the scaled side-on overpressure (Fig. 10.28)

$$P_S = 0.04 \tag{10.107}$$

$$P_{s\text{-}o} = 0.04 \cdot P_o = 4053\,\text{Pa} \tag{10.108}$$

Step 7—Calculation of the scaled impulse

$$I_S = 0.029 \tag{10.109}$$

$$I = \frac{I_s \cdot \sqrt[3]{P_o{}^2 \cdot E}}{a_o} = \frac{0.029 \cdot \sqrt[3]{101{,}325^2 \cdot 4{,}559{,}625}}{340} = 30.5\,\text{Pas} \tag{10.110}$$

The wave duration is:

$$t_w = 2 \cdot \frac{I}{P_{s\text{-}o}} = 2 \cdot \frac{30}{4053} \cdot 1000 \cong 15\,\text{ms} \tag{10.111}$$

Step 8—Evaluation of fragments impact

A reasonable estimation of the maximum kinetic energy can be made according to Baum (1984):

$$E_k = 0.2 \cdot \frac{P_1 \cdot V}{\gamma - 1} = 0.2 \cdot \frac{1{,}013{,}250 \cdot 15}{1.4 - 1} = 7{,}599{,}375\,\text{J} \tag{10.112}$$

10.12 Boiling Liquid Expansion Vapour Explosion

10.12.1 Background

BLEVE is defined by CCPS (2005) as:

> *A type of rapid phase transition (RPT) in which a liquid contained above its atmospheric boiling point is rapidly depressurised, causing a nearly instantaneous transition from liquid to vapour with a corresponding energy release. A BLEVE of flammable material is often accompanied by a large aerosol fireball, since an external fire impinging on the vapour space of a pressure vessel is a common cause. However, it is not necessary for the liquid to be flammable to have a BLEVE occur.*

This definition specifies that the only one requirement for a BLEVE to take place is the storage of the liquid at a temperature higher than its atmospheric boiling point. Furthermore, flammability is not necessary. Many causes can result in a liquid suddenly facing a pressure, typically atmospheric pressure, at which it cannot exist in a liquid phase. Regardless of the cause, loss of containment, operational error, or inadvertent discharge can be the cause of a BLEVE. The longest BLEVE case history includes LPG vessels impinged by external fire, which typically fail at 600°C and at an internal pressure that is around 15 bar. The following diagram describes the expansion curve of metastable subcooled carbon dioxide under pressure (Benintendi, 2014) (Fig. 10.29).

Once subcooled carbon dioxide has reached the saturation state B, it would be expected to isothermally and isobarically travel along the line BE. In reality, any liquid takes some time to accommodate to a new thermodynamic condition, depending on its relaxation time.

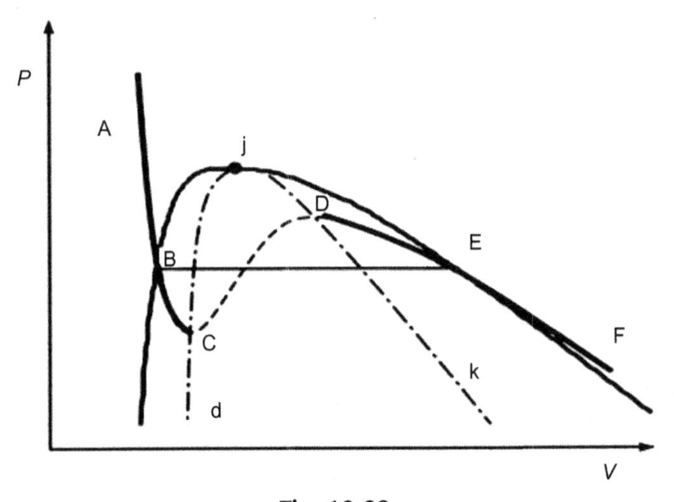

Fig. 10.29
Expansion curve of metastable subcooled carbon dioxide (Benintendi, 2014).

Carbon dioxide will move as a superheated liquid from B to C across a pressure and temperature field, where it should vapourise. This is possible until the superheat temperature limit is reached, which is a characteristic of the liquid. After this point, vapourisation starts, and the violence of the expansion depends also on the substance liquid to vapour expansion ratio.

10.12.2 BLEVE Scenarios

Table 10.23 includes practical data relating to substances which potentially undergo BLEVEs, along with their scenarios.

10.12.3 BLEVE Phases

The following simplified sequence identifies the typical phases of a BLEVE (Fig. 10.30).

10.12.4 BLEVE TNO Model

A simple model to describe the effects of BLEVE has been presented by Pietersen and Huerta (1984) and has shown a good accordance with the data related to Mexico City incident.

Table 10.23 BLEVE scenarios

Substance	Storage Conditions	Liquid to Vapour Expansion Factor	Fireball	Driving Force for BLEVE
LPG	P=8–10 bar Ambient temperature	270	Yes	Pressure difference
LNG	P=1 atm, −161°C	620	YES	Temperature difference (rapid phase transition)
Light hydrocarbons until pentane (Hasegawa and Kato, 1978)	Vapour pressure at ambient temperature	–	YES	–
Carbon dioxide	Above triple point	450	No	Pressure difference
Ammonia	–	840	YES	Pressure difference
Chlorine	–	–	NO	Pressure difference
Water	–	–	NO	Pressure difference Temperature difference
Cryogenic gases	Cryogenic	–	NO	Temperature difference Pressure difference

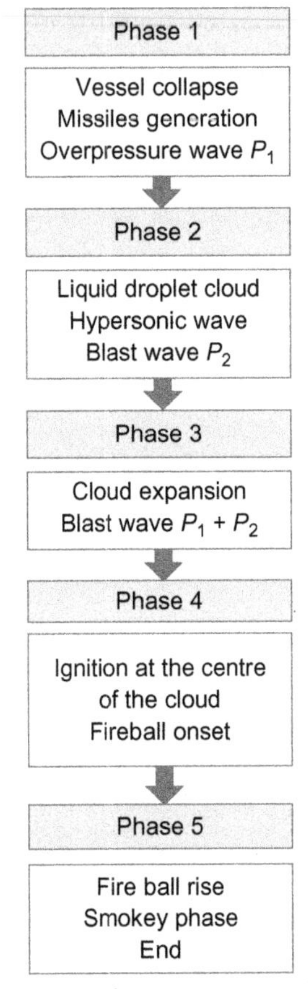

Fig. 10.30
Flow chart of BLEVE phases.

Diameter of BLEVE

$$D_b = 6.49 \cdot M_b^{0.325} \tag{10.113}$$

with:

- D_b, diameter of BLEVE, m
- M_b, amount of the material in the fireball, kg

Example 10.23

LPG is stored in a 10 m diameter sphere at ambient temperature (20°C). Pressurised liquid consists of 75% propane and 25% *n*-butane (mass fraction). Liquid occupies 50% of the geometric volume. Find the fireball diameter if BLEVE occurs.

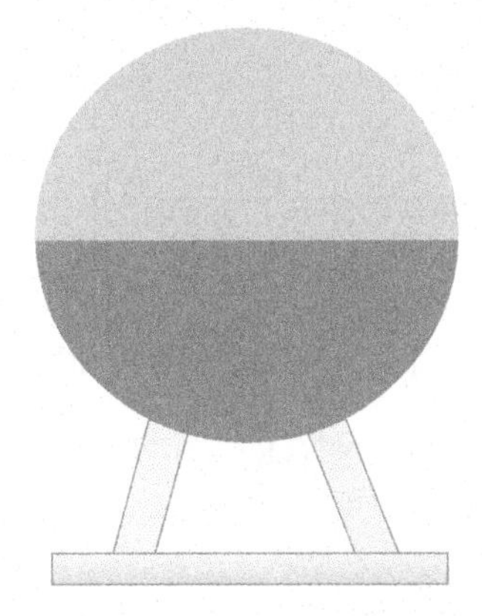

Solution

Propane

T_a (K)	D (m)	$MW_{propane}$	γ	x_m	Vapour Pressure P_p (20°C) (MPa)	Liquid Density (kg/m³)
293.16	10	44.1	1.13	0.75	0.83	580

***n*-Butane**

T_a (K)	D (m)	$MW_{n\text{-butane}}$	γ	x_m	Vapour Pressure P_b (20°C) (MPa)	Liquid Density (kg/m³)
293.16	10	58.1	1.18	0.25	0.208	604

Liquid and vapour volume:

$$V_v = V_L = \frac{2}{3} \cdot \pi \cdot R^3 = \frac{2}{3} \cdot \pi \cdot 5^3 = 261.7\text{m}^3 \tag{10.114}$$

Liquid mass:

$$M_L = 261.7(580 \cdot 0.75 + 604 \cdot 0.25) = 153{,}356\text{kg} \tag{10.115}$$

Vapour mass:

Mass of vapour can be calculated applying Raoult's law:

$$P_v = P_{vp}^o x_{mol\text{-}p} + P_{vp}^o x_{mol\text{-}b} \tag{10.116}$$

where P_v is vapour headspace pressure, and P_{vp}^o and P_{vb}^o are, respectively, propane and butane vapour pressures.

Liquid molar fractions are:

$$x_{\text{mol-p}} = \frac{\frac{0.75}{44.1}}{\frac{0.75}{44.1} + \frac{0.25}{58.1}} = 0.798 \tag{10.117}$$

$$x_{\text{mol-b}} = 1 - 0.798 = 0.202 \tag{10.118}$$

$$P_{\text{v}} = P^{\text{o}}_{\text{vp}} x_{\text{mol-p}} + P^{\text{o}}_{\text{vb}} x_{\text{mol-b}} = 0.798 \cdot 0.83 + 0.202 \cdot 0.208 = 0.704\,\text{MPa} \tag{10.119}$$

Vapour molar mass is:

$$n_{\text{v}} = \frac{P_{\text{v}} \cdot V_{\text{v}}}{R \cdot T} = \frac{704{,}000 \cdot 261.7}{8314 \cdot 293.16} \cong 75\,\text{kmoles} \tag{10.120}$$

$$n_{\text{vp}} = \frac{0.798 \cdot 0.83 \cdot 261.7}{8314 \cdot 293.16} \cong 71\,\text{kmoles} \tag{10.121}$$

$$n_{\text{vb}} \cong 4\,\text{kmoles} \tag{10.122}$$

Vapour mass is:

$$M_{\text{V}} = 71 \cdot 44.1 + 4 \cdot 58.1 \cong 3363\,\text{kg} \tag{10.123}$$

Total mass is:

$$M_{\text{b}} = M_{\text{L}} + M_{\text{V}} \cong 153{,}356 + 3363 = 156{,}719\,\text{kg} \tag{10.124}$$

$$D_{\text{b}} = 6.49 \cdot M_{\text{b}}^{0.325} = 6.49 \cdot 156{,}719^{0.325} = 316.7\,\text{m} \tag{10.125}$$

Duration of Fireball

$$t_{\text{b}} = 0.852 \cdot M_{\text{b}}^{0.26} \tag{10.126}$$

where t_{b} is in seconds.

- M_{b}, amount of the material in the fireball, kg.

Example 10.24

Find the duration of fireball for the previous example.

Solution

$$t_{\text{b}} = 0.852 \cdot 156{,}719^{0.26} = 19.1\,\text{s} \tag{10.127}$$

Heat radiation

The heat radiation Q_H (kW/m^2) to a target the same as that presented for pool and jet fire:

$$Q_H = F_v \cdot SEP_{av} \cdot \tau \tag{10.128}$$

where

- F_v is the overall view factor.
- SEP_{av} is the surface emissive power.
- τ is the transmissivity.

The values to be used in the previous formula area the following:

$$F_V = 0.25 \cdot \left(\frac{D_b}{x}\right)^2 \tag{10.129}$$

$$SEP_{av} = 184 \text{kW/m}^2 \tag{10.130}$$

The transmissivity can be calculated as:

$$\tau = P_o \cdot (P_w \cdot x)^{-0.09} \tag{10.131}$$

- P_o is 2.02 (N/m^2)$^{0.09}$$\cdot$m$^{0.09}$
- P_w is partial water vapour pressure of water in air at a relative humidity RH, in N/m^2:

$$P_w = 1013.25 \cdot RH \cdot \exp\left(14.4114 - \frac{5238}{T_a}\right) \tag{10.132}$$

Example 10.25

Find the heat radiated from the fireball of the previous example to a target at 500 m from fire edge. Relative humidity is 50%.

Solution

$$F_V = 0.25 \cdot \left(\frac{316.7}{500}\right)^2 = 0.1 \tag{10.133}$$

$$P_w = 1013.25 \cdot RH \cdot \exp\left(14.4114 - \frac{5238}{T_a}\right) = 1013.25 \cdot 50 \cdot \exp\left(14.4114 - \frac{5,238}{293.16}\right) = 1599 \text{Pa} \tag{10.134}$$

$$\tau = P_o \cdot (P_w \cdot x)^{-0.09} = 2.02 \cdot (1599 \cdot 500)^{-0.09} = 0.59 \tag{10.135}$$

$$Q_H = F_v \cdot SEP_{av} \cdot \tau = 0.1 \cdot 184 \cdot 0.59 = 10.856 \text{kW/m}^2 \tag{10.136}$$

10.12.5 BLEVE Pressure Wave

The pressure wave caused by a BLEVE can be calculated according to acoustic volume source concepts developed by Lighthill (1978). This approach has been implemented by Van den Berg et al. (2004) for BLEVE modelling, and by Benintendi and Rega (2014) for RPT modelling.

The van den Berg equation states:

$$\frac{\Delta P}{P_o} = \frac{\gamma}{2 \cdot \pi \cdot x \cdot u_s^2} \cdot \frac{2 \cdot V \cdot F \cdot \alpha}{\Delta t^2} \tag{10.137}$$

where

- F is the flash fraction.
- α is the expansion factor.
- x is the distance from the source, m.
- u_s is the ambient speed of sound, m/s.
- γ is the specific heat ratio.
- V is the volume, m^3.
- P_o is the atmospheric pressure, Pa.
- ΔP is the overpressure at x and after time Δt, s.

Example 10.26

Analyse the overpressure generated by the BLEVE of the previous example at a distance of 50, 100, and 500 m from the source, assuming a flash fraction equal to 50%. (Strictly speaking, in this example the geometric volume should be intended as full of liquid) (Fig. 10.31).

Solution

V (m^3)	P_o (Pa)	γ	u_s (m/s)	F	α
523.4	101, 325	1.14	343	0.5	260

10.12.6 Fragments Generated by BLEVE

The total number of fragments generated by BLEVE N_F is assumed to be a function of vessel size, according to Holden and Reeves (1985). The following formula can be used with a range of validity from 700 to 2500 m^3:

$$N_F = -3.77 + 0.0096 \cdot V \tag{10.138}$$

where V is the vessel capacity (m^3).

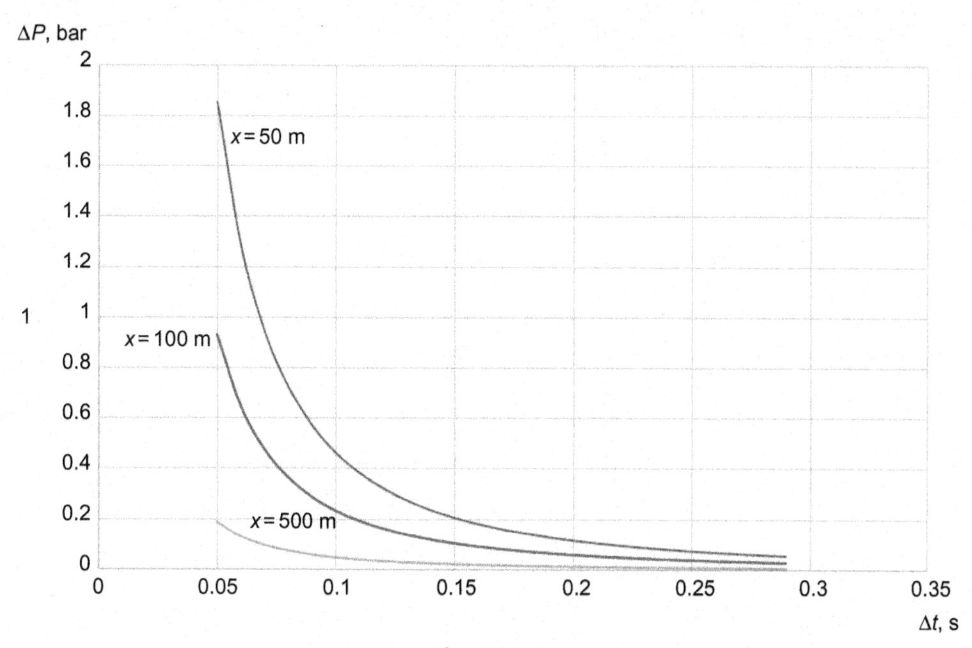

Fig. 10.31
Pressure wave of BLEVE for Example 10.26.

Example 10.27

Find the number of fragments produced by the BLEVE of a vessel with a volume of 1250 m^3.

Solution

$$N_F = -3.77 + 0.0096 \cdot 1250 \cong 8 \tag{10.139}$$

10.13 Rapid Phase Transition

LNG is stored on-shore and off-shore at a temperature which is much lower than ambient temperature, and this is the reason why, in the case of spillage and contact with water, a very rapid vapourisation can result, whose consequences can be catastrophic. This phenomenon, known as RPT, is a potentially explosive vapourisation of LNG in case of a sudden contact, mainly with water, generally sea-water. This phenomenon is very similar to the BLEVE of LPG, with the difference that the driving force in this case is the significant amount of heat transferred from sea water to LNG, which is stored at $-161°C$. Important aspects of LNG RPT are the following:

- LNG has a very great expansion factor, between 600 and 620.
- RPT is particularly likely and significant when concentrations of heavier than methane hydrocarbons are high, either because of LNG composition or because of its storage age.

- RPT is possible only provided that the liquid jet fragmentation does not promote full vapourisation prior to reaching the water phase.

10.13.1 Modelling

Benintendi and Rega (2014) developed a simple model, based on Lighthill's acoustic volume source concept (1978), to describe the overpressure behaviour following a RPT. This equation is very similar to the van den Berg equation presented for the BLEVE, and assumes liquid is fragmented into:

$$\frac{\Delta P}{P_o} = \frac{\gamma N_b}{2 \cdot \pi \cdot x \cdot u_s^2} \cdot \frac{d^2 V_b}{dt^2} \tag{10.140}$$

with:

- x is the distance from the source, m.
- u_s is the ambient speed of sound, m/s.
- γ is the specific heat ratio.
- V is the total volume, m^3.
- V_b is the bubble volume, m^3.
- N_b is the number of liquid bubbles of volume V_b.
- P_o is the atmospheric pressure, Pa.
- ΔP is the overpressure at x and after time Δt, s.

It is:

$$\frac{d^2 V}{dt^2} = N_b \frac{d^2 V_b}{dt^2} = N_b \frac{4}{3} \pi \frac{d^2 R_b^3}{dt^2} \tag{10.141}$$

Here R_b is bubble radius.

The derivative in the second term of this equation can be solved with Mikic's equation (Mikic et al., 1970):

$$\frac{d^2 R_b^3}{dt^2} = \frac{A_1 \times (A_2 + A_3 \times A_4)}{(A \times B)^2} \tag{10.142}$$

with:

$$A_1 = \frac{8}{9}\left(-t^{+3/2} + (1+t^+)^{3/2} - 1\right) \tag{10.143}$$

$$A_2 = \left[2\left(1.5t^{+1/2} - 1.5(1+t^+)^{1/2}\right)\right]^2 \tag{10.144}$$

$$A_3 = \frac{0.75}{\sqrt{t^+ + 1}} - \frac{0.75}{\sqrt{t^+}} \tag{10.145}$$

$$A_4 = -t^{+3/2} + (t^+ + 1)^{3/2} - 1 \tag{10.146}$$

and t^+ is a scaled time.

Example 10.28

Analyse the overpressure generated by the RPT for LNG, assuming methane data and values equal to 1 and 25 m^3.

Solution

Solving Mikic equation, values of function:

$$\frac{d^2 V_b}{dt^2} = f(t) \tag{10.147}$$

will be calculated. Substituting in the overpressure equation, the following diagram will be obtained (Fig. 10.32).

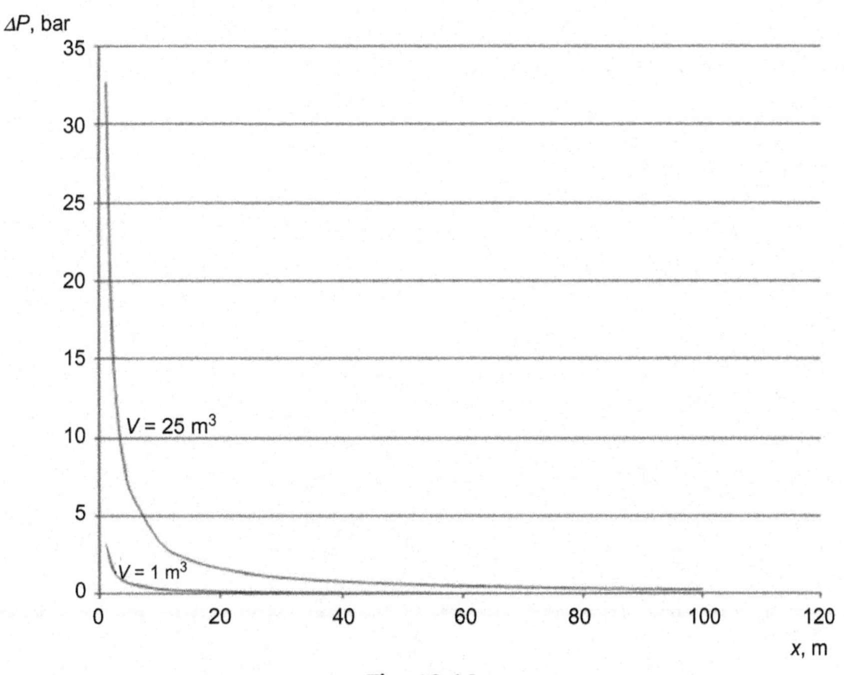

Fig. 10.32
Pressure wave of RPT for Example 10.27.

10.14 Thermal Runway

10.14.1 Definitions

Thermal runaway

Thermal runaway can be defined as:

> *A temperature and (typically) pressure rise in a chemical reactor which can no longer be recovered by the system and escalates to an out-of-control situation resulting in loss of containment, fire, and possibly explosion, unless the system is able to safely relief the escalation.*
>
> ***(Benintendi, 2013)***

The phenomenon is also known as *Runaway Reaction*, whose evidence is provided by CCPS (2005):

> *Runaway Reaction is a thermally unstable reaction system which shows an accelerating increase of temperature and reaction rate. A reaction that is out of control because the rate of heat generation by an exothermic chemical reaction exceeds the rate of cooling available.*

Time to thermal runaway

> *An estimation of the time required for an exothermic reaction, in an adiabatic container, (that is, no heat gain or loss to the environment), to reach the point of thermal runaway.*
>
> ***(ASTM E1445, 2015)***

10.14.2 Causes of Thermal Runaway

Thermal runaway is a loss of control phenomena triggered by a chemical reaction. Consequently, it belongs to reactive hazards. Typical causes for thermal runaway are:

- Bad design
- Wrong scale-up
- Wrong or misoperation
- Off spec chemicals
- Unintentional/intrinsic chemistry
- Cooling system failure
- Management of change

Example 10.29

An example of this approach is provided for the polymerisation of vinyl chloride monomer (VCM) in aqueous suspension. VCM is insoluble in water, so polymerisation in suspension is carried out with small VCM drops dispersed in a continuous medium, made up of water, which removes the heat of reaction and generates regular polymerisation in isothermal and controlled conditions. A block diagram, a general flow diagram, and a basic P&ID have been included, respectively, in Figs. 10.33–10.35. Fresh and recycled VCM are loaded along with additives and fluids necessary

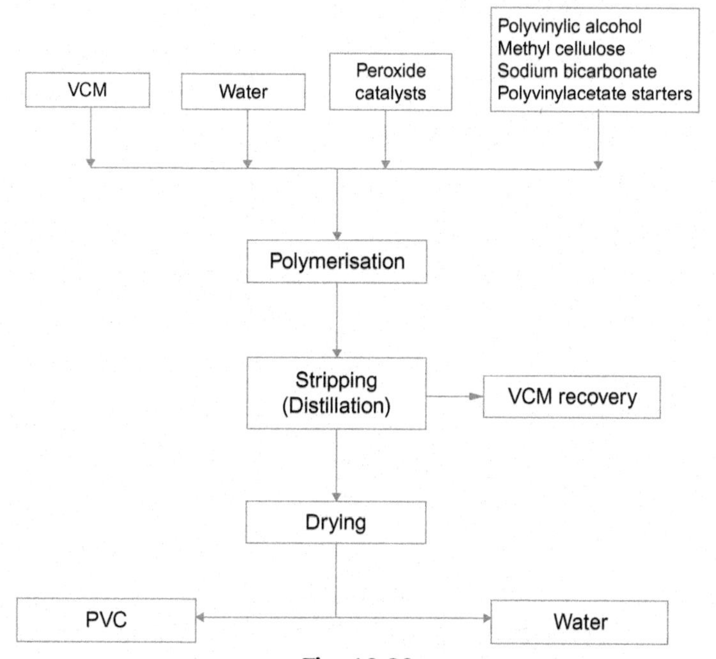

Fig. 10.33
Block diagram for Example 10.29.

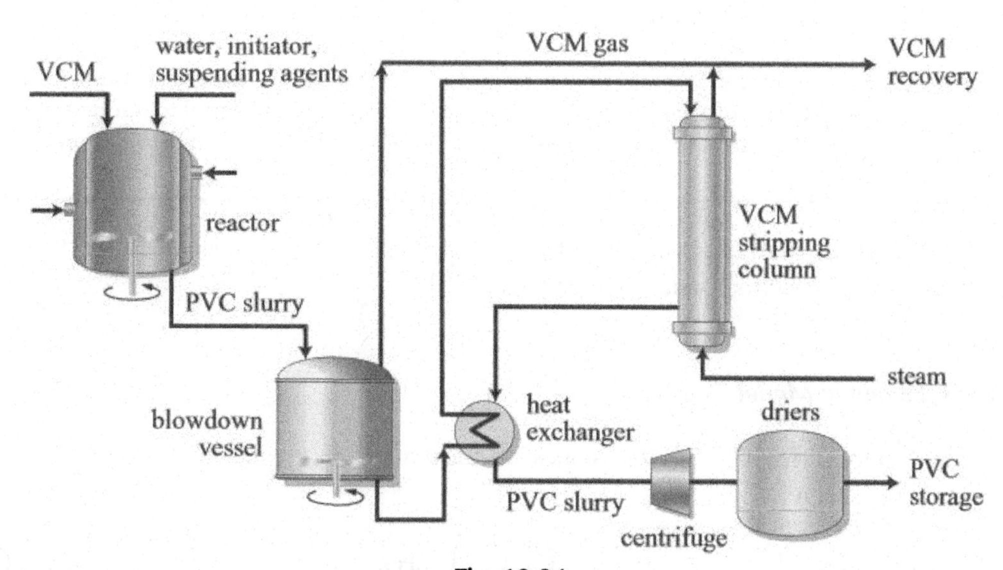

Fig. 10.34
VCM polymerisation in aqueous suspension—block diagram (Encyclopaedia Treccani, 2008).

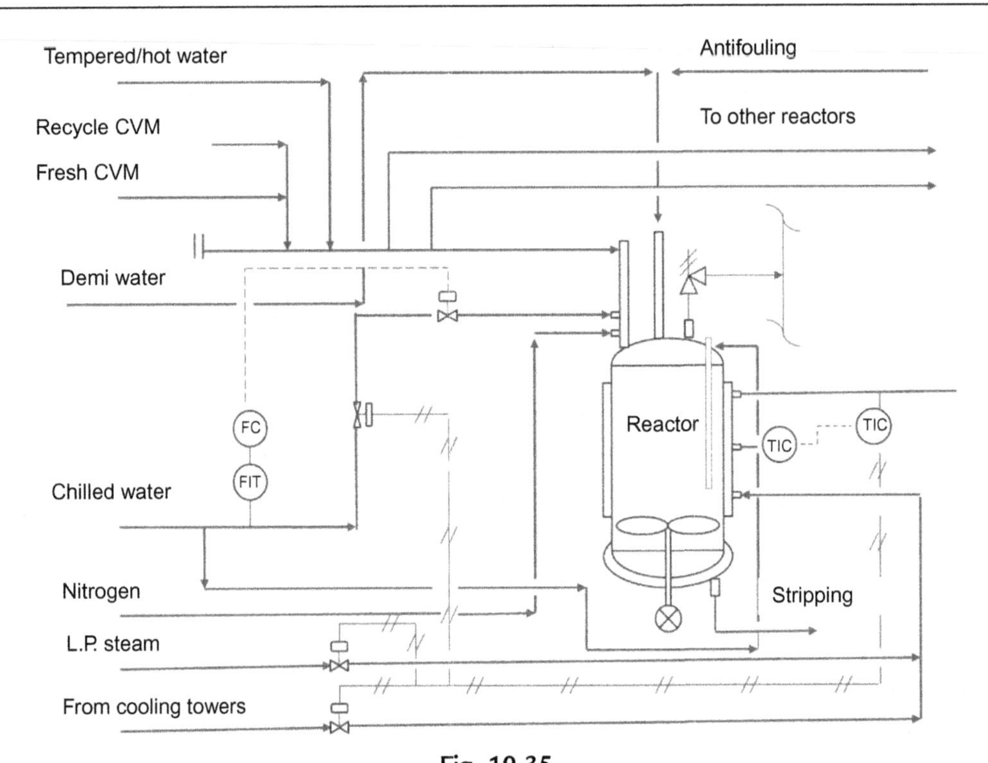

Fig. 10.35
VCM polymerisation in aqueous suspension—P&ID (Benintendi and Alfonzo, 2013).

to modify surface properties, to prevent fouling, and to control reaction temperature, which is exothermic. Chilled water and low-pressure steam are used to cool down and warm up the process until a steady state is reached. Reaction heat is removed by a jacket heat exchanger under temperature control. As long as the reaction proceeds, cooling requirements change, depending on kinetics and VCM concentration, as shown in Fig. 10.36. If the reaction is out of control, it can be stopped by the addition of a specific inhibitor, typically a bisphenol methanol solution. A step-by-step reactor operating description has been included in Table 10.24, which provides process and exothermic data.

A VCM polymerisation weakness analysis has been developed and included in Table 10.25, in relationship with the thermal runaway outcome.

10.14.3 Semenov Model

The Semenov model (Bradley, 1969) assumes chemical reactions to take place according to the Arrhenius mechanism:

$$r_{\text{rate}} = k \times \left(e^{-\frac{E_a}{R \times T}}\right) \times \Pi_i C_I^{n_i} \tag{10.148}$$

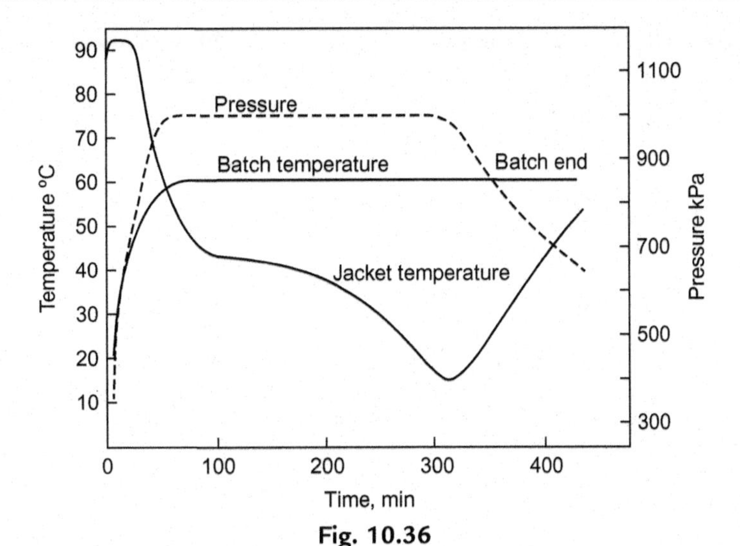

Fig. 10.36
VCM polymerisation temperature-time diagram (Encyclopaedia Treccani, 2008).

where

- k is the pre-exponential factor.
- E_a is the energy of activation.
- C_i is the concentration of component $-i$.
- n_i is the reaction order with respect to component $-i$.

If the reaction is exothermic, the reaction heat is developed:

$$H_r = r_{rate} \cdot \Delta H_r = k \cdot \left(e^{-\frac{E_a}{R \times T}} \right) \cdot \Pi_i C_i^{n_i} \tag{10.149}$$

On this basis, the qualitative heat generation rate curve is depicted in Fig. 10.37.

We suppose the system is thermally regulated by a cooling water plant, which is assumed to remove generated heat according to the linear formula:

$$H_c = U_A \cdot (T - T_o) \tag{10.150}$$

where

- U_A is the overall heat transfer coefficient related to the relative exchange area (J/s K).
- T_o is the cooling water temperature.

A steady state process will imply (Fig. 10.38):

$$H_r = H_c \tag{10.151}$$

Table 10.24 Step-by-step process description for Example 10.29 (Benintendi and Alfonzo, 2013)

#	STEP	Temperature (°C)	Duration (min)	
1	Cold demi water and hydro soluble ingredients load	30	10	
2	Recycle VCM load		10	
3	Fresh VCM load		20	
4	Hot demi water load	90	15÷18	
5	Peroxide catalyst load on demi water line		5	
6	Warm up to reaction temperature	60		
7	Polymerisation under mixing and reactor jacket cooling		480	Exothermic reaction (22.9 kcal/mole). Cooling is carried out using cooling tower or chilled water. Process is stopped at 85% conversion because reaction rate is very low after this conversion. Pressure is 8–9 ate. Thermal runaway and overpressure are controlled with chilled water or with a reaction stopper (bisphenol methanol solution). A rupture disk is installed with set pressure at 13.2 barg.
8	Stripping and separation			

Representing both the functions on a diagram against temperature, we can see that two working points can exist, at which the two heat rates are equal. Notably, points A and B are identified (Fig. 10.39).

If we disturb the system, removing it from the two working points, we can see:

- Point A is a stable point, so the system will have the ability to revert to t.

Table 10.25 VCM polymerisation weakness analysis for Example 10.29 (Benintendi and Alfonzo, 2013)

Reactive Threat	Protection	Weakness	Consequence	Safeguards and Recommendations
Polymerisation reaction heat	Cooling water	Loss of cooling water	- Thermal runaway - Overpressure	Cooling water failure analysis Adequate source redundancy
		Cooling water pump failure	- Thermal runaway - Overpressure	Adequate redundancy and failure analysis
		Temperature control failure	- Thermal runaway - Overpressure - Low reaction rate	Voting and SIL assessment
		Water jacket heat exchanger failure due to corrosion, equipment damage, other causes	- Thermal runaway - Overpressure - Reaction mixture leakage	Material management Control and inspection
		Water jacket heat exchanger loop fouling	- Thermal runaway - Overpressure	Chemical injection Maintenance programme
		More pumps running	- Low reaction rate	Production issue
		TCV fails closed	- Thermal runaway - Overpressure	Alarm management Failure analysis
		TCV fails open	- Low reaction rate	Production issue
		L.P. steam fed by mistake during reaction phase	- Thermal runaway - Overpressure	Failure analysis Alarm management
	Quenching	Loss of chilled water	- Thermal runaway - Overpressure	Failure analysis Alarm management Adequate redundancy
		Temperature control failure	- Thermal runaway - Overpressure - Low reaction rate	Failure analysis Alarm management Adequate redundancy

Continued

Table 10.25 VCM polymerisation weakness analysis for Example 10.29—cont'd

Reactive Threat	Protection	Weakness	Consequence	Safeguards and Recommendations
	Reaction inhibition	Loss of bisphenol	- Thermal runaway - Overpressure - Reaction stoppage	Operating procedure Operating barriers Multiple control Training
		Temperature control failure	- Thermal runaway - Overpressure - Reaction stoppage	Alarm management Failure analysis Voting and SIL assessment
		Substance oxidation	- Reactive with peroxides forming CO and CO_2	Loading procedure Mixing and distribution control

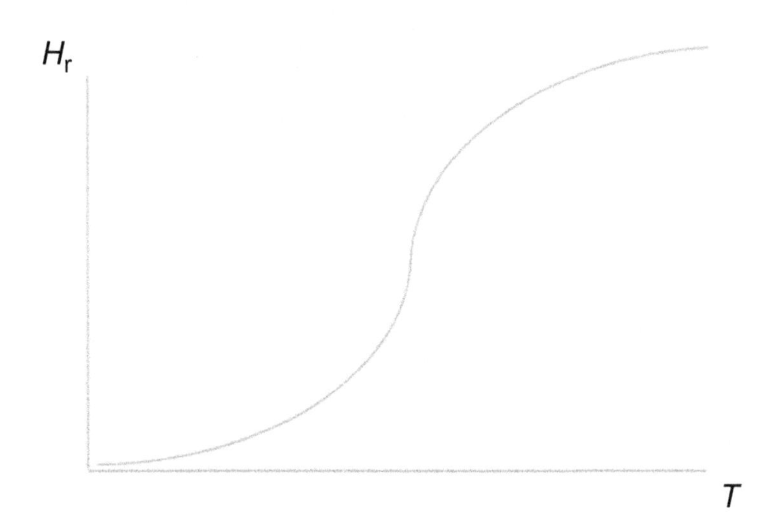

Fig. 10.37
Heat generation rate for Semenov model.

- Point B poses an instability risk: any situations above the temperature corresponding to point A and below the temperature corresponding to point B would take the system to point A; however, if point B is exceeded, the system temperature would increase with no intrinsic chance to recover. B is a nonreturn point for the system, which would undergo thermal runaway.

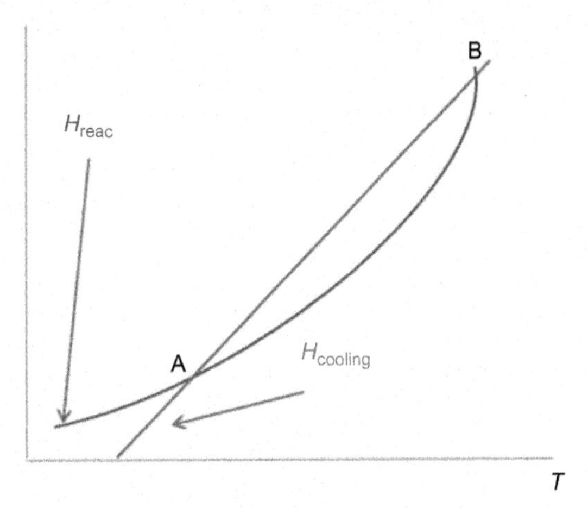

Fig. 10.38
Semenov diagram.

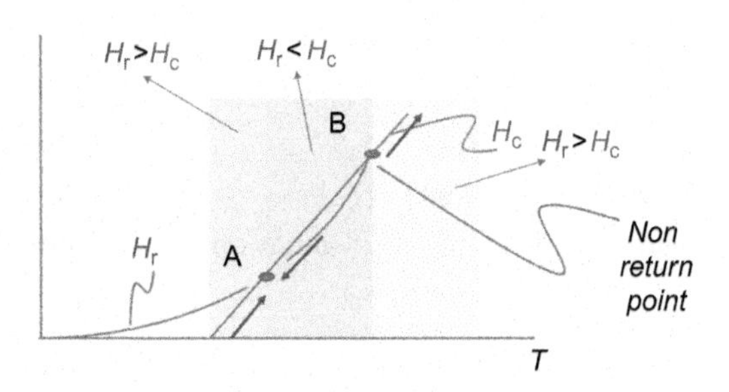

Fig. 10.39
Stability points on Semenov diagram.

Operational scenarios could result in thermal instability.

1. Significant variation of the overall heat transfer coefficient (Fig. 10.40).
   Some factors could reduce U, for example, fouling or scaling, or the exchange area A. Being:

$$U_A = U \cdot A = tg\alpha \tag{10.152}$$

   This would change the slope of the cooling line to an angle α', excluding any chance for it to cross the heat generation line.
2. Cooling water temperature rise (Fig. 10.41)
   Once again, generated heat would be greater than removed heat.

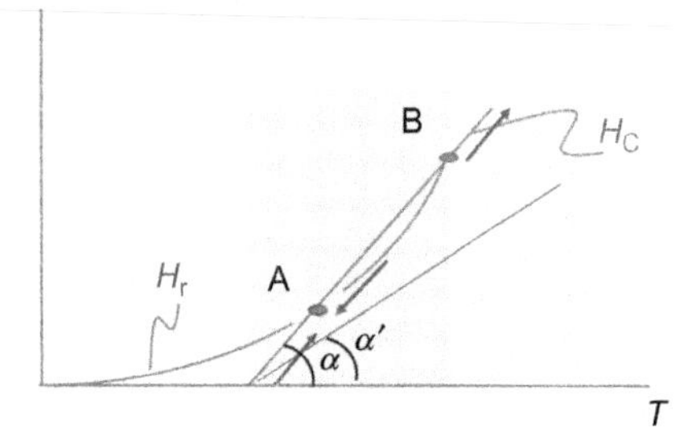

Fig. 10.40
Effect of thermal exchange properties variation on Semenov diagram.

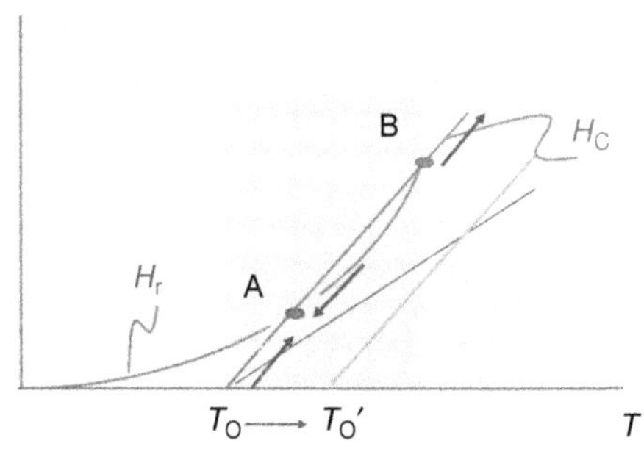

Fig. 10.41
Effect of cooling water temperature variation on Semenov diagram.

Example 10.30

A 1 M (1 mole/L) solution of methyl methacrylate (MM) monomer undergoes a polymerisation reaction, in a 1 L laboratory reactor, which produces 54,418 J/mole. The reaction rate is given by the following equation:

$$r_r = k \cdot \exp\left(\frac{-E_a}{R \cdot T}\right) \cdot C_{MM} \tag{10.153}$$

with:

- r_r = reaction rate, moles/s L
- k, rate constant, 1.5×10^7 L/mole s
- E_a, activation energy, 66,000
- R, gas constant, 8.314 J/mole K
- C_{MM}, MM molar concentration, 1 M

The reactor is cooled by means a heat exchanger with $U_A = 0.7$ J/s. The chilled water inlet temperature is 287 K.

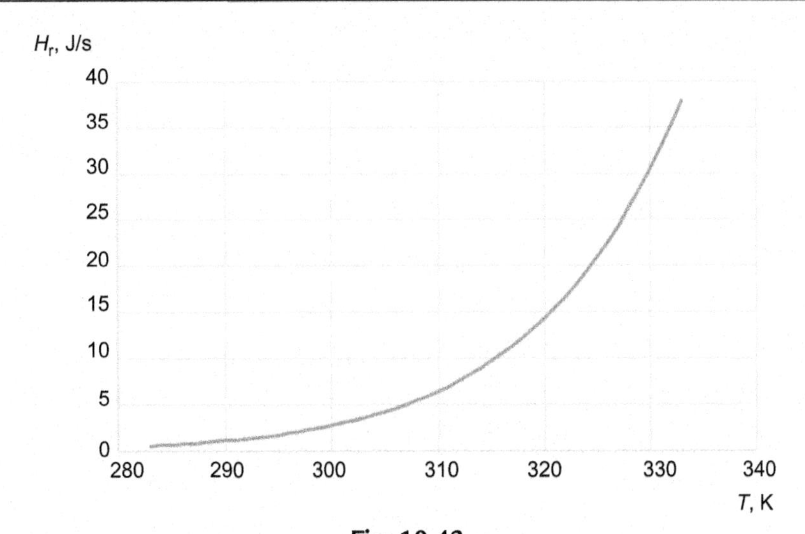

Fig. 10.42
Reaction heat generation curve for Example 10.30.

Find:

- The temperature corresponding to points A (stable) and B (nonreturn)
- The inlet cooling water critical temperature
- The critical coefficient U_A ($U \times A$)

Solution

$$H_r = 15{,}000{,}000 \cdot \exp\left(\frac{-66{,}000}{8.314 \cdot T}\right) \cdot 1 \cdot 54{,}418 \tag{10.154}$$

where H_r is expressed in J/s (Fig. 10.42).

The points A and B can be obtained equating the heat values and solving for T (Fig. 10.43):

$$H_r = 15{,}000{,}000 \cdot \exp\left(\frac{-66{,}000}{8.314 \cdot T}\right) \cdot 1 \cdot 54{,}418 = 0.7 \cdot (T - 287) \tag{10.155}$$

Solving:

- $T_A = 288.5$ K
- $T_B = 329.5$ K

The critical chilled water temperature can be found imposing:

$$H_r = 15{,}000{,}000 \cdot \exp\left(\frac{-66{,}000}{8.314 \cdot T_t}\right) \cdot 1 \cdot 54{,}418 = 0.7 \cdot (T_t - T_o) \tag{10.156}$$

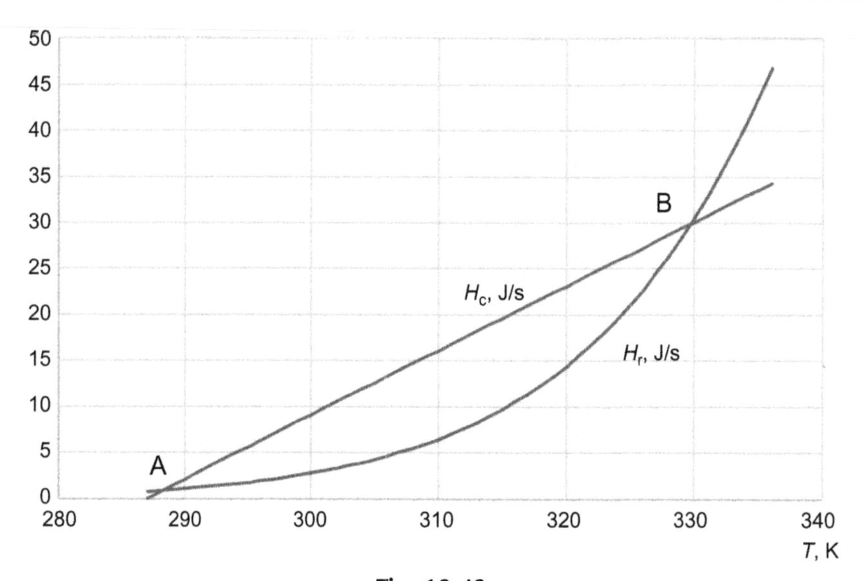

Fig. 10.43
Semenov diagram for Example 10.30.

We have two unknown temperatures, T_{oc} and T_t. Another equation is needed. Equating the slope of the H_c line to the derivative of H_r with respect to temperature:

$$\frac{dH_r}{dT}(T_t) = \frac{66{,}000}{8.314 \cdot T_t^2} \cdot 15{,}000{,}000 \cdot \exp\left(\frac{-66{,}000}{8.314 \cdot T_t}\right) \cdot 1 \cdot 54{,}418 = 0.7 \tag{10.157}$$

Solving:

$$T_t = 313.65\text{K} \tag{10.158}$$

Replacing in the first equation:

$$T_o = 301\text{K} \tag{10.159}$$

The last question includes two unknown variables, U_r and T_t (Fig. 10.44).

$$\frac{dH_r}{dT}(T_t) = \frac{66{,}000}{8.314 \cdot T_t^2} \cdot 15{,}000{,}000 \cdot \exp\left(\frac{-66{,}000}{8.314 \cdot T_t}\right) \cdot 1 \cdot 54{,}418 = U_r \tag{10.160}$$

$$H_r = 15{,}000{,}000 \cdot \exp\left(\frac{-66{,}000}{8.314 \cdot T_t}\right) \cdot 1 \cdot 54{,}418 = U_r \cdot (T_t - 287) \tag{10.161}$$

Solving (Fig. 10.45):

$$U_r = 0.22\text{J/s} \tag{10.162}$$

$$T_t = 299\text{K} \tag{10.163}$$

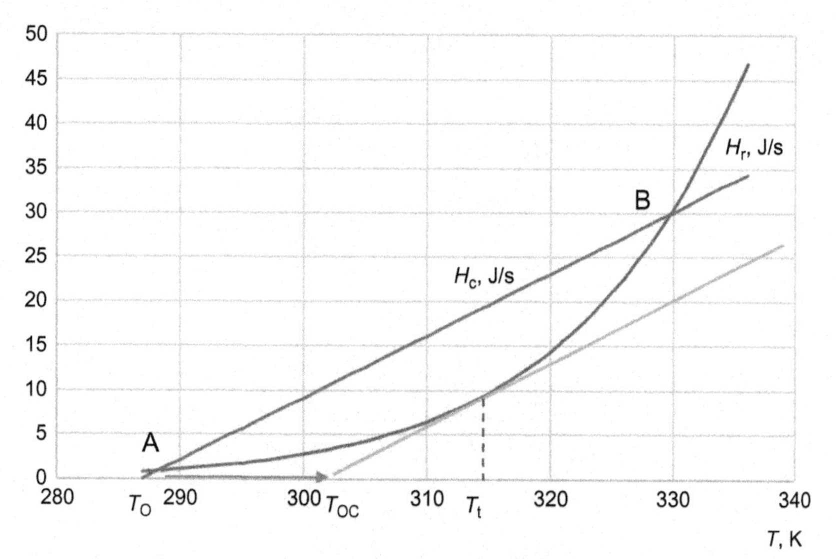

Fig. 10.44
Modified Semenov diagram for Example 10.30.

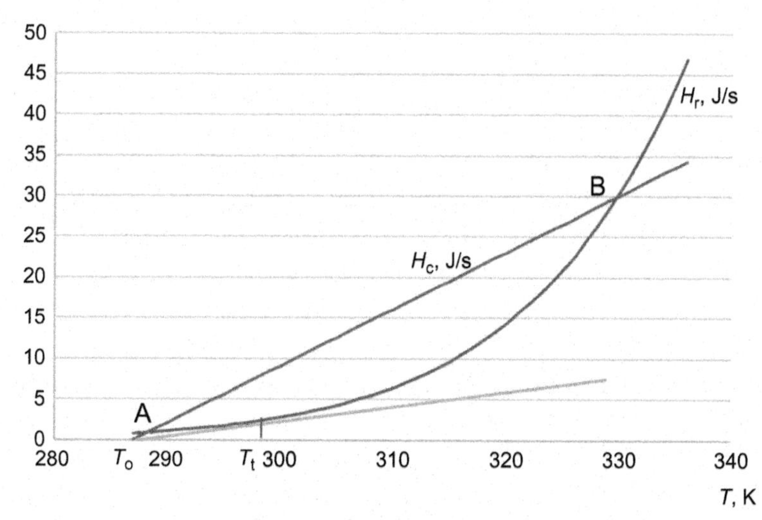

Fig. 10.45
Modified Semenov diagram for Example 10.30.

10.15 Design and Operational Issues: Hot Spots

Example 10.31

Pentane is isomerised with hydrogen in a plug flow reactor. The reaction takes place at a very high temperature and pressure. Reaction heat is removed by means of cooling pipes, arranged as

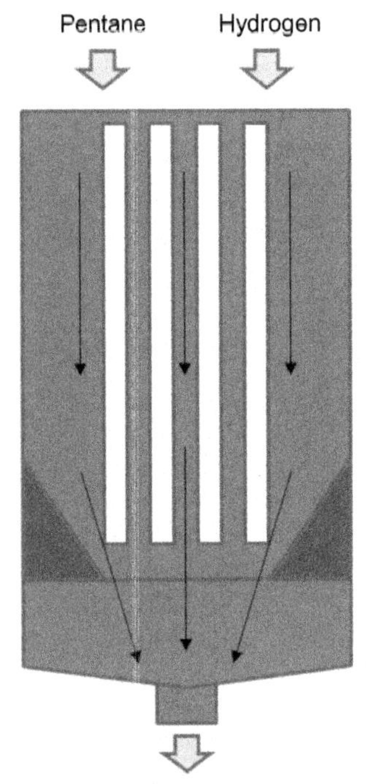

Fig. 10.46
PFR for pentane isomerisation.

illustrated in the next figure. Products are collected from the centre of the reactor. Discuss the scenario.

Solution
The system presents some defective aspects:

1. The cooling system is not properly distributed within the reacting mass. Reaction heat is not efficiently removed from the fraction close to the wall.
2. The products are collected from the centre of the bottom side. This promotes the formation of hot spots (Fig. 10.46).

Hot spots could be generated (red zones), and the system could undergo thermal runaway. A more adequate system is depicted in Fig. 10.47.

10.16 Pressure Piling

10.16.1 Definition

In a compartmented system in which there are separate but interconnected volumes, the pressure developed by the deflagration in one compartment causes a pressure rise in the unburned gas in the interconnected compartment, so that the elevated pressure in the

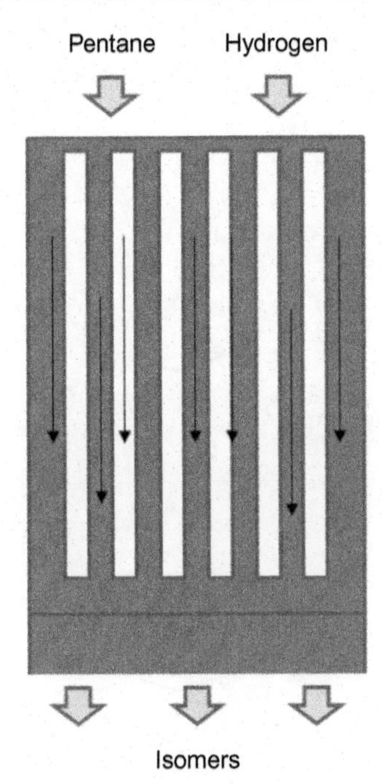

Fig. 10.47
PFR for pentane isomerisation modified design.

> *latter compartment becomes the starting pressure for a further deflagration. This effect is known as pressure piling, or cascading.*
>
> *(CCPS, 2005)*

Two interconnected vessels containing a flammable mixture are initially at the same pressure P_o (A). One of the two vessels is ignited, causing the pressure to rise to a predictable overpressure $X \cdot P_o$, X theoretically being a number ranging from 7 to 10, practically from 4 to 5 (B) (Fig. 10.48).

A pressure wave travels to the second vessel, where the flammable mixture is ignited at an initial pressure which is X time greater than P_o. Consequently, its final pressure P_2 will be roughly X^2 time P_o.

10.16.2 Factors Influencing Pressure Piling

Secondary pressure rise factors are:

- Primary vessel to secondary vessel volume ratios smaller than 1.
- Large diameters of the interconnecting pipe; values smaller than 0.1 m are reported as preventing pressure piling from developing significantly.

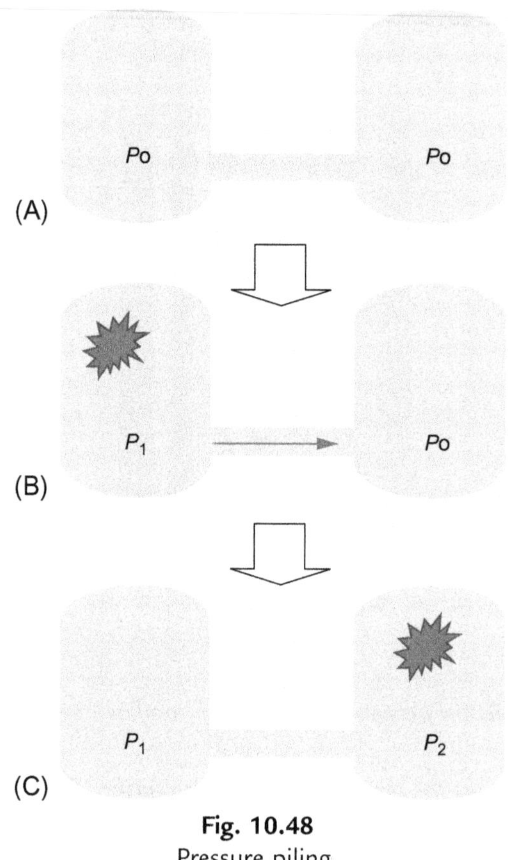

Fig. 10.48
Pressure piling.

10.16.3 Modelling

Holbrow et al. (1999) have proposed a methodology to evaluate whether or not the pressure-piling is significant between interconnected vessels, and which is the overpressure attained in the secondary containment. The method is strictly valid for dusts, but, due to the full similarity of explosions of gases and dusts in containments, an order of magnitude of the maximum overpressure can reasonably be obtained. If V_1 is the primary vessel volume, assumed to be greater than V_2, the volume of the secondary vessel, a compression factor (CF) can be calculated by means of the following diagram, obtained from Holbrow (Fig. 10.49).

The final explosion pressure attained in the vessel 2 is given by the product:

$$P_{2\text{max}} = \text{CF} \cdot P_{2\text{o}} \qquad (10.164)$$

where $P_{2\text{o}}$ is the explosion pressure of a stand-alone secondary vessel. This method is applicable, provided that:

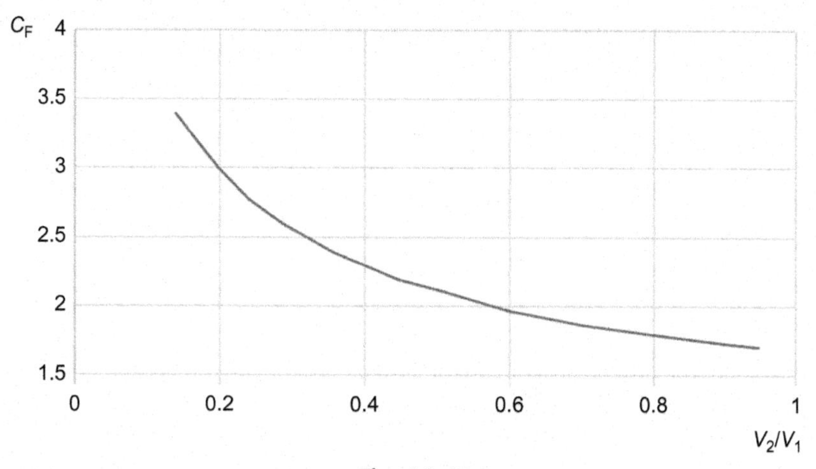

Fig. 10.49
Compression factor.

- Vessels volume are smaller than 20 m^3.
- Connecting pipe is longer than 5 m and shorter than 15 m.
- Explosion pressure of single vessels is not greater than 15 bar.
- Primary vessel volume includes also that of interconnecting pipe.

Example 10.32

Two interconnected vessels, whose volumes are respectively 12 m^3, including the interconnecting pipe volume, and 7 m^3, contain an ethane and air stoichiometric mixture. Pressure piling is caused by the ignition of gas in the first vessel at 1 bar and 20°C. Find the final pressure in the second vessel.

Solution

The stoichiometric reaction is the following:

$$C_2H_6 + 3.5O_2 + 3.5 \cdot 79/21N_2 \rightarrow 2CO_2 + 3H_2O + 3.5 \cdot 79/21N_2 \tag{10.165}$$

In Chapter 9 it was stated that practical values of final pressure in single vessels is generally assumed to range between 4 and 5 bar. If adiabatic conditions are considered, indicated respectively with P_{o1}, T_{o1}, and P_1, T_1 is the initial and final pressure and temperature of larger vessel 1:

$$P_1 = P_{o1} \cdot \frac{n_1}{n_{o1}} \cdot \frac{T_1}{T_{o1}} \tag{10.166}$$

where n_{1o} and n_1 are respectively the number of kilomoles of reactants and products. On the basis of stoichiometry, and assuming that T_1 coincides with the adiabatic flame temperature of ethane, it will be:

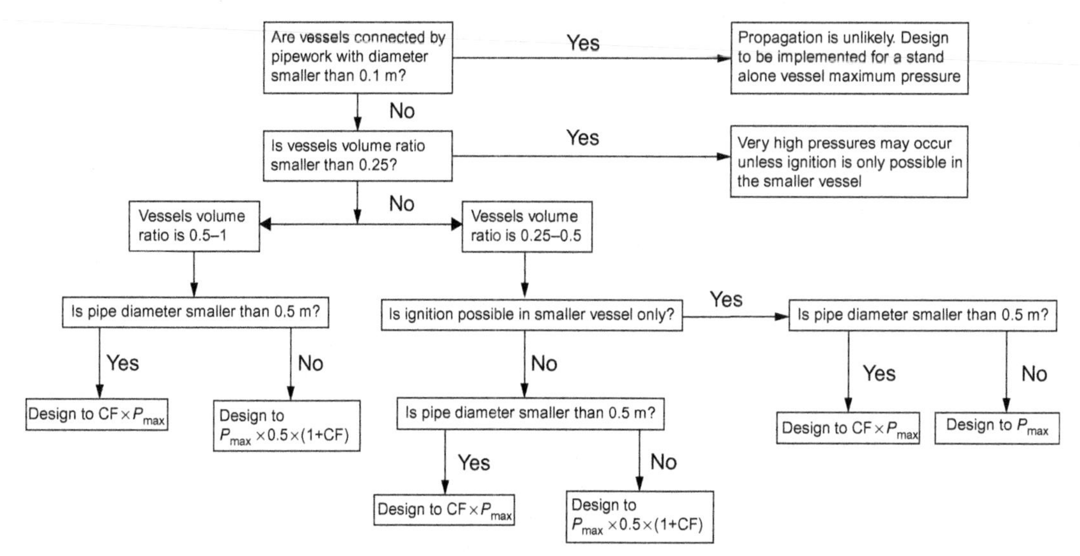

Fig. 10.50
Flow chart for design pressure selection in interconnected vessels.

$$P_1 = P_{o1} \cdot \frac{n_1}{n_{o1}} \cdot \frac{T_1}{T_{o1}} = 1 \cdot \frac{\left(5 + 3.5 \cdot \frac{79}{21}\right)}{\left(4.5 + 3.5 \cdot \frac{79}{21}\right)} \cdot \frac{1987 + 273.16}{293.16} = 7.92\,\text{bar} \tag{10.167}$$

Entering the diagram with $V_2/V_1 = 0.583$, it results in:

$$CF = 2 \rightarrow P_2 = 2 \cdot P_{\text{max}} = 2 \cdot 7.92 = 15.84\,\text{bar} \tag{10.168}$$

Holbrow et al. (1999) have provided a flow chart for design pressure selection in interconnected vessels (Fig. 10.50).

10.17 Drag Loads

Gas explosion causes overpressure and high-speed gas flows. Buildings are subject to a variation of pressure distribution along their length, so that a significant pressure difference exists on opposite sides. This doesn't happen for small cross section objects, such as piping, beams, and columns, for which pressure spatial and time distribution is practically the same on the opposite side. However, the dynamic pressure related to the explosion wind behind the blast wave can be significant and possibly destructive. Drag loads have been introduced in Chapter 5 along with the Baker's formula:

$$F_D = A \cdot C_D \cdot \frac{1}{2} \cdot \rho \cdot u^2 \tag{10.169}$$

where

- C_D is the drag coefficient, depending on the shape of the obstacle, on the Reynolds number and on the local Mach number.
- ρ is the fluid density.
- u is the large scale fluid velocity, whose distribution close to the obstacle may significantly vary.
- A is the maximum cross sectional area of the object in a plane normal to u.

Baker's formula is valid for obstacles with a reference size or diameter lower than 0.3 m. Drag coefficients against the obstacle shape have been included in Chapter 5.

A practical approach for estimating the drag pressure has been provided by Yasseri (2002) for off-shore structures, which consists of a graphical relationship between drag load and overpressure. The graph has been recalculated here (Fig. 10.51).

According to UKOOA (2006), when detailed information is not available, a ratio of drag load to overpressure load corresponding to an exceedance frequency of 10^{-4}/y equal to 1/3 can be used.

A practical formula for estimating F_D is given by TNO-Green Book (1992):

$$f_D \cdot = C_D \cdot \frac{5}{2} \cdot \frac{P_s^2}{7 \cdot P_o + P_s} \rightarrow F_D = f_D \cdot L \cdot D \qquad (10.170)$$

where L is the structural element length (m) and D is the diameter or transversal dimension (m).

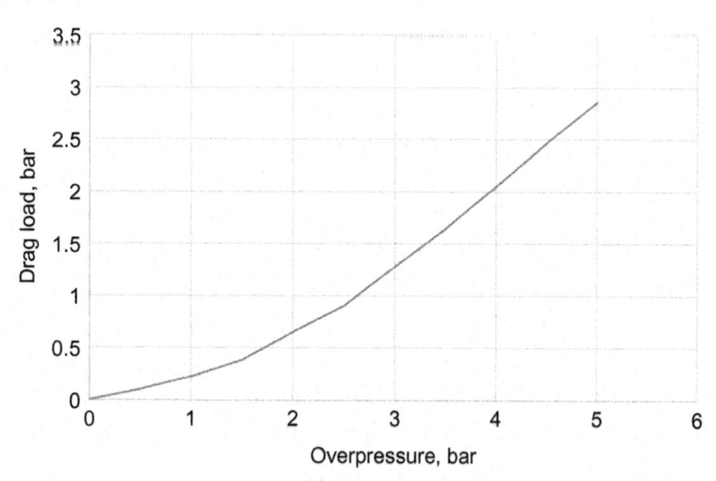

Fig. 10.51
Relationship between drag and overpressure.

Example 10.33

A pressure wave engulfs a 200 mm diameter pipe 15 m long. The peak side-on pressure is 3 psi. Find the blast load.

Solution

The diameter is smaller than 0.3 m, so the only one significant load is the drag load. P_s is equal to 20,684.3 Pa. The drag load is:

$$F_D = f_d \cdot L \cdot D = C_D \cdot \frac{5}{2} \cdot \frac{P_s^2}{7 \cdot P_o + P_s} \cdot L \cdot D = 1.2 \cdot \frac{5}{2} \cdot \frac{20,684.3^2}{7 \cdot 101,325 + 20,684.3} \cdot 15 \cdot 0.2 = 5275 \text{N} \tag{10.171}$$

10.18 Loads on Buildings

In Chapter 5 basic definitions of physical actions of pressure, and shock waves on buildings, have been provided, as well as some practical formulas to calculate the pressure impact on the basis of the side-on peak pressure P_s. In this section some indications about overall loads on buildings are included.

TNO-Green Book (1992) provides the following formula for the calculation of the wave-front velocity, based on the peak side-on overpressure P_s value that is included in the QRA findings:

$$u_{wf} = u_s \cdot \sqrt{1 + \frac{6 \cdot P_s}{7 \cdot P_o}} \tag{10.172}$$

where

- u_s is air sound speed, 340 m/s.
- P_s is the peak side-on overpressure.
- P_o is the ambient pressure.

As the pressure wave reaches the front wall of a building, the overpressure will coincide with the reflected pressure. A rigorous value of the perpendicularly reflected overpressure can be calculated by the following formula (TNO, Green Book, 1992):

$$P_r = 2 \cdot P_s \cdot + \frac{(\gamma + 1) \cdot P_s^2}{(\gamma - 1) \cdot P_s + 2 \cdot \gamma \cdot P_o} \tag{10.173}$$

where P_o is the atmospheric pressure, and γ is the specific heat ratio. Assuming $\gamma = 1.4$ for air, we can follow the behaviour of reflected overpressure with P_s (Fig. 10.52).

The reflected overpressure will decrease from the reflected pressure to an overpressure value, which is the summation of the peak side-on pressure and the dynamic pressure:

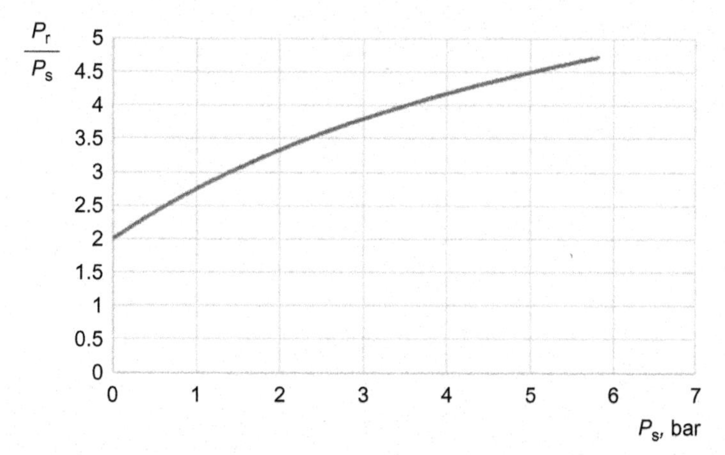

Fig. 10.52
Behaviour of reflected overpressure with peak side-on overpressure.

$$P_s + F_D = P_s + C_D \cdot \frac{5}{2} \cdot \frac{P_s^2}{7 \cdot P_o + P_s} \tag{10.174}$$

The time taken to reach this overpressure can be calculated by the formula:

$$t_s = \frac{3 \cdot L}{u_{wf}} \tag{10.175}$$

where L is a characteristic dimension of the surface. For the building front with height H and width B, it will be H or ½ B, whatever is smaller. The time for pressure wave and for the maximum overpressure ($\cong 0.7\ P_s$) to reach the back face is, respectively:

$$t_b = \frac{B}{u_{wf}} \tag{10.176}$$

$$t_{bmax} = \frac{4 \cdot L}{u_{wf}} \tag{10.177}$$

where B is the third building side.

Example 10.34

Determine the blast loadings for a substation engulfed by a peak side-on pressure is 3 psi. Its dimension are: $H=5$, $W=15$ m, $B=7$ m.

Solution

Reflected overpressure:

$$P_r = 2 \cdot P_s \cdot + \frac{(\gamma+1) \cdot P_s^2}{(\gamma-1) \cdot P_s + 2 \cdot \gamma \cdot P_o} = 2 \cdot 20{,}684.3 \cdot + \frac{(1.4+1) \cdot P_s^2}{(1.4-1) \cdot 20{,}684.3 + 2 \cdot 1.4 \cdot 101{,}325}$$
$$= 44{,}885.3\,\text{Pa} \tag{10.178}$$

Dynamic overpressure:

$$f_d \cdot = C_D \cdot \frac{5}{2} \cdot \frac{P_s^2}{7 \cdot P_o + P_s} = 1.05 \cdot \frac{5}{2} \cdot \frac{20,684.3^2}{7 \cdot 101,325_o + 20,684.3} = 1,538.5\,\text{Pa} \tag{10.179}$$

Wave front velocity:

$$u_{wf} = u_s \cdot \sqrt{1 + \frac{6 \cdot P_s}{7 \cdot P_o}} = 340 \cdot \sqrt{1 + \frac{6 \cdot 20,684.3}{7 \cdot 101,325}} = 368.5\,\text{s} \tag{10.180}$$

$$t_s = \frac{3 \cdot L}{u_{wf}} = \frac{3 \cdot 5}{368.5} = 0.04\,\text{s} \tag{10.181}$$

$$t_b = \frac{B}{u_{wf}} = \frac{7}{368.5} = 0.019\,\text{s} \tag{10.182}$$

$$t_b \max = \frac{4 \cdot L}{u_{wf}} = \frac{4 \cdot 5}{368.5} = 0.053\,\text{s} \tag{10.183}$$

10.19 Blast Resistant Structures in the QRA Framework

The identification of the peak side on overpressures, along with the positive duration, provides all necessary data to ensure that affected buildings, stick-built structures, and equipment are blast resistant.

10.19.1 Blast Resistant Buildings

Baker's diagram presented in Chapter 5 shows that overpressure is not sufficient to determine the damage zone for buildings. Duration or impulse are required. The effects of overpressures on buildings and main equipment have been presented by Clancey (1972) and has been reported in Table 10.26.

Buildings are assessed in terms of their potential blast-resistant requirements on the basis of three main criteria:

- occupancy
- safety-related functions
- asset integrity

The first two categories are the most stringent, the third one depends on company asset management policies and ALARP limits. Occupied buildings are designed in compliance with international standards, which generally apply the ALARP as Low approach. The UK ALARP reference source is the publication Reducing Risk, Protecting People, which includes and

Table 10.26 Blast effects on structures and equipment (Clancey, 1972)

Blast Effect	psi
Annoying noise (137 dB), if of low frequency (1–15 Hz)	0.02
Occasional breaking of large glass windows already under strain	0.03
Loud noise (143 dB); sonic boom glass failure	0.04
Breakage of windows, small, under strain	0.10
Typical pressure for glass failure	0.15
Safe distance (probability 0.95 no serious damage beyond this value)	0.30
Missile limit	
Some damage to house ceiling; 10% window glass broken	
Limited minor structural damage	0.40
Large and small windows usually shattered; occasional damage to window frames	0.50–1.0
Minor damage to house structures	0.70
Partial demolition of houses, made uninhabitable	1.00
Corrugated asbestos shattered	1.0–2.0
Corrugated steel or aluminium panels, fastenings fail, followed by buckling	
Wood panels (standard housing), fastening fail, panels blown in	
Steel frame of clad building slightly distorted	1.30
Partial collapse of walls and roofs of houses	2.00
Concrete or cinder block walls, not reinforced, shattered	2.00–3.00
Lower limit of serious structural damage	2.30
50% destruction of brickwork of house	2.50
Heavy machines (3000 lb) in industrial building suffer little damage	3.00
Steel frame building distorted and pulled away from foundations	
Frameless, self-framing steel panel building demolished	3.00–4.00
Rupture of oil storage tanks	
Cladding of light industrial buildings ruptured	4.00
Wooden utilities poles (telegraph, etc.) snapped	5.00
Tall hydraulic press (40,000 lb) in building slightly damaged	
Nearly complete destruction of houses	5.00–7.00
Loaded train wagons overturned	7.00
Brick panels, 8–12 in. thick, not reinforced, fail by shearing or flexure	7.00–8.00
Loaded train boxcars completely demolished	9.00
Probable total destruction of buildings	10.00
Heavy (7000 lb) machine tools moved and badly damaged	
Very heavy (12,000 lb) machine tools survived	
Limit of crater lip	300

justifies some statistically-based acceptability limits. Different ALARP criteria are adopted in Europe and Asia. Safety-related functional buildings can be substations, control room and FARs, whose integrity is required to properly ensure the implementation of shut down and safety-related process functions.

In the safety engineering practice, a building risk assessment (BRA) is carried out with the aim to compare the relevant findings of the QRA to pre-defined limits.

10.19.2 BRA Data and Design of Blast Resistant Buildings

Overpressure

An overpressure threshold is typically assumed to distinguish affected from unaffected buildings.

The Chemical Industries Association (2010) presents the following benchmark values below for which no specific safety measures are required (Table 10.27).

ASCE (2010) provides a general picture of typical assessed buildings and of corresponding minimum blast overpressures. A size range of 30–120 m is reported along, with a design overpressure range from 1.5 to 15 psi.

1 psi at a given frequency could be assumed as the threshold overpressure for blast resistant buildings.

Cumulative and exceedance frequency

An explosive outcome is the result of some Loss of Containment scenarios, which are based on specific assumptions such as hole size, weather conditions, and process data. Consequently, on a specific location, a number of different overpressures and durations/impulses can be found against the related so-called cumulative frequency, which is the combined result of various probabilistic factors. In the QRA frame, this produces a list of peak side-on overpressures and durations/impulses for a specific building, where increasing overpressure is associated with decreasing cumulative frequencies.

For a given overpressure, cumulative frequency is defined as the summation of all frequencies related to all calculated overpressure greater than and equal to it.

An example is provided in the following tables. Table 10.28 shows 10 hypothetical UVCE scenarios relevant to a specified building located in a specified unit. For each scenario the predicted overpressures and frequency is listed. The variety of the scenarios depends on the complexity of the approach in terms of release locations, ignition sources, hole sizes, weather conditions, and chemical compositions.

Table 10.28 is sorted out so that peak overpressure values are listed in decreasing order, as shown in Table 10.29.

Table 10.27 Bench mark Fire and blast limits (CIA, 2010)

Fire and Blast Effect	Unit	Threshold
Overpressure	mbar (psi)	30 (0.5)
Thermal radiation	kW/m^2	6.3

Table 10.28 Hypothetical UVCE scenarios

Scenario	Location	Building	Overpressure (psi)	Frequency (y^{-1})
1	Unit-015	Control Room	7.5	4E−05
2	Unit-015	Control Room	4.0	3E−05
3	Unit-015	Control Room	7.0	3E−05
4	Unit-015	Control Room	2.0	3E−04
5	Unit-015	Control Room	1.5	1E−03
6	Unit-015	Control Room	1.0	3E−03
7	Unit-015	Control Room	5.5	4E−05
8	Unit-015	Control Room	3.0	3E−05
9	Unit-015	Control Room	4.5	3E−05
10	Unit-015	Control Room	1.7	3E−04

Table 10.29 Hypothetical UVCE scenarios sorted out with peak overpressures values listed in decreasing order

Scenario	Location	Building	Overpressure (psi)	Frequency (y^{-1})
1	Unit-015	Control Room	7.5	4E−05
3	Unit-015	Control Room	7.0	3E−05
7	Unit-015	Control Room	5.5	4E−05
9	Unit-015	Control Room	4.5	3E−05
2	Unit-015	Control Room	4.0	3E−05
8	Unit-015	Control Room	3.0	3E−05
4	Unit-015	Control Room	2.0	3E−04
10	Unit-015	Control Room	1.7	3E−04
5	Unit-015	Control Room	1.5	1E−03
6	Unit-015	Control Room	1.0	3E−03

The curve generated on the semilog plan is called the exceedance curve, and it shows the behaviour of cumulative frequencies against the peak on-side overpressure. At every overpressure, the corresponding frequency is the sum of all the frequencies corresponding to that overpressure and to the greater ones (Fig. 10.53).

An *exceedance frequency* is selected on this diagram, which is defined as the frequency in correspondence of which the design overpressure will be adopted for the building. It is common practice to select a return period of 10,000 years (10^{-4}/y). This frequency is proposed by various institutions, such as the UKOOA/HSE (2006), the Chemical Industries Association (2010), and DNV.

Example 10.35

Find the design peak overpressure for the building tabulated in Tables 10.28 and 10.29.

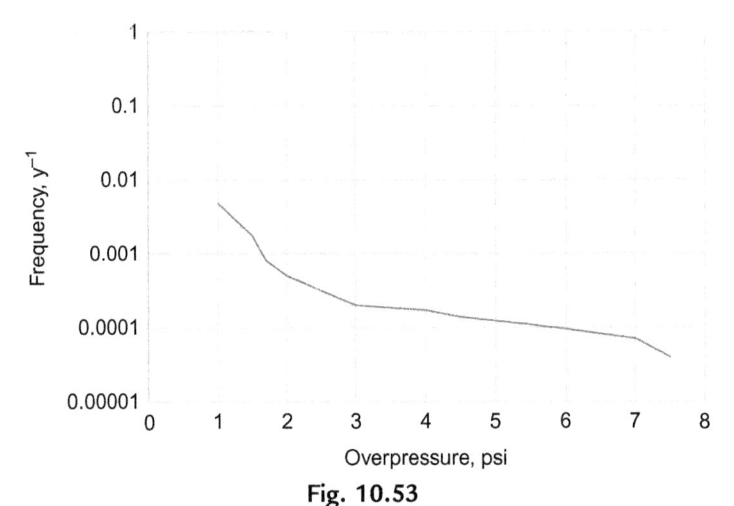

Fig. 10.53
Exceedance frequency curve for UVCE hypothetical scenarios.

Solution

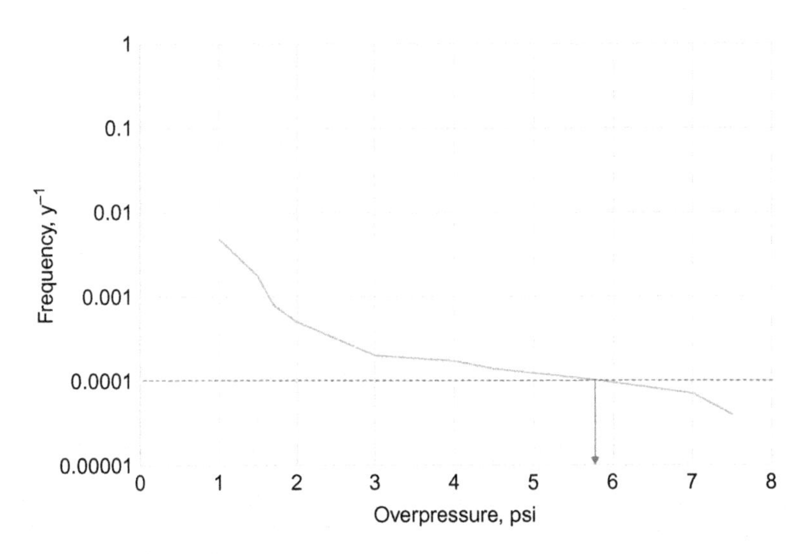

Drawing the exceedance frequency, this line crosses the exceedance curve at 5.75 psi.

Iso-contours charts

An useful way to show the behaviour of damage indices from data calculated in the QRA are the iso-charts, which consists of frequency contours drawn on the plot plan related to a specific damage index, such as individual risk of death due to thermal radiation, toxicity, and overpressure. Likewise, isobaric contours are produced to show the frequency behaviour for a given constant peak side overpressure (Fig. 10.54).

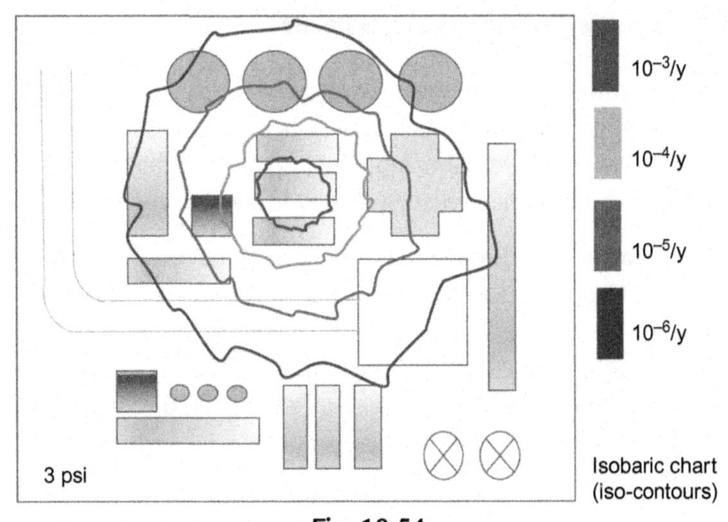

Fig. 10.54
Isobaric chart with iso-frequency contours.

It is important to take note that no design data can be collected from the contours. Even if the generation system is the same as that used for the QRA findings data, however:

- the level of approximation of the graphical representation is not so accurate to enable collection of design data.
- no information is provided about pressure wave duration or impulse.

Tabulated Design Accidental Loads (DAL)

The correct approach to produce design data for blast-resistant buildings is to list them along with the corresponding peak on-side overpressure and maximum duration and/or impulse. The typical approach consists of the tabulating of cumulative frequency data, as shown in Table 10.30. The design values will be the ones related to the exceedance frequency, 10^{-4}/y.

Table 10.30 Typical tabulated blast data for buildings

Location	Building	Overpressure (psi)	Max Duration (ms)	Max Impulse (psi ms)	Cumulative Frequency (y^{-1})
Unit-015	Shelter	1.0	297	148	2.67E−04
Unit-015	Shelter	1.5	290	216	1.00E−04
Unit-015	Shelter	2.0	283	283	2.29E−05
Unit-015	Shelter	3.0	96.0	145	2.29E−05

10.19.3 Building Damage Probability Assessment

Buildings designed according to the exceedance cumulative frequency blast data can be assumed as undamaged if an explosion occurs. However, in some instances, it is needed to assess what is the expected probability in relationship with a specific damage level. This is the case if the blast design criteria are not implemented, and a building, for any reasons, is not blast-rated.

This evaluation can be carried out through the probit functions presented by TNO-Green Book (1992), and provided in Chapter 6 for minor and major damage and collapse zones. Notably:

Minor damage

$$V = \left(\frac{4600}{P_s}\right)^{3.9} + \left(\frac{110}{i_s}\right)^{5.0} \rightarrow \mathrm{Pr} = 5 - 0.26 \times \ln V \tag{10.184}$$

Major damage

$$V = \left(\frac{17500}{P_s}\right)^{8.4} + \left(\frac{290}{i_s}\right)^{9.3} \rightarrow \mathrm{Pr} = 5 - 0.26 \times \ln V \tag{10.185}$$

Collapse

$$V = \left(\frac{40{,}000}{P_s}\right)^{7.4} + \left(\frac{460}{i_s}\right)^{11.3} \rightarrow \mathrm{Pr} = 5 - 0.22 \times \ln V \tag{10.186}$$

These three zones and the calculated probabilities are consistent with the impulse-overpressure regions included in the Baker's diagram (Baker et al., 1983) (Fig. 10.55).

Example 10.36

A nonblast-resistant building is affected by an overpressure of 2.25 psi (15,513 Pa), an impulse of 1901 Pa s, and at a frequency of 10^{-4}/y. Find the minor, major, and collapse damage probability. The explosive event probability is assumed to be 1.

Solution

Minor damage

$$V_1 = \left(\frac{4600}{15{,}513}\right)^{3.9} + \left(\frac{110}{1901_s}\right)^{5.0} \rightarrow \mathrm{Pr}_1 = 5 - 0.26 \times \ln V_1 = 6.23 \tag{10.187}$$

Major damage

$$V_2 = \left(\frac{17500}{15{,}513}\right)^{8.4} + \left(\frac{290}{1{,}901}\right)^{9.3} \rightarrow \mathrm{Pr}_2 = 5 - 0.26 \times \ln V_2 = 4.73 \tag{10.188}$$

Collapse (Table 10.31)

$$V_3 = \left(\frac{40{,}000}{15{,}513}\right)^{7.4} + \left(\frac{460}{1901}\right)^{11.3} \rightarrow \mathrm{Pr}_3 = 5 - 0.22 \times \ln V_3 = 3.457 \tag{10.189}$$

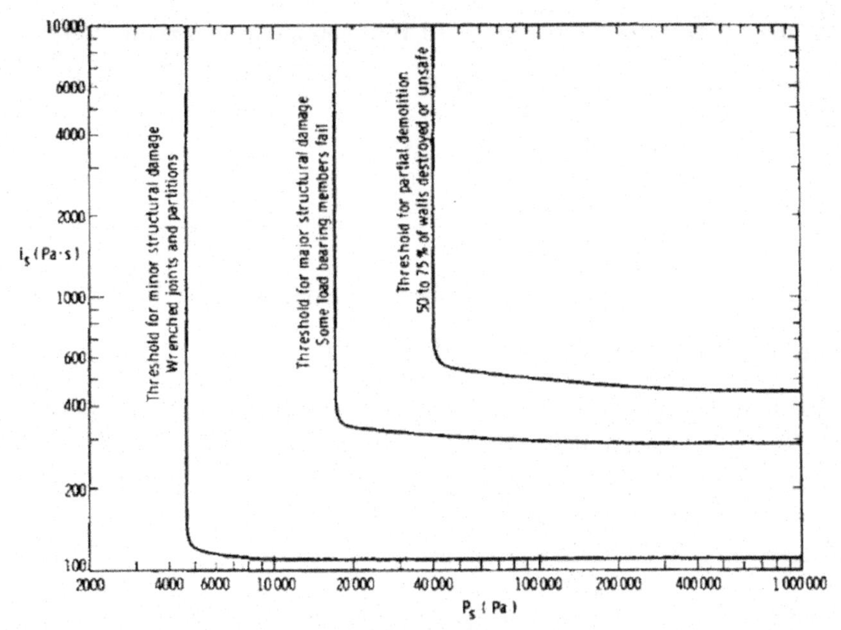

Fig. 10.55
Baker's diagram.

Table 10.31 Damage probability for building of Example 10.36

Frequency (y^{-1})	ΔP (psi)	ΔP (Pa)	Duration (ms)	Impulse (Pa s)	V_1	Pr_1	Minor Damage Probability (%)
1E-04	2.25	15,513	245	1901	0.0087	6.23	89
Frequency (y^{-1})	ΔP (psi)	ΔP (Pa)	Duration (ms)	Impulse (Pa s)	V_2	Pr_2	Major Damage Probability (%)
1E-04	2.25	15,513	245	1901	2.75	4.73	39.5
Frequency (y^{-1})	ΔP (psi)	ΔP (Pa)	Duration (ms)	Impulse (Pa s)	V_3	P_3	Collapse Probability (%)
1E-04	2.25	15,513	245	1901	1106.75	3.457	6

10.19.4 Blast Effects on Process Equipment

The evaluation of the damage of explosion on some main process equipment can be carried out by means of probit functions, following the same approach presented for the estimation of minor damage, major damage, and collapse of nonblast rated buildings.

Salzano and Cozzani (2004) have specified the following coefficients of a probit function related to damage to vessels and equipment caused by overpressure:

$$\mathrm{Pr} = k_1 + k_2 \cdot \ln(\Delta P) \tag{10.190}$$

Table 10.32 Coefficients of Eq. (10.190) Salzano and Cozzani (2004)

Equipment	k_1	k_2
Atmospheric tanks	−18.96	2.44
Pressure vessels	−42.44	4.33
Elongated equipment	−28.07	3.16
Small equipment	−17.79	2.18

The coefficients for specific equipment have been included in Table 10.32:

In the probit equation the overpressure is in Pascal unit.

Overpressure data specific for process equipment and damage typology have been presented by Stephens (1970) and have been included in Table 10.33.

10.20 Hazardous Area Classification

The identification and calculation of the size of areas where flammable atmospheres can be present at a given frequency or probability is an important design tool. The European ATEX framework is a comprehensive approach, aiming to control ignition probability of electrical and nonelectrical equipment through two main steps:

- The quantification of the size of flammable atmospheres due to vapours, gases, mists, and dusts.
- The adoption of electrical and nonelectrical equipment which are designed to be used within the specified hazardous areas.

The first task belongs to the process safety engineer's commitment, which is in charge to identify and quantify hazardous areas. A reference book providing guidance about the methodology to quantify hazardous areas was released by Cox et al. (1990). In the engineering practice, standards are used depending on the specific regulatory frame. Notably:

Europe

- EN 60079-10-1:2015 Explosive atmospheres. Classification of areas. Explosive gas atmospheres.
- EN 60079-10-2:2015 Explosive atmospheres. Classification of areas. Explosive dust atmospheres.

United Kingdom

- EI Model code of safe practice Part 15: Area classification for installations handling flammable fluids, Energy Institute.

Table 10.33 Overpressure data specific for process equipment and damage typology (Stephens, 1970)

Equipment	Damage Estimates Based on Overpressure for Process Equipment (Stephens, 1970)																							
	Overpressure (psi)																							
	0.5	1.0	1.5	2.0	2.5	3.0	3.5	4.0	4.5	5.0	5.5	6.0	6.5	7.0	7.5	8.5	9.0	9.5	10	12	14	16	18	20
Control house steel roof	A	C	D				N																	
Control house concrete roof	A	E	P	D			M																	
Cooling tower	B			F			O																	
Tank: cone roof		D				K							U											
Instrument cubicle			A			LM						T												
Fired heater				G	I					T														
Chemical reactor				A				I					P				T							
Filter				H					F									V		T				
Regenerator						I				IP					T									
Floating roof tank						K							U											D
Reactor: cracking							I							I						T				
Pipe supports							P					SO												
Utilities: gas metre									Q															
Utilities: electronic									H						I				T					
Electric motor										H							I							V
Blower										Q									T					
Fractionation column											R			T										
Horizontal pressure vessel												PI					T							
Utilities: gas regulator												I							MQ					
Extraction column													I						V	T				

Continued

Table 10.33 Overpressure data specific for process equipment and damage typology—cont'd

Equipment	Damage Estimates Based on Overpressure for Process Equipment (Stephens, 1970)																							
	Overpressure (psi)																							
	0.5	1.0	1.5	2.0	2.5	3.0	3.5	4.0	4.5	5.0	5.5	6.0	6.5	7.0	7.5	8.5	9.0	9.5	10	12	14	16	18	20
Steam turbine															I					M	S			V
Heat exchanger															I		T							
Sphere vessel																I					I	T		
Vertical pressure vessel																				I	T			
Pump																				I		V		

A. Windows and gauges broken.
B. Louvers fall at 0.2–0.5 psi.
C. Switchgear is damaged from roof collapse.
D. Roof collapses.
E. Instruments are damaged.
F. Inner parts are damaged.
G. Brick cracks.
H. Debris—missile damage occurs.
I. Unit moves and pipes break.
J. Bracing falls.
K. Unit uplifts (half tilted).
L. Power lines are severed.
M. Controls are damaged.
N. Block walls fall.
O. Frame collapses.
P. Frame deforms.
Q. Case is damaged.
R. Frame cracks.
S. Piping breaks.
T. Unit over turns or is destroyed.
U. Unit uplifts (0.9 tilted).
V. Unit moves on foundation.

- IGEM/SR/25 Hazardous area classification of natural gas installations, The Institution of Gas Engineers and Managers, Edition 2.

USA

- NFPA 497: Recommended Practice for the Classification of Flammable Liquids, Gases, or Vapours, and of Hazardous (Classified) Locations for Electrical Installations in Chemical Process Areas.
- Recommended Practice 500 for Classification of Locations for Electrical Installations at Petroleum Facilities Classified as Class I, Division I, and Division 2, Third Edition.
- API Recommended Practice 505 for Classification of Locations for Electrical Installations at Petroleum Facilities Classified as Class I, Zone 0, Zone 1, and Zone 2, 1st Edition, November 1997.

Russia
GOST R 51330.9-99, Electrical apparatus for explosive gas atmospheres. Part 10. Classification of hazardous areas.

10.20.1 EN 60079-10-1

Definitions
Explosive atmosphere
A mixture with air, under atmospheric conditions, of flammable substances in the form of gas, vapour, dust, fibres, or flyings which, after ignition, permits self-sustaining propagation.

Hazardous area (on account of explosive gas atmospheres)
An area in which an explosive gas atmosphere is, or may be expected to be, present, in quantities which require special precautions for the construction, installation, and use of equipment.

Source of release
A point or location from which a gas, vapour, mist or liquid may be released into the atmosphere so that an explosive gas atmosphere could be formed.

Release rate
A quantity of flammable gas, vapour, or mist emitted per unit time from the source of release.

Continuous grade of release
Release which is continuous, or is expected to occur frequently or for long periods.

Primary grade of release
Release which can be expected to occur periodically, or occasionally, during normal operation.

Secondary grade of release
Release which is not expected to occur in normal operation and, if it does occur, is likely to do so only infrequently and for short periods.

Ventilation
The movement of air, and its replacement with fresh air, due to the effects of wind, temperature gradients, or artificial means (for example, fans or extractors).

Fundamental equations

The standard EN 60079-10-1 provides the full methodology for carrying out the hazardous area classification. The fundamental equations are included here.

Release rate of liquid

$$\frac{dG}{dt} = S \cdot \sqrt{2 \cdot \rho_L \cdot \Delta P} \tag{10.191}$$

where

- $\frac{dG}{dt}$ is the release rate of liquid, kg/s.
- S is the opening surface area, m^2.
- ρ is the liquid density, kg/m^3.
- ΔP is the pressure difference across the opening, Pa.

Release rate of gas or vapour choked flow

$$\frac{dG}{dt} = S \cdot P \cdot \sqrt{\frac{\gamma \cdot \mathrm{WM}}{R \cdot T}} \cdot \left(\frac{2}{\gamma + 1}\right)^{\frac{\gamma+1}{2 \cdot (\gamma - 1)}} \tag{10.192}$$

where

- $\frac{dG}{dt}$ is the release rate, kg/s.
- S is the opening surface area, m^2.
- P is the pressure difference inside the container, Pa.
- WM is the gas or vapour molecular weight.
- T is the absolute temperature inside the container, K.
- R is the ideal gas constant, 8314 J/kmole K.
- γ is the gas or vapour specific heat ratio.

Release rate of gas or vapour nonchoked flow

$$\frac{dG}{dt} = S \cdot P \cdot \sqrt{2 \cdot \frac{\mathrm{WM} \cdot \gamma}{R \cdot T \cdot (\gamma - 1)} \cdot \left[1 - \left(\frac{P_o}{P}\right)\right]^{(\gamma-1)/\gamma}} \left(\frac{P_o}{P}\right)^{1/\gamma} \tag{10.193}$$

where

- $\frac{dG}{dt}$ is the release rate of liquid, kg/s.
- S is the opening surface area, m^2.
- P_o is the atmospheric pressure, Pa.
- P is the pressure difference inside the container, Pa.
- WM is the gas or vapour molecular weight.
- T is the absolute temperature inside the container, K.
- R is the ideal gas constant, 8314 J/kmole K.
- γ is the gas or vapour specific heat ratio.

The velocity of gas at the discharge opening is equal to the speed of sound, which may be calculated with the following formula:

$$u_w = \sqrt{\frac{R \cdot T \cdot \gamma}{\text{WM}}} \tag{10.194}$$

A hole size of 0.25 mm^2 is normally adopted for gas fittings such as flanges, screwed fittings, and joints. For specific applications, a larger value up to 2.5 mm^2 is selected (HSE, 2008).

The hypothetical volume V_z

According to the EN 60079-10-1, an hypothetical volume V_z represents the volume over which the mean concentration of flammable gas or vapour will typically be either 0.25 or 0.5 times the LEL, depending on the value of a safety factor, k. The standard states that V_z is not necessarily related to the size of the hazardous area, but is only intended to assist in assessing the degree of the ventilation.

V_z is defined as:

$$V_Z = \frac{f}{C} \cdot \left(\frac{dV}{dt}\right)_{\min} = \frac{f}{C} \cdot \left(\frac{dG}{dt}\right)_{\max} \frac{T}{k \cdot \text{LEL}_m \cdot 293} \tag{10.195}$$

where

- f is the efficiency of the ventilation ranging from 1 (ideal situation) to 5 (impeded air flow).
- C is the number of fresh air changes per unit time, s^{-1}, 0.03 for an open air situation.
- K is the safety factor.
- LEL$_m$ is the lower explosion limit, kg/m$_3$.
- T is the ambient temperature, K.
- $\left(\frac{dV}{dt}\right)_{\min}$ is the minimum volumetric flow rate of fresh air (volume per time, m^3/s);

It must be considered that V_z is not representative of the hazardous volume, because its shape is influenced by the ventilation, and no information is included in V_z as to the cloud location. Furthermore, the concentration in a significant part of the cloud is lower than the LEL, so V_z is generally greater than the flammable cloud. The volume V_z is used for the rating of the ventilation as high, medium, or low for each grade of release, and this will define the zone.

Example 10.37

Gaseous hydrogen is released in open air from a 10 barg and 20°C pipeline to atmosphere through a flange hole of surface area of 5 mm^2. Find the hypothetical volume V_z, assuming that ventilation is very efficient.

Solution

T (K)	P (Pa)	P_o (Pa)	γ	WM	S (mm^2)	LEL$_m$ (kg/m^3)
293.16	1,100,000	100,000	1.41	2	5	0.0033

Critical pressure is:

$$P_C = P_o \cdot \left(\frac{\gamma+1}{2}\right)^{\frac{\gamma}{\gamma-1}} = 100{,}000\left(\frac{2{,}41}{2}\right)^{\frac{1.41}{0.41}} = 190{,}000\,\text{Pa} \tag{10.196}$$

Flow is choked:

$$\left(\frac{dG}{dt}\right)_{max} = S \cdot P \cdot \sqrt{\frac{\gamma \cdot WM}{R \cdot T}} \cdot \left(\frac{2}{\gamma+1}\right)^{\frac{\gamma+1}{2\cdot(\gamma-1)}}$$
$$= 5 \cdot 10^{-6} \cdot 1{,}100{,}000 \cdot \sqrt{\frac{1.41 \cdot 2}{8314 \cdot 293.16}} \cdot \left(\frac{2}{1.41+1}\right)^{\frac{2.41}{2\cdot(1.41-1)}} = 0.0034\,\text{kg/s} \tag{10.197}$$

$$V_Z = \frac{f}{C} \cdot \left(\frac{dG}{dt}\right)_{max} \frac{T}{k \cdot LEL_m \cdot 293} = \frac{1}{0.03} \cdot 0.0034 \cdot \frac{293}{0.5 \cdot 0.0033 \cdot 293} = 68.7\,\text{m}^3 \tag{10.198}$$

10.21 Inerting

Inerting is a technique to keep a flammable gas outside its flammable range. An inert gas is added so that a fraction of the flammable gas is replaced, and its final concentration is outside the flammable composition range. Specific definitions of this operation are provided by the NFPA 69 (2014).

10.21.1 Definitions

Inert gas
A gas that is noncombustible and nonreactive.

Purge gas
An inert or a combustible gas that is continuously or intermittently added to a system to render the atmosphere nonignitable.

Inerting
A technique by which a combustible mixture is rendered nonignitible by adding an inert gas or a noncombustible.

Blanketing
The technique of maintaining an atmosphere that is either inert or fuel-enriched in the vapour.

Batch purging
This method includes siphon, vacuum, pressure, and venting into to atmosphere.

Continuous purging
This method includes fixed-rate application and variable-rate or demand application.

Siphon purging
In this method, equipment might be purged by filling it with liquid and introducing purge gas into the vapour space to replace the liquid as it is drained from the enclosure.

Vacuum purging
In this method, equipment that normally operates at reduced pressure, or in which it is practical to develop reduced pressure, might be purged during shutdown by breaking the vacuum with purge gas.

Pressure purging
In this method, enclosures might be purged by increasing the pressure within the enclosure by introducing purge gas under pressure and, after the gas has diffused, venting the enclosure to the atmosphere.

Sweep-through purging
This method involves introducing a purge gas into the equipment at one opening and letting the enclosure content escape to the atmosphere through another opening, thus sweeping out residual vapour. The following points should be noted:

- The total quantity required might be less than that for a series of steps of pressure purging.
- Four to five volumes of purge gas are sufficient to almost completely displace the original mixture, assuming complete mixing.

Fixed-rate purging
This method involves the continuous introduction of purge gas into the enclosure at a constant rate, which should be sufficient to supply the peak requirement in order that complete protection is provided, and a corresponding release of purge gas and whatever gas, mist, or dust has been picked up in the equipment.

10.21.2 Calculations

The determination of the amount required to reduce a flammable gas concentration below a given value over time can be approached, assuming perfect mixing between the fuel and the inert gas. The final formula is:

$$C_f(t) = C_o \cdot \exp\left(-\frac{Q}{V} \cdot t\right) \tag{10.199}$$

In practical applications, perfect mixing is an underconservative assumption. It's recommended to account for the nonideal scenario including, in the equation, a mixing efficiency coefficient K. According to the worst scenario, NFPA 69 recommends $K = 0.2$. This results in a factor of 5 for the time required by a given inert gas flow rate to lead a fuel concentration below a given value:

$$C_f(t) = C_o \cdot \exp\left(-K \cdot \frac{Q}{V} \cdot t\right) \tag{10.200}$$

Example 10.38

A 10 m^3 vessel contains 10% hydrogen (volume). Calculate the nitrogen flow rate required to reduce hydrogen concentration below its LEL (4%) in 10 min.

$$Q_{N_2} = -\ln\left(\frac{C_{LEL}}{C_o}\right) \cdot \frac{V}{K \cdot t} = -\ln\left(\frac{4}{10}\right) \cdot \frac{10}{0.2 \cdot 10} = 4.5\,m^3/min \tag{10.201}$$

10.21.3 Explosion Protection by Limiting Oxidant Concentrations

Dilution of a pure fuel gas with nitrogen can prevent flammable mixtures from occurring. However, sometimes air is inadvertently supplied, sometimes the supply of pure nitrogen is very expensive, and optimisation of nitrogen consumption is important. In all cases, it is very useful to follow the inerting operation on flammability diagrams, like the one included in Fig. 10.56.

Line BA is the place where air and fuel are mixed at different concentrations. Point A represents pure air, and points C and D represent the LEL and the UEL in air, respectively. A flammable zone can be recognised at oxygen concentrations lower than 21%. At point E the LEL coincides with the UEL and with the extreme point of the flammable zone.

If F is the starting point for the inertisation, some options can be considered:

- Feeding air would force the system moving across the line FA, and the flammable zone would be attained.

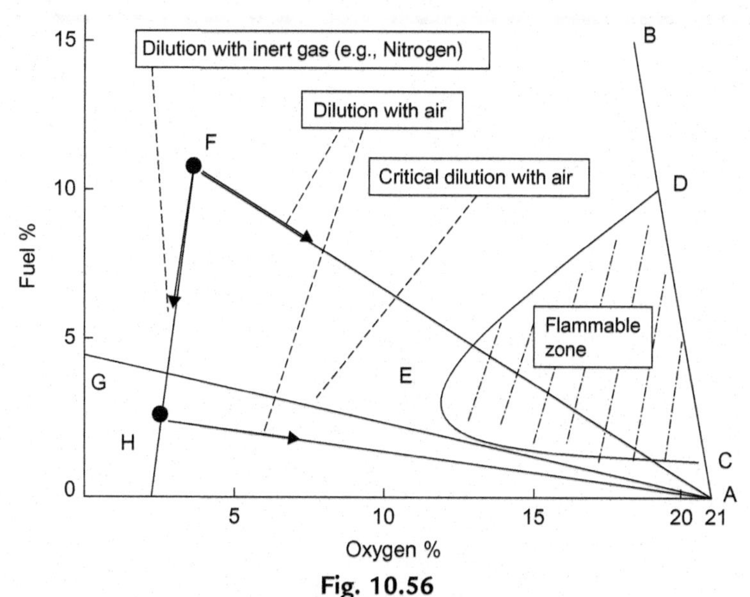

Fig. 10.56
Flammable and inert zone for fuel purging.

- Feeding nitrogen across the line FH, and starting supplying air, would prevent this system entering the flammable zone, for example through line HA.
- A critical line is defined, above which the flammable zone is crossed: this is line GA.

The adoption of flammability diagrams like this allows one to optimise nitrogen consumption. However, it has to be considered that this is an ideal situation, based on a lumped parameters assumption, whereas local concentration gradients can cause flammable mixtures to exist.

Limited Oxidant Concentration values have been presented in Chapter 3. However, they have to be considered as theoretical values; for practical applications, safety margins are applied. NFPA 69 recommends the following approach, based on whether or not the oxygen in a controlled atmosphere is continuously or noncontinuously monitored:

Oxygen concentration is continually monitored

- A safety margin of at least 2 vol% below the worst credible case LOC shall be maintained.
- The LOC shall be less than 5%, in which case the equipment shall be operated at no more than 60% of the LOC.

Oxygen concentration is not continually monitored

- The oxygen concentration shall be designed to operate at no more than 60% of the LOC, or 40% of the LOC if the LOC is below 5%.
- The oxygen concentration shall be checked on a regularly scheduled basis.

Example 10.39

A storage vessel is filled with 100% air, and it has to be ensured that, if methane is fed into the tank, oxygen concentration is properly reduced by supplying nitrogen. The storage tank cannot be pressurised, and oxygen is not continually monitored. Adopt the most appropriate inerting system and find a relationship between them.

Solution

The storage tank is atmospheric, so sweep-through purging will be implemented. Solving the mass balance for nitrogen, the following transient equation applies, assuming perfect mixing and a lumped parameters system (Chapter 3, Reactor Schemes) (Fig. 10.57):

$$V \cdot \frac{dC_{N2}}{dt} = Q_{N2} \cdot (C_{N2o} - C_{N2}) \tag{10.202}$$

where

- V is the volume of the storage tank.
- Q_{N2} is nitrogen flow rate.
- C_{N2} is nitrogen concentration at time t, both inside the storage tanks and in the outlet stream.
- C_{N2o} is the inlet nitrogen concentration in the, 100%.

To integrate the equations, two other values are needed:

- Nitrogen concentration at time $t=0$, 79%
- Nitrogen concentration corresponding to the final controlled atmosphere oxygen concentration.

From the table included in Chapter 3, we see that LOC for methane is 10%. However, NFPA 69 states that maximum oxygen concentrations should not be greater than 6%.

Integrating:

$$C_{N2}(t) = 100\% + (79\% - 100\%) \cdot \exp\left(-\frac{Q_{N2}}{V} \cdot t\right) \tag{10.203}$$

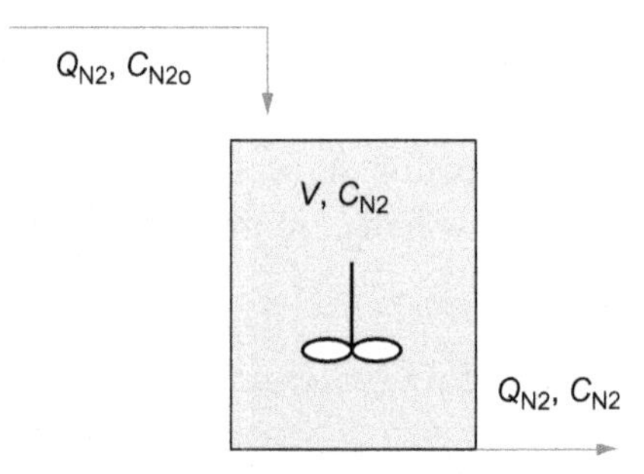

Fig. 10.57
CSTR.

This equation will be solved assuming $C_{N2}(t) = 100\% - 6\% = 94\%$ and specifying either the flow rate or the time.

In this example, perfect missing has been assumed. However in practical applications, an efficiency of 20% (0.2), as shown in the previous example, should be adopted.

References

Abdel-Gayed, R.G., Bradley, D., 1977. Dependence of turbulent burning velocity on turbulent Reynolds number and ratio of flaminar burning velocity to RMS turbulent velocity. In: International Symposium on Combustion, vol. 166. The Combustion Institute, Pittsburgh, PA, pp. 1725–1735.

API Recommended Practice 581, 2016. Risk-Based Inspection Methodology, third ed. American Petroleum Institute.

American Society of Civil Engineers, 2010. Design of Blast-Resistant Buildings in Petrochemical Facilities Second Edition. Task Committee on Blast-Resistant Design of the Petrochemical Committee of the Energy Division of the American Society of Civil Engineers.

ASTM E1445, 2015. Standard Terminology Relating to Hazard Potential of Chemicals.

Baker, W.E., Cox, P.A., Westine, P.S., Kulesz, J.J., Strehlow, R.A., 1983. Explosion Hazards and Evaluation. Elsevier Scientific Publishing Company.

Baker, Q.A., Tang, M.J., Scheier, E.A., Silva, G.J., 1994. Vapor cloud explosion analysis. In: 28th Loss Prevention Symposium. American Institute of Chemical Engineers, Atlanta, GA, New York.

Baker, Q.A., Doolittle, C.M., Fitzgerald, G.A., Tang, M.J., 1997. Recent development in the Baker-Strehlow VCE analysis methodology. In: 31st Annual Loss Prevention Symposium.

Bartknecht, W., 1981. Explosions. Course Prevention Protection. Springer-Verlag.

Bartknecht, W., 1993. Explosions-Schutz: Grundlagen und Anwendung. Springer-Verlag, Berlin.

Baum, M.R., 1984. The velocity of missiles generated by the disintegration of gas pressurized vessels and pipes. Trans. ASME 106, 362–368.

Benintendi, R., 2013. Course on Basic Process Safety Calculations. Foster Wheeler, Reading, United Kingdom.

Benintendi, R., Alfonzo, J., 2013. Identification and analysis of the key drivers for a systemic and process-specific reactive hazard assessment (RHA) methodology. In: 2013 International Symposium. Mary Kay O'connor Process Safety Center, College Station, TX.

Benintendi, R., 2014. Non-equilibrium phenomena in carbon dioxide expansion. Process Safety Environ. Protect. 92 (1), 47–59.

Benintendi, R., Rega, S., 2014. A unified thermodynamic framework for LNG rapid phase transition on water. Chem. Eng. Res. Des. 92, 3055–3071.

Bodurtha, F.T., 1980. Industrial Explosion Prevention and Protection. McGrawHill Book Company.

Boot, H., van der Voort, M.M., 2013. Determining VCE damage zones using the GAME relations and explosion regions. In: 9th Global Congress of Process Safety, AIChE, Spring Meeting.

Bradley, J.N., 1969. Flame and Combustion Phenomena. Barnes & Noble.

Britter, R.E., McQuaid, J., 1988. Workbook on the Dispersion of Dense Gases, Health and Safety Executive (Open source). http://www.hse.gov.uk/research/crr_pdf/1988/crr88017.pdf.

Britton, L.G., 2000. Using heats of oxidation to evaluate flammability hazards. Process Safety Prog. 21 (1), 1–24.

Cardillo, P., n.d. Le Esplosioni di Gas, Vapori e Polveri, Stazione Sperimentale per i Combustibili, Milan (Italy).

Cates, A.T., 1991. Fuel gas explosion guidelines. In: Int. Conf. Fire and Explosion Hazards Inst. Energy.

CCPS, 1999. Guidelines for Consequence Analysis of Chemical Releases.

CCPS, 2005. Glossary of Terms. American Institute of Chemical Engineers.

CCPS, 2010. Guidelines for Vapor Cloud Explosion, Pressure Vessel Burst, BLEVE, and Flash Fire Hazards, second ed. John Wiley & Sons.

Chemical Industries Association, 2010. Guidance for the Location and Design of Occupied Buildings on Chemical Manufacturing Sites, third ed.
Clancey, V.J., 1972. Diagnostic features of explosion damage. In: Sixth International Meeting of Forensic Sciences, Edinburgh.
Cox, W., Lees, F.P., Ang, M.L., 1990. Classification of Hazardous Locations, Institution of Chemical Engineers. Inter-Institutional Group on the Classification of Hazardous.
Craven, A.D., Greig, T.R., 1968. The Development of Detonation Over-Pressures in Pipelines IChemE Symposium Series No. 25.
Dugdale, D., 1985. An Introduction to Fire Dynamics. Wiley.
Encyclopaedia Treccani, 2008. Polymeric materials/12.5 Polyvinyl Chloride—Volume II/Refining and Petrochemicals, http://www.treccani.it.
Englund, S.M., 1990. Chemical Hazard Engineering Guidelines, Corporate Loss Prevention. The Dow Chemical Company.
European Norm, 2007. EN 14994. Gas Explosion Venting Protective Systems.
Fenning, R.W., 1926. Gaseous combustion at medium pressures. Phil. Trans. R. Soc. Lond. Ser. A 225.
Harvey, B.H., 1979. Second Report of the Advisory Committee on Major Hazards. Stationery Office, London.
Hasegawa, K., Kato, K., 1978. Study of the fireball following steam explosion of *n*-pentane. In: 2nd Int. Symp. Loss Prevention, Heidelberg, pp. 297–305.
Hess, K., Leuckel, W., Stoeckel, A., 1973. Ausbildung von Explosiblen Gaswolken bei Uberdachentspannung und MaBnahmen zu deren Vermeidung. Chem. Ing. Technol. 45 (73), 5323–5329.
Holbrow, P., Lunn, G.A., Tyldesley, A., 1999. Explosion venting of bucket elevators. J. Loss Prevent. Process Ind. 12, 373–383.
Holden, P.L., Reeves, A.B., 1985. Fragment hazards from failures of pressurised liquefied gas vessels. In: IChemE Symposium Series No. 93: Assessment and Control of Major Hazards, Rugby, UK, pp. 205–220.
HSE, 2008. Area Classification for Secondary Releases From Low Pressure Natural Gas Systems. Health and Safety Executive.
Karnesky, J., Damazo, J., Shepherd, J.E., Rusinek, A., 2010. Plastic response of thin-walled tubes to detonation. In: ASME 2010 Pressure Vessels and Piping Conference: Volume 4 Bellevue, Washington, USA, July 18–22.
Kinsella, K., 1993. A rapid assessment methodology for the prediction of vapour cloud explosion overpressure. Society of Loss Prevention in the Oil, Chemical and Process Industries in Health, Safety and Loss Prevention in the Oil, Chemical and Process Industries, Butterworth-Heinemann, pp. 200–211.
Lewis, B., von Elbe, G., 1961. Combustion, Flames and Explosions of Gases, second ed. Academic Press, NY.
Lighthill, J., 1978. Waves in Fluids. Cambridge University Press, Cambridge.
Mannan, S. (Ed.), 2012. Lees' Loss Prevention in the Process Industries, fourth ed. Butterworth-Heinemann.
Mercx, W.P.M., van den Berg, A.C, van Leeuwen, D., 1998. Application of correlations to quantify the source strength of vapour cloud explosions in realistic situations. Final Report for the Project: GAMES. TNO Report PML 1998-C53.
Mikic, B.B., Rohsenow, W.M., Griffith, P., 1970. On bubble growth rate. Int. J. Heat Mass Transfer 13, 657–666.
National Fire Protection Association, 2013. NFPA 68. Standard on Explosion Protection by Deflagration Venting.
National Fire Protection Association, 2014. NFPA 69. Standard on Explosion Prevention Systems.
National Fire Protection Association, 2016. NFPA 67. Guide on Explosion Protection for Gaseous Mixtures in Pipe Systems.
Pfreim, H., 1941. Forsch Arb. Geb. IngWes., 1941,12,148: Idem ibid., 244.
Pierorazio, A.J., Thomas, J.K., Baker, Q.A., Ketchum, D.E., 2005. An update to the Baker-Strehlow-Tang vapor cloud explosion prediction methodology flame speed table. Process Saf. Prog. 24 (1), 59–65.
Pietersen, C., Huerta, S., 1984. Analysis of the LPG Incident in San Juan Ixhuatepec. Mexico City. 19 November, TNO Report No. 85-0222: 8727-3325, The Hague.
Puttock, J., 1995. Fuel gas explosion guidelines—the congestion assessment method. In: 2nd European Conference on Major Hazards On- and Off- Shore, Manchester.
Puttock, J., 2001. Developments for the congestion assessment method for the prediction of vapour-cloud explosions. In: 10th International Symposium on Loss Prevention and Safety Promotion in the Process Industries, Stockholm.

Salzano, E., Cozzani, V., 2004. Blast wave damage to process equipment as a trigger of domino effects. In: Center for Chemical Process Safety—19th Annual International Conference: Emergency Planning Preparedness, Prevention & Response.

Shepherd, J.E., Krok, J.C., Lee, J.L., 1997. Jet A Explosion Experiments: Laboratory Testing. Pasadena, CA (California Institute of Technology. Explosion Dynamics Laboratory Report FM97-5), pp. 27–28.

Simmonds, W.A., Cubbage, P.A., 1960. The design of explosion reliefs for industrial drying ovens. In: Symposium on Chemical Process Hazards, IChemE.

Singh, J., 1994. Gas explosion in interconnected vessels: pressure piling. In: Institution of Chemical Engineers, Hazards XII, European Advances in Process Safety, Symposium Series No 134, EFCE Event No 504, EFCE Publication, No 104, pp. 195–212. ISBN 0 85 295 327 5.

Stephens, M.M., 1970. Minimizing Damage from Nuclear Attack, Natural and Other Disasters. The Office of Oil and Gas. Department of the Interior, Washington.

Tang, M.J., Cao, C.Y., Baker, Q.A., 1996. Blast effects from vapor cloud explosions. In: International Loss Prevention Symposium, Bergen, Norway.

Tang, M.J., Baker, Q.A., 1999. A new set of blast curves from vapor cloud explosion. Process Safety Prog. 18(3).

TNO, 1992. Methods for the Determination of Possible Damage, CPR 16E, Green Book.

UKOOA, 2006. Fire and Explosion Guidance. Part 2: Avoidance and Mitigation of Fires. Final Draft.

UKOOA/HSE, 2006. Update Guidance for Fire and Explosion Hazards, MSL Engineering Consortium, CTR 103, Explosion Hazard Philosophy.

Van Buijtenen, C.J.P., 1980. Calculation of the amount of gas in the explosive region of a vapour cloud released in the atmosphere. J. Hazard. Mater. 3, 201–220.

Van den Berg, A.C., van der Voort, M.M., Weerheijm, M.M., Versloot, N.H.A., 2004. Expansion-controlled evaporation: a safe approach to BLEVE blast. J. Loss Prevent. Process Ind. 17, 397–405.

Van den Bosch, C.J.H, Duijm, N.J., 2005, Methods for the Calculation of Physical Effects. Yellow Book, C.J.H. van den Bosch, R.A.P.M. Weterings.

Weibull, H.R.W., 1968. Pressures recorded in partially closed chambers at explosion of TNT charges. Ann. N. Y. Acad. Sci. 152 (1), 357–361.

Yasseri, S., 2002. Response of Secondary Systems to Explosion, FABIG Newsletter Article No. R454, FABIG Newsletter Issue No. 33.

Zeeuwen, J. P., Wiekema, B.J, 1978. The measurement of relative reactivities of combustible gases. In: Conference on Mechanisms of Explosions in Dispersed Energetic Materials.

Further Reading

Eggen, J.B.M.M., 1998. GAME: Development of Guidance for the Application of the Multi-Energy Method, Health and Safety Executive.

HSE, 2001. Reducing Risk, Protecting People. HSE's Decision-Making Process.

NORSOK, 2001. Z-013-Risk and Emergency Preparedness Analysis. Norwegian Technology Centre.

Woodward, J.L., 1998. Estimating the Flammable Mass of a Vapor Cloud. CCPS, AIChE.

Worldwide Refinery Processing Review, 2010. Causes of Temperature Runaway in C4 and C5/C6 Isomerization Units, Hydrocarbon Publishing, First Quarter.

CHAPTER 11

Dust Explosions

...in collecting flour, a sudden fall of a great quantity took place, followed by a thick cloud, which immediately caught on fire and caused a violent explosion

(Count Morozzo 1785)

11.1 Powders Classification

Powders are solid materials belonging to the three categories: pellets, granules and dusts.

Pellets: Solid particles having a diameter greater than 2 mm according to 10 *U.S. Standard Sieve* and typically greater than 3 mm.
Granules: Solid particles having a diameter ranging from 0.42 to 2 mm.
Dusts: Solid particles having a diameter ranging from 0.42 mm and 1 μm.
Suspended fines smaller than 1 μm are called fumes.
Combustible dusts: According to NFPA 654, dusts traditionally have been defined as a material 420 μm diameter or smaller (capable of passing through a U.S. No. 40 standard sieve). Combustible particulates with an effective diameter of less than 420 μm should be deemed to fulfil the criterion of the definition.
Explosive dusts: According to Field (1982), explosions of 500 μm (0.5 mm) diameter dusts is highly unlikely, even if not impossible.

11.2 Dust Explosions Thermodynamics

Dust explosions are very similar to gas explosions. Depending on the containment configuration, if any, a pressure rise can take place. An ignition source (BS EN 1127) typically causes a primary explosion, which provides the energy and promotes airborne dusts, resulting in a secondary explosion, which is generally much more serious than the primary. Case history generally refers to the secondary explosive event. Dust explosions are typically deflagrative events, and are very infrequently detonations. In open systems, the ignition of a powder cloud results predominantly in thermal effects, whereas dust ignition in containments may have significant overpressure consequences. This has been summarised in the next flow chart (Fig. 11.1).

Process Safety Calculations. https://doi.org/10.1016/B978-0-08-101228-4.00011-3

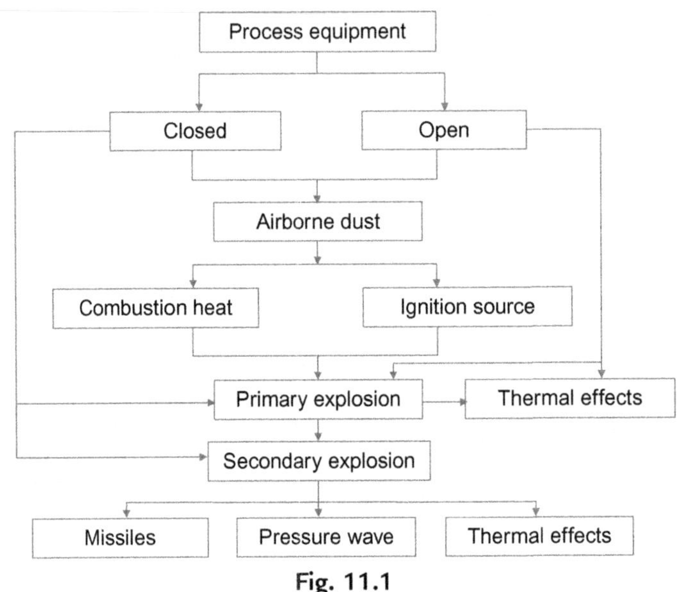

Fig. 11.1
Flow chart of dust explosions.

11.2.1 Semi-Confined Explosion

Semi-confined explosion has been presented in Chapter 10.

11.2.2 Confined Explosion

As for gases, confined explosions result in maximum pressure generation. The pressure rise curve is depicted in Fig. 11.2.

After the ignition, a pressure rise to P_1 is detected, also due to air ingress in the cloud. The faster the explosion kinetics, the more negligible this value, compared to peak overpressure P_{max}. Downstream of P_{max}, overpressure drops to zero. The following value:

$$\left[\frac{dP}{dt}\right]_{max} = tg\alpha \tag{11.1}$$

identifies the maximum pressure rise. As for gases, P_{max} and $\left[\frac{dP}{dt}\right]_{max}$ define the violence of the explosion: the former is an indicator of the thermodynamic potential of the dust, the latter of the kinetic potential.

Maximum pressure

As demonstrated for gases, the maximum theoretical overpressure inside a containment does not depend on the volume and can be estimated as:

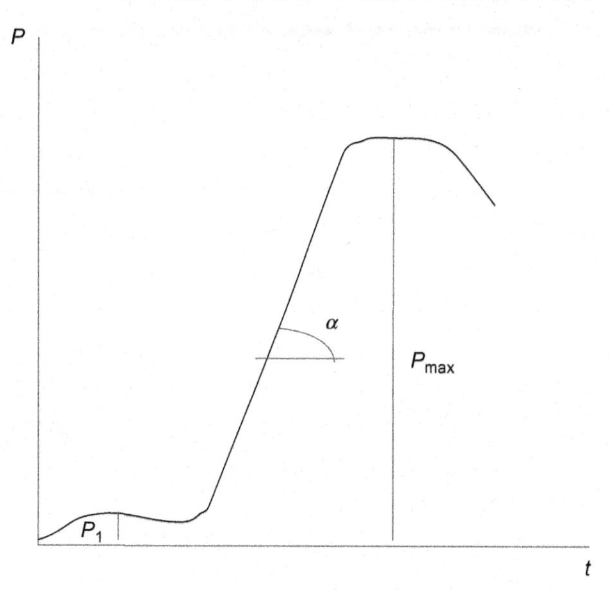

Fig. 11.2
Pressure rise in confined explosion.

$$P_1 = P_o \cdot \frac{n_1 \cdot T_1}{n_o \cdot T_o} \tag{11.2}$$

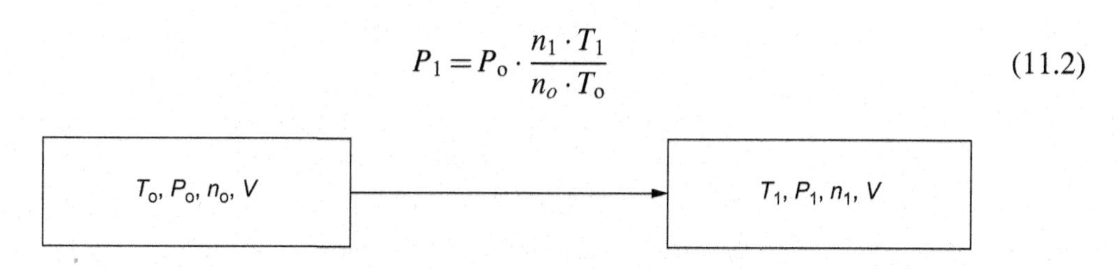

and is attained around the stoichiometric dust to air ratio.

Example 11.1

Brown coal can be roughly assumed to have 50% mass composed of carbon (including also hydrogen and sulphur thermal contribution). The remaining part is water and ash. The temperature is 20°C at atmospheric pressure.

Solution

The stoichiometric reaction is:

$$C + O_2 + 79/21 N_2 \rightarrow CO_2 + 79/214 N_2 \tag{11.3}$$

1 kmol of carbon requires about 5 kmol of air. In terms of mass, 12 kg of carbon (50% mass). Overall, 24 kg of brown coal require $5 \times 22.414 \times 293/273.16 \cong 120$ m^3 of air. This is equivalent to $24,000/120 \approx 200$ g/m^3 of brown coal. The maximum overpressure should be found at this concentration. According to Wieman (1987), this is verified for initial pressures from 1 to 4 bar abs.

The maximum pressure can be calculated assuming adiabatic combustion.

Example 11.2

Find the maximum overpressure for the previous dust.

Solution

Assuming a lower heating value equal to 18 MJ/kg (4300 kcal/kg), and considering an average value of air specific heat at constant volume:

$$M_p \cdot H_p = (M_p + M_a) \cdot c_v \cdot (T_c - T_{amb}) \quad (11.4)$$

where:

- M_p is brown coal mass, 24 kg.
- H_p is the lower heating value, 4300 kcal/kg.
- M_a is stoichiometric air mass, 145 kg.
- T_c is the maximum (adiabatic) combustion temperature.
- T_{amb} is ambient temperature, 293.16.
- c_v is air average specific heat at constant volume $= c_p - R = 0.2$ kcal/kg K.

Solving:

$$T_c = 2760\,\text{K} \quad (11.5)$$

And:

$$P_{max} = P_o \cdot \frac{n_c . T_c}{n_o \cdot T_{amb}} = \frac{5 \cdot 2760}{6 \cdot 293.16} \cong 8\,\text{atm} \quad (11.6)$$

This value is very close to the one reported by Cardillo (n.d.).

If experimental data are missing, dust explosion thermodynamic characteristics can be estimated on the basis of the elementary composition and thermal data.

11.2.3 Pressure Piling

Pressure piling has been dealt with in Chapter 10. The methodology presented for gases is specific for dusts.

11.2.4 Pressure Rise Rate

Cube root law presented for gases is identically applicable to dusts, replacing the explosion constant k_{ST} with k_G:

$$\left(\frac{dP}{dt}\right)_{max} \cdot V^{1/3} = k_{ST} \quad (11.7)$$

Subscript 'ST' stands for the German *staub*, dust, the first experiment having been carried out by the Swiss scientist Wolfgang Bartknecht.

The evidence that the maximum overpressure does not change with volume changes, whereas maximum pressure rise does, can be explained by assuming that pressure increases according to the kinetic advancement of flame front are:

$$\frac{P(t) - P_o}{P_{max} - P_o} = \frac{r(t)}{R} \tag{11.8}$$

where R is the spherical containment radius, $r(t)$ is the spherical flame transient radius, and $P(t)$ the corresponding transient pressure.

This equation can be written as:

$$\frac{P(t) - P_o}{P_{max} - P_o} = \frac{S_b \cdot t}{R} \tag{11.9}$$

where S_b is the flame front velocity. Differentiating:

$$\frac{dP(t)}{dt} = (P_{max} - P_o) \cdot \frac{S_b}{R} = (P_{max} - P_o) \cdot \frac{\rho_u}{\rho_b} \cdot \frac{S_u}{R} \cong (P_{max} - P_o) \cdot \frac{P_{max}}{P_o} \cdot \frac{S_u}{R} \tag{11.10}$$

where S_u is the laminar burning velocity, and ρ_u and ρ_b, respectively, the unburnt and burnt gas density. It is worth noting that the derivative is a constant and:

$$k_{ST} = \left(\frac{dP(t)}{dt}\right)_{max} \cdot V^{1/3} = \left(\frac{4}{3} \cdot \pi\right)^{1/3} \cdot (P_{max} - P_o) \cdot \frac{P_{max}}{P_o} \cdot S_u \tag{11.11}$$

NFPA 68 (2013) has proposed the following classification for dusts, on the basis of the explosion constant k_{ST}. Typical materials have been selected by OSHA (2007) (Table 11.1):

The explosion constant k_{ST}, like k_G, is determined experimentally in standard conditions. A screening evaluation of the dust classification can be done by means of the previous equation.

Its value is affected by many variables, such as process conditions, moisture, particle size, size distribution, concentration, and, last but not least, ignition energy. Moreover, turbulence is a very important factor, which is not easily accounted for in the calculation.

Table 11.1 Classification of explosive dusts (NFPA 68, 2013; OSHA, 2007)

Dust Explosion Class	k_{ST}, bar m/s	Characteristic	Typical Material
St 0	0	No explosion	Silica
St 1	0–200	Weak explosion	Powdered milk, charcoal, sulphur, sugar, zinc
St 2	200–300	Strong explosion	Cellulose, wood flour, PPMA
St 3	>300	Very strong explosion	Anthraquinone, aluminium, magnesium

Its effect is the production of a variable P_{max} and a different value of the laminar burning velocity. The latter is typically tabulated with reference to standard conditions (300 K), but is strongly dependent on the unburnt gas and on pressure. A general formula is the following (Landry et al., n.d.):

$$S_u = S_{uo} \cdot \left(\frac{T_u}{300}\right)^{1.85} \cdot (P)^{-0.45} \tag{11.12}$$

where S_{uo} is the laminar burning velocity corresponding to the standard conditions.

If one knows the laminar burning velocity in a condition 1:

$$S_{u1} = S_{uo} \cdot \left(\frac{T_{u1}}{300}\right)^{1.85} \cdot (P_1)^{-0.45} \tag{11.13}$$

and wants to knows the laminar burning velocity in a condition 2:

$$S_{u2} = S_{uo} \cdot \left(\frac{T_{u2}}{300}\right)^{1.85} \cdot (P_2)^{-0.45} \tag{11.14}$$

taking ratios of both members of these equations:

$$\frac{S_{u2}}{S_{u1}} = \left(\frac{T_{u2}}{T_{u1}}\right)^{1.85} \cdot \left(\frac{P_2}{P_1}\right)^{-0.45} \tag{11.15}$$

applying Gay Lussac's law:

$$\frac{S_{u2}}{S_{u1}} = \left(\frac{P_2}{P_1}\right)^{1.85} \cdot \left(\frac{P_2}{P_1}\right)^{-0.45} = \left(\frac{P_2}{P_1}\right)^{1.4} \tag{11.16}$$

Knowing a given k_{ST} value, the one related to another state can be calculated, provided that P_{max} is known:

$$k_{ST2} = k_{ST1} \cdot \frac{\left(P_{2max}{}^{3.4} - P_{2max}{}^{2.4}\right)}{\left(P_{1max}{}^{3.4} - P_{1max}{}^{2.4}\right)} \tag{11.17}$$

Example 11.3

Check the previous formula on the basis of the following data:

Dust	Volume, m^3	P_{max}, bar g	k_{ST}, bar m/s
Wheat starch	0.0012	7.95	68
Wheat starch	0.0012	10.2	128

Solution

$$k_{ST2} = k_{ST1} \cdot \frac{\left(P_2 max^{3.4} - P_2 max^{2.4}\right)}{\left(P_1 max^{3.4} - P_1 max^{2.4}\right)} = 68 \cdot \frac{\left(11.2^{3.4} - 11.2^{2.4}\right)}{\left(8.95^{3.4} - 8.95^{2.4}\right)} = 130 \text{bar m/s} \tag{11.18}$$

that is very close to 128.

Example 11.4

Check the previous formula on the basis of the following data (van Wingerden):

Dust	Volume, m^3	P_{max}, bar g	k_{ST}, bar m/s
Cornstarch	0.020	8.65	160
Cornstarch	0.020	7.9	114

Solution

$$k_{ST2} = k_{ST1} \cdot \frac{(P_2 max^{3.4} - P_2 max^{2.4})}{(P_1 max^{3.4} - P_1 max^{2.4})} = 114 \cdot \frac{(9.65^{3.4} - 9.65^{2.4})}{(8.9^{3.4} - 8.9^{2.4})} = 152 \text{bar m/s} \quad (11.19)$$

that is very close to 160.

11.2.5 Characteristics of Explosivity

Minimum ignition energy (MIE)

Ignition energy for dusts is greater also by orders of magnitude than ignition energy of gases. MIEs are typically tabulated for dust clouds and dust layers.

Minimum explosive concentration (MEC)

It is the minimum dust concentration in a cloud which, after the ignition, is able to propagate and sustain combustion. Unlike the UEL for gases, no maximum explosive concentration is identified.

Minimum ignition temperature (MIT)

It is the minimum temperature at which a dust cloud or layer is ignited.

Limiting oxidant concentration (LOC)

It is the equivalent for dusts of the LOC defined for gases.

Ignition severity (IS) and explosion severity (ES)

These are two parameters, introduced by the US Bureau of Mines (USBM), to have an indication of the intrinsic hazardous behaviour of a dust.

$$\text{IS} = \frac{(T_C \cdot E \cdot C)_1}{(T_C \cdot E \cdot C)_2} \quad (11.20)$$

where:

- T_C is the ignition temperature.
- E is the MIE.
- C is the MEC.

Table 11.2 Ignition sensitivity and explosion severity (Jacobson et al., 1961)

Hazard Level	Ignition Sensitivity	Explosion Sensitivity
Weak	<0.2	<0.5
Medium	0.2–1	0.5–1
Strong	1–5	1–2
Severe	>5	>2

Table 11.3 Ignition sensitivity and explosion severity for some dusts (Jacobson et al., 1961)

IS and ES for Some Dusts (USBM)		
Dust	Ignition Sensitivity	Explosion Sensitivity
Coffee	0.1	0.1
Egg white	<0.1	0.2
Flour	2.1	1.8
Grain	2.8	3.3
Milk	1.6	0.9
Sugar	4.0	2.4

And:

$$\mathrm{ES} = \frac{(P_{\max} \cdot dP/dt_{\max})_1}{(P_{\max} \cdot dP/dt_{\max})_2} \tag{11.21}$$

where subscripts 1 and 2 refer, respectively, to Pittsburgh coal and to the dust. According to Jacobson et al. (1961) (Tables 11.2 and 11.3):

11.2.6 Parameters of Influence

The following factors influence the onset and the characteristics of explosive events for dusts.

Chemistry

Chemical identity of dusts strongly affects both the energetic potential (thermodynamics), and the modality and rate at which the explosion takes place (kinetics). Table 11.4 shows how the energetic potential changes depending on the single substance (Eckhoff, 2003).

Being:

$$k_{\mathrm{ST}} = \left(\frac{dP(t)}{dt}\right)_{\max} \cdot V^{1/3} = \left(\frac{4}{3} \cdot \pi\right)^{1/3} \cdot (P_{\max} - P_o) \cdot \frac{P_{\max}}{P_o} \cdot S_u \tag{11.22}$$

it will be obtained:

Dust	P_{max}, bar g	k_{ST}, bar m/s
Milk	8.1	90
Cocoa bean shell	8.1	68

Table 11.4 Energetic potentials of dusts (Eckhoff, 2003)

Substance	Oxided to	kJ/kmol O_2
Calcium	Calcium oxide	1270
Magnesium	Magnesium oxide	1240
Aluminium	Aluminium oxide	1100
Silica	Silica oxide	830
Chromium	Chromium oxide (III)	750
Zinc	Zinc oxide	700
Iron	Iron oxide (III)	530
Copper	Copper oxide	300
Sugar	Carbon dioxide and water	470
Corn	Carbon dioxide and water	470
Polyethylene	Carbon dioxide and water	390
Sulphur	Sulphur dioxide	300

It is worth noting that, against the same energetic potential, a significant difference of the explosion constant is due to a likely different value of the laminar burning velocity, which is a parameter of kinetic nature. The existence of specific functional groups in an organic molecule can provide a useful indication of the expected flammability and explosivity. Abbot (1988) has provided the following list of chemical groups which can be related to an explosive tendency:

- NO_2 or $—ONO_2$
- N=N or N—N
- NX_2 (f.i. NCl_2)
- C=N
- $OClO_2$
- O—O or O—O—O
- C≡C
- Light metallic atoms not stably bound to C or to some organic radicals

According to Field (1982), groups COOH, OH, NH_2, C=N, C≡N, and N=N increase the explosion hazard, whereas molecules containing halogens Cl, Br, and F have a low ignition tendency.

Size

The smaller the size, the higher the explosion likelihood. Organic molecules are deemed to proceed according to the following process in a series (Eckhoff, 2003). According to it, a pyrolytic mechanism generates a gas atmosphere, and this is why dust explosion presents so many similarities with gas explosions.

$$\text{Devolatilisation} \rightarrow \text{Gas} - \text{phase mixing} \rightarrow \text{Gas} - \text{phase combustion}$$

According to the rate-limiting step theory, the overall rate of a process is increased if the slowest one is increased. For dust explosions, it is experimentally proved that the smaller the dust size, the higher the explosion rate and violence. Consequently, the pyrolysis is the rate-limiting step, so the overall process is controlled by the gas transfer from the particle, that is described by the following general equation:

$$W = k \cdot A_v \cdot V \cdot [C_S(T) - C_\infty] \tag{11.23}$$

where:

- k is the transfer coefficient.
- A_V is the surface to volume ratio.
- C_S is the gas concentration at the particle surface.
- C_∞ is the gas concentration in the gas bulk.

For a given mass, A_v significantly increases by reducing the particle size to a certain limit, below which no further rate increase is registered, due to fact that kinetics becomes the rate-limit step.

Humidity

Humidity is a very important factor to control the explosion risk.

a. Water content is a significant *absorbing reservoir* of ignition energy and combustion energy, through heating and evaporation.
b. Water vapour is mixed with pyrolised gas, reducing the mixture reactivity.
c. Water causes particle aggregation.
d. Dusts' electrical conductivity is increased, significantly reducing charge build-up and static electricity occurrence.

Practical data are the following:

- According to Field (1982), 15% is the upper limit for dust to undergo ignition.
- For practical calculations, a lower limit of 30% should be adopted.

Turbulence

Turbulence affects:

- The onset of the primary explosion.
- The occurrence of the secondary disruptive explosion.
- The delay between dispersion and ignition which, in turn, affects the explosion strength.

Oxygen contents

Like for gases, oxygen contents affect the occurrence and the strength of dust explosions. LOC data have been tabulated by standards (e.g., NFPA 69).

Initial cloud temperature

As temperature increases, MEC and MIE decrease. For the latter, Glarner (1984) has found a common point of convergence of 0.088 mJ at 1000°C. This suggests that, if a MIE is known at a given temperature, all others can be found by interpolation.

Initial cloud pressure

Effects of initial cloud pressure can be theoretically predicted considering the nature of P_{max} and dP/dT_{max}. For P_{max}, assuming $n_c \approx n_o$:

$$P_{max} = P_o \cdot \frac{T_c}{T_{amb}} \tag{11.24}$$

The behaviour of this is represented by a line whose slope is related to the adiabatic combustion to ambient temperature ratio (Fig. 11.3):

Behaviour of dP/dT_{max} can be easily predicted as:

$$k_{ST} = \left(\frac{dP(t)}{dt}\right)_{max} \cdot V^{1/3} = \left(\frac{4}{3} \cdot \pi\right)^{1/3} \cdot (P_{max} - P_o) \cdot \frac{P_{max}}{P_o} \cdot S_u \tag{11.25}$$

Replacing the pressure ratio:

$$\left(\frac{dP(t)}{dt}\right)_{max} \cdot V^{1/3} = \left(\frac{4}{3} \cdot \pi\right)^{1/3} \cdot P_o \cdot \left(\frac{T_c}{T_{amb}} - 1\right) \cdot \frac{T_c}{T_{amb}} \cdot S_u \tag{11.26}$$

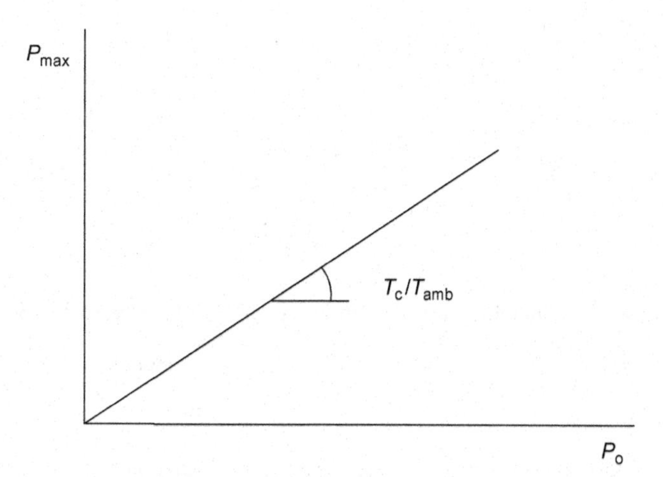

Fig. 11.3
Effect of initial cloud pressure.

A linear growth of the Maximum Pressure Rise Rate (MPRR) with initial pressure is expected, which is experimentally verified.

Glow temperature

This is a very important parameter which applies to the dust layers. Its value is lower than the MIT. It is the temperature at which a dust layer can smoulder, igniting a dust cloud. By definition, T_{glow} is defined as the lowest temperature at which a dust layer 5 mm thick is ignited. Glow temperature is a reference parameter to test ATEX rated equipment for use in presence of combustible dusts (EN 61241).

Hybrid mixtures

Explosive mixtures are called hybrid if they contain both gases and dusts. The final effect is a significant reduction of MIE and MEC. Examples of hybrid mixtures are mixtures of methane and coal dust, in air. A particular case of hybrid mixture is the one where the added component is an inert material. This will result in the reduction of the tendency to explode by the original mixture. An example provided by Eckhoff has been included in Table 11.5.

Electrical resistivity

Electrical resistivity is related to the capacity of dusts to build up and release static electrical charges, which can easily ignite flammable clouds. Furthermore, its value affects selection and certification of materials and equipment for use in the presence of combustible dusts (ATEX directives). Typically, industrial dusts are weak electrical conductors, so static electricity charges are easily accumulated across the various processes: grinding, mixing, micronising, filtration, drying, and pneumatic transport. In Chapter 9 static electricity has been discussed. Relaxation time, that is, time required for a charge to dissipate 37% of its initial charge, is a general parameter to predict the behaviour of a dust in this respect (ref. Chapter 9). Dusts with a relaxation time τ greater than 1 s, or with a resistivity greater than 10^{10} Ω m, are to be considered as hazardous. In Table 11.6, relaxation time and resistivity of some dusts have been included:

Table 11.5 Inerting of sugar dust by $NAHCO_3$ (Eckhoff 2003)

Explosive Dust		Inert Dust		
Material	Diameter (μm)	Material	Diameter (μm)	Minimum Inert Mass to Prevent Ignition (%)
Sugar	30	$NaHCO_3$	35	50

Table 11.6 Resistivity and relaxation time for some dusts

Dust	Resistivity, Ω m	τ, s
Iron	10^4	10^{-7}
Aluminium	10^7	10^{-4}
Corn	$10^7 \div 10^9$	10^{-4}–10^{-2}
Sugar	10^{11}	1
Milk	$10^{11} \div 10^{13}$	1–100
Nylon	10^{13}	2–200
Polyethylene	$>10^5$	$>10^4$

If the electrical resistivity is known, a *rest time* can be calculated to make sure that most parts of the charge has been released ($t \geq 5 \times \tau$), with:

$$\tau = \varepsilon_0 \times \varepsilon \times \rho \tag{11.27}$$

where:

- $\varepsilon_0 = 8.85 \times 10^{-12}$ F/m
- ε = dielectric constant
- ρ = resistivity

The standard 61241-2-2

The standard BS EN 61241-2-2 (*Electrical Apparatus for Use in the Presence of Combustible Dust Part 2: Test Methods Section 2: Method for Determining the Electrical Resistivity of Dust in Layers*) states that a dust is conductive if its resistivity is smaller than 10^3 Ω m. Values greater than 10^8 Ω m identify a very high tendency to charge build-up.

NFPA 77

Powders are divided into the following three groups:

- Low-resistivity powders having volume resistivities of up to 10^8 Ω m, including metals, coal dust, and carbon black
- Medium-resistivity powders having volume resistivities between 10^8 Ω m and 10^{10} Ω m, including many organic powders and agricultural products
- High-resistivity powders having volume resistivities above 10^{10} Ω m, including organic powders, synthetic polymers, and quartz

Ignition sources

The full list of all possible ignition sources is provided by BS EN 1127-1 and has been described in detail in Chapter 10. In the following list the likely ignition sources have been highlighted in red, the unlikely in green. It has to be considered that static electricity is by far the most frequent ignition sources for dusts (Table 11.7).

Table 11.7 Ignition sources and their effectiveness likelihood for dusts

Ignition Source	Likelihood
Hot surfaces	Likely
Flames and hot gases	Likely
Mechanically generated sparks	Likely
Electrical apparatus	Very likely
Stray electric currents, cathodic corrosion protection	Likely
Static electricity	Very likely
Lightning	Likely
Radio frequency electromagnetic waves (10^4 Hz to $3 \cdot 10^{11}$ Hz)	Unlikely
Radio frequency electromagnetic waves ($3 \cdot 10^{11}$ 10^4 Hz to $3 \cdot 10^{15}$ Hz)	Unlikely
Ionizing radiations	Unlikely
Ultrasonics	Unlikely
Adiabatic compression and shock waves	Weakly likely
Exothermic reactions, including self-ignition of dusts	Likely

11.3 Explosion Inherent Hazard

See Fig. 11.4.

11.4 Dusts Explosivity Data

Explosive data for organic dusts have been collected from several sources (Mannan, 2005; VDI, 2002; Eckhoff, 2003), and included in Table 11.8.

11.5 Venting Devices

The function and the operating principles of venting panels have been discussed in Chapter 10. Calculation models depend on aspect ratio. Notably:

- Systems with aspect ratio (AR) $L/D = 1 \div 20 \rightarrow$ calculation codes
- Aspect ratio (AR) > 20: Semiempirical methods

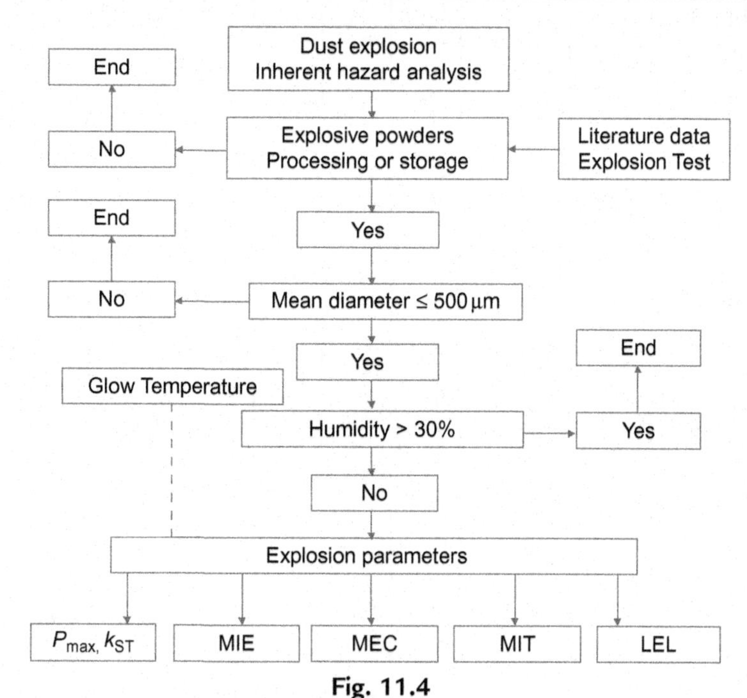

Fig. 11.4
Flow chart—explosion hazard analysis of dusts.

11.5.1 Venting Devices for Items With AR 1 → 20

The calculation can be carried out according to validated standards (Tables 11.9 and 11.10):

- NFPA 68—*Standard on explosion protection by deflagration venting*
- EN 14491—*Dust explosion venting protective systems*
- *VDI 3673—Pressure venting of dust explosions*

11.6 Equipment an Plants for Dust Processing

The equipment configuration of dust processing, in the light of the explosion risk assessment, can ideally be classified into two main categories:

- Vessel-type equipment
- Pipe-like equipment

11.6.1 Vessel-Type Equipment

Equipment and structures showing an aspect ratio (length to diameter ratio) not very far from 1:

- Tanks
- Short silos

Table 11.8 Dusts explosivity data

Dust	Particle Distribution Weight % < Size (μm)							Median (μm)	Dust Cloud Explosion Tendency						Dust Layers		
									1 m^3 or 20 L Vessel			Mod. H.			VDI	DIN	
	500	250	125	71	63	32	20		C_{min} (g/m^3)	P_{max} (bar(g))	k_{ST} (bar m/s)	Explos. <63 μm (Class)	T_{min} (°C) G.G.	T_{min} (°C) BAM	MIE (mJ)	Glow Temp. (°C)	Flam. <250 μm (Class)
Cotton, wool, peatland																	
Cotton			98	72		38	25	44	(100)	7.2	24	St. 1	560			350	3
Cellulose			92	71		20	3	51	60	9.3	66		500		250	380	5
Wood dust				90		47	7	33					500		100	320	
Wood dust	58		57	55		43	39	80					480		7	310	
Wood dust (chipboard)				70		30		43	60	9.2	102	St. 2	490			320	3
Peat (15% moisture)			84	58		26	3	58	60	10.9	157	St. 1	480			320	4
Peat (22% moisture)			82	65		40	15	46	125	8.4	69	St. 1	470			320	
Peat (31% moisture)			87	76		43	20	38	125	8.1	64		500			320	
Peat (41% moisture)			88	76		40	18	39	No Ignition				500			315	
Peat (from bottom of sieve)			78	48		22		74	125	8.3	51	St. 1	490			310	4
Peat (dust deposit)				66		33	11	49	60	9.5	144		360			295	
Food dusts																	
Gravy powder (21% starch)					100					5.1	12			500	>1000		
Citrus pellets					100				60	7.7	39			460	250		

Dextrose, ground			100		94	71		22				St. 2					2
Dextrose				38		5	4	80	60	4.3	18	St. 1	500			570	3
Fat/whey mixture	76		11	3				330		7.0	23		450		180	410	5
Fat powder (48% fat)		100	75		24	7		92	30	6.4	20	St. 2					2
Dough					100									430	>100		
Fishmeal	68		23		12			320	125	7.0	35		530				
Fructose (from filter)	99		39	17				150	60	9.0	102		430		<1	Melts	
Fructose	92		15					200	60	7.0	28		440		180	440q	
Fructose	81							400	125	6.4	27		530		>4000	Melts	
Barley grain dust	79	51	25		8	3		240							100		4
Dough					100				125	7.7	83			400			
Oats grain dust	64		24		8			295	750	6.0	14	St. 1				350	3
Wheat grain dust				48		30		80	60	9.3	112	St. 1				290	3
Wheat grain dust	100	81	50		322	25		125				St. 2					3
Coffee (from filter)				100		99	89	<10	60	9.0	90		470			>450	
Coffee (refined)					100					6.8	11			460	>500		4
Cocoa bean shell dust					100				125	8.1	68				>250		4
Cocoa/sugar mixture	53		20					500	125	7.4	43	St. 1	580			460	2
Potato granulate					100					6.4	21			440	>250		3
Potato flour			86	53		26	17	65	125	9.1	69		480			>450	
Lactose (from filter)				83		60	47	22	125	6.9	29		450		80	>450	

Continued

Table 11.8 Dusts explosivity data—cont'd

Dust	Particle Distribution Weight % < Size (μm)							Median (μm)	Dust Cloud Explosion Tendency						Dust Layers		
									1 m^3 or 20 L Vessel			Mod. H.			VDI	DIN	
	500	250	125	71	63	32	20		C_{min} (g/m^3)	P_{max} (bar(g))	k_{ST} (bar m/s)	Explos. <63 μm (Class)	T_{min} (°C) G.G.	T_{min} (°C) BAM	MIE (mJ)	Glow Temp. (°C)	Flam. <250 μm (Class)
Lactose (from cyclone)				97		70	41	23	60	7.7	81	St. 2	520			>450	3
Maize seed waste (9% moisture)	68	97	40		23	16		165	30	8.7	117			440	>10		
Milk powder			34	18				165	60	8.1	90		460		75	330	
Milk powder	98		15	8				235	60	8.2	75		450		80	320	
Milk powder (low fat, spray dried)	100	100	99		60	17		46	30	7.5	109				>100		
Milk powder (low fat, spray dried)				30				88	60	8.6	83	St. 1	520			330	2
Whey fat emulgator	62		7	2				400		7.2	38		450		90	420	5
Olive pellets					100				125	10.4	74			470	>1000		
Rice flour					100				60	7.4	57			360	>100		
Rye flour			94	76		58	15	29		8.9	79		490			>450	
Soy bean flour				85		63	50	20	(200)	9.2	110	St. 1	620			280	2
Potato starch					100				30	7.8	43			420	>1000		
Potato starch				100		50	17	32		(9.4)	(89)		520		>3200	>450	2
Maize starch				99		98	94	<10		10.2	128		520		300	>450	2
Maize starch				94		81	60	16	60	9.7	158	St. 1	520			440	2

Rice starch (hydrolyzed)				29		15		120	60	9.3	190	(St. 2)	480			555	5
Rice starch				99		74	54	18					470		90	390	3
Rice starch				86		62	52	18		10.0	190	(St. 2)	530			420	3
Wheat starch						84	50	20	60	9.8	132	(St. 2)	500			535	3
Tobacco			81	64		29		49		4.8	12		470			280	
Tapioca pellets				61		42		44	125	9.0	43	St. 1	(450)			290	4
Tea (6% moisture)					100				30	8.1	68			510			≥3
Tea (black, from dust collector)			64	48		26	16	76	125	8.2	59	St. 1	510			300	4
Meat flour			69	52		31	21	62	60	8.5	106	St. 1	540			>450	2
Wheat flour								50					500		540	>450	
Wheat flour			97	60		32	25	57	60	8.3	87		430			>450	
Dough					100									400	>100		
Wheat flour 550				60		34	25	56	60	7.4	42		470		400	>450	
Milk sugar				99		92	77	10	60	8.3	75		440		14	Melts	5
Milk sugar				98		64	32	27	60	8.3	82	St. 1	490			460	2
Sugar (icing)				88		70	52	19					470			>450	
Other natural products																	
Cotton seed expellers	66		24	10			245	125	7.7	35	St. 1	(480)				350	3
Blood flour			93	61		27	5	54	60	9.4	85		610			>450	1
Hops, malted	52		14	9				490		8.2	90		420			270	
Linen (containing oil)	63		21					300		6.0	17		(440)			230	

Continued

Table 11.8 Dusts explosivity data—cont'd

Dust	Particle Distribution Weight % < Size (μm)							Median (μm)	Dust Cloud Explosion Tendency: 1 m³ or 20 L Vessel			Mod. H.			Dust Layers: VDI	DIN	
	500	250	125	71	63	32	20		C_{min} (g/m³)	P_{max} (bar(g))	k_{ST} (bar m/s)	Explos. <63 μm (Class)	T_{min} (°C) G.G.	T_{min} (°C) BAM	MIE (mJ)	Glow Temp. (°C)	Flam. <250 μm (Class)
Lycopodium					100	91								410		280	
Grass dust	96		26					200	125	8.0	47		470			310	
Walnut shell powder									(100)			St. 1					4
Pharmaceuticals, cosmetics, pesticides																	
Acetyl salicylic acid					100				15	7.9	217	(St. 2)		550			2(5)
Ascorbic acid, L(+)-				93		75	61	14	60	6.6	48	(St. 2)	490			Melts	2(2)
Caffeine					100				30	8.2	165	(St. 2)		>550		Melts	2(5)
Intermediate and auxiliary products																	
Casein				99		65	40	24	30	8.5	115		560			>450	
Methyl cellulose				96		87	30	22		10.0	157		400		12	380	
Methyl cellulose				100		69	10	29	60	10.0	152		400		105	>450	5
Methyl cellulose				93		37	12	37	30	10.1	209		410		29	450	5
Ethyl cellulose				66		40		40		8.1	162		(330)			275	
D (−)-Mannite				61		24	13	67	60	7.6	54	St. 1	460			Melts	2
Pectin			86	61		21		59	60	9.5	162		460			300	
Salicylic acid									(30)			(St. 2)					2(5)
Tartaric acid	100	5	1					480				St. 1					2

Table 11.9 Venting panel calculation parameters

Code	Venting Panel Calculation Parameters	Notes
NFPA 68	P_{max}, P_{stat}, P_{red}, k_{ST}, V, L, D	L/D < 8
EN 14491	P_{max}, P_{stat}, P_{red}, k_{ST}, V, L, D	L/D < 20
VDI 3673	P_{max}, P_{stat}, P_{red}, k_{ST}, V, L, D	L/D < 20

Table 11.10 Comparison between EN 14491 and NFPA 68

Parameters	EN 14491	NFPA 68
Vessel volume, m^3	$0.1 < V < 10{,}000$	$0.1 < V < 10{,}000$
P_{stat} (bar)	$0.1 < P_{stat} < 0.75$	$P_{stat} < 0.75$
P_{red} (bar)	$P_{stat} < P_{red} < 2$ Recommended $P_{red} > 0.12$	L/D < 20
P_{max} (bar)	$5 < P_{max} < 10$ for $10 < k_{ST} < 300$ bar m s^{-1} $5 < P_{max} < 12$ for $300 < k_{ST} < 800$ bar m s^{-1}	$5 < P_{max} < 12$ for $10 < k_{ST} < 800$ bar m s^{-1}
Aspect ratio (L/D)	1 < L/D < 20	L/D < 8
Turbulence	-	All air velocity components lower than 20 m/s
Atmospheric conditions	80 kPa < Atmospheric pressure < 110 kPa −20°C < temperature < 60°C 5% < RH < 85% Oxygen = 20.9 + 0.2%	Initial pressure = 1 bar absolute + 0.2 bar
Further conditions	No venting ducts Venting effectiveness factor = 1	No venting ducts Panel inertia < 40 kg/m^2

- Spray dryers
- Cyclones
- Bag filters
- Dust collectors
- Batch reactors
- Batch mixers
- Circumferential vibrating feeders
- Hammer mills

11.6.2 Pipe-Type Equipment

- Tall silos
- Pneumatic transport ducts
- Screw conveyors
- Closed belt conveyors

- Plug mixers
- Axial vibrating feeders
- Closed bucket elevators

11.7 HAZID of Equipment and Process Scenarios

In this section a general hazard identification for the most frequent dust processing equipment is presented.

11.7.1 Pneumatic Transport

Pneumatic transport of solids is a high risk context. Should powders have appropriate size and characteristics, the following ignition sources and causes have to be accounted for, along with the associated risk (Tables 11.11 and 11.12).

11.7.2 Bucket Elevator

Most of the risk scenarios illustrated for the pneumatic transport also apply to the bucket elevator. In Table 11.13, just the specific hazards are presented.

11.7.3 Screw Conveyor

Most of the risk scenarios illustrated for the previous equipment also apply to the screw conveyor. Internal explosion results in the axial propagation of the flame. According to the NFPA 654, a baffle plate installed at regular spans would choke the explosion (Figs 11.5 and 11.6).

11.7.4 Silos and Dust Collectors

These enclosures present most of the hazards identified for the previous equipment. In addition, these containments:

Table 11.11 Ignition causes and sources (Cenelec, 2003)

Ignition Cause	Ignition Source
Friction and contact between particles and duct	Static electricity
High temperature due to grinding operations	Temperature
Presence of metals	Mechanical sparks
Potential difference between insulated masses and conductive bodies	Static electricity
	Stray currents
Autoignition due to micro-organism or thermal instability (Frank Kamenetskii)	Chemical reaction

Table 11.12 Explosion HAZID for pneumatic transport

Explosion HAZID for Pneumatic Transport		
Ignition Source	**Cause**	**Prevention Measure**
Any	Dust concentration C_D within the explosivity range	As far as possible UEL $< C_D <$ LEL even if UEL is not always defined
Any	Airborne dust at indoor plants	Pressing instead than sucking fans
Static electricity	Unearthed equipment and structural parts	Earthing and grounding according to Cenelec 50404/NFPA 77
Static electricity	Lack or defective equipotential bonding	Earthing and grounding according to Cenelec 50404/NFPA 77
Static electricity	Ratio of the Charge to Mass Ratio (CMR) to the mass flow resulting in an inadequate steaming current	Optimisation of these parameters Adoption of antistatic materials or with discharge voltage (electric breakdown) $<$ 4 kV as per EN 13463
Mechanical sparks Static electricity	Fire due to friction in narrow radius bends	Avoidance of narrow bends
Mechanical sparks Static electricity	Deposit of solid material promoting fire and explosion	Adequate speed (15–20 min/s) Provision of openings to facilitate frequent housekeeping Dust supply only after full air flow feeding Air supply stop only after dust removal
Electrical equipment Mechanical equipment	Thermal and electrical ignition	Adoption of ATEX rated equipment
Mechanical sparks	Presence of foreign materials	Adoption of filtration/separation of foreign materials
Mechanical sparks	Improper materials such as rust, magnesium, titanium, zirconium, and inappropriate alloys	Compliance with EN 13463
Autoignition/Chemical reactions	High temperature or thermally instable materials	Semenov/Frank Kamenetskii analysis Loading procedure

- can be subject to buckling collapse (Chapter 5) with consequent fire and explosion. Adequate breathing valves are to be provided to maintain the atmospheric pressure.
- can be ignited by mechanical sparks caused by falling structural elements. Regular maintenance can prevent this hazard.
- could inadvertently be loaded with burning or ignited materials. Prevention measures are:

1. Provision of CO detectors
2. Inertisation
3. Free falling minimisation

Table 11.13 Explosion HAZID for bucket elevator

Explosion HAZID for Bucket Elevator		
Ignition Source	**Cause**	**Prevention Measure**
Static electricity Friction between contact parts High temperature	Choked leg jogging	Choked leg jogging prevention procedure
Static electricity Friction between contact parts High temperature	Damaged bearings Bearings overheating due to misalignment Chains/pulleys slipping Friction on side inspection doors Dust compression in the space gaps	Antislippery devices Temperature detectors Bearings check Bearings low working rating Lube oil check Gaps minimisation
Adiabatic compression Shock wave	Pressure piling from silos or closed belt conveyors	Venting panels design according to NFPA 68/VDI 3673/EN 14491
Any	Dust spreading through the elevator casing potentially resulting in a primary explosion	Dust-tight casing, head and tail as per NFPA 61
Any	Missiles from inspection doors fragment	Inspection doors as resistant as casing

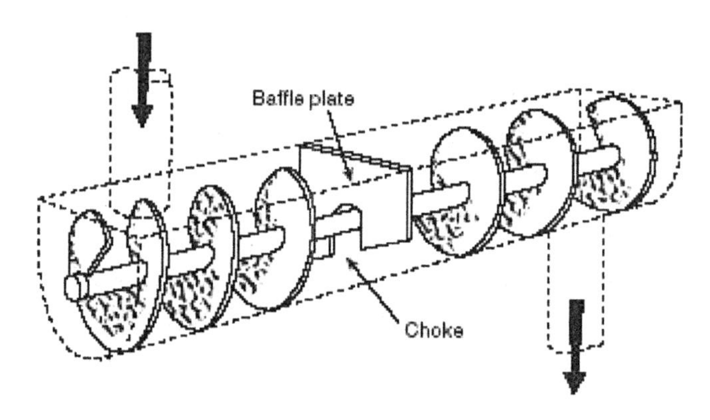

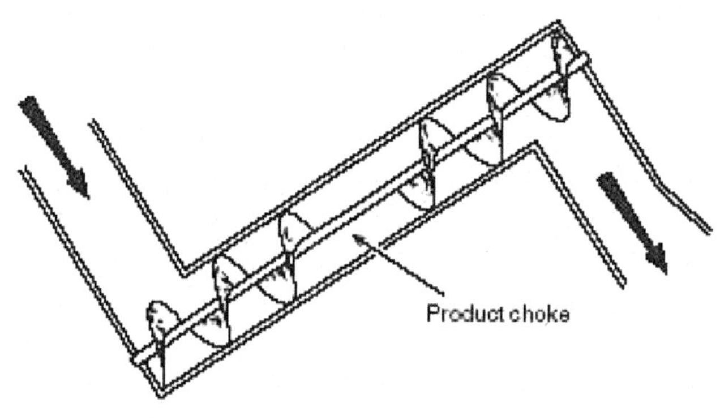

Fig. 11.5
Screw conveyor–Baffle plate for explosion choking (NFPA 654).

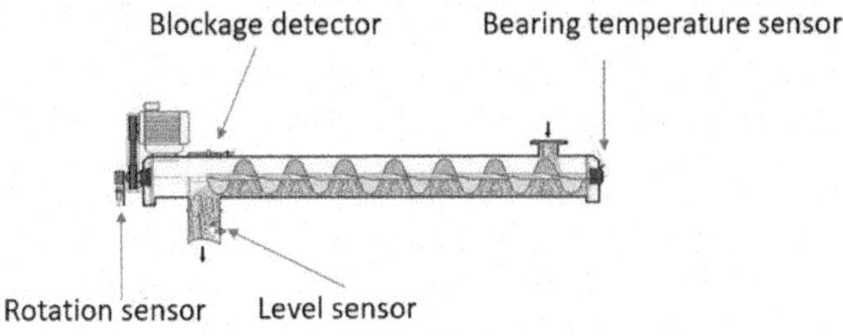

Fig. 11.6
Screw conveyor—safeguards for explosion prevention.

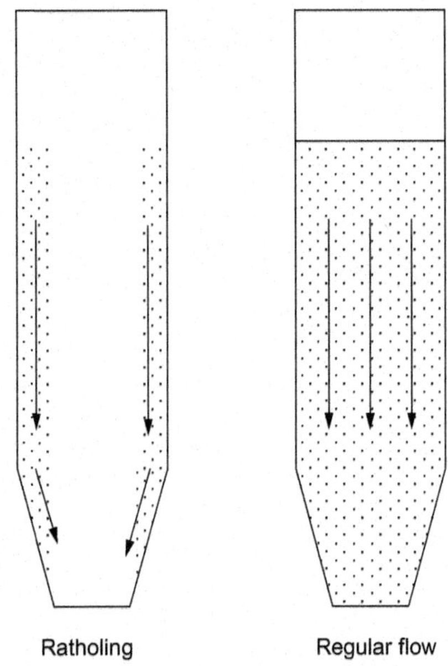

Fig. 11.7
Ratholing and regular flow.

4. Moisturing
5. Prevention of funnel flow/ratholing

Ratholing is the packed accumulation of dust on the silo walls causing a sudden release of the material with a possible electrical charge release and subsequent ignition. Fig. 11.7 illustrate ratholing and the correct flow.

Ratholing can be avoided by preventing funnel flow and by implementing an adequate silo diameter to discharge hole diameter ratio.

Inertisation

Fig. 11.8 shows the typical instrumented system for an inertised silo.

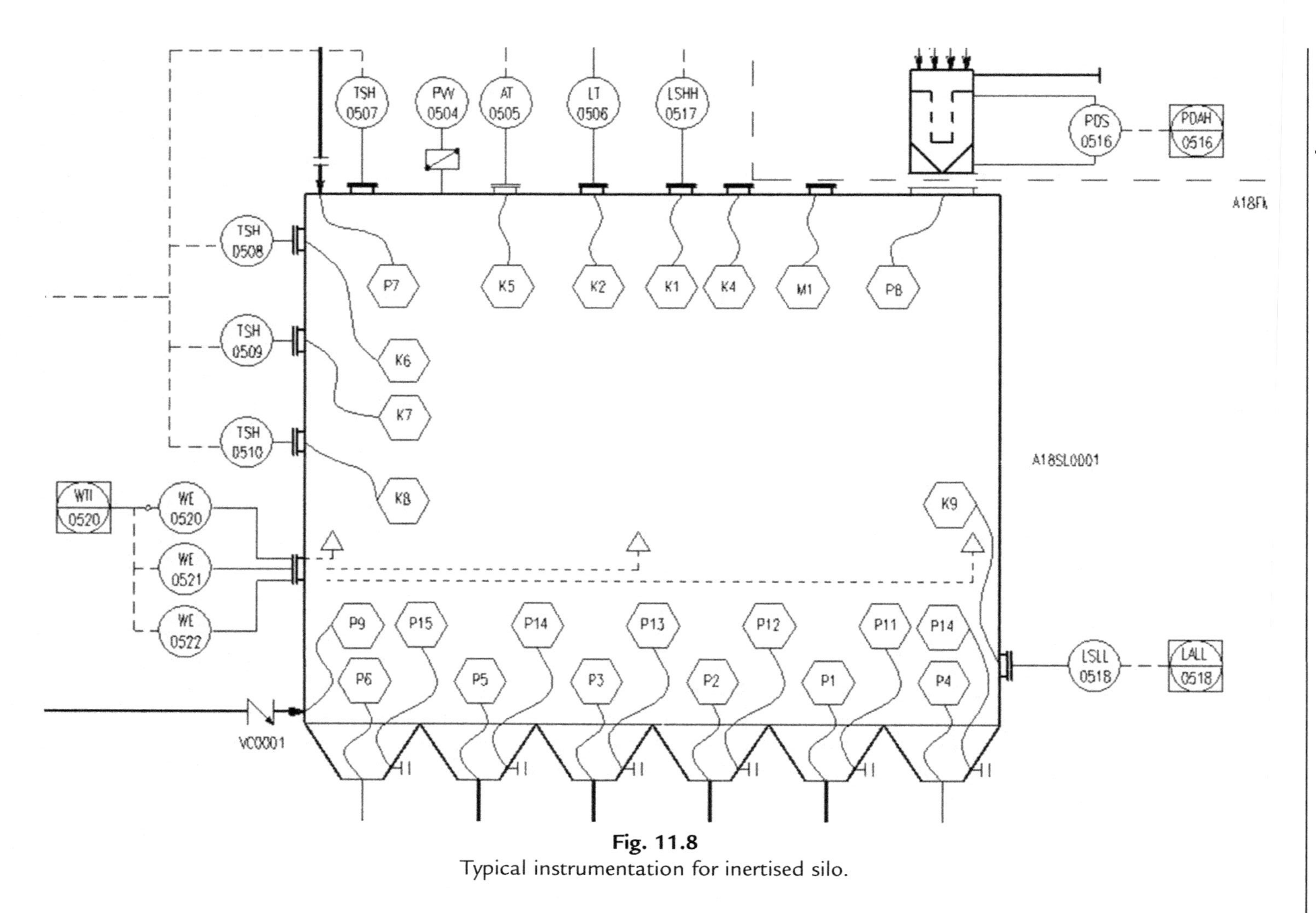

Fig. 11.8
Typical instrumentation for inertised silo.

Table 11.14 Cause and effect analysis of explosion in spray driers (UK 1972 1986)

Dust	Cause	Effect
Milk	The spinner of the motor shaft was ejected inside the dryer, causing explosive sparks.	The dryer was protected by venting panels. The downstream section was damaged.
Milk	Damaged o-rings allowed dust transfer, and the accumulation which underwent autoignition, triggering an explosion.	The dryer was protected by venting panels. Fire
Milk	Friction caused by contact between a plate and the top of the disk resulted in melted steel drops which ignited the dust.	The dryer was protected by venting panels. Fire Loss of the tightness at the dryer base. Flames vented through venting panels damaged the building roof.
Cheese	Ignition of dust coating on air heater pipe.	Deformation of steel fluid bed. Partial burning of steel belt.
Milk	Smouldering dust in the fine recycle pipe due to a three way valve failure	Intervention of rupture disks. Other plastic parts ruptured supporting the relief.
Milk	Ignition of dust coating around the fines return pipe.	Explosion
Milk	Excessive oil amount at the atomiser filled the dust accumulated on the vault, which caught on fire.	Fire
Milk	Dust autoignition due to poor housekeeping	Panels burst and damage of two cyclones
Coconut milk	Dust accumulation at the fluidised bed until an obstruction occurred. After removal, ignited mass entered the bed and ignited the mass.	Damage to the bed and to the ducts. Flames crossed panels and reached walls, windows and ceilings
Milk	Dust releases from the dryer bottom ignited by hot air. Primary explosion uplifted a steel plate which ignited the dust and caused secondary explosion.	Overturning of dryer roof. Relief of the explosion inside the building

11.7.5 Spray Drier

The spray dryer is a complex very common equipment used in the food, pharmaceutical, and chemical industry. Most of the hazards analysed in the previous sections are identically applicable. In Table 11.14, there is a summary of case history data relating to incidents and near misses that occurred in the United Kingdom from 1972 to 1986.

Spray dryer: lesson learnt

1. Dust autoignition is very frequent, and accumulation is likely due to the dynamic nature of the dryer. Glow temperature of 5 mm thick milk layer is 330°C, but it is lower as thickness decreases.

2. High speed moving parts are present, so friction and impact have to be considered both in the design and in the operation.
3. Explosions have occurred due to static electricity mainly in the conical zone when humidity has been removed.
4. A typical fire and explosion event is triggered by autoignition of dust layers or by mechanical friction/impacts. The fall of burning materials towards the conical section provokes the explosion.

11.8 Quantified Risk Assessment

Example 11.5

Find the risk of explosion of a system consisting of a bucket elevator which delivers dust material into silos by adopting a Fault Tree Analysis. An ignition temperature of 165°C has been assumed, along with a self ignition temperature.

Solution

The calculation has been performed considering, for the sake of simplicity, just probabilities. A rigorous approach would require conversion of probabilities into frequency assuming a Poisson distribution (Chapter 6). For the purpose of this example, the findings of the calculations have been assumed as acceptable (Fig. 11.9).

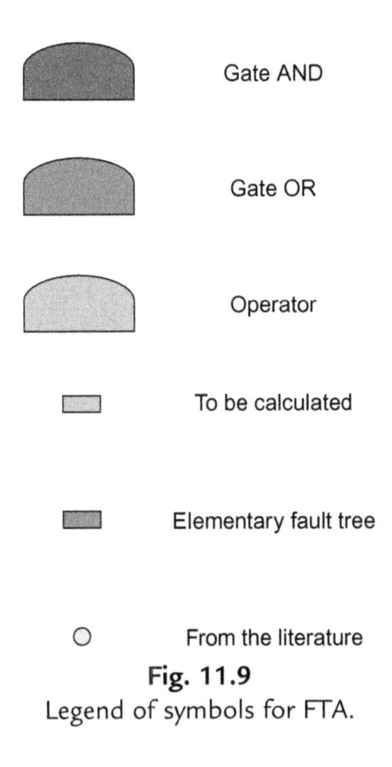

Fig. 11.9
Legend of symbols for FTA.

Bucket Elevator: Fault Tree Analysis (FTA)

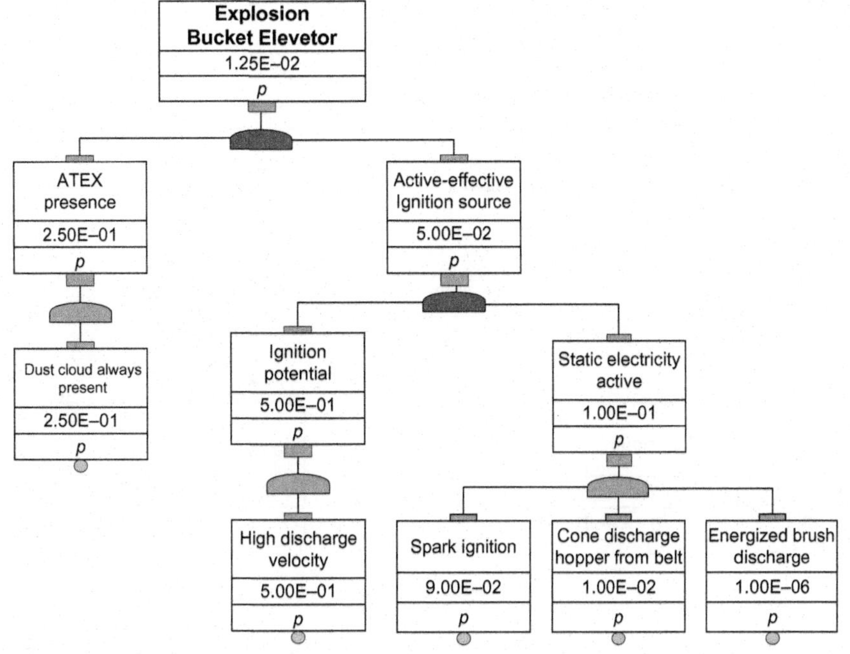

Root cause 1: static electricity

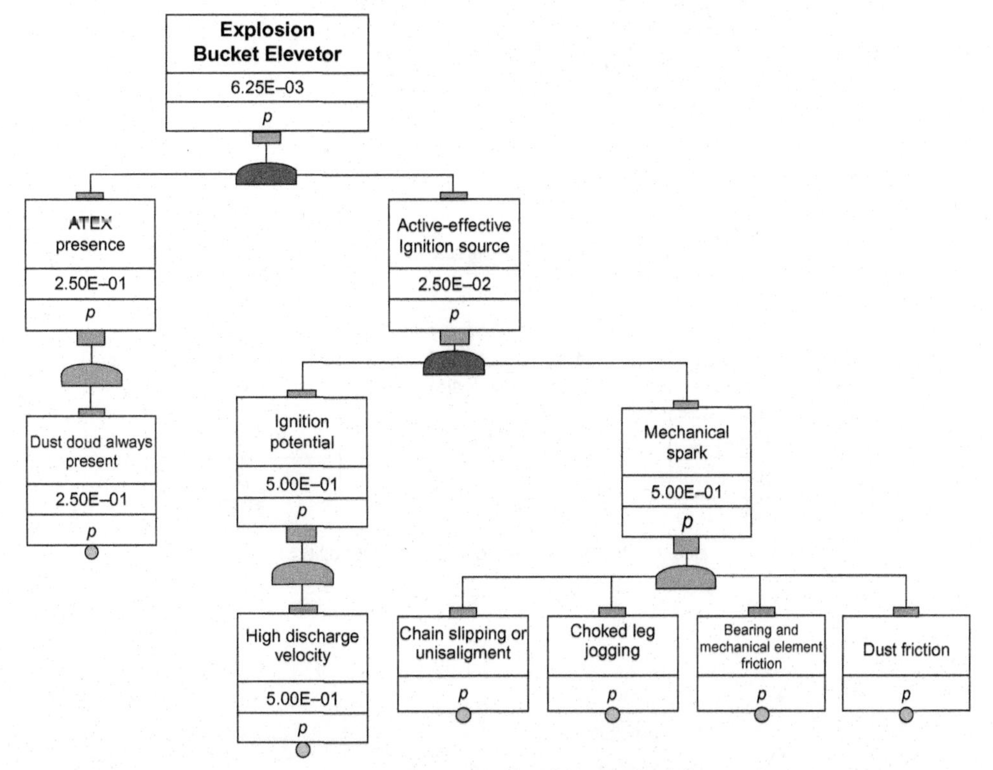

Root cause 2: mechanical spark

Bucket Elevator: Fault Tree Analysis (FTA)

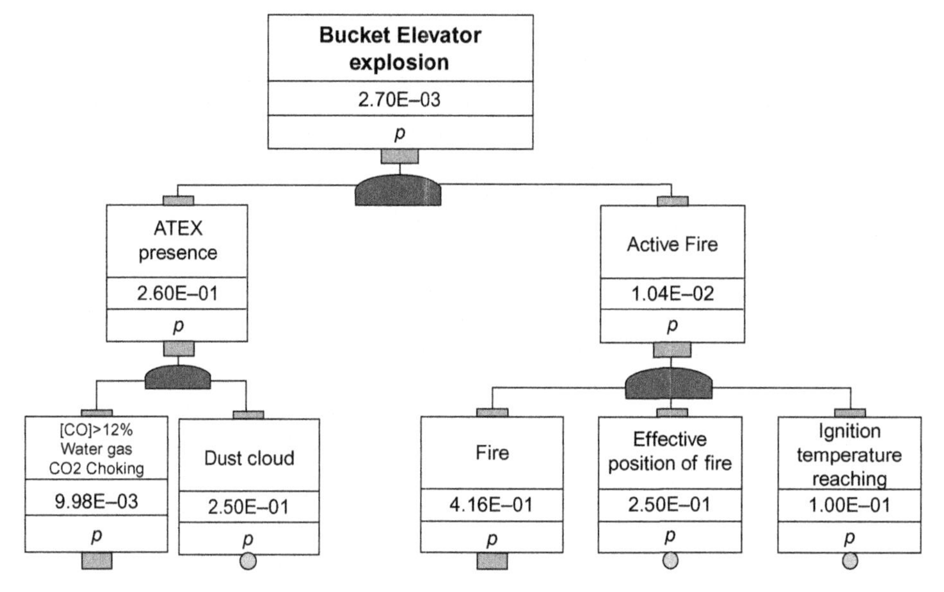

Root cause 3: Fire

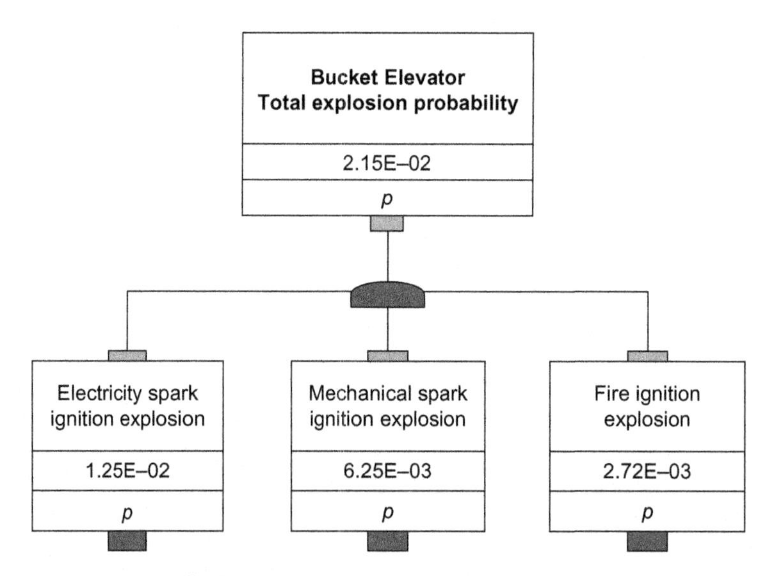

Bucket elevator: overall explosion probability

Final Silo: Fault Tree Analysis (FTA)

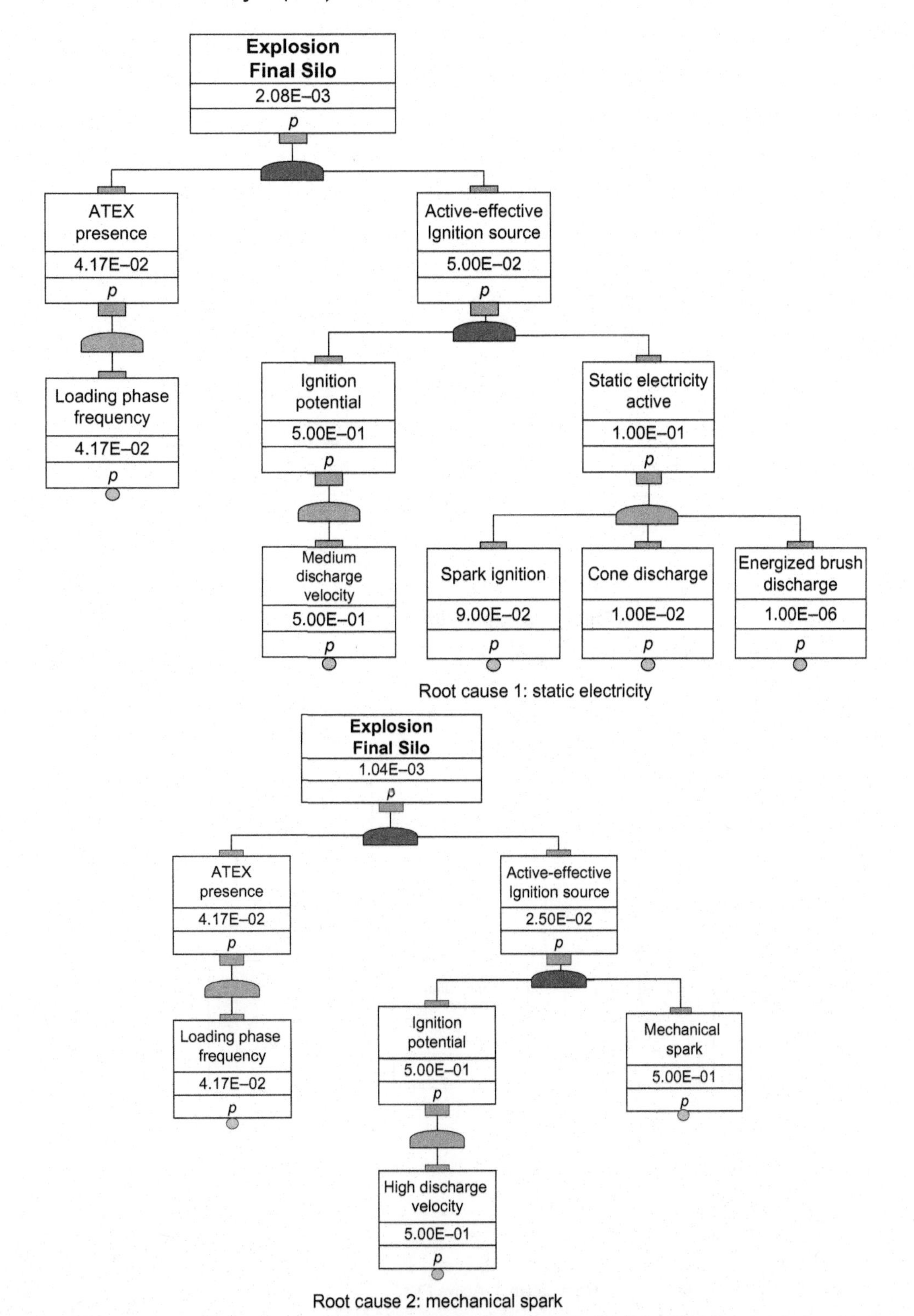

Final Silo: Fault Tree Analysis (FTA)

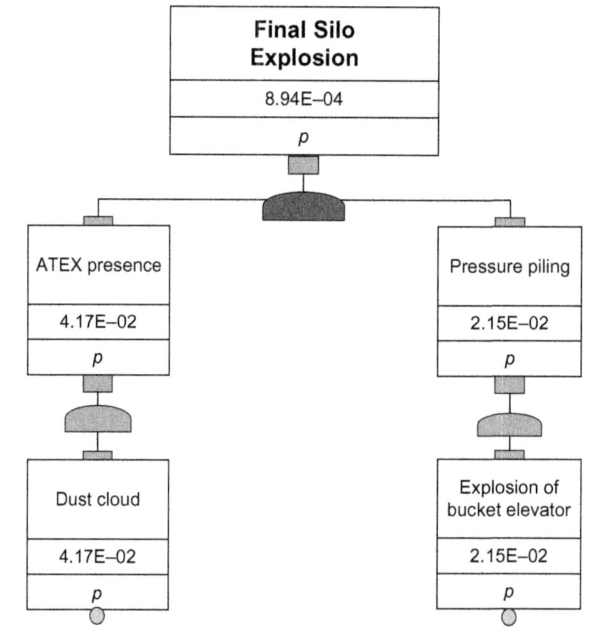

Root cause 3: pressure piling

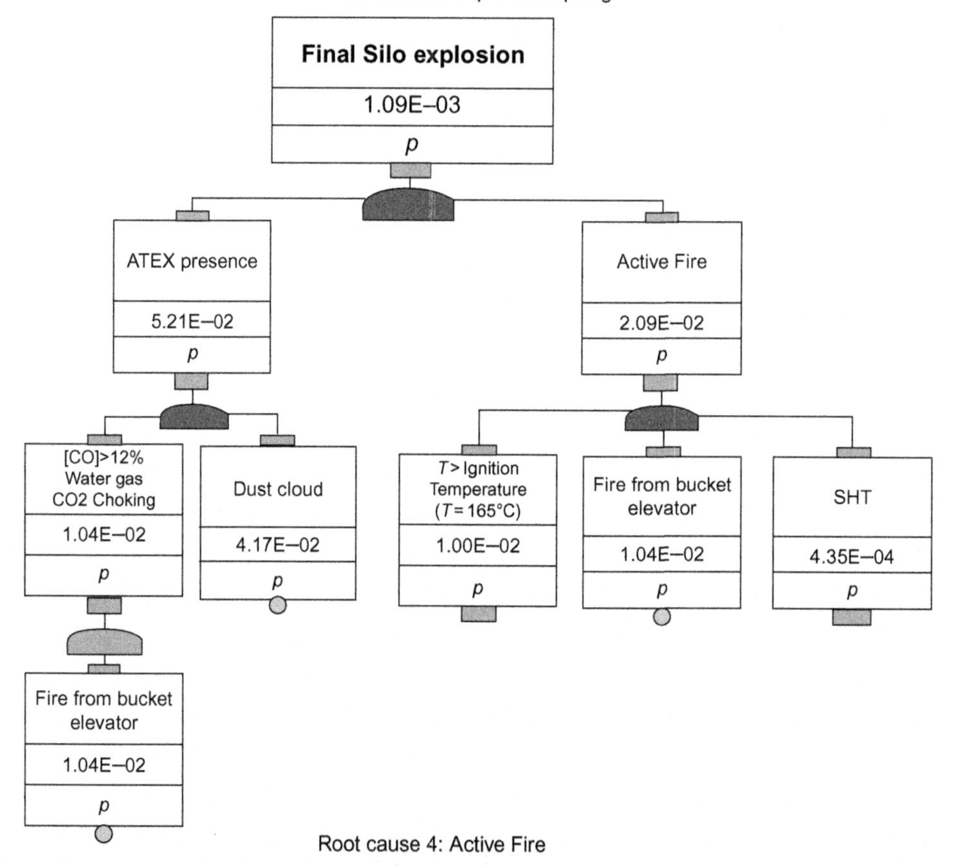

Root cause 4: Active Fire

Final Silo: Fault Tree Analysis (FTA)

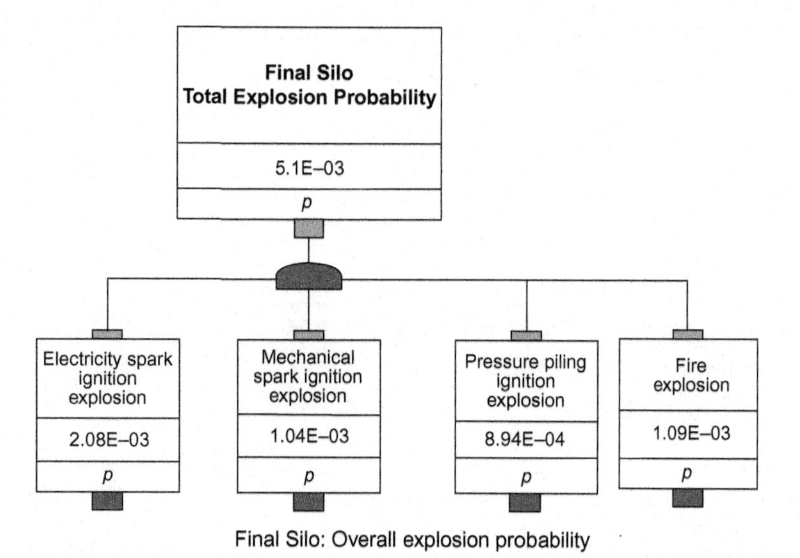

Final Silo: Overall explosion probability

11.9 Risk Assessment in the ATEX Framework

The ATEX directives 100 and 137 require that, as for gases, workplaces where explosive dusts can be present must be classified, and that electrical and non electrical equipment are suitable to be used in hazardous areas and certified accordingly. In addition, an Explosion Protection Document (EPD) has to be prepared, where explosion risks are shown to have been analysed and quantified. This requirement is common for gases and dusts: for the former, a large literature and validated criteria are available, as discussed in Chapter 10. For dusts, which are highly non-ideal systems, a limited number of criteria and guidelines are available. An overview of the equipment explosion risk assessment for starch dust in the ATEX framework has been presented in this section.

11.9.1 Reference Standards

- BS EN 1127-1, 2011 Explosive Atmospheres—Explosion Prevention and Protection. Part 1: Basic Concepts and Methodology
- BS EN 15198:2007 Methodology for the risk assessment of non-electrical equipment and components for intended use in potentially explosive atmospheres

11.9.2 Risk Assessment and ATEX Compliance

Example 11.6

Explosion risk assessment and ATEX conformity check for a vibrating screen.

Solution

Figs 11.10 and 11.11 present the flow chart methodology for the risk assessment collected from the standard BS EN 15198. *Methodology for the risk assessment of non-electrical equipment and components for intended use in potentially explosive atmospheres:*

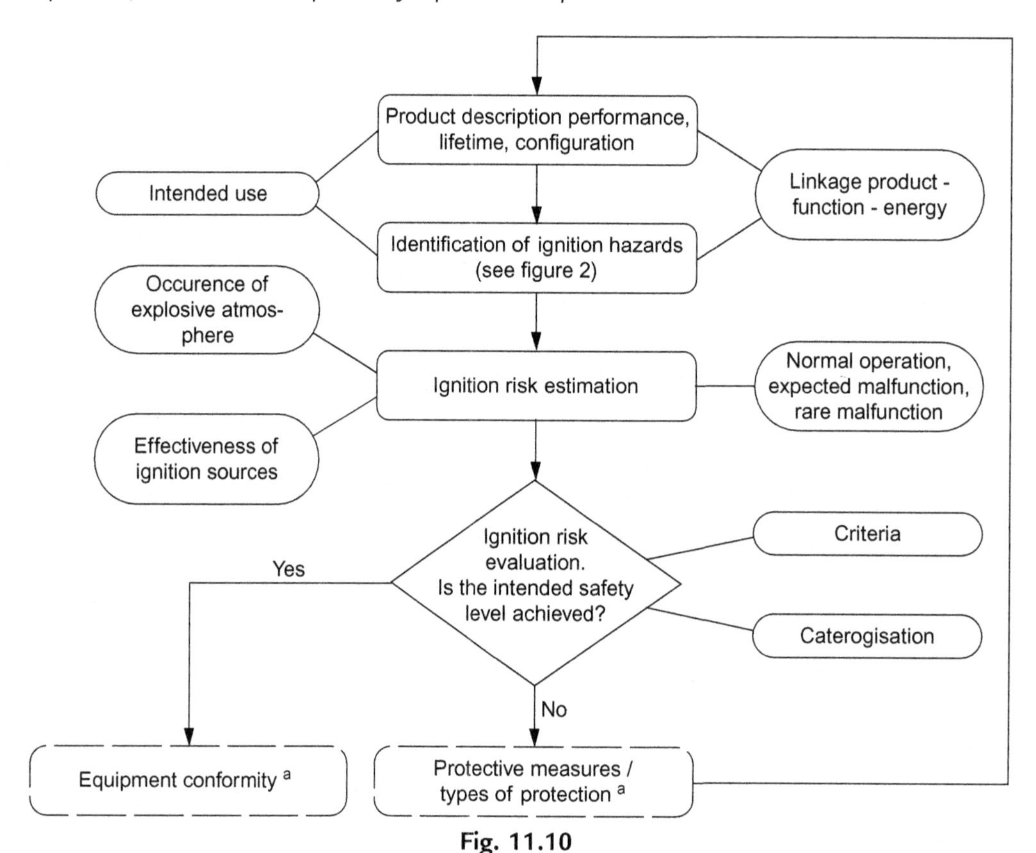

Fig. 11.10
Risk assessment according to BS EN 15198.

Block 1—Zoning

The *zoning* is defined on the basis of frequency at which an explosive atmosphere can be present. The following reference could be considered (Table 11.15) (CEI 31–35, 2014):

Block 2—Likelihood and Occurrence

The real presence of explosive atmosphere in relationship with operating cycles, regeneration, and downtime will be evaluated.

Block 3—Ignition Sources and Energy

Table 11.16 shows ignition sources likelihood and locations.

Block 4—Ignition Hazard Assessment

The ignition hazard assessment can be carried out on the basis of the data included in Table 11.16.

Ignition Risk Assessment
(EN 15198 par. 5.4)

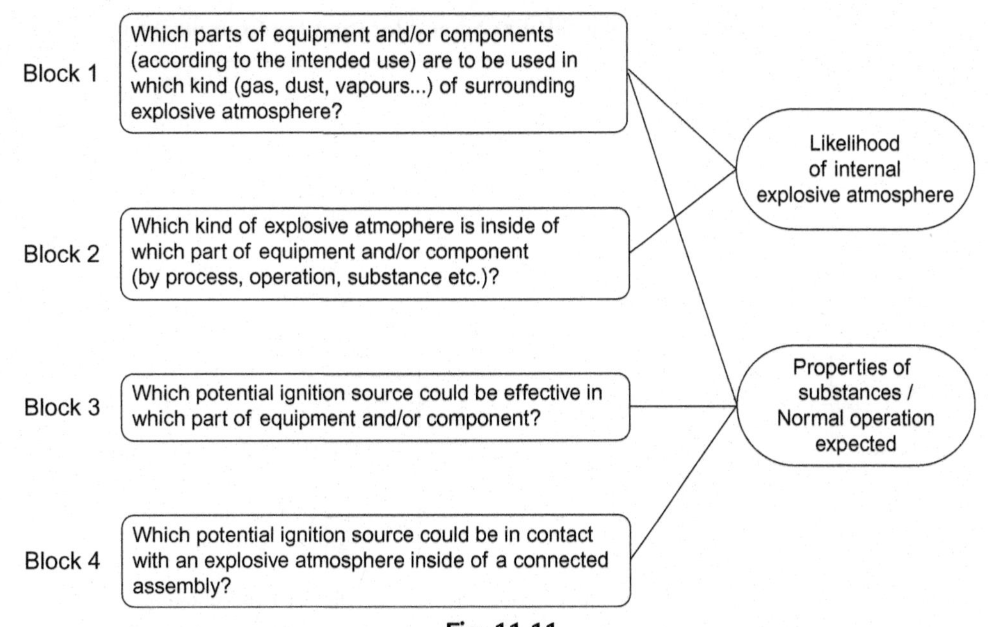

Fig. 11.11
Risk Ignition hazard assessment according to BS EN 15198.

Table 11.15 Criteria for zoning (standard CEI 31–35, 2014)

Zone	Likelihood of Atmosphere Explosive Within 365 Days	Duration Within 365 Days
20	$P>0.1$	More than 1000 hr
21	$0.1>P>0.001$	Between 10 and 1000 h
22	$0.001>P>0.00001$	Between 0.1 and 10 h

The rationale of the ignition risk assessment is based on the following considerations.

a. Bearings
   The roller bearing failure rate is 5 failures/10^6 h (Mannan, Lees, 2005). This failure can be converted into a probability through Poisson's equation:

$$P=1-e^{-\lambda \cdot t} \tag{11.28}$$

b. Belt skidding
   Belt skidding can result in surfaces overheating. The assumed failure rate is 40/10^6 h (Mannan, Lees, 2005).
c. Electric motor
   The electric motor has a failure rate of 10/10^6 h (Mannan, Lees, 2005).
d. Earthing failure/loss of equipotentiality
   This failure is related to the charge build-up and release due to static electricity. Data from IChemE suggest a specific failure rate of 21 failures/10^6 h. The check time is undefined, however the compliance of the whole assembly with the standard BS-EN-13463-5 and the suitability of its use in classified areas identify a negligible ignition risk.

   If the whole assembly complies with the standard BS-EN-13463-5, and is suitable to be used in classified areas, identify a negligible ignition risk.

Table 11.16 Ignition sources likelihood and location for Example 11.6

Ignition Source	Likelihood	Location	Causes
Hot surfaces	Yes in anomalous situations	Bearings	- Defective/lack of lubrication; temperature can exceed 90°C - Damaged bearings - Bearing misalignment
		Belts skidding	- Defective belt tightening - Worn belts - Worn pulleys
Flame and hot gases	No	–	–
Mechanical sparks	Yes in anomalous situations	Bearings	- Defective/lack of lubrication; temperature can exceed 90°C - Damaged bearings - Bearing misalignment
		Belts skidding	- Defective belt tightening - Worn belts - Worn pulleys
Electrical equipment	Yes, in anomalous conditions	Motor Instruments	- Short circuit - Lack of insulation
Stray currents	Yes, in anomalous conditions	Motor	- Lack of insulation - Defective equipotentiality
Static electricity	Yes	Screening	- Defective/lack of earthing
Lightning	No	–	–

All other ignition sources are not applicable.

These data are summarised in Table 11.17.

The vibrating screen presents a generic failure rate potentially, causing an ignition event equal to 76 failures/10^6 h, which corresponds to a yearly failure rate of 0.67 y^{-1}.

Ignition Risk Assessment
(EN 15198 par. 5.5)

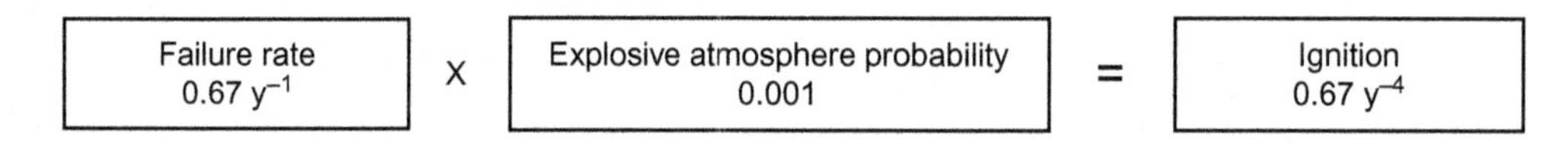

Categorisation
(EN 15198 par. 5.6)

Table 11.17 Ignition risk for Example 11.6

Failure Rate	Failure Scenario	Explosive Atmosphere Probability	Ignition Risk (Failure/Operation Time)
$5 \times 10^6\ h^{-1}$	Bearings wear Lack of lubrication	$P = 10^{-3}$	$P = 10^{-6}$
$40 \times 10^6\ h^{-1}$	Belt skidding	$P = 10^{-3}$	$P = 10^{-5}$
$10 \times 10^6\ h^{-1}$	Motor failure	$P = 10^{-3}$	Low risk due to compliance to construction safety EN 13463-5
$21 \times 10^6\ h^{-1}$	Static electricity Stray currents	$P = 10^{-3}$	Low risk due to compliance to (ohmic) construction safety EN 13463-5

Categorisation and acceptability analysis have been carried out according to the method proposed by Markowski and Mujumdar (2004), based on the TNT equivalence and relating the findings to the explosion constant k_{ST} and to the explosion class, which is St = 1 for starch. A *Consequence Severity Index* C_s can be obtained from Table 11.18.

Table 11.18 Determination of the Consequence Severity Index (Markowski and Mujumdar, 2004)

Dust Hazard Class	Equivalent TNT—Scale of Explosion Index, C_i									
	1–5 kg		5–50 kg		50–100 kg		100–500 kg		>500 kg	
	In-door	Out-door	In-door	Out-door	In-door	Out-door	In-door	Out-door	In-door	Out-door
C_{sev}	Consequence Severity Index C_S -									
St 0	I	I	I	I	I	I	I	I	I	I
St 1	II	II	III/IV	III	IV	IV	V	V	V	V
St 2	II	II	III/IV	III/IV	IV/V	IV/V	V	V	V	V
St 3	III	II	IV	IV	V	V	V	V	V	V

Considering the first limit of 5 kg equivalent TNT, it will correspond to the following starch mass:

$$M_{starch} = \frac{M_{TNT} \cdot E_{TNT}}{E_C \cdot \eta} \tag{11.29}$$

with:

- M_{starch} equivalent starch equivalent mass, kg
- M_{TNT}, TNT selected mass, 5 kg
- E_c, starch combustion heat, 17,000 kJ/kg
- E_{TNT}, TNT combustion heat, 4500 kJ/kg

It is found that 5 kg TNT corresponds to 26.5 kg of starch. The explosive atmosphere for this amount is easily calculated, assuming a LEL of 60 g/m^3 (Eckhoff, 2003):

$$V = \frac{17.3}{0.06} = 290\,m^3$$

This value is much bigger than the overall volume of the screen, so the selected range in Table 11.18 is adequate. Entering with St = 1:

$$C_s = II$$

Table 11.19 relates C_s to the explosion frequency.

Table 11.19 Risk ranking (Markowski and Mujumdar, 2004)

Frequency F^n_{ws} [1/year]	Consequence Category C_s				
	I (Negligible)	II (Minor)	III (Medium)	IV (Major)	V (Catastrophic)
A (10^0–10^{-1})	TA	TNA	NA	NA	NA
B (10^{-1}–10^{-2})	TA	TNA	TNA	NA	NA
C (10^{-2}–10^{-3})	A	TA	TNA	TNA	NA
D (10^{-3}–10^{-4})	A	TA	TA	TNA	TNA
E (10^{-4}–10^{-5})	A	A	TA	TA	TNA
F (10^{-5}–10^{-6})	A	A	A	TA	TA

Table 11.20 Safeguards from ignition for Example 11.6

Ignition Source/Cause	Ignition Prevention
Bearings wear Bearings lube oil loss Bearings damage	Pt 100 (maximum temperature threshold) Regular check Instruction manual ATEX protection by control of ignition source 'b' (BS EN 13463-5)
Mechanical coupling friction Belt skidding	ATEX protection by control of ignition source 'c' (BS EN 13463-6) DC breaking ATEX protection by control of ignition source 'c' (BS EN 13463-6)
Belt combustion Static electricity Stray current Loss of insulation Loss of equipotentiality	Belt materials compliance with EN 13478 ATEX protection by control of ignition source 'b' and 'c' (BS EN 13463-5/6) Dust conductivity control–Enclosure IP 64 ATEX protection by control ignition source 'b' and 'c' (BS EN 13463-5/6)
Motor overheating Overcurrent	PTC thermistor for the detection of: - Phase failure - Tension deviation - High temperature DC breaking ATEX protection by control ignition source 'b' and 'c' (BS EN 13463-5/6)
Compression of dust in interstitial gaps	ATEX protection by control of ignition source 'c' (BS EN 13463-6)

It is:

- A—Acceptable risk
- TA—Acceptable risk if ALARP
- TNA—Tolerable risk but improving actions required
- NA—Unacceptable risk

Ignition Risk Assessment (Safeguards)
(BS EN 13463-1, par 5.2) (Table 11.20).

11.10 Analysis of an Incident: Imperial Sugar Company Explosion

On February 7, 2008 a catastrophic series of explosions occurred at the Port Wentworth Imperial Sugar Company site, Georgia, resulting in 14 fatalities and 36 injured people. The incident has been analysed on the basis of the data collected from the U.S. Chemical Safety Board (U.S. CSB) Investigation Report (2009).

11.10.1 The Site

Imperial Sugar Company factory is located at Port Wentworth, Georgia. The aerial picture in Fig. 11.12 shows the site at the time of the incident (U.S. CSB, 2009).

The circled area is related to the silos complex and to the packing building. On the low side the raw sugar buildings are visible.

Fig. 11.12
Imperial Sugar Company aerial view before the incident (U.S. CSB, 2009).

Table 11.21 Thermal and granulometric characteristic of substances (U.S. CSB, 2009)

Material	Moisture Content (% wt)	Mean Particle Size (μm)	P_{max} (bar)	k_{ST} (bar m/s)	MEC (g/m^3)	MIE (mJ)
Cornstarch	11.5	10	8.5	189	105	10–30
Powdered sugar	0.5	23	7.5	139	95	10–30
Granulated sugar (as received)	0.1	Not determined	5.2	35	115	>1000
Granulated sugar (sieved to < 500 μm)	0.1	286	6.0	56	115	>1000

11.10.2 The Substances

In Table 11.21 the substances included in the production cycle have been listed, along with the explosivity parameters (U.S. CSB, 2009). A fraction of 3% of cornstarch was mixed with refined granulated sugar to prevent packing prior to being fed to the mills for powder sugar production.

11.10.3 Analysis of Intrinsic Explosivity Parameters

See Fig. 11.13.

11.10.4 Production Flow Chart

See Fig. 11.14.

11.10.5 Process and Equipment

Silos complex

Refined sugar was supplied to silo No. 3 of a three silos battery through a steel belt conveyor located at the top of the battery. From here, through an aerobelt system and the west elevator, sugar was sent to silos No. 1 and 2, where it went on towards mixing and packaging (Fig. 11.15).

Belt conveyors

In Fig. 11.16, one of the belt conveyors located at the top of the silos complex, early nineties. A large amount of dust is visible.

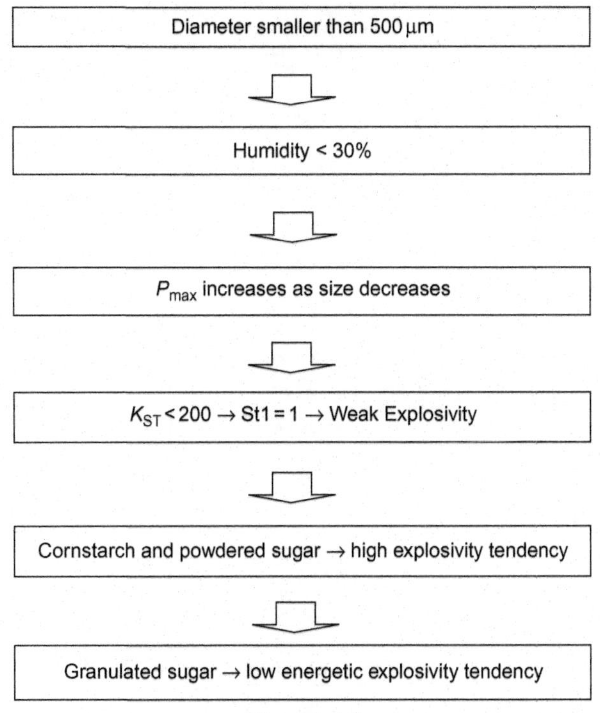

Fig. 11.13
Analysis of explosive parameters.

The final products were contaminated for many years with foreign fragments and debris, so in 2007 the company provided the equipment with frames and steel panels. No venting panels were installed.

Bucket elevators

Bucket elevators delivered sugar at different levels. East elevator, which transferred material from silo 3 to silos 1 and 2, and west elevator, which supplied material to packaging building, are recognisable in Fig. 11.15.

Screw conveyors

Some screw conveyors were also used to transfer the product (Fig. 11.17).

Bosch packing building

This was a four storeys building, located northward to the silos complex, used for packaging. Specifically:

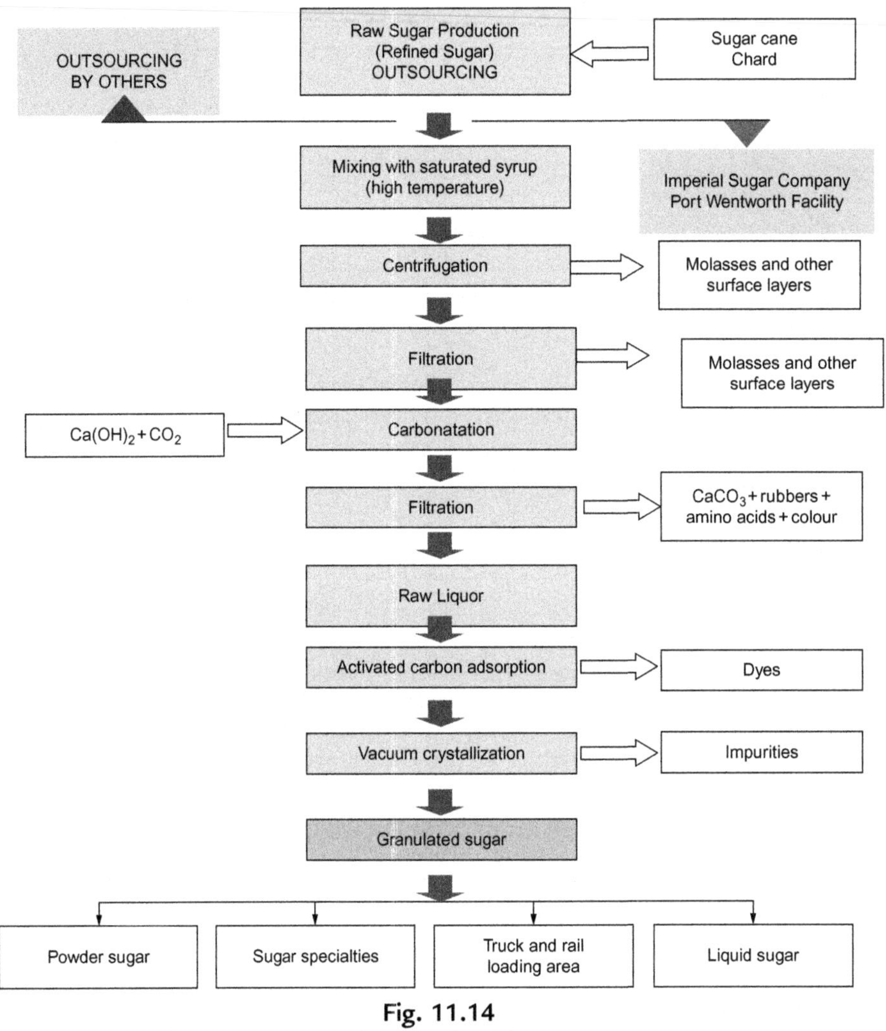

Fig. 11.14
Production flow chart.

- granulated sugar was sent to the top side through screw conveyors, and to the screen of the so-called *hummer room* through belt elevators.
- after screening, sugar was sent through screw conveyors to hoppers located at third floor above packaging machines.
- belt conveyers transferred the packed products to the palletiser rooms.

South packing building

Like the previous one, this was a four storeys building, southward to the silos complex. The equipment for the production of granulated sugar and soft brown sugar was located on

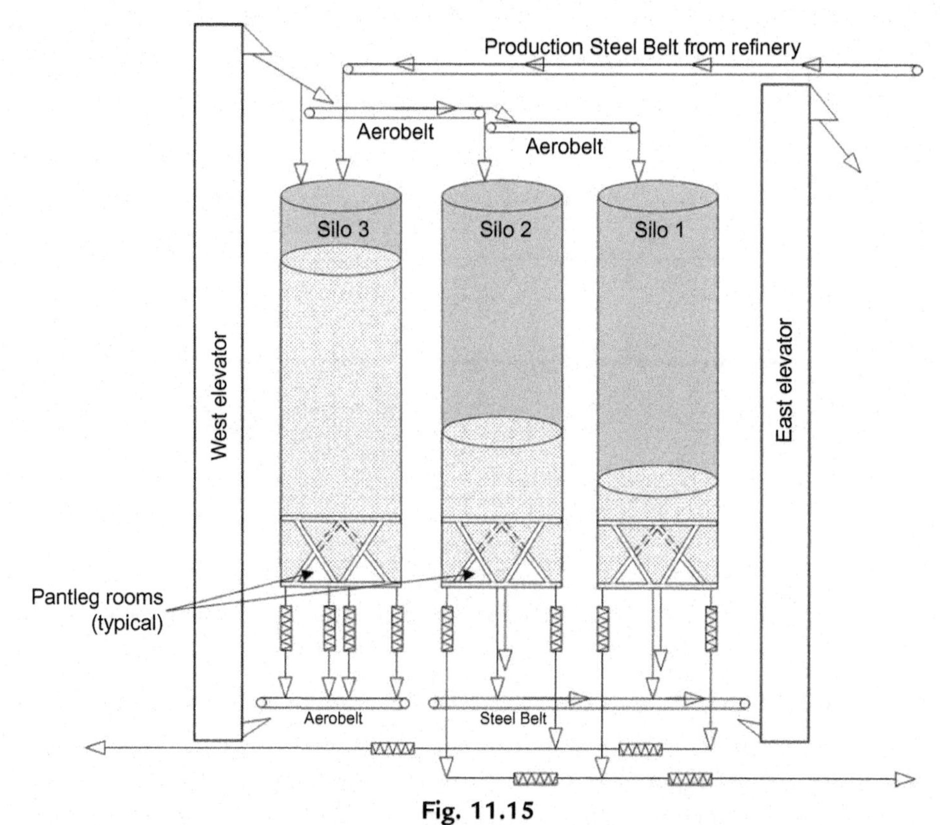

Fig. 11.15
Imperial Sugar Company arrangement of silos (U.S. CSB, 2009).

Fig. 11.16
Imperial Sugar Company—steel belt conveyor (U.S. CSB, 2009).

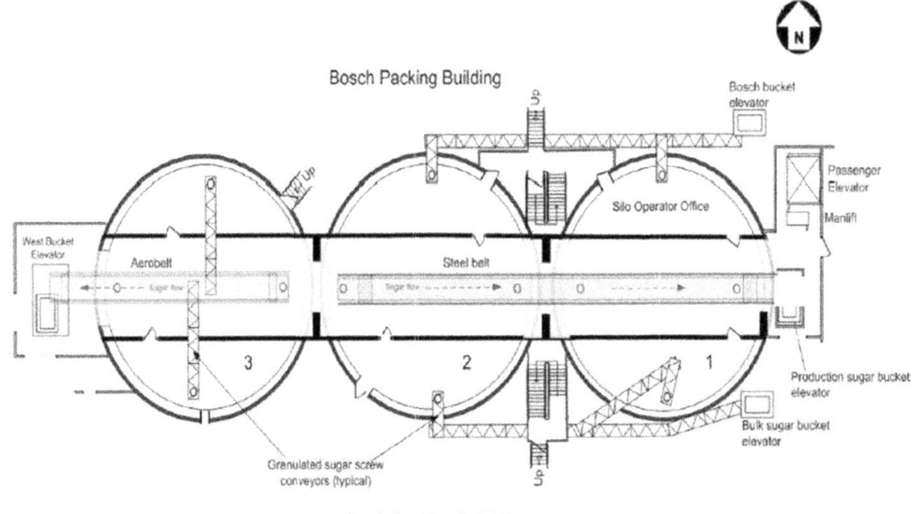

Fig. 11.17
Imperial Sugar Company—belts, screw conveyors, and elevators are visible (U.S. CSB, 2009).

the second and third floors. Screw conveyors and elevators transferred granulated sugar to the hoppers located above the packaging machines. Belt and roller conveyors carried the product to the palletiser room, some of them through large walls and floor openings located to the west. The plant for the powder sugar production was located on the fourth floor. A pneumatic transport line transferred starch above the mills, where it was mixed to sugar. The mills were connected to the dust collectors, which received the suspended powder.

Bulk sugar building

Approximately located westwards to the silos complex, not very far from the Refinery buildings, it was provided with refined granulated sugar by means of bulk sugar bucket elevators and screw conveyors. Finally, the product was loaded on railway carriages and trailers (Fig. 11.18).

11.10.6 The Explosion: Description and Consequences Assessment

Primary explosion

The technical investigation has shown that the primary explosion probably occurred within the silos complex tunnel, possibly at the centre of the belt. The steel panels installed in 2007 had been uprooted (Figs 11.19 and 11.20).

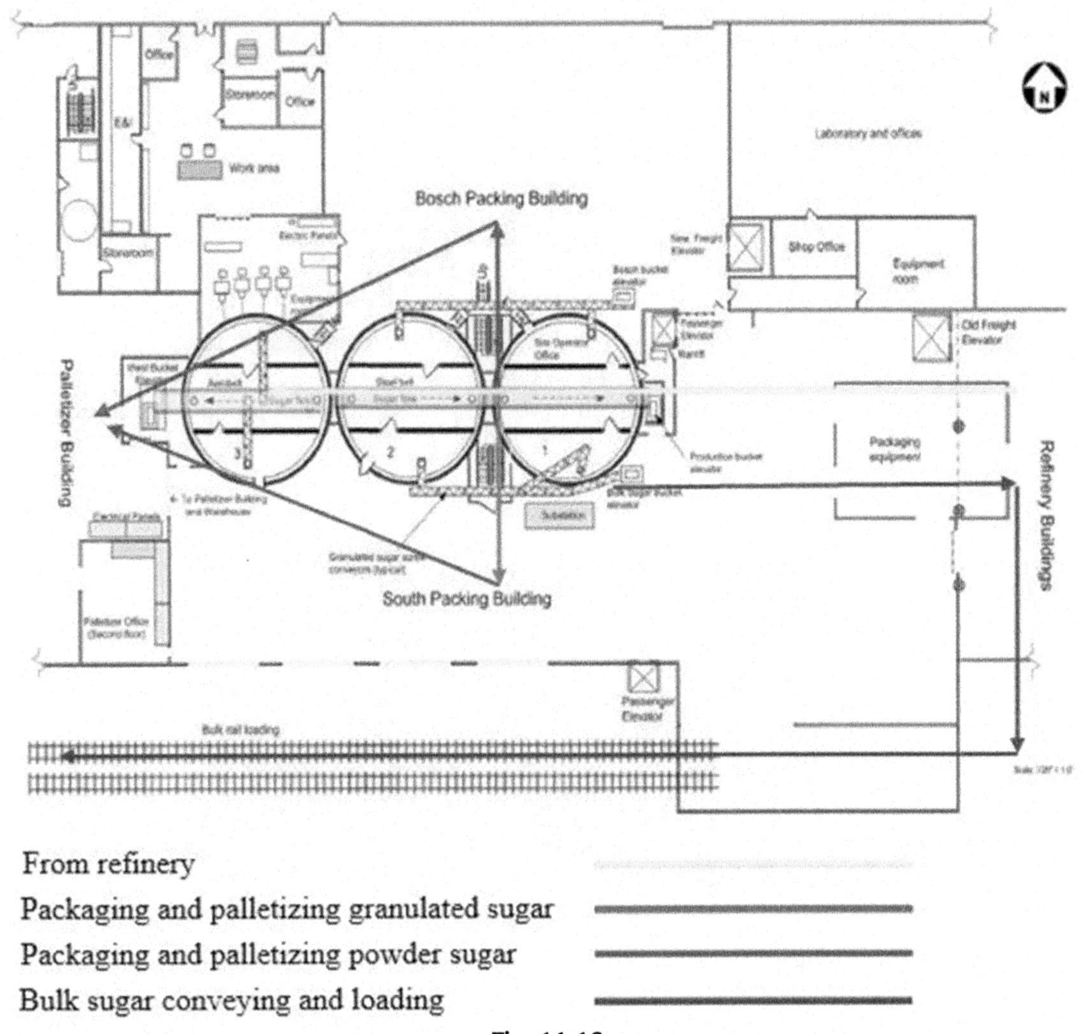

Fig. 11.18
Imperial Sugar Company: lay out and production flow line (U.S. CSB, 2009).

In the conveyor end the explosion has demolished the panels frame. Afterwards, the pressure wave moved between silos 1 and 2, reaching the packing building stairwell and destroying the brick containment (Fig. 11.21).

The causes: the fuel

Some days before the incident, the maintenance team had removed some materials which were in the way of the discharge opening of silo 1 to the belt. For this purpose, some metallic rods had been used. During the cleaning operation, according to the personnel's

Fig. 11.19
Imperial Sugar Company. Damage of belt conveyor: *arrows* show panels uprooted by explosion (U.S. CSB, 2009).

Fig. 11.20
Imperial Sugar Company. Panels supporting frames (U.S. CSB, 2009).

declarations, silo No. 2 had continued to feed materials to the belt, upward to silo 1. It is very likely that the packed material was small enough to go through silo 1 discharge channels to the belt, but too large to go through the free gap between the discharge channel and the belt. The conclusion was that it possibly created a kind of barrier that was effective in stopping the material coming from silo 2, that therefore went out from the belt prior to reaching the silo 1 discharge, and completely covered the internal parts between the bearings (Fig. 11.22).

The causes: Ignition sources

Sugar powder covered structural elements and components inside the belt, such as proximity switches and bearings. The former, precisely electrical sparks, were excluded because it was ATEX rated. The latter were considered one of the causes, also according to

Fig. 11.21
Imperial Sugar Company. Brick panels destroyed by the explosion (U.S. CSB, 2009).

Fig. 11.22
Imperial Sugar Company. Discharge duct from silo No 1. The *arrow* shows the material disengagement door.

the declaration of operators, which reported on the high temperature of its components. The court of enquiry examined the ignition features of sugar, and the conclusion was that the MIT decreases as the contact time with the heat source increases. Specifically, the belt closed configuration, which prevented natural ventilation, and the frequent smouldering material formation may have caused the ignition. Moreover, the *Ignition Sensitivity* and the *Explosion Severity* were high, notably 4 e 2.4 according to the USMB. The court has excluded any friction between metallic parts. Surprisingly, static electricity is not mentioned.

Sugar glow temperature

The picture of ignition causes confirms the critical aspects of *glow temperature* in the dust-layer's spontaneous combustion. Dust layering above the conventional 5 mm suggests very low layer autoignition temperatures. The standard IEC 61241-2-1 shows a minimum glow temperature of a 5 mm sugar layer at 360°C. According to Fig. 11.23, it can be seen that temperature, as layer thickness increase, drops down to 100°C for a layer 5 cm thick.

Consequence assessment

The primary explosion occurred around the centre of the belt, 24 metres long, located inside the silos tunnel. Specifically, on the basis of the inspection findings, three points have been identified (Fig. 11.24):

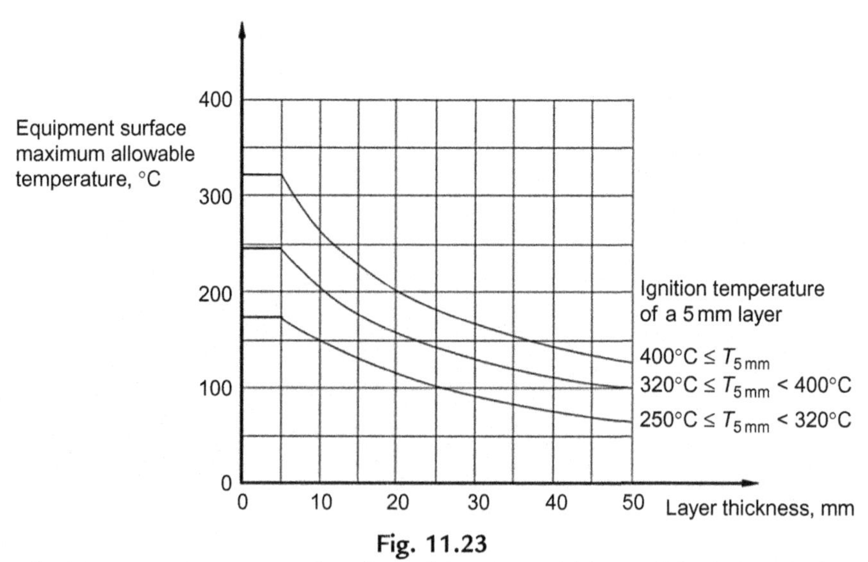

Fig. 11.23
Ignition temperature as a function of dust layer thickness (EN 61241-2-1).

- *Point 1*: along the belt, the pressure wave has uprooted the panels almost completely.
- *Point 2*: at the belt end the panel frames are not in place anymore.
- *Point 3*: at the Packing Building stairwell the pressure wave has destroyed the masonry structure.

The specific blast effects can be collected from Clancey (1972) (Table 11.22).

The overpressure can be estimated as:

$$\Delta P = \rho_o \cdot u_S \cdot v_F \tag{11.30}$$

where:

- ΔP is the overpressure, Pa.
- ρ_o is the unburnt gas density at the ambient conditions, $kg/m^{3.}$
- υ_S is the sound speed of burnt gas, m/s.

Considering the same values:

- $\Delta P = 13172, 17225, 20265$ Pa (Clancey)
- ρ_o (air density at 20°C) = 1.2 kg/m^3
- $\upsilon_S = 800$ m/s (sound speed assumed at 1500 K)
- v_F is flame velocity

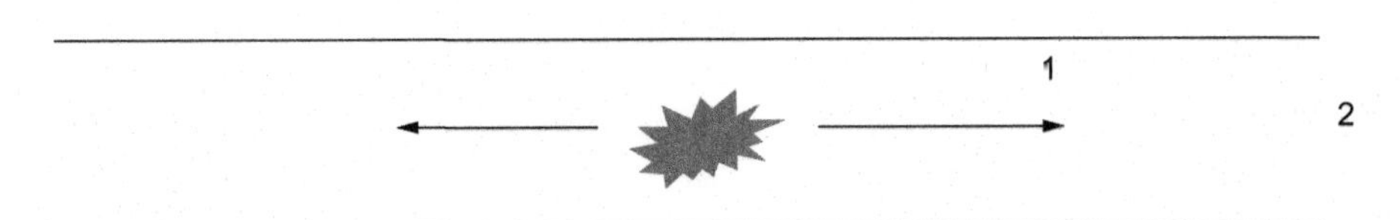

Fig. 11.24
Explosion epicentre and propagation.

Table 11.22 Overpressure on the structural elements of points 1, 2 and 3 (Clancey, 1972)

Location	Effect	Overpressure (atm)
1	Failure of corrugated metal panels; housing wood panels are blown in	0065 ÷ 0.13
2	Distortion of steel frame buildings; may pull away from their foundations	0.2
3	50% destruction of home brickwork	0.17

Replacing:

The order of magnitude of the calculated flame speeds is consistent with the maximum values of agricultural dust flame speed for upward propagation in ducts, where the maximum flame speeds were about 20 m/s (Table 11.23). It can be concluded that the thermo-fluid dynamic scenario occurring inside the belt is consistent with the turbulent flame regime of an enclosed system with the end partially closed. It is worth noting the growing value of the speeds from

Table 11.23 Overpressures and calculated flame speeds flame speeds (Eq. 11.30)

Location	Overpressure (Pa)	Flame Speed (m/s)
1	13,172	13.7
2	20,265	21.1
3	17,225	17.9

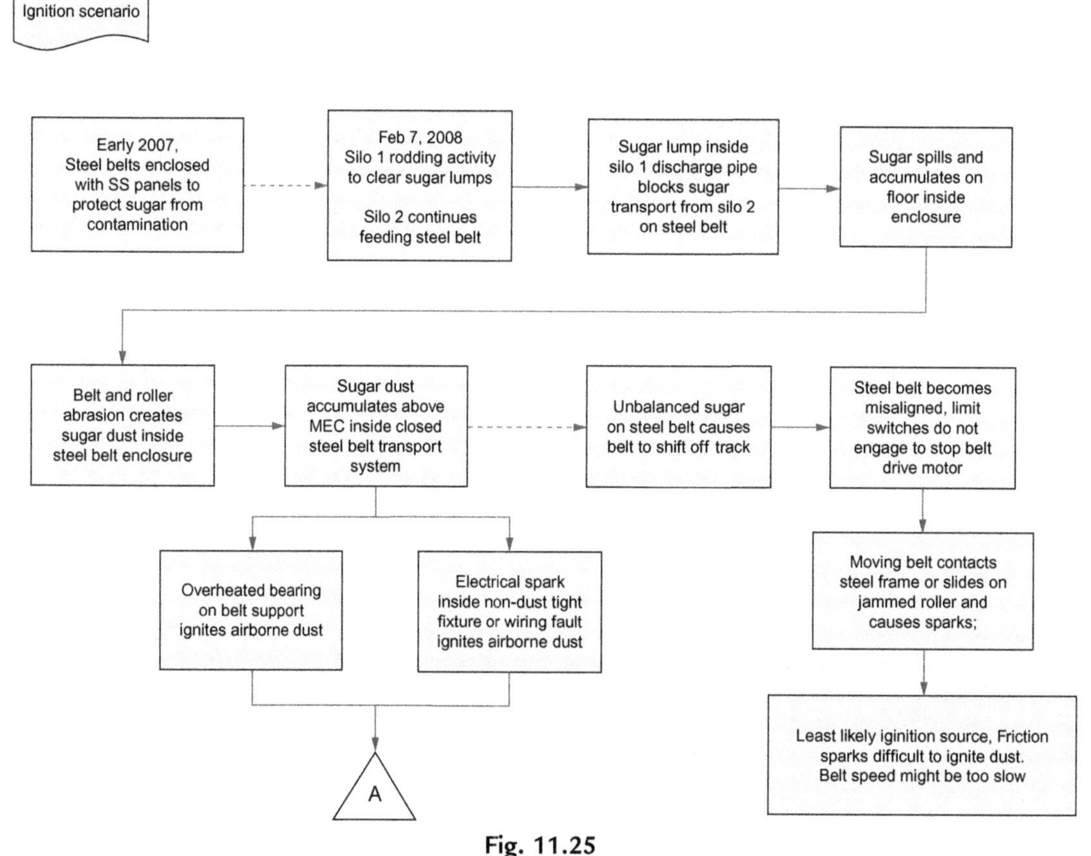

Fig. 11.25
Imperial Sugar Company. Ignition scenario (U.S. CSB, 2009).

point 1 to point 2 and the decreased value at the stairwell. The ignition scenario is shown in Fig. 11.25 (U.S. CSB, 2009).

Secondary explosions

Primary explosion, due to logistic causes and energetic circumstances, propagates towards all the plant, sustained and strengthened by sugar powder spread everywhere by the moving pressure wave. In Figs 11.26–11.35, the sequential explosion flow charts and the related layouts have been included (U.S. CSB, 2009).

11.10.7 Root Causes

Figs 11.36–11.38 are self-explanatory regarding the incident causes. Most parts of the photographic information is dated months earlier than the explosion. In the United Stated, through the standard NEC 70, and in Europe through the ATEX standards, hazardous area classification due to the presence of explosive dusts was mandatory, along with the obligation

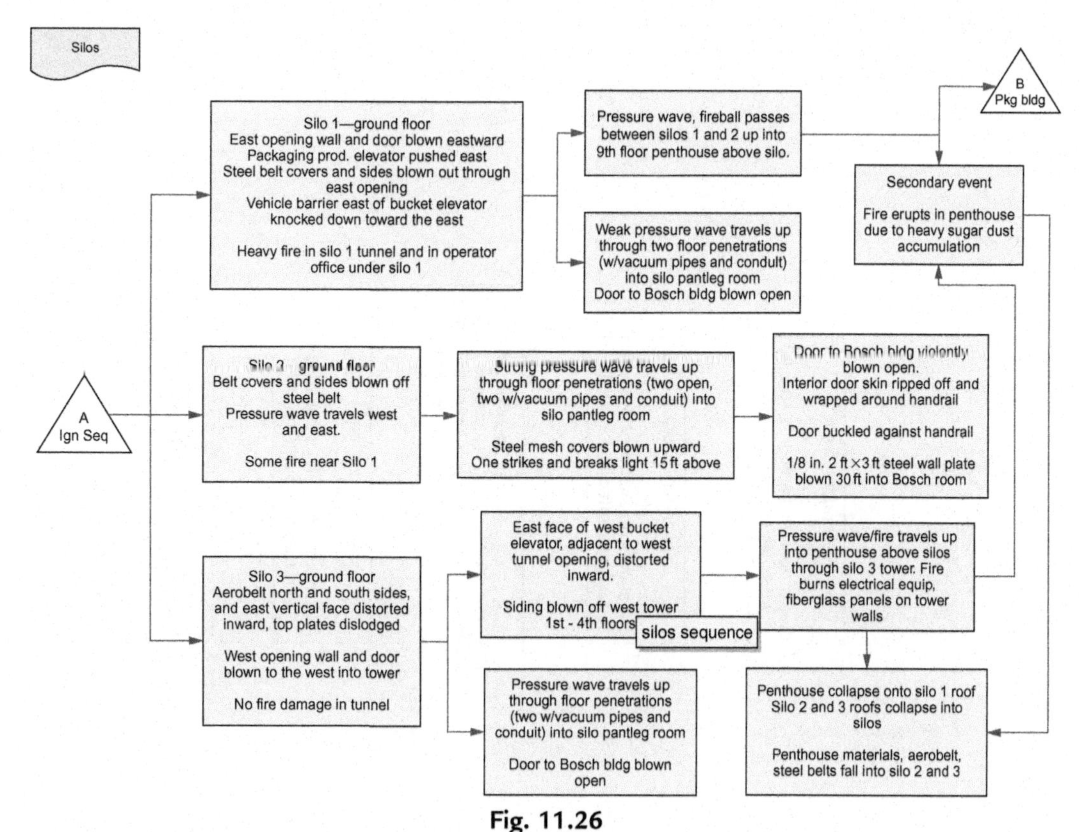

Fig. 11.26
Imperial Sugar Company. Bosch building explosion—Flow chart 1/2 (U.S. CSB, 2009).

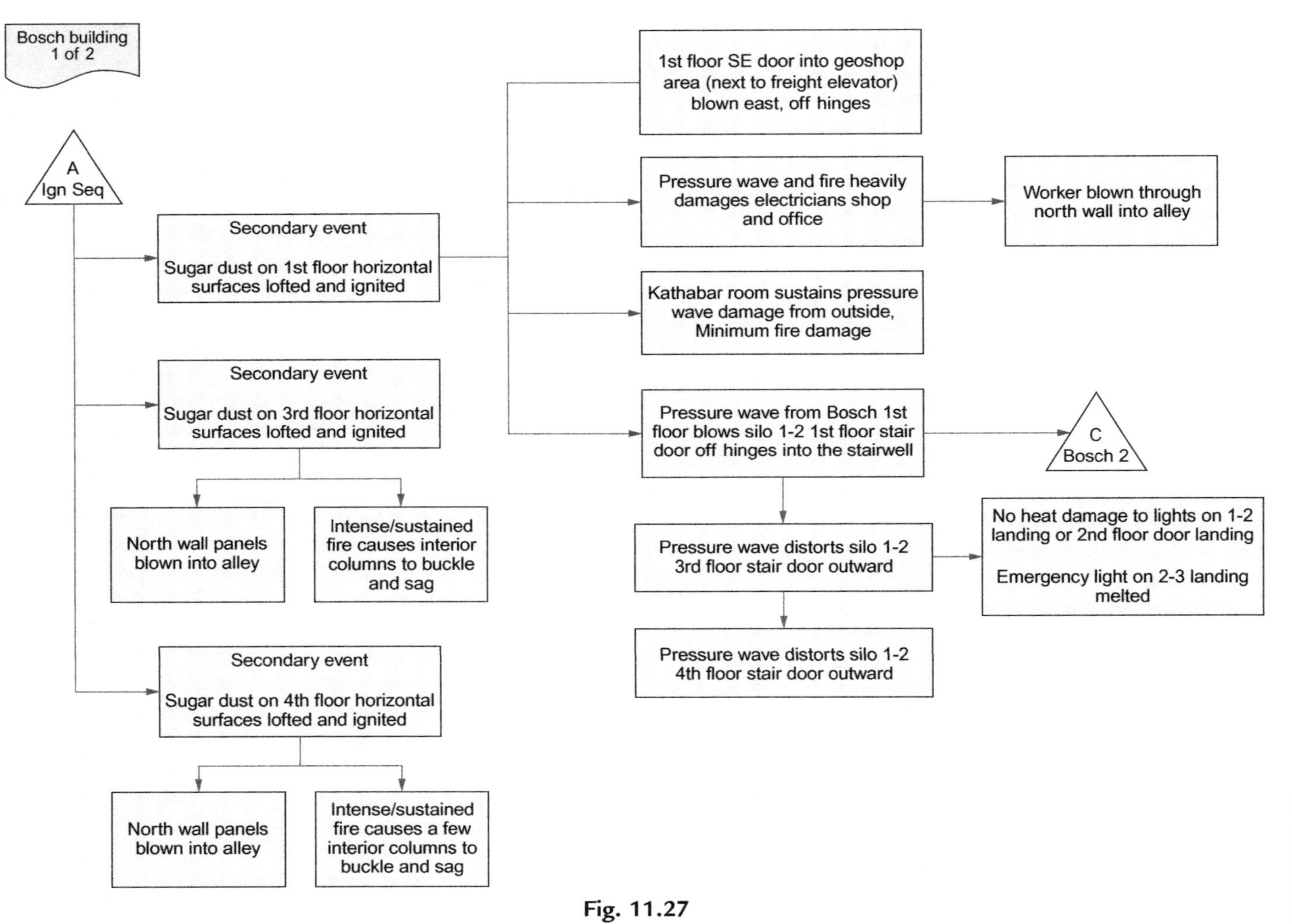

Fig. 11.27
Imperial Sugar Company. Bosch building explosion—Flow chart 2/2 (U.S. CSB, 2009).

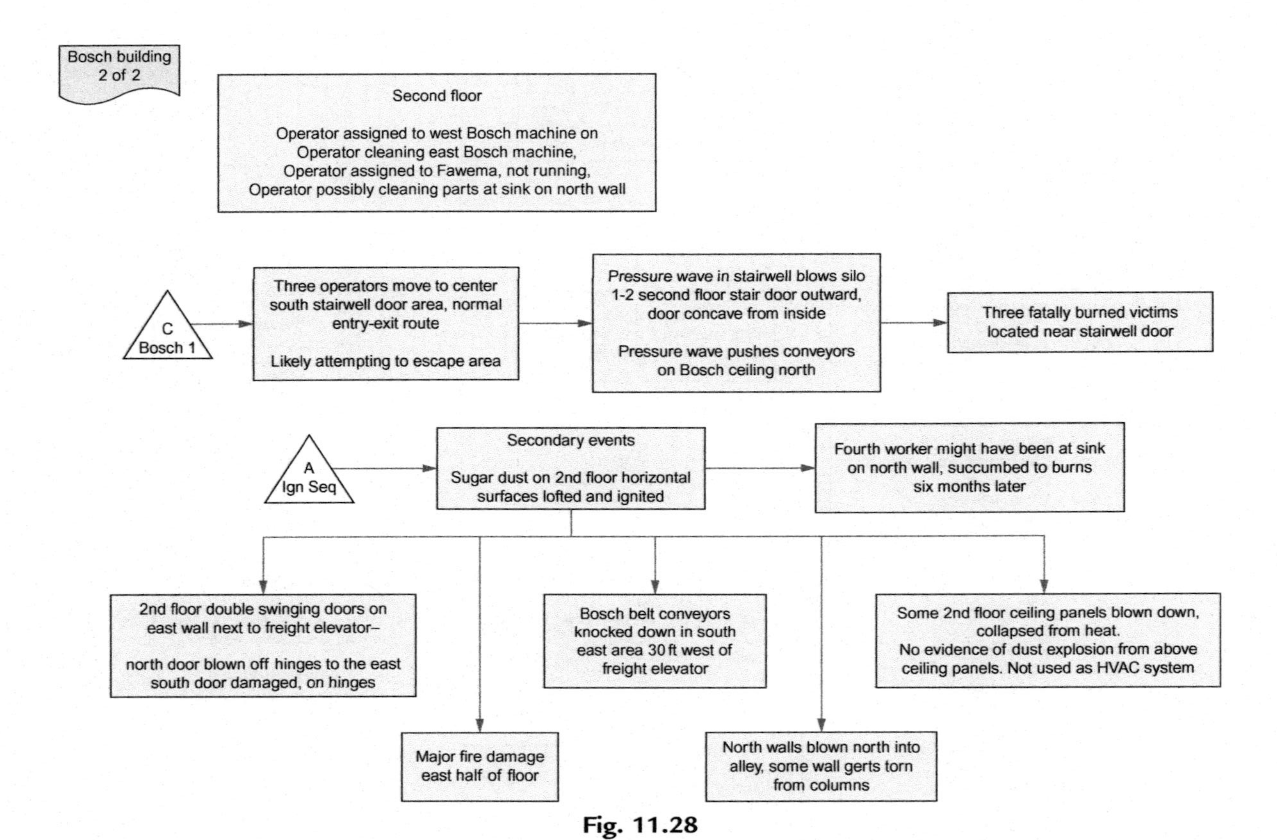

Fig. 11.28
Imperial Sugar Company. Pelletizer room explosion (U.S. CSB, 2009).

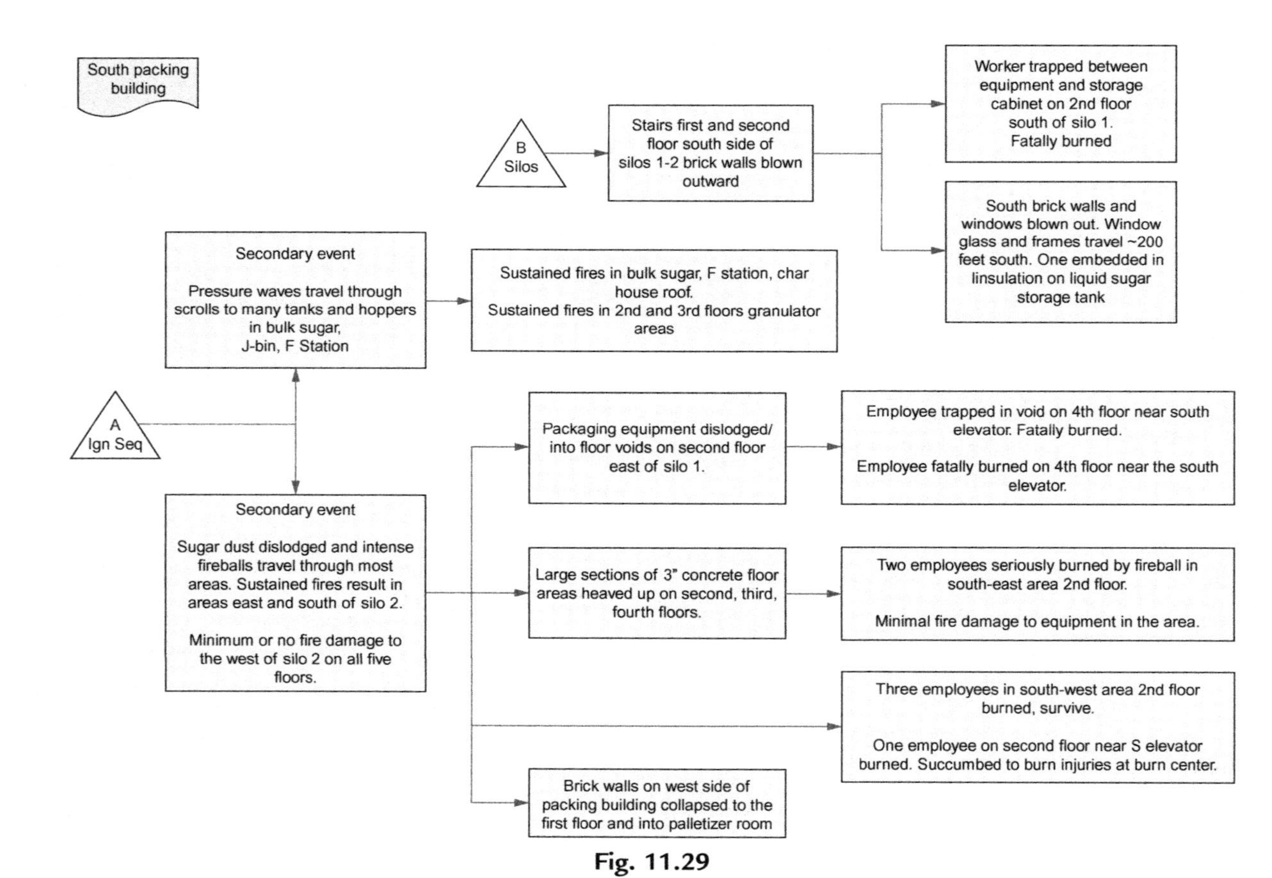

Fig. 11.29
Imperial Sugar Company South packing building explosion (U.S. CSB, 2009).

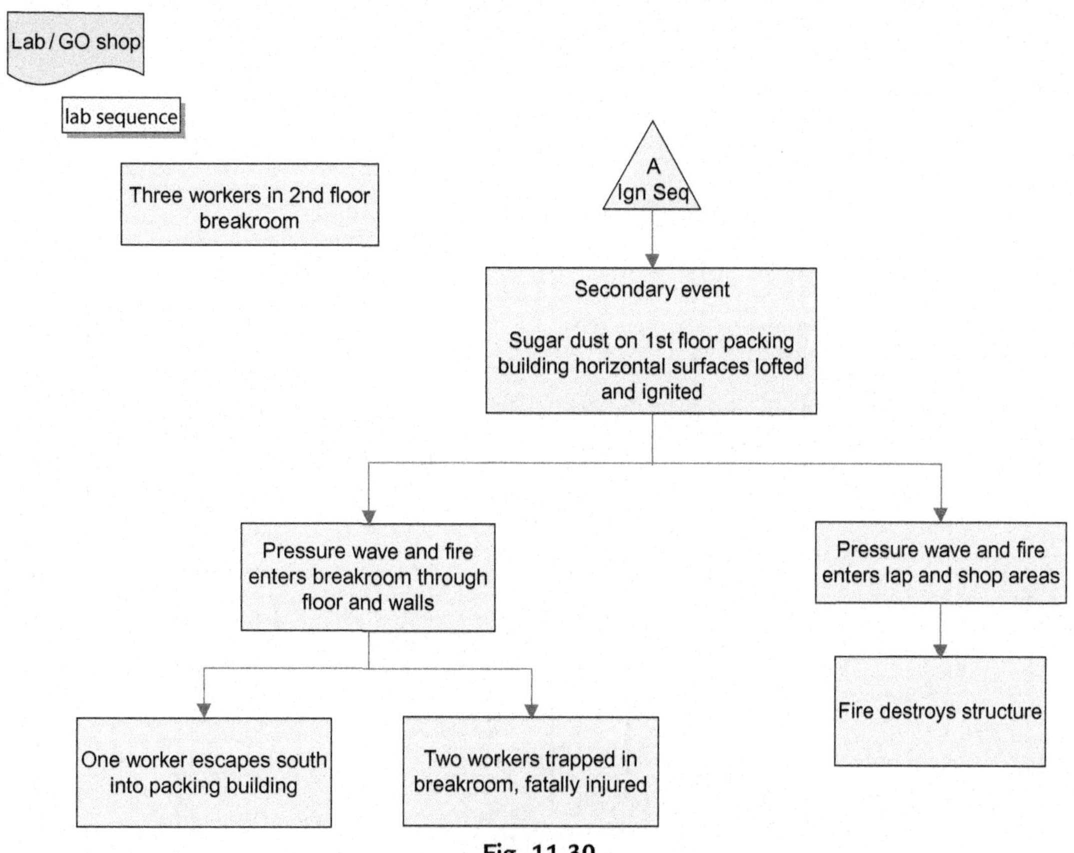

Fig. 11.30
Imperial Sugar Company Lab/GO explosion (U.S. CSB, 2009).

to install certified equipment and to implement efficient housekeeping. The following testimony on the incident was released by John S. Bresland, Chairman and Chief Executive Officer U.S. Chemical Safety Board (2008):

> *Madam Chairman, this accident was preventable. Testimony of John S. Bresland Senate HELP Committee July 29, 2008. Combustible dust is an insidious workplace hazard when it accumulates on surfaces, especially elevated surfaces. A wide range of common combustible materials can explode in finely powdered form, including wood, coal, flour, sugar, metals, plastics, and many chemicals and pharmaceuticals. At Imperial Sugar, a catastrophic explosion resulted from massive accumulations of combustible sugar dust on surfaces throughout the packaging plant. My written testimony details what we have learned to date about the accident. Let me summarize a few key points. The photographs on the easel were taken in September and October 2006 at different locations inside the sugar packaging building at Imperial's Savannah refinery. They confirm the existence of substantial dust accumulations on various ducts, motors, switch boxes, and processing equipment. These accumulations – ranging in depth from an inch or two up to several feet – far exceed the NFPA-recommended limit of 1/32 of an inch. Witnesses told the CSB that large*

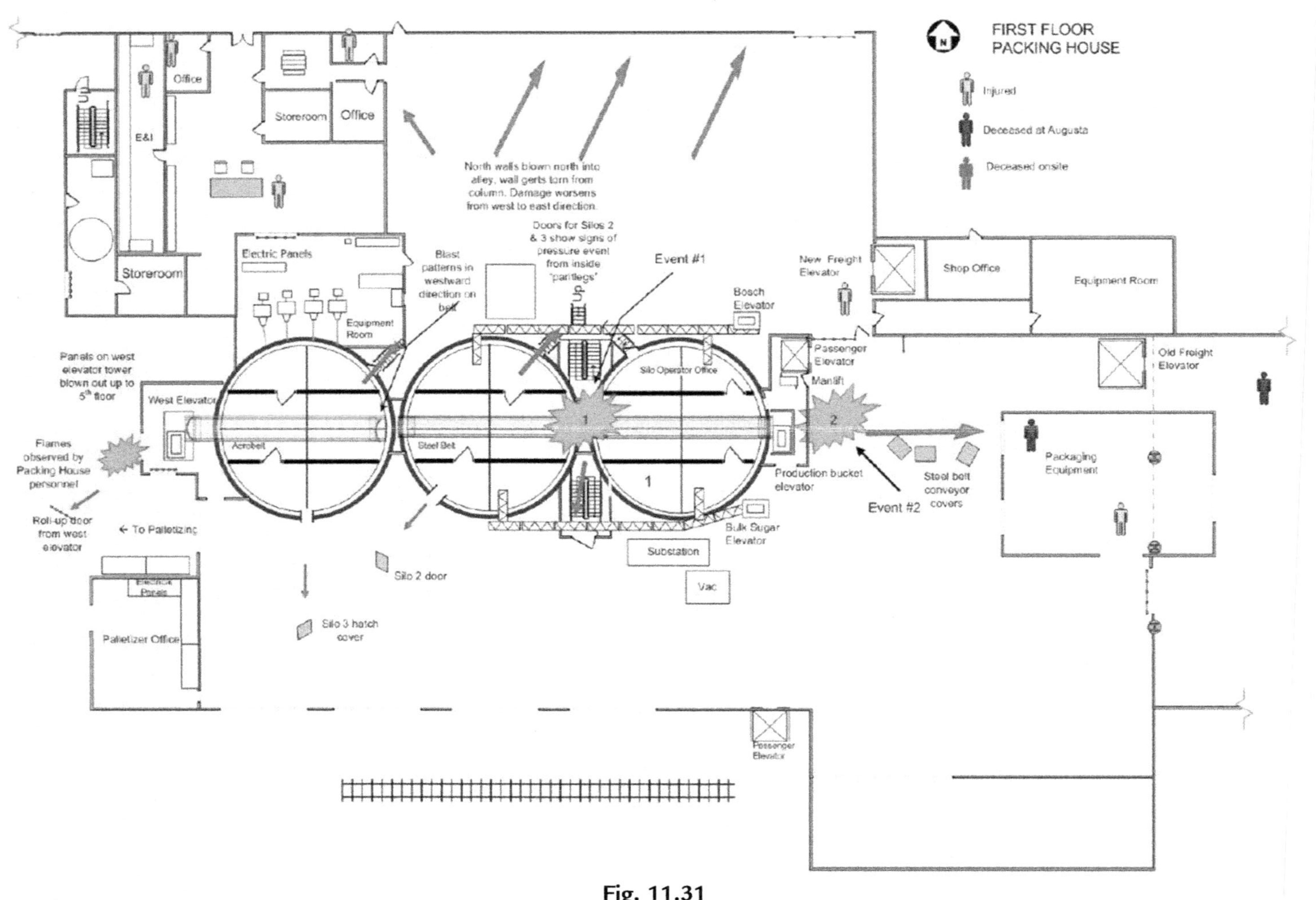

Fig. 11.31
Imperial Sugar Company. Explosion in packing buildings. First floor (U.S. CSB, 2009).

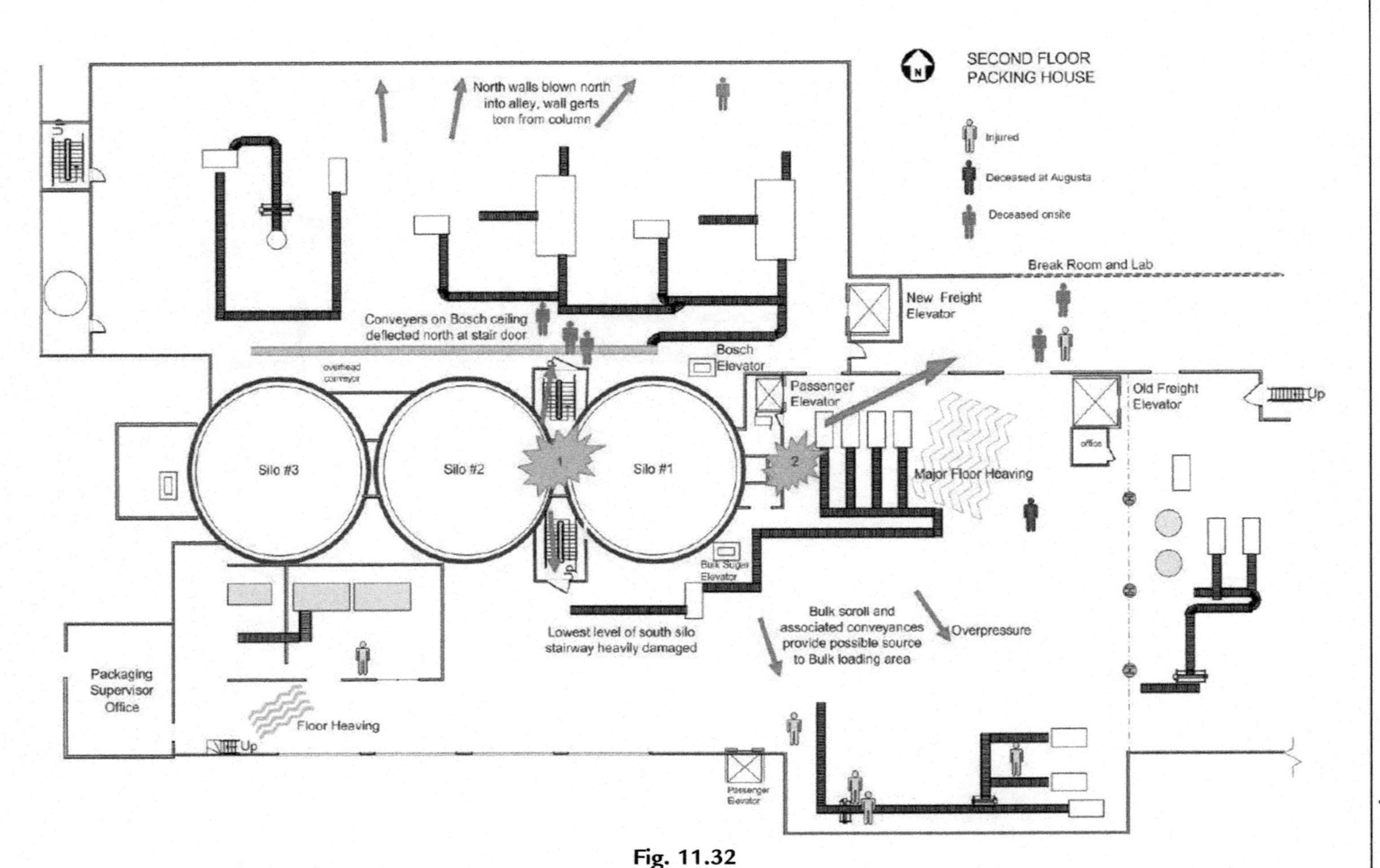

Fig. 11.32
Imperial Sugar Company. Explosion in packing buildings. Second floor (U.S. CSB, 2009).

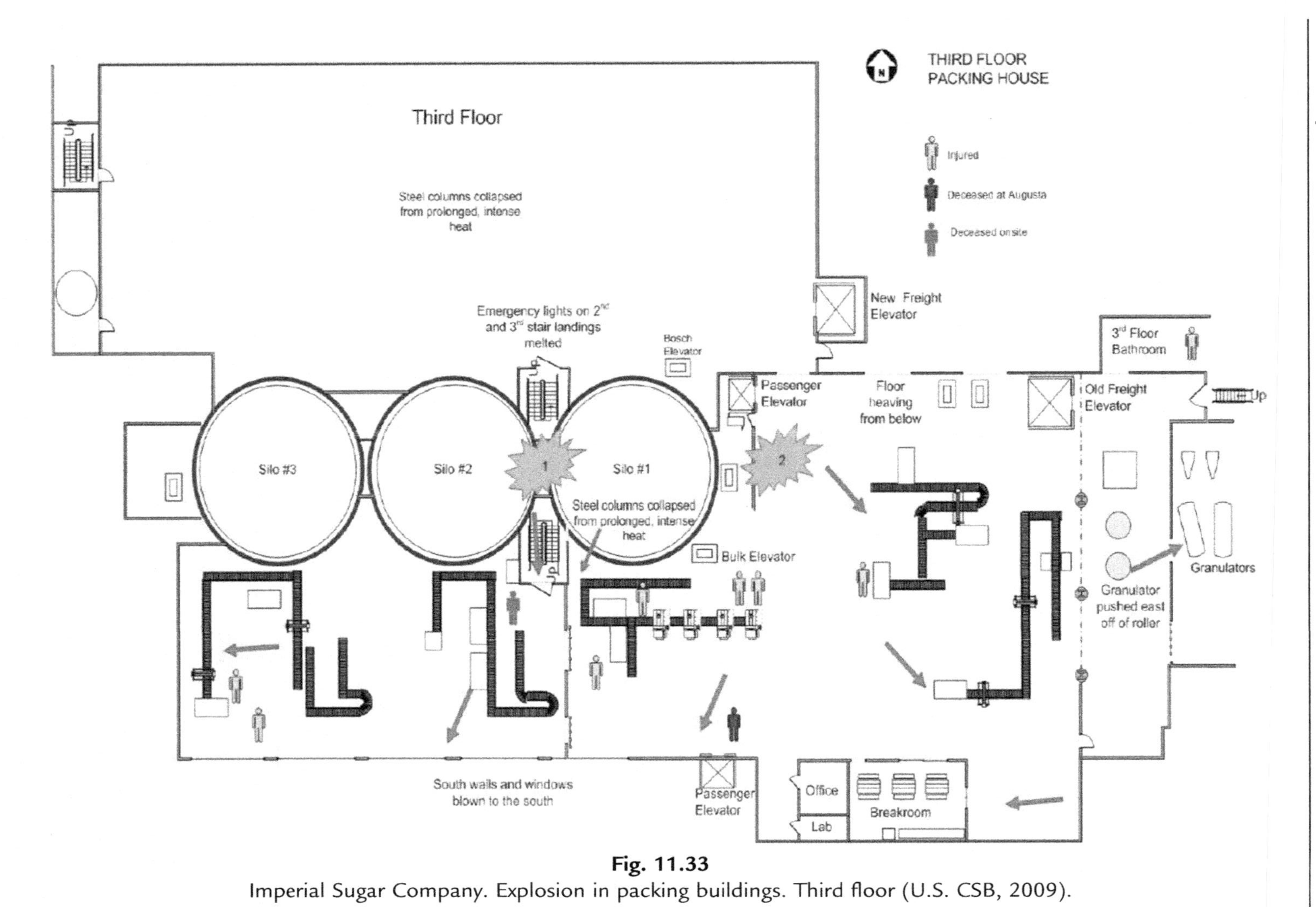

Fig. 11.33
Imperial Sugar Company. Explosion in packing buildings. Third floor (U.S. CSB, 2009).

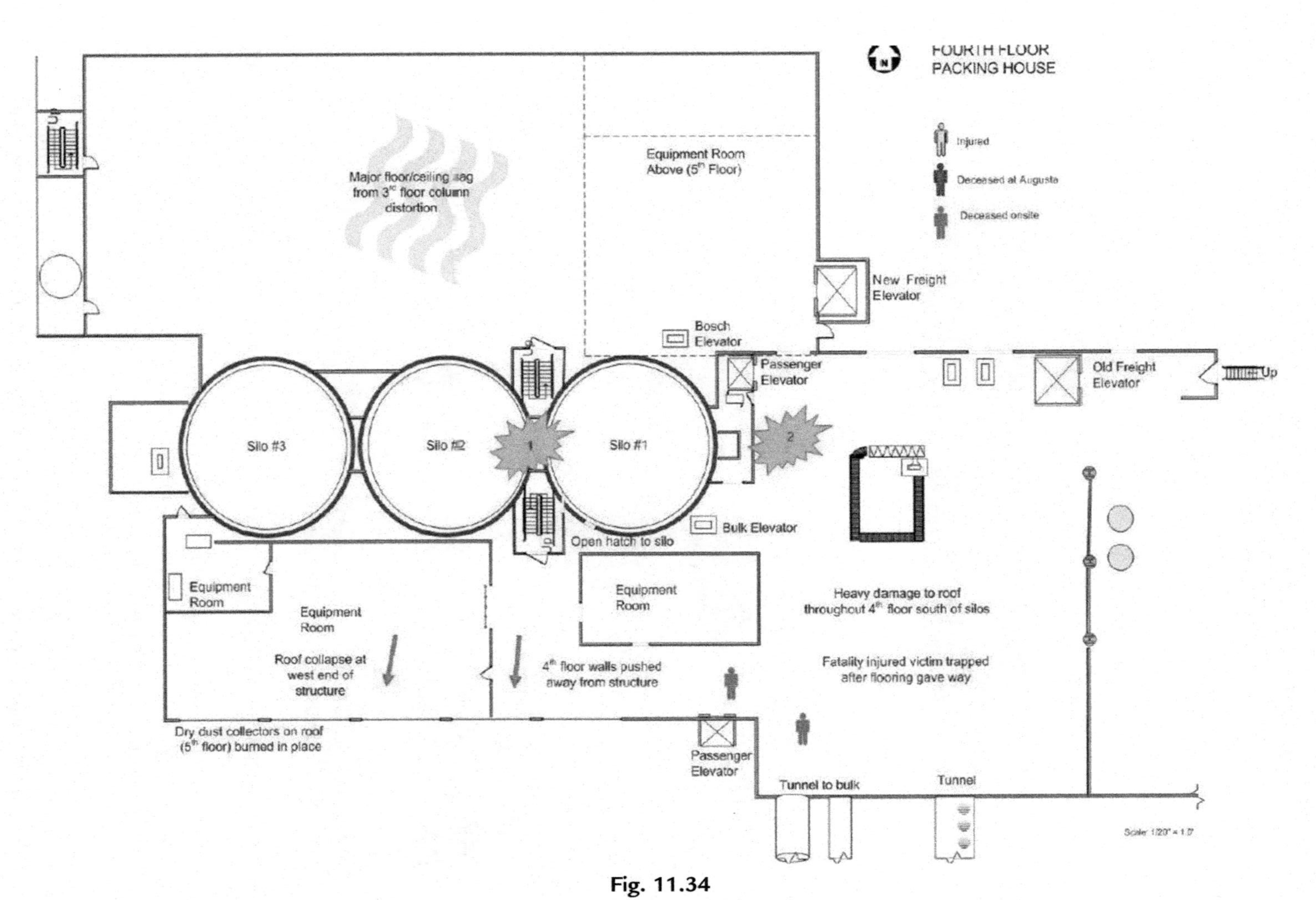

Fig. 11.34
Imperial Sugar Company. Explosion in packing buildings. Fourth floor (U.S. CSB, 2009).

Fig. 11.35
Imperial Sugar Company. Silos of west elevator and south packing building after the explosion (U.S. CSB, 2009).

Fig. 11.36
Imperial Sugar Company. Dust layers on process equipment. March–May 2006 (U.S. CSB, 2009).

accumulations of dust were present until the day of the explosion. According to an employee, near the powder mills, sugar accumulated on the floor to a "mid-leg" height. We were told that airborne sugar in this room made it difficult for workers to see each other. We obtained documents indicating that certain parts of Imperial's milling process were releasing tens of thousands of pounds of sugar per month into the work area. Based on our evidence, Imperial did not have a written dust control program or a program for using safe dust removal methods. And the company lacked a formal training program to educate its workers about combustible dust hazards.

Fig. 11.37
Imperial Sugar Company. Dust layers on process equipment. September–October 2006 (U.S. CSB, 2009).

Fig. 11.38
Imperial Sugar Company. Dust layers on a screw conveyor (U.S. CSB, 2009).

References

Abbot, J.A., 1988. Survey of dust fires and explosions in UK 1979–1984: dust explosion prevention and protection-latest development. In: Proceeding of the International Meeting Organised by IBC/BMHB. Ascot, Berkshire, UK British Material Handling Board.

Benintendi, R., 2004. Identificazione e sviluppo di criteri applicativi per la valutazione del rischio di esplosione nelle attività di deposito e trasformazione di polveri impiegate nei processi dell'industria agroalimentare. Progetto di Studio e Ricerca proposto ai sensi del D.M 29 dicembre 2003, Ministero del lavoro e delle politiche sociali.
BS EN 1127–1, 2011. Explosive atmospheres- explosion prevention and protection. Part 1: basic concepts and methodology.
BS EN 15198, 2007. Methodology for the risk assessment of non-electrical equipment and components for intended use in potentially explosive atmospheres.
BS EN 61241-0, 2006. Electrical apparatus for use in the presence of combustible dust. General requirements.
BS EN 61241-2-2, 1996. Electrical apparatus for use in the presence of combustible dust. Test methods. Method for determining the electrical resistivity of dust in layers.
Cardillo, P. Le Esplosioni di Gas, Vapori e polveri. Paolo Cardillo - Stazione sperimentale per i Combustibili.
CEI 31–35, 2014. Atmosfere esplosive. Guida alla classificazione dei luoghi con pericolo di esplosione per la presenza di gas in applicazione della Norma CEI EN 60079–10-1 (CEI 31–87).
CENELEC CLC/TR 50404, 2003. Electrostatics—Code of practice for the avoidance of hazards due to static electricity.
Clancey, V.J., 1972. Diagnostic features of explosion damage. In: Sixth International Meeting of Forensic Science, Edinburgh, England.
Eckhoff, R.K., 2003. Dust Explosions in the Process Industries, third ed. Gulf Professional Publishing, Elsevier.
EN 14491, 2012. Dust explosion venting protective systems.
Field, P., 1982. Dust Explosions. Elsevier, Amsterdam Fire Research Station, Borehamwood, Hertfordshire.
Glarner, T., 1984. Mindestzündenergie-Einfluss der Temperatur. VDI Ber. 494, 109–118. VDI-Verlag GmbH, Düsseldorf.
Jacobson, M., Nagy, J., Cooper, A.R., Ball, F.J., 1961. Explosibility of agricultural dusts. Report of Investigation 5753, US Bureau of Mines.
Landry, E., Samson, O., Pajot, F., Halter, F., Foucher, Mounaïm-Rousselle, C. Experimental Study of Spark-Ignition Engine Combustion Under High Pressure and high dilution, PSA Peugeot Citroën, Direction de la Recherche et Innovation Automobile and Laboratoire de Mécanique et Energétique, Ecole Poytechnique de l'Université d'Orléans.
Mannan, S., 2005. Lees' Loss Prevention in the Process Industries, third ed. Elsevier, Butterworth-Heinemann, Burlington, MA.
Markowski, A.S., Mujumdar, A.S., 2004. Layer of Protection Analysis for Industrial Drying Risk Assessment. In: Proceedings of the 14th International Drying Symposium (IDS 2004) São Paulo, Brazil, 22–25 August, pp. 760–766.
National Fire Protection Association, 2013a. NFPA 68. Standard on Explosion Protection by Deflagration Venting. NFPA.
National Fire Protection Association, 2013b. NFPA 654. Prevention of Fire and Dust Explosions from the Manufacturing, Processing, and Handling of Combustible Particulate Solids. NFPA.
National Fire Protection Association, 2014. NFPA 69. Standard on Explosion Prevention Systems. NFPA.
National Fire Protection Association, 2017. NFPA 61. Standard for the Prevention of Fires and Dust Explosions in Agricultural and Food Processing Facilities. NFPA.
OSHA CPL 03–00-006, 2007. Combustible dust national emphasis program.
U.S. Chemical Safety Board Investigation, 2009. Sugar dust explosion and fire. Investigation Report No. 2008–05-I-GA.
VDI 3673, 2002. Pressure venting of dust explosions.
Wieman, W., 1987. Influence of temperature and pressure on the explosion characteristics of dust/air and dust/air/inert gas mixtures. In: Cashdollar, Hertzberg, (Eds.), Paper in Industrial Dust Explosions, ASTM STP 958. American Society for Testing and Materials, Philadelphia, PA, pp. 33–34.

Further Reading

BS EN 1346, 2011. Non-electrical equipment intended for use in potentially explosive atmospheres.

PART 3

Quantitative Risk Assessment

CHAPTER 12

Quantitative Risk Assessment

There are more things in heaven and earth, Horatio, than are dreamt of in your philosophy.

W. Shakespeare. Hamlet (1.5.167-8)

12.1 Definitions

12.1.1 Risk Indices (RI)

Risk Indices are single numbers, or tabulations of numbers, which are correlated to the magnitude of risk (CCPS, 1999).

12.1.2 Location Specific Individual Risk (LSIR)

LSIR is the risk (yearly frequency) of death or serious injury to a person in the vicinity of a hazard, assuming that the person is there 365 days a year, 24 hours a day. LSIR is given in terms of year^{-1}. CCPS define the following notations to calculate the LSIR:

$$\text{LSIR}_{x,y} = \sum_{i=1}^{n} \text{LSIR}_{x,y,i} \tag{12.1}$$

where:

$\text{LSIR}_{x,y}$ = chances of fatality per year at geographical location x, y
$\text{LSIR}_{x,y,i}$ = chances of fatality per year at geographical location x, y from incident outcome case i (chances of fatality per year, or year^{-1})
n = the total number of incident outcome cases considered in the analysis

And:

$$\text{LSIR}_{x,y,i} = f_i \cdot P_{x,y,i} \tag{12.2}$$

where:

f_i = frequency of incident outcome case i (year^{-1})
$P_{x,y,i}$ = probability that incident outcome case i will result in a fatality at location x, y.

Process Safety Calculations. https://doi.org/10.1016/B978-0-08-101228-4.00012-5

However, it has to be considered that the probability of death P_i due to the exposure for a time t_i to a specific concentration C_i of a specific substance i at a given location x, y can be calculated on the basis of the probit function described in Chapter 6:

$$\mathrm{LSIR}_{x,y,i} \leftarrow P_{x,y,i} \leftarrow Pr_{x,y,i} = a + b \cdot \ln\left(C^{n}_{x,y,i} \cdot t_{x,y,i}\right) \tag{12.3}$$

where $\mathrm{LSIR}_{x,y,i}$ is obtained from $P_{x,y,i}$ applying Poisson's equation described in Chapter 6. This approach is different from the one generalised by the CCPS.

It is worth noting that at a specific location it is:

$$\mathrm{LSIR}_{x,y} = \mathrm{LSIR}_{x,y,\mathrm{TOX}} + \mathrm{LSIR}_{x,y\mathrm{TH}} + \mathrm{LSIR}_{x,y\mathrm{OP}} \tag{12.4}$$

The overall LSIR at a specific location is the sum of the LSIRs at the sum location due to the contribution of toxic, thermal, and overpressure impact on people. Of course the overpressure damage thresholds and the corresponding probabilities included here are other than those considered for buildings and equipment. The probit functions to calculate the probability of death due to each specific hazard have been presented in Chapter 6.

LSIR values can be tabulated, and are usually depicted, on iso-frequency contours like those done for blast loads, as shown in the next drawings. This can be done for the overall risk and for the risk associated to each single hazard (Fig. 12.1).

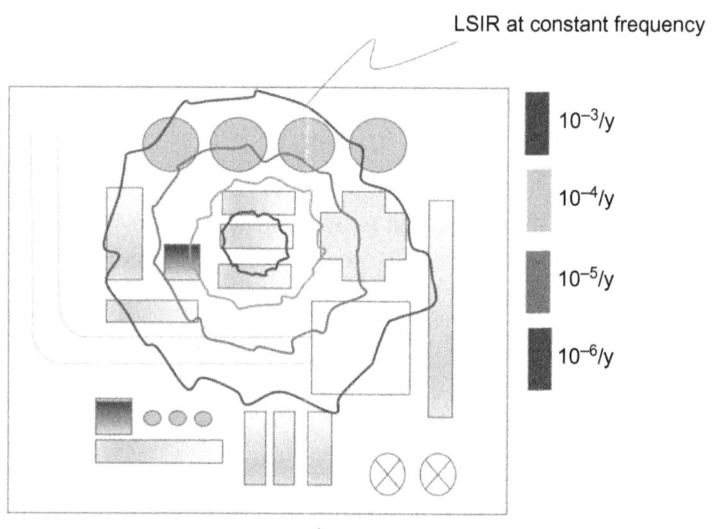

Fig. 12.1
LSIR at constant frequency contours.

12.1.3 Individual Risk Per Annum (IRPA)

The risk (yearly frequency) of death or serious injury to a person in the vicinity of a hazard considering the fraction of the year really spent by the person is:

$$\mathrm{IRPA} = f \times \mathrm{LSIR} \tag{12.5}$$

where f is the yearly fraction spent by a person on-site. Frequency contours are also drawn for IRPA values, based on manning distribution.

12.1.4 Maximum Individual Risk

Individual risk to the person(s) exposed to the highest risk in an exposed population.

12.1.5 Average Individual Risk

The average of all individual risk estimates over a defined population.

$$\mathrm{LSIR_{AV}} = \frac{\sum_{x,y} \mathrm{LSIR}_{x,y} \cdot \Pi_{x,y}}{\sum_{x,y} \Pi_{x,y}} \tag{12.6}$$

where $\prod_{x,y}$ is number of people at location x,y.

12.1.6 Potential Loss of Life (PLL) or Rate of Death (ROD)

The expected number of fatalities within a specific population per year

On-shore

$$\mathrm{PLL} = \frac{\mathrm{IRPA} \cdot \mathrm{POS}}{f_{\mathrm{OS}}} = \mathrm{LSIR} \cdot \mathrm{POS} \tag{12.7}$$

Off-shore

$$\mathrm{PLL} = \frac{\mathrm{IRPA} \cdot \mathrm{POB}}{f_{\mathrm{OB}}} = \mathrm{LSIR} \cdot \mathrm{POB} \tag{12.8}$$

where

- POS is number of people on-site.
- POB is number of population on-board.
- f_{OB} and f_{OB} are the respective fractions of spent time.

12.1.7 Fatal Accident Rate (FAR)

A measure of individual risk expressed as the estimated number of fatalities per 10^8 exposure hours (roughly 1000 employee working lifetimes):

$$FAR = LSIR_{AV} \cdot 11{,}415 \tag{12.9}$$

where 11,415 is the ratio of 10^8 to the hours in 1 year (8760). Kletz (1977) has listed the following FARs for some activities (Table 12.1).

Example 12.1

Hydrogen sulphide is released from the two following units at the given frequency, and the gas clouds reach the zones occupied by the identified people. Calculate the risk indices.

Incident	Frequency (y^{-1})	Impacted zones	Populations	H_2S concentration (ppm)	Exposure time (min)
1	10^{-4}	A	5	1200	7
	10^{-5}	B	10	1000	5
2	10^{-4}	A	5	1100	6
	10^{-5}	B	10	1400	9

Solution

Probit equation for hydrogen sulphide (ppm, min) (Van den Bosch, Duijm, Yellow Book, 2005)

$$Pr = -10.833752 + \ln\left(C^{1.9} \cdot t\right) \tag{12.10}$$

Table 12.1 Fatal accident rate for various activities (Kletz, 1977)

Activity	FAR (fatalities/10^8 exposure hours)
British industry (all)	4
Clothing and footwear manufacture	0–15
Vehicle manufacture	1–3
Timber, furniture, and so on	3
Metal manufacture, ship building	8
Agriculture	10
Coal mining	12
Railway shunters	45
Construction erectors	67
Staying at home (men 16–65)	1
Travelling by train	5
Travelling by car	7

Specifically:

$$Pr_{1A} = -10.833752 + \ln\left(1200^{1.9} \cdot 7\right) = 4.5 \quad (12.11)$$

$$Pr_{1B} = -10.833752 + \ln\left(1000^{1.5} \cdot 5\right) = 3.9 \quad (12.12)$$

$$Pr_{2A} = -10.833752 + \ln\left(1100^{1.9} \cdot 6\right) = 4.2 \quad (12.13)$$

$$Pr_{2B} = -10.833752 + \ln\left(1400^{1.9} \cdot 9\right) = 5.1 \quad (12.14)$$

Percentage probability from probit (Eq. 12.11).

%	0	1	2	3	4	5	6	7	8	9
0	–	2.67	2.95	3.12	3.25	3.36	3.45	3.52	3.59	3.66
10	3.72	3.77	3.82	3.87	3.92	3.96	4.01	4.05	4.08	4.12
20	4.16	4.19	4.23	4.26	4.29	4.33	4.36	4.39	4.42	4.45
30	4.48	4.50	4.53	4.56	4.59	4.61	4.64	4.67	4.69	4.72
40	4.75	4.77	4.80	4.82	4.85	4.87	4.90	4.92	4.95	4.97
50	5.00	5.03	5.05	5.08	5.10	5.13	5.15	5.18	5.20	5.23
60	5.25	5.28	5.31	5.33	5.36	5.39	5.41	5.44	5.47	5.50
70	5.52	5.55	5.58	5.61	5.64	5.67	5.71	5.74	5.77	5.81
80	5.84	5.88	5.92	5.95	5.99	6.04	6.08	6.13	6.18	6.23
90	6.28	6.34	6.41	6.48	6.55	6.64	6.75	6.88	7.05	7.33
–	0.0	0.1	0.2	0.3	0.4	0.5	0.6	0.7	0.8	0.9
99	7.23	7.37	7.41	7.46	7.51	7.58	7.65	7.75	7.88	8.09

Percentage probability from probit (Eq. 12.12).

%	0	1	2	3	4	5	6	7	8	9
0	–	2.67	2.95	3.12	3.25	3.36	3.45	3.52	3.59	3.66
10	3.72	3.77	3.82	3.87	3.92	3.96	4.01	4.05	4.08	4.12
20	4.16	4.19	4.23	4.26	4.29	4.33	4.36	4.39	4.42	4.45
30	4.48	4.50	4.53	4.56	4.59	4.61	4.64	4.67	4.69	4.72
40	4.75	4.77	4.80	4.82	4.85	4.87	4.90	4.92	4.95	4.97
50	5.00	5.03	5.05	5.08	5.10	5.13	5.15	5.18	5.20	5.23
60	5.25	5.28	5.31	5.33	5.36	5.39	5.41	5.44	5.47	5.50
70	5.52	5.55	5.58	5.61	5.64	5.67	5.71	5.74	5.77	5.81
80	5.84	5.88	5.92	5.95	5.99	6.04	6.08	6.13	6.18	6.23
90	6.28	6.34	6.41	6.48	6.55	6.64	6.75	6.88	7.05	7.33
–	0.0	0.1	0.2	0.3	0.4	0.5	0.6	0.7	0.8	0.9
99	7.23	7.37	7.41	7.46	7.51	7.58	7.65	7.75	7.88	8.09

Percentage probability from probit (Eq. 12.13).

%	0	1	2	3	4	5	6	7	8	9
0	–	2.67	2.95	3.12	3.25	3.36	3.45	3.52	3.59	3.66
10	3.72	3.77	3.82	3.87	3.92	3.96	4.01	4.05	4.08	4.12
20	4.16	4.19	4.23	4.26	4.29	4.33	4.36	4.39	4.42	4.45
30	4.48	4.50	4.53	4.56	4.59	4.61	4.64	4.67	4.69	4.72
40	4.75	4.77	4.80	4.82	4.85	4.87	4.90	4.92	4.95	4.97
50	5.00	5.03	5.05	5.08	5.10	5.13	5.15	5.18	5.20	5.23
60	5.25	5.28	5.31	5.33	5.36	5.39	5.41	5.44	5.47	5.50
70	5.52	5.55	5.58	5.61	5.64	5.67	5.71	5.74	5.77	5.81
80	5.84	5.88	5.92	5.95	5.99	6.04	6.08	6.13	6.18	6.23
90	6.28	6.34	6.41	6.48	6.55	6.64	6.75	6.88	7.05	7.33
–	0.0	0.1	0.2	0.3	0.4	0.5	0.6	0.7	0.8	0.9
99	7.23	7.37	7.41	7.46	7.51	7.58	7.65	7.75	7.88	8.09

Percentage probability from probit (Eq. 12.14).

%	0	1	2	3	4	5	6	7	8	9
0	–	2.67	2.95	3.12	3.25	3.36	3.45	3.52	3.59	3.66
10	3.72	3.77	3.82	3.87	3.92	3.96	4.01	4.05	4.08	4.12
20	4.16	4.19	4.23	4.26	4.29	4.33	4.36	4.39	4.42	4.45
30	4.48	4.50	4.53	4.56	4.59	4.61	4.64	4.67	4.69	4.72
40	4.75	4.77	4.80	4.82	4.85	4.87	4.90	4.92	4.95	4.97
50	5.00	5.03	5.05	5.08	5.10	5.13	5.15	5.18	5.20	5.23
60	5.25	5.28	5.31	5.33	5.36	5.39	5.41	5.44	5.47	5.50
70	5.52	5.55	5.58	5.61	5.64	5.67	5.71	5.74	5.77	5.81
80	5.84	5.88	5.92	5.95	5.99	6.04	6.08	6.13	6.18	6.23
90	6.28	6.34	6.41	6.48	6.55	6.64	6.75	6.88	7.05	7.33
–	0.0	0.1	0.2	0.3	0.4	0.5	0.6	0.7	0.8	0.9
99	7.23	7.37	7.41	7.46	7.51	7.58	7.65	7.75	7.88	8.09

Incident	Incident frequency (y^{-1})	Probability of fatality	Frequency of fatality (y^{-1})
1	10^{-4}	0.31	3.1×10^{-5}
	10^{-5}	0.14	1.4×10^{-6}
2	10^{-4}	0.22	2.2×10^{-5}
	10^{-5}	0.54	5.4×10^{-6}

$$LSIR_1 = 3.1 \times 10^{-5} + 2.2 \times 10^{-5} = 5.3 \times 10^{-5} y^{-1};$$
$$LSIR_2 = 1.4 \times 10^{-6} + 5.4 \times 10^{-6} = 6.8 \times 10^{-6} y^{-1} \quad (12.15)$$

$$\mathrm{LSIR_{AV}} = \frac{\sum_{x,y} \mathrm{LSIR}_{x,y,i} \cdot \Pi_{x,y}}{\sum_{x,y} \Pi_{x,y}}$$

$$= \frac{5 \times 3.1 \times 10^{-5} + 10 \times 1.4 \times 10^{-6} + 5 \times 2.2 \times 10^{-5} + 10 \times 5.4 \times 10^{-6}}{15} \tag{12.16}$$

$$= \frac{3.33 \times 10^{-4}}{15} = 2.2 \times 10^{-5} \mathrm{y}^{-1}$$

$$\mathrm{PLL} = \mathrm{LSIR_{AV}} \cdot \mathrm{POS} = 2.2 \times 10^{-5} \mathrm{y}^{-1} \times 15 = 3.3 \times 10^{-4} \mathrm{y}^{-1} \tag{12.17}$$

$$\mathrm{FAR} = \mathrm{LSIR_{AV}} \cdot 11{,}415 = 0.25 \text{ fatalities}/10^8 \text{ hours} \tag{12.18}$$

Example 12.2

For the following installation, find the shift turn-over necessary to have an IRPA of $0.5 \times 10^{-3}\, \mathrm{y}^{-1}$.

Installation	Shifts	IRPA (y^{-1})
Onshore	2/day	10^{-3}

Solution

$$\mathrm{LSIR} = \frac{\mathrm{IRPA}_1}{fos_{-1}} = \frac{10^{-3}}{0.5} = 0.002 \mathrm{y}^{-1} \tag{12.19}$$

$$fos_{-2} = \frac{\mathrm{IRPA}_2}{\mathrm{LSIR}} = \frac{0.5 \times 10^{-3}}{0.002} = 0.25 \tag{12.20}$$

In order to reduce the IRPA 4 shifts a day, 6 hours each are required.

12.1.8 ALARP

It is the acronym for as low as reasonably practicable, and was introduced into the British regulation in 1974. Disproportion between costs and benefits is a key driver for the model:

> "Provision and maintenance of plant and systems of work that are, so far as is reasonably practicable, safe and without risks to health."

The acronym included in the original act is SFAIRP (so far as is reasonably practicable).

12.1.9 Safety Integrity

It is the likelihood that a safety-related system will achieve its required safety functions under all the stated conditions within a specified period of time (CCPS, 2005).

12.1.10 Safety Integrity Level (SIL)

It is the discrete level (one out of four) for specifying the safety integrity requirements of the safety instrumented functions to be allocated to the safety instrumented systems. Safety integrity level 4 has the highest level of safety integrity; safety integrity level 1 has the lowest (IEC, 61511, 2016).

12.2 ALARP Model

The conceptual and historical background of the ALARP model, as well as the rationale of its various worldwide variations, are not included in this book. The reader can refer to the keystone publications issued by the Health and Safety Executive (Ball and Floyd 1998; HSE, 2001). In this section, guidance on the UK ALARP model has been provided as a support to the quantitative risk assessment (QRA) studies.

12.2.1 ALARP Limits for Workers and Public

In Figs. 12.2 and 12.3 the three bands of the ALARP model are shown both for workers and public.

12.2.2 ALARP Limits for Land-Use Planning

In Table 12.2 the limits of the ALARP model for land-use planning are shown both for nonvulnerable and vulnerable people.

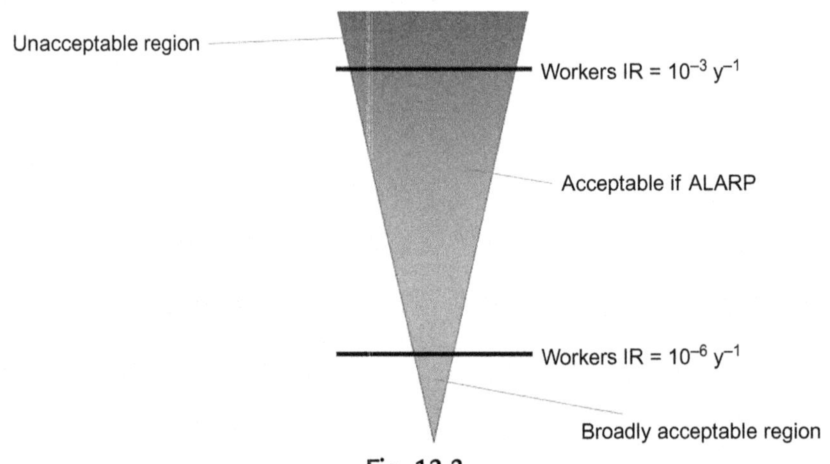

Fig. 12.2
ALARP model bands for workers (HSE, 2001).

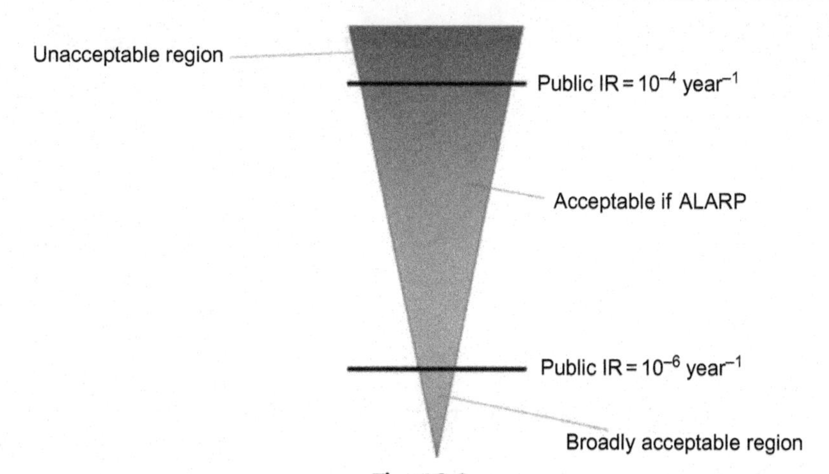

Fig. 12.3
ALARP model bands for public (HSE, 2001)

Table 12.2 ALARP limits for land-use planning (HSE)

Risk band	Fatalities for nonvulnerable public (y^{-1})	FAR (fatalities/10^8 exposure hours)
Intolerable	$IR > 10^{-5}$	$IR > 10^{-6}$
Tolerable If ALARP	$10^{-5} \geq IR \geq 10^{-6}$	$10^{-6} \geq IR \geq 3.3\ 10^{-7}$
Broadly acceptable	$IR < 10^{-6}$	$IR < 3.3\ 10^{-7}$

12.2.3 FN *Curves*

The *FN* curves are a means of presenting the behaviour of fatalities versus the cumulative frequency for societal risk. Each F, N couple of points of the curve represent the frequency F corresponding to N or more fatalities. An *FN* curve is constructed as shown in the following examples.

Example 12.3

Table 12.3 includes 10 incident scenarios relating to a number of fatalities occurred at the given frequency f. Construct the *FN* curve.

Solution
The table is sorted out so that fatality values are listed in decreasing order, and the frequencies corresponding to N or more are cumulated, as shown in Table 12.4, where the cumulative frequency F_i for the number of fatalities N_i is the sum of the single frequencies f_i relative to all the scenarios with $N > N_i$.

The *FN* curve has been depicted in Fig. 12.4.

Table 12.3 Incident scenarios and related frequencies for Example 12.3

Scenario	N	Frequency f (y^{-1})
1	10	4E−03
2	70	3E−06
3	24	7E−04
4	20	9E−04
5	14	7E−03
6	18	5E−03
7	41	1E−05
8	48	9E−04
9	6	9E−03
10	23	3E−04

Table 12.4 Incident scenarios of Table 12.3 sorted out by decreasing number of fatalities

Scenario	N	Cumulative frequency F (y^{-1})
2	70	3.00E−06
8	48	9.00E−04
7	41	9.10E−04
3	24	1.60E−03
10	23	1.90E−03
4	20	2.80E−03
6	18	7.80E−03
5	14	1.48E−02
1	10	1.88E−02
9	6	2.78E−02

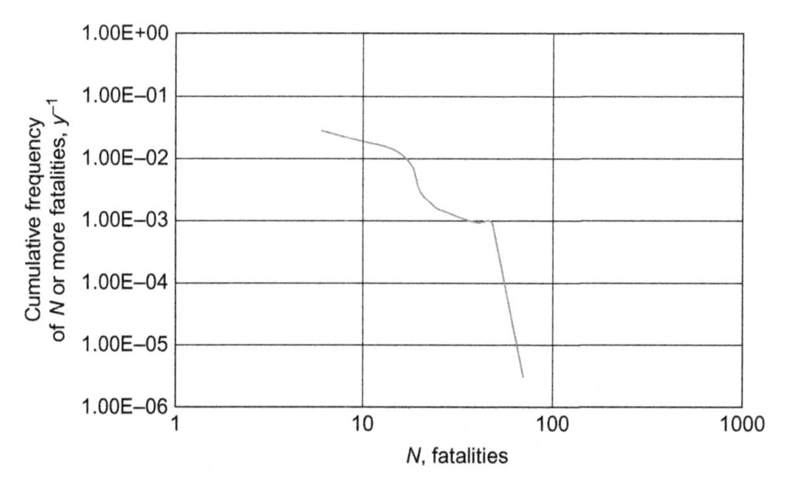

Fig. 12.4
Exceedance frequency curve for Eample 12.3.

The UK Health and Executive, with specific reference to societal risk, has proposed the following partition for the three zones of the ALARP model (Fig. 12.5).

It is concluded that the scenario included in this example does not comply with the UK HSE ALARP limits for societal risk as highlighted in the orange zone, where the FN curve exceeds the ALARP band (Fig. 12.6).

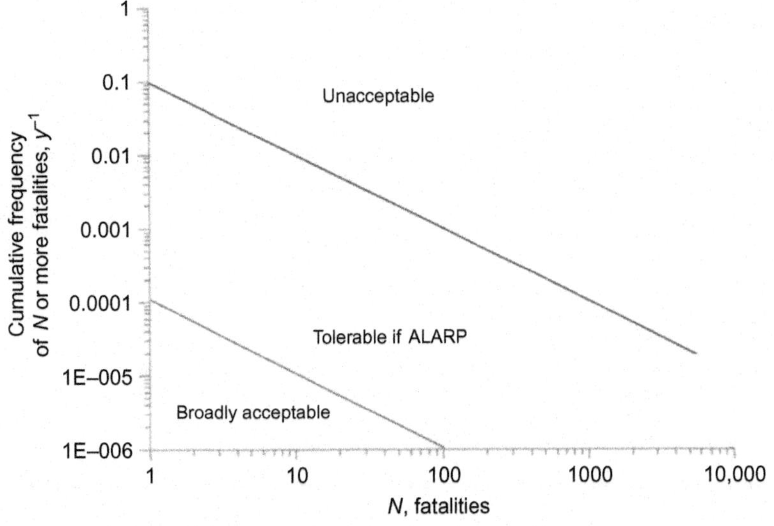

Fig. 12.5
FN curves ALARP bands (HSE).

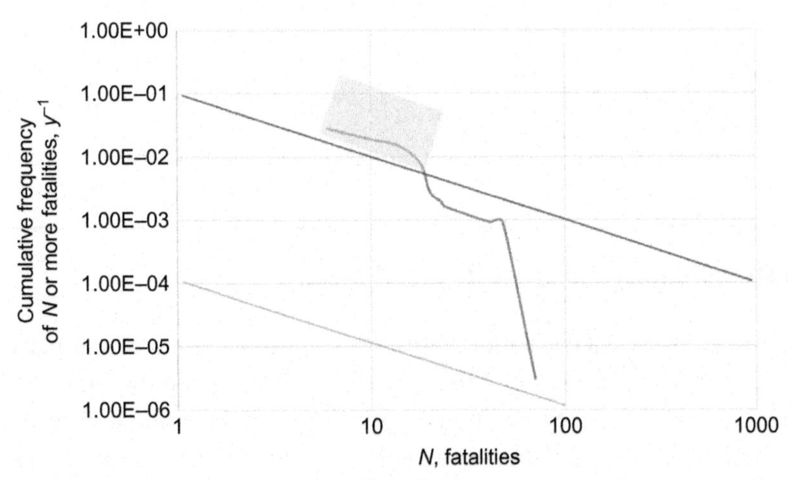

Fig. 12.6
Exceedance frequency curve and ALARP bands for Example 12.6.

12.3 PLL From FN Curve

The ALARP lines of the FN diagram mathematically define two PLLs, notably:

- the PLL corresponding to the upper line
- the PLL corresponding to the lower line

According to the definition of cumulative frequency, for the upper line:

$$f(N) = F(N) - F(N+1) \tag{12.21}$$

For the red (upper) line:

$$F(N) = \frac{1}{10 \times N} \tag{12.22}$$

By definition of PLL:

$$\mathrm{PLL_{upper}} = \int_{1}^{N_{max}} f(N) \cdot dN = 0.1 \times \int_{1}^{N_{max}} \left[\frac{1}{N} - \frac{1}{N+1}\right] \cdot dN = 0.1 \times [\,\ln(N) - \ln(N+1)]_{1}^{N_{max}} \tag{12.23}$$

The following example shows the calculation procedure.

Example 12.4

Find the upper ALARP PLL for the previous example with a maximum number of fatalities equal to 70.

Solution

$$\begin{aligned}\mathrm{PLL_{upper}} &= \int_{1}^{70} f(N) \cdot dN = 0.1 \times \int_{1}^{70} \left[\frac{1}{N} - \frac{1}{N+1}\right] \cdot dN = 0.1 \times [\,\ln(N) - \ln(N+1)]_{1}^{70} \\ &= 0.1 \times [\,\ln(70) - \ln(71)] - 0.1 \times [\,\ln(1) - \ln(2)] = 0.0679\mathrm{y}^{-1}\end{aligned} \tag{12.24}$$

12.4 ALARP Demonstration—ICAF Method

The central band of the ALARP model identifies a conditional zone where individual risk and FN region are acceptable, provided the adopted risk mitigation measures are such that a disproportion would exist between additional costs and corresponding expected benefits in terms of safety improvement. A validated technique to determine whether or not a risk reduction is reasonably practicable is the ICAF (Implied Cost of Avoiding a Fatality).
A possible ICAF formula is the following:

$$\mathrm{ICAF} = \frac{\mathrm{COST}}{\mathrm{PLT} \times \Delta\mathrm{PLL}} \tag{12.25}$$

where:

- COST is the cost of implementing measure.
- PLT is plant life time.
- ΔPLL is the PLL reduction

An example of the application of the ICAF method has been shown here, along with some typical parameters adopted with the method.

Example 12.5

PLL of $1.5 \times 10^{-3}\, y^{-1}$ has to be reduced to one third. The plant time life is 30 years. Find the ICAF ALARP ZONE for workers if the value of a statistical fatality is 1,000,000 US Dollars (HSE, 2002).

Solution

According to the HSE (2002), the lower and upper border lines of the disproportion factors of the ICAF and ALARP zones are respectively 1 and 10 (Fig. 12.7).

The COST corresponding to the disproportion factor = 1 is:

$$ICAF_L \times PLT \times \Delta PLL = 1{,}000{,}000 \times 30 \times (1.5 - 0.5) \times 10^{-3} = 30{,}000\$ \qquad (12.26)$$

The COST corresponding to the disproportion factor=1 is:

$$ICAF_U \times PLT \times \Delta PLL = 10{,}000{,}000 \times 30 \times (1.5 - 0.5) \times 10^{-3} = 300{,}000\$ \qquad (12.27)$$

This is an example. It is worth noting that the financial rules should be stated by the corporate policy. The ALARP model is also applied in the execution of safety studies and for the acceptable design of plants and buildings. In these cases, semiquantitative methods such as ALARP matrices can be used to establish the level of compliance and check if a safer approach has to be adopted. A typical example is provided by the standard BS EN 1473:2016, where consequences and frequencies are related, and the three ALARP model bands are identified for safety, environment, and asset (Tables 12.5–12.9).

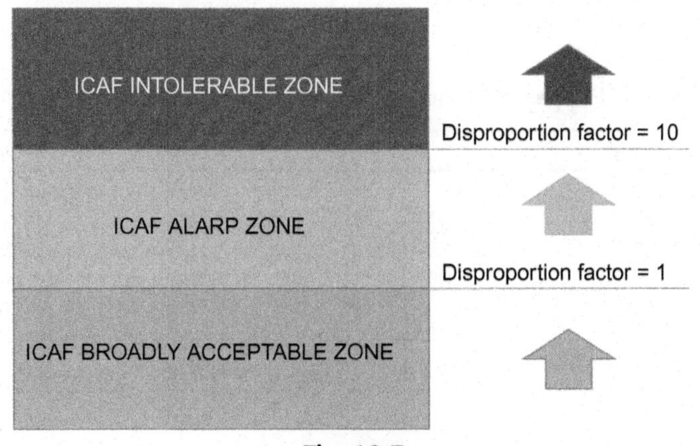

Fig. 12.7
ICAF bands for Example 12.5.

Table 12.5 BS Class of consequences (people) (BS EN 1473:2016)

Classes of consequences for hazard assessment (people)						
	Criteria unit	**Class 1**	**Class 2**	**Class 3**	**Class 4**	**Class 5**
Fatalities	Dead persons	>10	1–10	0	0	0
Accident with loss time	Injured persons	>100	11–100	2–10	1	0
Release of hydrocarbons	Tons	>1000	10.01–100	1.01–10	0.1–1	<0.1

Table 12.6 Class of consequences (environment) (BS EN 1473:2016)

Classes of consequences for hazard assessment (environment)						
	Criteria unit	**Class 1**	**Class 2**	**Class 3**	**Class 4**	**Class 5**
Release of substance harmful to the environment	Impact on environment	Massive effect	Major effect	Localised effect	Minor effect	No effect

Table 12.7 BS Class of consequences (assets) (BS EN 1473:2016)

Classes of consequences for hazard assessment (assets)						
	Criteria unit	**Class 1**	**Class 2**	**Class 3**	**Class 4**	**Class 5**
Release of flammable or corrosive substance	Impact on assets	Extensive damage	Major damage	Localised effect	Minor effect	No effect

Table 12.8 Ranges of frequency of occurrence (BS EN 1473:2016)

Range	Frequency of occurrence – *f*
1	More than once in 10 years
2	10 years >*f*> 100 years
3	100 years >*f*> 1000 years
4	1,000 years >*f*> 10,000 years
5	10,000 years >*f*> 100,000 years
6	100,000 years >*f*> 1000,000 years
7	More than once in 1000,000 years

Table 12.9 Risk ranking matrix (BS EN 1473:2016)

Risk assessment matrix for hazard assessment						
Risk		Consequence class				
Frequency	Cumulative frequency y^{-1}	5	4	3	2	1
Range 1	> 0.1	2	2	3	3	3
Range 2	0.1 to 0.01	1	2	2	3	3
Range 3	0.01 to 0.001	1	1	2	2	3
Range 4	0.001 to 10^{-4}	1	1	1	2	2
Range 5	10^{-4} to 10^{-5}	1	1	1	1	2
Range 6	10^{-5} to 10^{-6}	1	1	1	1	1
Range 7	> 10^{-6}	1	1	1	1	1

Legend
Level 1 - Acceptable
Level 2 - Acceptable if ALARP
Level 3 - Unacceptable

12.5 Parts Count

The typical QRA studies just deal with Loss of Containment (LOC) scenarios, assuming that any other process upsets, such as thermal runaway and reactive hazard assessment are covered in safety studies, notably HAZID, HAZOP, FMECA. Although this is not completely correct, LOC analysis requires the identification of all release sources, the calculation of releases, the definition of frequencies, and the combination of all data. Release models have been presented in Chapter 8; practical formulas have been included in this section (DNV) along with the description of the parts count technique.

12.5.1 QRA Release Models

Gas releases

The following simplified formula for initial release of gases is based on the assumptions $\gamma = 1.31$ and $C_D = 0.85$:

$$W_g = 1.4 \times 10^{-4} \times D_e^{\,2} \cdot \sqrt{\rho_g \cdot P_g} \tag{12.28}$$

where:

- W_g is the initial mass flow rate (kg/s).
- D_e is the hole diameter (mm).

- ρ_g is the initial gas density (kg/m^3).
- P_g is the initial pressure (barg).

Liquid releases

The following simplified formula for initial release of liquids is based on the assumptions that $C_D = 0.61$, and static head is negligible:

$$W_L = 2.1 \times 10^{-4} \times D_e^2 \cdot \sqrt{\rho_L \cdot P_L} \quad (12.29)$$

where:

- W_L is the initial mass flow rate (kg/s).
- D_e is the hole diameter (mm).
- ρ_L is the initial gas density (kg/m^3).
- P_L is the initial pressure (barg).

Two phase releases

The following simplified formula for initial release of liquids is based on the assumptions that $C_D = 0.61$ and static head is negligible:

$$W_{gL} = \frac{\text{GOR}}{\text{GOR} + 1} W_g + \frac{1}{\text{GOR} + 1} W_L \quad (12.30)$$

where:

- W_{gL} is the initial mass flow rate (kg/s).
- GOR is the gas to oil ratio (kg of gas/kg of liquid).

12.5.2 Transient Analysis

The previous equations are related to the initial mass flow rate. A real system will undergo the following three release phases:

1. At time = 0 initial constant mass flow is released. This is typically choked flow.
2. Depending on the detection system and the emergency procedure the equipment segment will be isolated. A flow decay will start.
3. Blow down of isolated segments will increase the decay slope, until the driving force is lost.

The three phases have been depicted in the following diagram (Fig. 12.8).

In this diagram:

- W_o is the mass flow rate at time = 0.
- W_B is the mass flow rate when blowdown starts.

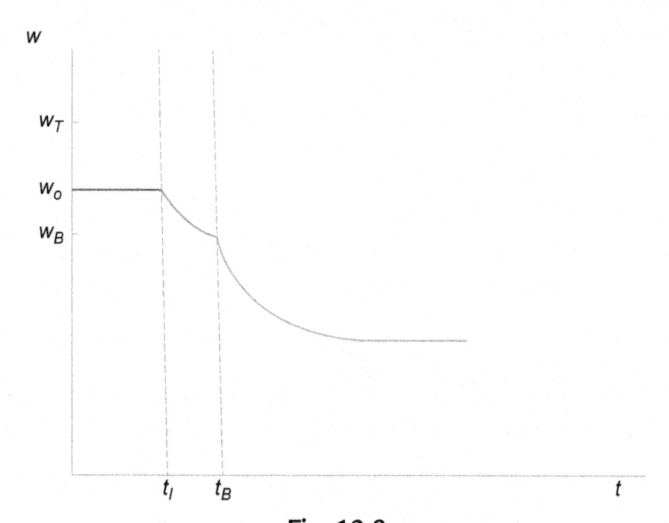

Fig. 12.8
Transient decay of mass flow.

- W_T is the total mass flow rate when blowdown starts from the BD valve and from the leak.
- t_I is the time of start of leak (s).
- t_B is the time of start of blowdown (s).

and

$$R_E = W_o \cdot t_I + I \cdot \left(1 - \frac{W_B}{W_o}\right) + M_B \cdot \frac{W_B}{W_T} \tag{12.31}$$

$$W_B = W_o \cdot \exp\left[\frac{-W_o \cdot (t_B - t_I)}{I}\right] \tag{12.32}$$

$$M_B = I \cdot \frac{W_B}{W_o} \tag{12.33}$$

$$\frac{W_B}{W_T} = \frac{D_e^2}{D_e^2 + \rho_d \cdot b^2} \tag{12.34}$$

with

- R_E = ESD-limited outflow (kg)
- I = inventory in isolated section (kg)
- M_B = mass remaining when blowdown starts (kg)
- ρ_d = density factor = 1 for gas and 2-phase releases, and 0.5 multiplied by the gas to liquid ratio for liquid releases
- b = blowdown valve diameter (mm)

A rigorous method accounting for long pipelines and vessels has been provided by Norris (1993).

Incident scenarios are developed splitting process units into isolatable segments and predicting the release frequencies of the single elements. This is done through the parts count technique, according to some criteria, such as the following:

1. Isolatable segments are identified on the P&ID.
2. Different phases are counted as separate segments, and the frequency of the common equipment (e.g. flash drum) is halved and partitioned.
3. Same size components, equipment, and pipes belonging to the same segment are grouped together.
4. Frequencies of same size items are summed.
5. Leak cumulative frequencies are developed to support the specific incident scenarios such as pool fire, jet fire, flash fire, explosion, and toxic cloud.

Leak frequencies can be collected from validated sources, such as:

- *Hydrocarbon Release (HCR) data base by the Health and Safety Executive*
- *API RP 581 Risk-Based Inspection Methodology, 2016*
- *OREDA Offshore Reliability Data Handbook, 2009*
- *TNO/VROM Guidelines for quantitative risk assessment*
- *RIVM Reference Manual Bevi Risk Assessments version 3.2, 2009*
- *Spouge, J.R., Funnemark, E., Process Equipment Failure frequencies, DNV, 2006*
- *Cox, A., Lees, F., Ang, M., Classification of Hazardous, 1990*
- *EGIG (European Gas Pipeline Incident Data Group) 2015*

Example 12.6

Develop the parts count for the vapour phase of the following section.

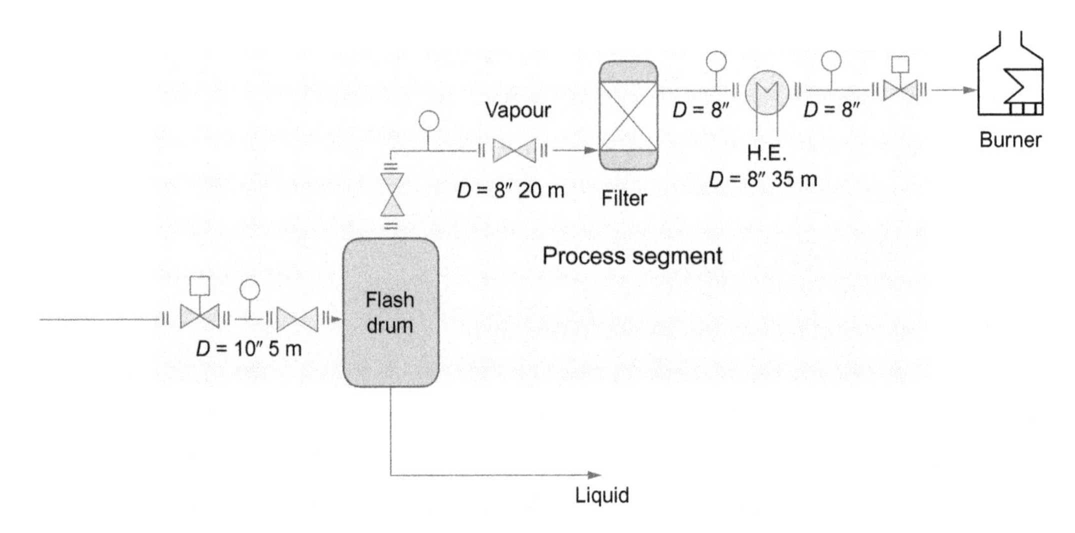

Solution

According to Spouge and Funnemark (2006) the leak frequency for an item with diameter D from a hole diameter greater than d can be calculated according to the expression:

$$f(d) = C \cdot (1 + a \cdot D^n) \cdot d^m + F_{rup} \tag{12.35}$$

with:

- f, frequency of leaks (per year or per metre year for pipes only) exceeding size d
- C and m, constants representing hole size distribution
- a and n, constants representing equipment size dependency
- D, item diameter mm
- d, hole diameter, mm (assumed as $\geq$10 mm)

Total leaks frequency of 8″ pipe (200 mm, 55 m)

Scenario	C	a	n	m	F_{rup}
Total leaks	3.7E−05	1000	−1.5	−0.74	3.0E−06

$$\begin{aligned} f \cdot L &= \left[C \cdot (1 + a \cdot D^n) \cdot d^m + F_{rup}\right] \cdot L \\ &= \left[3.7\text{E}-05 \cdot \left(1 + 1000 \cdot 200^{-1.5}\right) \cdot 10^{-0.74} + 3.0\text{E}-06\right] \cdot (20 + 35) = 6.66\text{E}-04\text{y}^{-1} \end{aligned} \tag{12.36}$$

Total leaks frequency of 10″ pipe (250 mm, 5 m)

$$\begin{aligned} f \cdot L &= \left[C \cdot (1 + a \cdot D^n) \cdot d^m + F_{rup}\right] \cdot L \\ &= \left[3.7\text{E}-05 \cdot \left(1 + 1000 \cdot 250^{-1.5}\right) \cdot 10^{-0.74} + 3.0\text{E}-06\right] \cdot 5 = 5.72\text{E}-05\text{y}^{-1} \end{aligned} \tag{12.37}$$

Total leaks frequency of 8″ flanges

Scenario	C	a	n	m	F_{rup}
Total leaks	7.5E−05	1.9E−04	1.5	−0.84	5.0E−06

$$\begin{aligned} f &= \left[C \cdot (1 + a \cdot D^n) \cdot d^m + F_{rup}\right] \\ &= \left[7.5-05 \cdot \left(1 + 1.9\text{E}-04 \cdot 200^{1.5}\right) \cdot 10^{-0.84} + 5.0\text{E}-06\right] = 2.4\text{E}-05\text{y}^{-1} \end{aligned} \tag{12.38}$$

Total leaks frequency of 10″ flanges

$$\begin{aligned} f \cdot L &= \left[C \cdot (1 + a \cdot D^n) \cdot d^m + F_{rup}\right] \\ &= \left[7.5-05 \cdot \left(1 + 1.9\text{E}-04 \cdot 250^{1.5}\right) \cdot 10^{-0.84} + 5.0\text{E}-06\right] = 2.69\text{E}-05\text{y}^{-1} \end{aligned} \tag{12.39}$$

Total leaks frequency of 8″ hand valves

Scenario	C	a	n	m	F_{rup}
Total leaks	7.0E−05	3.4−05	2.0	−0.76	1.0E−05

$$\begin{aligned} f &= \left[C \cdot (1 + a \cdot D^n) \cdot d^m + F_{rup}\right] \\ &= \left[7.0 - 05 \cdot \left(1 + 3.4\text{E} - 05 \cdot 200^{2.0}\right) \cdot 10^{-0.76} + 1.0\text{E} - 05\right] = 1.88\text{E} - 04\text{y}^{-1} \end{aligned} \tag{12.40}$$

Total leaks frequency of 10″ hand valves

$$\begin{aligned} f &= \left[C \cdot (1 + a \cdot D^n) \cdot d^m + F_{rup}\right] \\ &= \left[7.0 - 05 \cdot \left(1 + 3.4\text{E} - 05 \cdot 250^{2.0}\right) \cdot 10^{-0.76} + 1.0\text{E} - 05\right] = 2.81\text{E} - 04\text{y}^{-1} \end{aligned} \tag{12.41}$$

Total leaks frequency of 8″ isolation valve

Scenario	C	a	n	m	F_{rup}
Total leaks	6.9E−04	0	0	−0.82	0

$$f = \left[C \cdot (1 + a \cdot D^n) \cdot d^m + F_{rup}\right] = \left[6.9 - 05 \cdot (1) \cdot 10^{-0.82}\right] = 1.04\text{E} - 04\text{y}^{-1} \tag{12.42}$$

Total leaks frequency of 10″ isolation valve

$$f = \left[C \cdot (1 + a \cdot D^n) \cdot d^m + F_{rup}\right] = \left[6.9 - 05 \cdot (1) \cdot 10^{-0.82}\right] = 1.04\text{E} - 04\text{y}^{-1} \tag{12.43}$$

Total leaks frequency of fittings (50″ hole)

$$f = \left[C \cdot d^m + F_{rup}\right] = \left[6.1 - 04 \cdot 50^{-0.80} + 0\right] = 2.67\text{E} - 05\text{y}^{-1} \tag{12.44}$$

Total leaks frequency of flash drum (50 mm hole)

$$f = \left[C \cdot d^m + F_{rup}\right] = \left[1.8 - 03 \cdot 50^{-0.43} + 0.0001\right] = 4.3\text{E} - 0\text{y}^{-1} \tag{12.45}$$

Total leaks frequency of heat exchanger (50 mm hole)

$$f = \left[C \cdot d^m + F_{rup}\right] = \left[4.7 - 03 \cdot 50^{-0.65} + 0\right] = 3.7\text{E} - 04\text{y}^{-1} \tag{12.46}$$

Total leaks frequency of filter (50 mm hole)

$$f = \left[C \cdot d^m + F_{rup}\right] = \left[3.7 - 03 \cdot 50^{-0.66} + 0\right] = 2.8\text{E} - 04\text{y}^{-1} \tag{12.47}$$

Parts count summary

	Item	Size	Length (m)	Frequency	*N*	Total frequency
1	Pipes	8″	55	6.66E−04	–	6.66E−04
2	Pipes	10″	5	5.72E−05	–	5.72E−05
3	Flanges (couples)	8″	–	2.40E−05	8	1.92E−04
4	Flanges (couples)	10″	–	2.69E−05	4	1.0E−04
5	Hand valves	8″	–	1.88E−04	3	5.6E−04
6	Hand valves	10″	–	2.81E−04	2	5.6E−04
7	Isolation valve	8″	–	1.04E−04	1	1.04E−04
8	Isolation valve	10″	–	1.04E−04	1	1.04E−04
9	Fittings	–	–	2.67E−05	4	1.06E−04
10	Flash drum	–	–	4.30E−04	0.5	2.15E−04
11	Heat exchanger	–	–	3.07E−05	1	3.07E−05
12	Filter	–	–	2.80E−04	1	2.80E−04

This table can then be sorted out so that release rate values are listed in decreasing order and, as done for the overpressures and for the FN diagram, the cumulative frequency curve is constructed.

References

Ball, D.J., Floyd, P.J., 1998. Societal risk, Crown Copyright. Available from Health & Safety Executive, Risk Assessment Policy Unit, 2 Southwark Bridge, London SE1 9HS.

BS EN 1473:2016 Installation and equipment for liquefied natural gas. Design of onshore installation.

CCPS, 1999. Guidelines for Chemical Process Quantitative Risk Analysis, second ed. American Institute of Chemical Engineers, Wiley.

CCPS, 2005. Glossary of Terms, American Institute of Chemical Engineers, http://www.emergencymgt.net/sitebuildercontent/sitebuilderfiles/ccpsHazards.pdf.

Health and Safety Executive, 2001. Reducing Risks, Protecting People – HSE's decision-making process, ISBN 0 7176 2151 0, http://www.hse.gov.uk/risk/theory/r2p2.pdf.

HSE- HID, 2002. Guidance on As Low As Reasonably Practicable (ALARP). Decisions in Control of Major Accident Hazards (COMAH). SPC/Permissioning/12.

IEC 61511–1:2016. Functional safety—Safety instrumented systems for the process industry sector—Part 1: Framework, definitions, system, hardware and application programming requirements.

Kletz, T.A., 1977. The Risk Equations—What Risks Should We Run? New Scientist 12, 320–325.

Norris III, H.L., 1993. Single-Phase or Multiphase Blowdown of Vessels or Pipelines, SPE-26565-MS. Society of Petroleum Engineers.

Spouge, J.R., Funnemark, E., 2006. Process Equipment Failure Frequencies. DNV.

Van den Bosch, C.J.H., Duijm, N.J., 2005. Methods for the Calculation of Physical Effects, Yellow Book. In: Van den Bosch, C.J.H., Weterings, R.A.P.M. (Eds.), http://content.publicatiereeksgevaarlijkestoffen.nl/documents/PGS2/PGS2-1997-v0.1-physical-effects.pdf.

Further Reading

DET NORSKE VERITAS. Quantified Risk Assessment. Failure Frequency Guidance. Det Norske Veritas as NO-1322 Høvik, Norway www.dnv.com.

Health and Safety at Work etc. Act 1974, http://www.legislation.gov.uk/ukpga/1974/37/pdfs/ukpga_19740037_en.pdf.

Index

Note: Page numbers followed by *f* indicate figures, and *t* indicate tables.

I

J

W